CORONAL MASS EJECTIONS

Cover illustration:

Drawing of the corona as it appeared to Temple at Torreblanca, Spain during the total solar eclipse of 18 July 1860 showing what may be the first observation of a CME (see Eddy, J. A.: 1974, *Astron. Astrophys.* **34**, 235).

Space Sciences Series of ISSI

Volume 21

The International Space Science Institute is organized as a foundation under Swiss law. It is funded through recurrent contributions from the European Space Agency, the Swiss Confederation, the Swiss National Science Foundation, and the University of Bern. For more information, see the homepage at http://www.issi.unibe.ch/.

CORONAL MASS EJECTIONS

Edited by

H. KUNOW
Christian-Albrechts-Universität zu Kiel,
Kiel, Germany

N. U. CROOKER
Boston University,
Boston MA, USA

J. A. LINKER
Science Applications MS C2,
International Corporation,
San Diego CA, USA

R. SCHWENN
Max-Planck Institut für Sonnensystemforschung,
Katlenburg-Lindau, Germany

R. VON STEIGER
International Space Science Institute (ISSI),
Bern, Switzerland

Reprinted from *Space Science Reviews*, Volume 123, Nos. 1–3, 2006

Springer

INTERNATIONAL SPACE SCIENCE INSTITUTE
Space Sciences Series of ISSI

A.C.I.P. Catalogue record for this book is available from the Library of Congress

ISBN: 978-0-387-45086-5

Published by Springer
P.O. Box 990, 3300 AZ Dordrecht, The Netherlands

Sold and distributed in North, Central and South America
by Springer,
101 Philip Drive, Norwell, MA 02061, U.S.A.

In all other countries, sold and distributed
by Springer,
P.O. Box 322, 3300 AH Dordrecht, The Netherlands

Printed on acid-free paper

Printed in the Netherlands

TABLE OF CONTENTS

FOREWORD

Coronal Mass Ejections are a spectacular an violent phenomenon of the solar atmosphere with repercussions throughout the entire heliosphere. They are a spectacular sight when seen to erupt from the Sun with the aid of a coronagraph such as LASCO on the Solar and Heliospheric Observatory SoHO. They are a violent phenomenon when arriving at Earth, pounding on our magnetosphere, and sometimes disrupting all kinds of achievements of the industrial and information age. CMEs have been with us ever since the existence of the solar system, yet only in the past century and a half they make themselves known to us in that way. They are a continuously observable phenomenon only since the Skylab and SoHO era, save for some very brief periods of solar eclipses, one of which is pictured on the front cover. The flare that was observed live through the telescope by Lord Carrington in 1859 led to a gigantic CME that, would it happen today, could easily cause a global blackout. Understanding CMEs is thus a first step in protecting ourselves from their potentially devastating effects.

This volume is the result of a series of workshops during the years 2000–2004 to study in detail origin, development, and effects of coronal mass ejections (CMEs). An international team of about sixty experimenters, ground observers, and theoreticians worked on interpreting the observations and developing new models for CME initiations, development, and interplanetary propagation. Under investigation were also effects on charged particles and related phenomena like energetic particle acceleration, interaction with ambient solar wind and other CMEs, as well as the internal structure of CMEs and its time variation. Fundamental questions concerning CMEs (e.g., CME initiation) and many detailed observations are still not understood. The workshops helped to jointly investigate these questions with scientists from all scientific areas involved.

The workshops were subdivided into eight working groups with always four of them held in parallel. Each participant attended two different working groups. While in the first four working groups (A-D) scientists from the same field discussed and described the topics from their own point of view, the second four (E-H) were topic-oriented with participants from all relevant areas attending. Their goal was to investigate all aspects of the phenomenon and to present a comprehensive interpretation. Occasionally this working scheme led to duplications in different working groups, however, this was intended and helped to further clarify the topic, especially in the case of conflicting statements.

The eight working group reports constitute the main body of the book. In addition, seven introductory chapters describe the state of knowledge prior to the first workshop and serve as introduction to the topics discussed later in more detail. The

Space Science Reviews (2006) 123: 1–2
DOI: 10.1007/s11214-006-9007-z

volume is rounded off with a historical overview to start with and with a paper on geoeffectiveness and a summary to conclude.

We are happy to complement with this volume an earlier ISSI book that has been conceived and compiled in a very similar manner. Volume 7 in the *Space Sciences Series of ISSI* was dealing with Corotating Interaction Regions (CIRs), which are shaping the heliosphere at times of solar minimum activity. CMEs, conversely, are an important phenomenon mainly at solar maximum activity. Thus the two volumes now form a nice pair covering the entire solar cycle.

It is a pleasure to thank all those who have contributed to this volume and to the workshops in general. First of all, we thank the authors for writing up their contributions, in particular the Working Group co-chairs for compiling the massive WG reports. All papers were peer reviewed by referees, and we thank the reviewers for their critical reports. We also thank the directorate and staff of ISSI for selecting this topic for a workshop and for their support in making it happen, in particular Roger M. Bonnet, Brigitte Fasler, Vittorio Manno, Saliba F. Saliba, Irmela Schweizer, and Silvia Wenger.

July 2006

H. Kunow, N. U. Crooker, J. A. Linker, R. Schwenn and R. von Steiger
ISSI, Hallerstrasse 6
CH-3012 Bern, Switzerland

A BRIEF HISTORY OF CME SCIENCE

DAVID ALEXANDER[1,*], IAN G. RICHARDSON[2] and THOMAS H. ZURBUCHEN[3]
[1]*Department of Physics and Astronomy, Rice University, 6100 Main St., Houston, TX 77005, USA*
[2]*The Astroparticle Physics Laboratory, NASA GSFC, Greenbelt, MD 20771, USA*
[3]*Department of AOSS, University of Michigan, Ann Arbor, MI 48109, USA*
*(*Author for correspondence: E-mail: dalex@rice.edu)*

(Received 15 July 2004; Accepted in final form 5 May 2005)

Abstract. We present here a brief summary of the rich heritage of observational and theoretical research leading to the development of our current understanding of the initiation, structure, and evolution of Coronal Mass Ejections.

Keywords: CMEs, corona, history

1. Introduction

The key to understanding solar activity lies in the Sun's ever-changing magnetic field. The potential role played by the magnetic field in the solar atmosphere was first suggested by Frank Bigelow in 1889 after noting that the structure of the solar minimum corona seen during the eclipse of 1878 displayed marked equatorial extensions, called 'streamers'. Bigelow (1890) noted that the coronal streamers had a strong resemblance to magnetic lines of force and proposed that the Sun must, in fact, be a large magnet. Subsequently, Henri Deslandres (1893) suggested that the forms and motions of prominences seen during solar eclipses appeared to be influenced by a solar magnetic field. The link between magnetic fields and plasma emitted by the Sun was beginning to take shape by the turn of the 20th Century. The epochal discovery of magnetic fields on the Sun by American astronomer George Ellery Hale (1908) signalled the birth of modern solar physics. This realization led to the modern emphasis on solar transient activity and its relationship to the solar magnetic field and its reconfiguration.

2. Historical Observations

The first terrestrial phenomena recognized to be of solar origin were geomagnetic disturbances. Colonel Sabine, in the middle of the 19th century (Sabine, 1852), noted that the frequencies of both geomagnetic storms and sunspots followed an 11-year cycle. The first step in associating geomagnetic storms with transient solar activity – what later became known as solar flares – rather than simply with the associated spot regions, was the memorable observations in 1859 by British amateur

Space Science Reviews (2006) 123: 3–11
DOI: 10.1007/s11214-006-9008-y

astronomers Richard Carrington and Richard Hodgson (Carrington, 1860). They independently witnessed a rapid intense flash of two bright ribbons on the Sun in visible light accompanied, essentially simultaneously, by a marked disturbance of the Earth's magnetic field detected at Kew Observatory in London. Some 17.5 hours later, one of the largest magnetic storms on record occurred. While Carrington was reluctant to suggest a physical connection between the visible event at the Sun and the geomagnetic storm, Balfour Stewart, the Director of Kew Observatory, claimed that they had caught the Sun in the act of producing a terrestrial event. The first systematic evidence for a flare-storm connection, however, didn't come until the work of Hale (1931) (see Cliver, 1994a,b, 1995 for a detailed history). Over a century and a half later, solar and space physicists are revisiting the remarkable event of 1859 in a concerted effort to apply 21st Century tools to model its solar and terrestrial effects (e.g. Tsurutani *et al.*, 2003).

The importance of the "chromospheric eruptions", as the early flares were known, for the Earth's space environment came through the study of these events and their apparent association with geomagnetic storms. Lindemann (1919) suggested that geomagnetic storms were caused by ejections of magnetically neutral matter from the Sun impacting the Earth's magnetic field several days afterwards, as illustrated in the top panel of Figure 1. The statistical association of large flares and

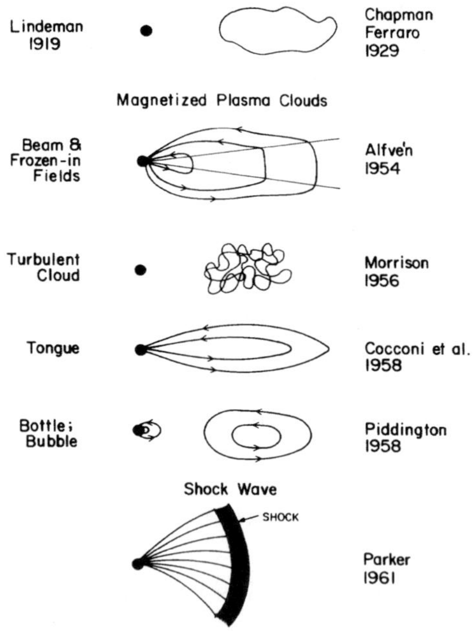

Figure 1. Early concepts of the structure of ICMEs, showing (from the top): unmagnetized material; a plasma cloud including frozen-in magnetic field loops; plasma including turbulent magnetic fields; a "tongue" of magnetic field loops rooted at the Sun; a disconnected "plasmoid" or "bubble"; and a shock wave ahead of a region of enhanced turbulence (Burlaga, 1991).

storms was solidified by Newton (1943), who surveyed all the large flares observed since 1892 and found a significant correlation between those flares and subsequent geomagnetic storms.

The expulsion of hydrogen was also observed near the time of peak intensity of the majority of bright flares. These emissions occurred in specific directions, usually along nearly vertical trajectories, and exhibited all the characteristics of the well-known eruptive prominences. The initial velocity of a mass expulsion was around 500 km/s and, while its H brightness was several times that of normal quiescent prominences, it was still much fainter than the flare emission itself. The physical relationship between solar flares and prominences dates back to the *disparition brusques* phenomena catalogued in the late 1940s by researchers at Meudon Observatory. The factors which cause this relationship are important since filament eruptions appear to have a role in many of the coronal transients that make up the most energetic solar activity. However, despite the fact that solar prominences have been observed for several hundred years, they were not thought to play a role in geomagnetic storms. A relationship was suggested by Greaves and Newton (1928); but Hale disagreed, pointing out three years later (Hale, 1931) that erupting prominences generally fall back to the Sun. The connection between prominence eruptions and geomagnetic storms, while hinted at by Newton (1936), was not fully appreciated until the work of Joselyn and McIntosh (1981).

It was pointed out by Kiepenheuer (1953) that the sudden disappearance of a prominence could result as the prominence rises into the corona with an increasing velocity that may eventually exceed the velocity of escape. This process was studied in detail with the conclusion that the ejected plasma is accelerated as it rises. Such studies were the precursors to present-day investigations into the relationship between filament eruptions and flares, and preceded by as much as three decades the discovery of coronal mass ejections.

Combined with the apparently clear association between geomagnetic disturbances and solar flares, the observed acceleration of material associated with prominence eruptions suggested a physical mediator for the transfer of energy from the solar atmosphere to the Earth's. Given the incontrovertible evidence for the existence of corpuscular radiation from the Sun, a major effort to detect the particles in transit was performed. Waldmeier (1941) and Ellison (1943) independently detected a strong asymmetry in the wings of the H emission line of flares. Ellison interpreted this as being due to the absorption by hydrogen atoms expelled in all directions from the flare site. This asymmetry was subsequently confirmed with spectrohelioscopes at observatories around the world. Ellison did caution, however, that: "While these asymmetric profiles provide the strongest possible evidence for the general expulsion of hydrogen during flares, we must await further work in order to prove that this constitutes the initial departure of the geomagnetic storm particles". Coinciding with large flares, sudden increases in cosmic ray intensity were occasionally detected (e.g., Forbush, 1946; Meyer *et al.*, 1956), suggesting that flares were also able to accelerate charged particles to energies in excess of

5 GeV. The notion that the particles could be accelerated en route did not occur to researchers at the time.

Early cosmic ray studies also provided evidence for ejections of material from the Sun that are related to geomagnetic storms, and strongly suggested that this material was magnetized. Decreases in the galactic cosmic ray intensity that accompany some storms were reported by Forbush (1937), and these were later explained by the exclusion of the cosmic rays from "magnetic bottles" formed when the ejection of highly-conducting coronal material drags solar magnetic fields into interplanetary space. Such bottles may remain connected to the Sun (Cocconi *et al.*, 1958) or be disconnected plasmoid-like structures (Piddington, 1958), as illustrated in Figure 1. An alternative, a turbulent cloud with tangled magnetic fields, also shown in Figure 1, was proposed by Morrison (1956).

Gold (1955) noted that many geomagnetic storms have remarkably abrupt onsets and suggested that shocks generated ahead of fast ejections cause the sudden onsets as they arrive at the Earth. The possibility that a large solar flare could drive a hydrodynamic blast wave to the Earth in 1–2 days was demonstrated by Parker (1961) (Figure 1). This idea was subsequently "confirmed" by a series of calculations on interplanetary shocks and was supported by observations of shocks which became available with the advent of in-situ measurements of the interplanetary plasma and magnetic fields in the space era (e.g., Sonnet *et al.*, 1964). Nevertheless, Hundhausen (1972) noted a number of apparent discrepancies between shock wave models and observations, expressing some reservations about the association between large flares and interplanetary shocks. Thus, by this time, one year prior to the launch of Skylab, the physics of storm-causing interplanetary shocks was understood but the shocks themselves could not be directly related to any coronal events.

While there had been indications of large, transient disturbances traveling through the Sun's outer corona in solar radio records and coronagraph observations from earlier unmanned spacecraft, it took the as-then unprecedented sensitivity of the Skylab coronagraph to put these observations in perspective. Skylab observations showed "gargantuan loops rushing outwards from the Sun at remarkable speeds" with the frequently observed "expulsion from the Sun of an eruption bigger than the disk of the Sun" (see Eddy, 1979, chapter 7). The first quantitative summary of the Skylab coronal disturbances (Gosling *et al.*, 1974) strongly indicated that these transients were the long-sought eruptions of coronal material required to produce the high-speed solar wind flows responsible for geomagnetic storms: measured speeds ranged from $<$100 km/s to $>$1200 km/s (Gosling *et al.*, 1976). These events came to be known by a variety of names such as "plasma clouds", "solar mass ejections", "mass ejection coronal transients", "coronal mass ejection events" and then simply "coronal mass ejections".

The detailed observations of CMEs by Skylab led Eddy (1974) to scour eclipse records for evidence of similar phenomena. The paucity of reports of such coronal transients was readily explained by the combination of the Skylab CME occurrence rate, the typical CME speed and the short duration of eclipse totality, resulting in

Figure 2. Drawing of the corona as it appeared to Tempel at Torreblanca, Spain during the total solar eclipse of 18 July 1860 showing what may be the first observation of a CME (see Eddy, 1974).

the expectation of one chance per century of capturing a CME during an eclipse. Despite these slim odds, Eddy (1974) found signs of a transient, very similar in form to the Skylab CMEs (see Figure 2) in a drawing of the Spanish eclipse of July 18, 1860, made by the Italian astronomer Gugliemo Tempel with supporting evidence from other observers. Other examples include a disconnection event from 16 April 1893 (Cliver, 1989) and a 3-part structure observation from an eclipse on 29 May 1919[1].

Following Skylab, several space-based coronagraphs were flown to study the transient Sun. The Solar Maximum Mission (SMM), launched in 1980, significantly advanced our knowledge of solar flares and coronal mass ejections. The nine years of SMM coronagraph observations resulted in a dramatic shift in the paradigm of the Sun-Earth interaction and brought CMEs to the fore of solar-terrestrial research. A complete summary of the contribution of SMM to our understanding of solar transients can be found in Strong *et al.* (1998). The theme of solar-terrestrial interactions continued into the 1990's with the launches of the Yohkoh and SOHO satellites. Observations by Yohkoh/SXT have demonstrated that CMEs typically produce a response in the hot corona even when this response does not include typical flare emissions. In particular, intriguing "dimmings" of the X-ray corona preceding arcade formation suggest that a significant volume (and mass) of gas is ejected from the flare site, consistent with coronagraph observations in white light. The quantitative relationship between this ejected mass and that seen in the CME, however, has yet to be established.

[1]Memoirs of the RAS, 64, plates 18 and 19, 1929; E. W. Cliver personal communication

Coronal mass ejections returned to the fore of solar activity research with the launch of SOHO in 1995. The combination of three white light coronagraphs, collectively known as LASCO, together with a full disk EUV imager (known as the Extreme Ultraviolet Imaging Telescope, EIT) has demonstrated the coronal consequences of these large-scale magnetic reconfigurations. While the characteristics of the CMEs observed by LASCO are similar to those observed in previous coronagraphs, there are several new aspects: (i) many CMEs are accompanied by a global response of the solar corona, (ii) many show acceleration to the edge of the LASCO field of view (32 R_s), (iii) partial disconnection is a frequent occurrence, (iv) CMEs are occurring more frequently than had been expected at solar minimum, and (v) CMEs undergo extensive internal evolution as they move outward. (see Howard *et al.*, 1997) In addition, LASCO has a greater ability to detect CMEs moving well out of the plane of the sky, in particular 'halo' CMEs which may be directed towards the Earth. The dimming events, discovered by Yokhoh, have been confirmed in EUV observations by EIT and also by the TRACE spacecraft. (e.g. Thompson *et al.*, 1998; Wills-Davey and Thompson, 1999)

CME research also extends to their interplanetary and heliospheric effects, with significant effort being devoted to identifying and measuring in-situ the characteristics of the material ejected into interplanetary space during CMEs. Such material was first identified in the early space era through regions of plasma with unusual characteristics, such as enhanced helium abundances (Hirshberg *et al.*, 1970) commencing a few hours following some interplanetary shocks. These regions extended over periods of $\sim$1 day, suggesting scale sizes of $\sim$0.2 AU, and were initially referred to by terms such as "shock driver", "piston", "plasma cloud", "solar mass ejection", and "ejecta", under the supposition that this plasma was the material ejected from the Sun that generated the shock. At the time of these first observations, it was assumed that the ejected material originated, or at least contained a component, from solar flares that was accelerated through some explosion, or piston process. Subsequent combined CME observations by coronagraphs and in-situ measurements made by spacecraft off the limbs of the Sun (e.g., Schwenn, R. 1983; Sheeley *et al.*, 1985; Lindsay *et al.*, 1999) or near the Earth (e.g., Webb *et al.*, 2000) have demonstrated the clear association (though not necessarily one-to-one, e.g., Cane *et al.*, 2000)) between CMEs launched in the general direction of the observing spacecraft and the subsequent detection of shocks and the related ejected material. The interplanetary manifestations of CMEs are currently frequently termed "Interplanetary Coronal Mass Ejections" (ICMEs), although this does imply an association with CMEs that is arguably not completely proven.

ICMEs are characterized by an array of signatures, most of which had been identified by the early 1980's with the exception of certain compositional signatures which are only observable under all solar wind conditions with the later generation of specialized instruments, such as the Solar Wind Ion Composition Spectrometer (SWICS) on the Advanced Composition Explorer (ACE) satellite. The in-situ signatures of ICMEs are summarized by Zurbuchen and Richardson (this volume).

It was also clear from early in-situ observations (e.g., Bryant *et al.*, 1962) that CME-driven shocks can accelerate particles as they move out through the heliosphere such that major solar energetic particle events include, and may even be dominated by, shock-accelerated particles (e.g., Cane *et al.*, 1988). See the papers in this volume by Cane and Lario, and by Klecker *et al.* for further discussion of this topic.

3. Theories

The observational developments, as in any scientific field, progressed hand-in-hand with theoretical considerations. The development of theoretical models of solar activity has as rich a history as the observational side of solar physics (see Alexander and Acton, 2001 for a more complete discussion of the early developments in flare theory). However, it was realized very early that most solar phenomena had something to do with the magnetic field and its variability. Consequently, the major improvements in our theoretical understanding of solar activity has come about through our ability to investigate the interplay between the plasma and the magnetic field. An important series of models worth mentioning briefly here appeared in the 1960s and 1970s. The first of these, by Carmichael in 1964, proposed that magnetic field lines high above the photosphere could be forced open by the solar wind. Developments of this line of thinking appeared from Sturrock and Coppi (1966), Hirayama (1974) and Kopp and Pneuman (1976) earning this class of the models the sobriquet of the CSHKP model. Since these early models, there have been major advancements in the development of theories to explain the initiation and evolution of solar eruptive transients (Forbes *et al.*, this volume). The development of theoretical models is a small but vibrant area of solar research and the synergy with observation only helps to improve the subtlety and relevance of the theoretical ideas.

4. Overview

As this volume indicates, the study of the formation and development of Coronal Mass Ejections at the Sun and their impact on the heliosphere is a burgeoning field of research with important consequences for our understanding of the Sun and its interaction with the interplanetary medium and planetary magnetospheres. The recent ubiquitous interest in Space Weather is a fitting testament to the heritage provided by the 150 year effort to understand the Sun-Earth connection.

There is still much to learn about solar eruptive events, but it is clear that flares, CMEs, and ICMEs are all important components of the Space Weather system. Studies of these phenomena will continue to drive our need to understand solar variability and increase our ability to predict these events and their potential terrestrial effects.

Acknowledgements

The authors would like to thank the E. W. Cliver and an anonymous referee for helpful comments on the manuscript. This work was patially supported by SHINE under NSF Grant ATM-0353345.

References

Alexander, D., and Acton, L. W.: 2001, The Active Sun, in *The Century of Space Science*, Chapter 46, Kluwer, Netherlands.

Bigelow, F. H.: 1890, Further Study of the Solar Corona. New Haven.

Bryant, D. A., Cline, T. L., Desai, U. D., and McDonald, F. B.: 1962, *J. Geophys. Res.* **67**, 4983.

Burlaga, L. F.: 1991, in: Schwenn, R. and Marsch, E. (eds.), *Physics of the Inner Heliosphere 2*, Springer-Verlag, Berlin and Heidelberg, p. 1.

Cane, H. V., Reames, D. V., and von Rosenvinge, T. T.: 1988, *J. Geophys. Res.* **93**, 9555.

Cane, H. V., Richardson, I. G., and St. Cyr, O. C.: 2000, *Geophys. Res. Lett.* **27**, 3591.

Carmichael, H.: 1964, A Process for Flares, in AAS/NASA Symposium on Physics of Solar Flares. NASA SP-50, Washington, DC. 451.

Carrington, R. C.: 1860, *MNRAS*, **20**, 13.

Chapman, S. and Ferraro, V. C. A.: 1929, *Mon. Not. Roy. Astron. Soc.* **89**, 470.

Cliver, E. W.: 1989, *Solar Phys.*, **122**, 319.

Cliver, E. W.: 1994a, *EOS*, **75**, (569), 574–575.

Cliver, E. W.: 1994b, *EOS*, **75**, (609), 612–613.

Cliver, E. W.: 1995, *EOS*, **76**, 75–83.

Cocconi, G., Gold, T., Greisen, K., Hayakawa, S., and Morrison, P.: 1958, *Nuovo Cimento* **8**, 161.

Deslandres, H.: 1893, *Comptes Rendus Acad. Sci. Fr.* **116**, 127.

Eddy, J. A.: 1974, *Astron. Astrophys.* **34**, 235.

Eddy, J. A.: 1979, A New Sun: The Solar Results from Skylab. NASA, SP-402, Washington, DC.

Ellison, M. A.: 1943, *MNRAS* **103**, 3.

Forbush, S. E.: 1937, *Phys. Rev.* **51**, 1108.

Forbush, S. E.: 1946, *Phys. Rev.* **70**, 771.

Gold, T.: 1955, in van de Hulst, J. C. and Burgers, J. M. (eds.), *Gas Dynamics of Cosmic Clouds*, North-Holland Publishing Co., Amsterdam, p. 103.

Gosling, J. T., Hildner, E., MacQueen, R. M., Munro, R. H., Poland, A. I., and Ross, C. L.: 1974, *J. Geophys. Res.*, **79**, 4581.

Gosling, J. T., Hildner, E., MacQueen, R. M., Munro, R. H., Poland, A. I., and Ross, C. L.: 1976, *Solar Phys.* **48**, 389.

Greaves, W. M. H. and Newton, H. W.: 1928, *MNRAS*, **89**, 84.

Hale, G. E.: 1908, *Astrophys. J.* **28**, 315.

Hale, G. E.: 1931, *ApJ* **73**, 37953.

Hirayama, T.: 1974, *Solar Phys.* **34**, 323.

Hirshberg, J., Alksne, A., Colburn, D. S., Bame, S. J., and Hundhausen, A. J.: 1970, *J. Geophys. Res.* **75**, 1.

Howard, R. A., *et al.*: 1997, in: Crooker, N., Joselyn, J. A. and Feynman, J., (eds.), *Geophysical Monograph* 99, p. 17.

Hundhausen, A. J.: 1972, Coronal Expansion and Solar Wind, Springer-Verlag, New York.

Joselyn, J. A., and McIntosh, P. S.: 1981, *J. Geophys. Res.* **86**, 4555.

Kiepenheuer, K. O.: 1953, in Kuiper, G.P., (ed.) *University Press*, Chicago, p. 322.
Kopp, R. A., and Pneuman, G. W.: 1976, *Solar Phys.*, **50**, 85.
Lindemann, F. A.: 1919, *Phil. Mag.* **38**, 669.
Lindsay, G. M., Luhmann, J. G., Russell, C. T., and Gosling, J. T.: 1999, *J. Geophys. Res.* **104**, 12,515.
Meyer, P., Parker, E. N., and Simpson, J. A.: 1956, *Phys. Rev.* **104**, 768.
Morrison, P.: 1956, *Phys. Rev.* **101**, 1397.
Newton, H. W.: 1936, *Observatory*, **59**, 51.
Newton, H. W.: 1943, *MNRAS*, **103**, 244.
Parker, E. N.: 1961, *Astrophys. J.*, **133**, 1014.
Piddington, J. H.: 1958, *Phys. Rev.* **112**, 589.
Sabine, E.: 1852, *Philos. Trans. R. Soc. London*, **142**, 103.
Schwenn, R.: 1983, *Space Sci. Rev.* **34**, 85.
Sheeley, N. R. Jr., *et al.*: 1985, *J. Geophys. Res.* **90**, 163.
Sonnet, C. P., Colburn, D. S., Davis, L., Smith, E. J., and Coleman, P. J.: 1964, *Phys. Rev. Lett.* **13**, 153.
Strong K. T. *et al.*: 1998, The Many Faces of the Sun. Springer-Verlag, New York.
Sturrock, P. A. and Coppi, B.: 1966, *ApJ* **143**, 3.
Thompson, B. J., Plunkett, S. P., Gurman, J. B., Newmark, J. S., St. Cyr, O. C., and Michels, D. J.: 1998, *Geophys. Res. Lett.* **25**, 2465.
Tsurutani, B. T., Gonzalez, W. D., Lakhina, G. S., and Alex, S.: 2003, *J. Geophys. Res.* **108**, 1268.
Waldmeier, M.: 1941, Ergebinisse und Probleme der Sonnenforschung. Becker and Erler, Leipzig.
Webb, D. F., Cliver, E. W., Crooker, N. U., St. Cyr, O. C., and Thompson, B. J.: 2000, *J. Geophys. Res.* **105**, 7,491.
Wills-Davey, M. J., and Thompson, B. J.: 1999, *Solar Phys.*, **190**, 467.

CORONAL MASS EJECTIONS: OVERVIEW OF OBSERVATIONS

H. S. HUDSON[1,*], J.-L. BOUGERET[2] and J. BURKEPILE[3]
[1]*Space Sciences Laboratory, University of California, Berkeley, CA 94720, USA*
[2]*Observatoire de Paris, Meudon, France*
[3]*High Altitude Observatory, Boulder, CO, USA*
*(*Author for correspondence: E-mail: hhudson@ssl.berkeley.edu)*

(Received 11 October 2004; Accepted in final form 7 March 2006)

Abstract. We survey the subject of Coronal Mass Ejections (CMEs), emphasizing knowledge available prior to about 2003, as a synopsis of the phenomenology and its interpretation.

Keywords: sun, corona, CMEs

1. Background

A Coronal Mass Ejection (CME) is "...an observable change in coronal structure that (1) occurs on a time scale of a few minutes and several hours and (2) involves the appearance and outward motion of a new, discrete, bright, white-light feature in the coronagraph field of view" (Hundhausen *et al.*, 1984; Schwenn, 1996). With a kinetic energy that may exceed 10^{32} ergs, it is one of the most energetic forms of solar activity. We believe a CME in essence to be the eruption of a magnetically closed volume of the lower and middle corona.[1] The CMEs are interesting in their own right; they also have substantial effects on the Earth's environment. In this chapter we give an overview of the CME phenomenon, touching on all of its manifestations – traceable now from the photosphere into the distant heliosphere as far as human exploration has extended. This chapter summarizes the basic knowledge available prior to 2003. Figure 1 shows representative examples.

Originally termed "coronal transients," CMEs entered the modern era (but Figure 1 also shows one historical observation) with the Skylab observations (Gosling *et al.*, 1974; Munro *et al.*, 1979). Detailed records from the P78-1 coronagraph (Howard *et al.*, 1985) provided an early comprehensive view, including the discovery of the "halo CME" (Howard *et al.*, 1982; see also Alexander *et al.*, 2006, this volume) now known to be mainly responsible for terrestrial effects.

The modern view of CMEs has broadened considerably as the result of observations made by instruments other than coronagraphs at visual wavelengths. The Chapman Conference of 1997 (Crooker *et al.*, 1997) provides an excellent set of papers covering both the classical and the newer material available then.

[1]In our usage the lower and middle corona are below and above, respectively, the projected height of a typical coronagraphic occulting edge.

Space Science Reviews (2006) 123: 13–30
DOI: 10.1007/s11214-006-9009-x

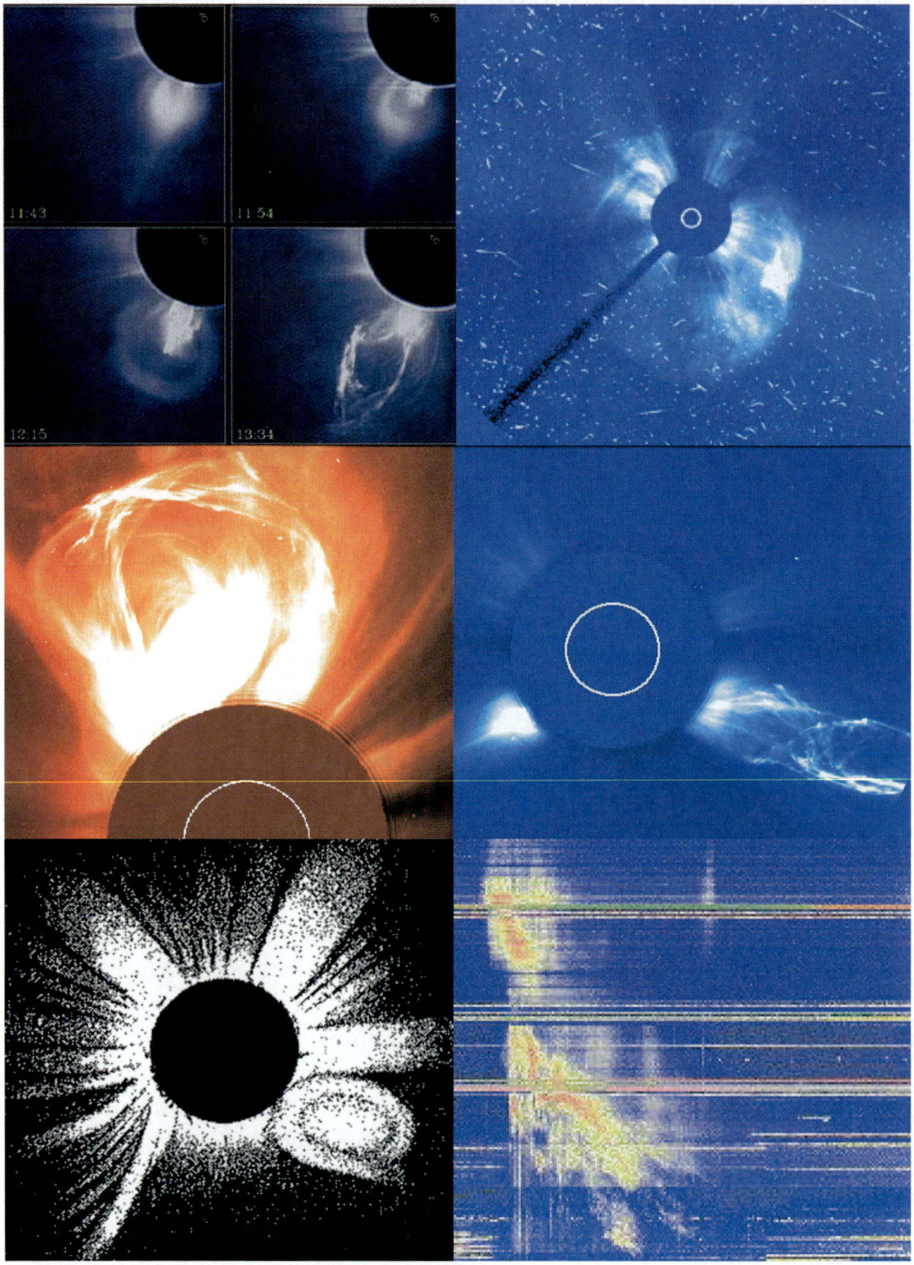

Figure 1. Six views of coronal mass ejections. Top: Prototypical "3-part CME" as observed by SMM; halo CME from LASCO. Middle: two views of flux-rope CMEs (LASCO). Bottom: Historical eclipse observation of possible CME; type II radio burst (Culgoora spectrogram).

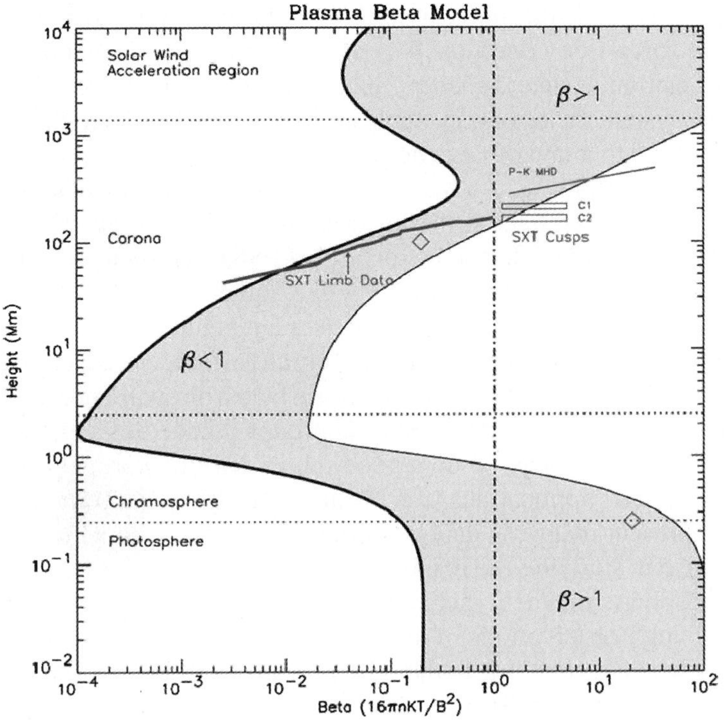

Figure 2. Survey of coronal plasma β, from Gary (2001), as a function of height above the photosphere. Note that this display ignores non-radial variation. A similar plot for Alfvén speed would show a radial decrease outward, followed by a rise to a local maximum in the upper corona, then a monotonic decline into the heliosphere.

The solar corona consists mainly of hot (10^6 K) and ionized plasma, bounded above by the solar wind and below by atmospheric layers at much lower temperatures. The magnetic field dictates the structure of the corona, according to its generally low plasma beta (the ratio of gas to magnetic pressure; see Figure 2). CME movies give the impression that a sector of the coronal field simply expands and opens out into the solar wind. It thus (temporarily, at least) must increase the open-field fraction of the photospheric field. The corona (to $10R_\odot$) contains 10^{18-19} g according to the semi-empirical models of Withbroe (1988). The mass content above $3R_\odot$, representative of the domain of coronagraphic observations, would not amount to 10^{15} g in the angular domain of a large CME, so that (as the images show) most of the CME mass typically originates in or below the lower corona.

Figure 2 shows estimates of the distribution of β with height (Gary, 2001); note that this survey ignores non-radial structure. Large local variations of plasma β occur in active regions because of the presence of dense loops. Our direct knowledge of the coronal magnetic field is extremely limited because of observational difficulties. As a result one must use representative ranges (as presented in Figure 2) or extrapolations from the photospheric Zeeman-splitting observations, usually based on the force-free condition $\nabla \times \mathbf{B} = \alpha\mathbf{B}$ (where α generally would be a function of position as determined by subphotospheric conditions). These extrapolations have systematic errors, the most obvious of which is that the photospheric observations refer to a layer that is not itself force-free.

In general the corona supports a system of currents, and so potential-field representations based upon data at the lower boundary cannot exactly represent the geometry. The "potential field source surface" (PFSS) method ingeniously sidesteps this problem (Altschuler and Newkirk, 1969; Schatten *et al.*, 1969), at least for the large-scale structure. In this approach one uses a potential representation from the photosphere out to an optimum spherical "source surface," almost universally now set at $2.5 R_\odot$. A fictitious current flows at this surface with such a distribution that the field external to it is strictly radial. Several groups pursue this practical approach, which (for example) appears to do a good job in defining coronal holes and open field for heliospheric applications (e.g., Wang *et al.*, 1996). Unfortunately it cannot be used to represent magnetic energy storage within the coronal domain itself, so it is of little use in studying the details of flare or CME evolution.

The photospheric magnetic field does not reflect CME occurrence in any obvious way, although observations of subtle flare effects do exist, especially in limb observations where a small tilt in the field may affect the line-of-sight component (Cameron and Sammis, 1999). This absence of strong effects is consistent with the general idea of coronal energy storage and release to explain the transients, but this conclusion must be understood more quantitatively. It is also consistent with the important idea (Melrose, 1995) that the vertical currents responsible for coronal magnetic energy storage must have their origin deep in the convection zone, and not vary appreciably during the transient.

CMEs usually come from active regions in close association with major solar flares, but they also can come from filament channels in the quiet Sun. The three-part structure for the quiet-Sun events, often associated with filament eruptions from the polar crown, can be directly identified with the appearance of a streamer cavity seen on the limb in white light or soft X-rays. Quiet-Sun events correspond to weak flare-like effects seen in chromospheric observations (Harvey *et al.*, 1986); such events often have slow, low-temperature soft X-ray emissions that do not produce recognizable GOES[2] signatures (e.g., Hudson *et al.*, 1995).

2. Techniques of Observation

CMEs are observed directly by white-light coronagraphs, mostly via photospheric light Thomson-scattered by coronal electrons. Eclipse images show coronal structure definitively well, and in spite of their infrequency have shown CMEs in rare historical cases (see Figure 1). Phenomena related to CMEs appear at virtually

[2]Geostationary Operational Environmental Satellite.

every observable wavelength (the "non-coronagraphic" observations; see Hudson and Cliver, 2001) as well as in many interplanetary signatures (e.g., Gosling, 1991).

2.1. Optical/UV

Bernard Lyot's invention of the coronagraph permitted time-series observations of changes in coronal structure. A coronagraph is a special-purpose telescope that images only the corona, suppressing the bright photosphere by either internal or external occultation; stray-light levels can now be reduced to the order of 10^{-15} of disk brightness at an elongation of 18° (Buffington *et al.*, 2003). The essential point of the visible-light observations is that they show the electron-scattered emission of the K-corona; the intensity thus determines the line-of-sight column density of the corona, which is optically thin outside prominences. The high temperature of the corona smears out the photospheric Fraunhofer line spectrum, but an emission-line component appears prominently at short wavelengths.

2.2. Radio

Within the vast spectral range of ground-based radio techniques (roughly 3×10^6 Hz to 10^{12} Hz) one finds a variety of emission mechanisms and observing techniques. The meter-decimeter wavelength ranges show us the corona mainly via coherent emission mechanisms; because these are bright at the plasma frequency one gets a rough measure of the density. At shorter wavelengths the optical depth decreases until at submillimeter wavelengths one sees right into the upper photosphere. Free-free emission can be detected from either over-dense coronal loops following flares or the quiet lower solar atmosphere; gyrosynchrotron radiation comes from high-energy electrons. Below about 10 MHz radio receivers in space allow us to study solar-wind phenomena as far down as the local plasma frequency at 1 AU, normally at $\sim 3 \times 10^4$ Hz.

2.3. EUV/X-ray

The EUV and X-ray wavelengths show us the K-corona directly in emission. The emissivity of the hot corona decreases rapidly at short wavelengths, but the extreme temperature dependence ($\propto e^{-h\nu/kT}$ in the limit) results in large image contrast for X-rays at $h\nu > kT$. Focusing optics (grazing incidence for soft X-rays to a few keV; normal incidence for narrow-band imaging longward of about 100 Å) with good angular resolution led to many discoveries. The first systematic X-ray and EUV observations were those from Skylab, and showed coronal holes, flares, CME-related ejecta and dimmings, and in general many counterparts of phenomena previously studied only at other wavelengths. The normal-incidence TRACE observations have revolutionized our views of coronal dynamics, owing to their high resolution (0.5″ pixels; see Handy *et al.*, 1999).

2.4. Interplanetary

The interplanetary data mostly consist of *in-situ* measurements of particles and fields, in which one characterizes the bulk parameters (speed, density, temperature, magnetic field) of the solar wind, plus the distribution functions and abundances (ionization states, elements, isotopes) within the plasma (Zurbuchen and Richardson, 2006, this volume; Wimmer *et al.*, 2006, this volume). These include solar energetic particles resulting ultimately from flares and CMEs; the interplanetary shock waves have a close association with CMEs (Sheeley *et al.*, 1985), and these shock waves cause SEP (Solar Energetic Particle) events (e.g., Reames, 1999; Klecker *et al.*, 2006, this volume; Cane and Lario, 2006, this volume). Most of the interplanetary observations are from near-Earth space, but Helios, Ulysses, and the Voyagers have now explored as far in as $0.3R_{\odot}$, out of the ecliptic plane, and out to the heliopause (Gazis *et al.*, 2006 his volume).

3. Coronagraphic Observations

3.1. White Light

CMEs are unambiguously identified in white light coronal observations as outward-moving density structures (Tousey, 1973; Gosling *et al.*, 1974). The rate at which they occur correlates well with the solar activity cycle (Webb and Howard, 1994); (St. Cyr *et al.*, 2000); their appearance does not significantly differ between sunspot minimum and sunspot maximum. CMEs often appear as a “three-part” structure comprised of an outer bright front, and a darker underlying cavity within which is embedded a brighter core as shown in Figure 1 (Hundhausen, 1987). The front may contain swept-up as well as primary material (Hildner *et al.*, 1975; Illing and Hundhausen, 1985). The cavity is a region of lower plasma density but probably higher magnetic field strength. The cores of CMEs can often be identified as prominence material on the basis of their visibility in chromospheric emission lines (Sheeley *et al.*, 1975; Schmieder *et al.*, 2002) and often appear to have helical structure.

In addition to the familiar 3-part CMEs, other types commonly occur – narrow CMEs and CMEs with clear flux-rope morphology, in particular (Howard *et al.*, 1985). Halo CMEs (Figure 1) have special properties resulting from projection effects (see Burkepile *et al.*, 2004).

Five different coronagraphs have contributed substantial information about CME properties in a statistical sense: those on Skylab, Solwind, SMM, and SOHO from space, and the MK3 coronagraph at Mauna Loa Solar Observatory. These instruments have different properties (sampling, radius of occulting edge, epoch of observation) but a consistent picture generally prevails. We can distinguish the observational properties of CMEs into morphological (geometry, kinematics) and

TABLE I
CMEs: average properties.

	MK3[a]	SMM[b]	Skylab[c]	Solwind[d]	LASCO[e]
Period of observation	1980–99	1980 1984–89	1973–1974	1979–1980 1984–85	1996– present
Field of view ($R_\odot$)	1.15–2.24	1.8–∼5	2–6	3–10	1.1–32
Angular Size (deg)	37	47	42	43	72
Speed (km/s)	390	349	470	460	424
Mass (g)		3.3×10^{15}	4.7×10^{15}	4.0×10^{15}	1.7×10^{15}
K. E. (erg)		6.7×10^{30}	3.1×10^{30}	3.4×10^{30}	4.3×10^{30}
P. E. (erg)		7.1×10^{30}	8.0×10^{30}		

[a] St. Cyr *et al.* (1999).
[b] Hundhausen (1993).
[c] Gosling *et al.* (1976), Rust (1979) and Hundhausen (1993).
[d] Howard *et al.* (1985) and Howard *et al.* (1986).
[e] St. Cyr *et al.* (2000) and Vourlidas *et al.* (2002).

physical (mass, energy) categories. For reference we quote the average properties from the different sources in Table I; these are roughly consistent among the different data sets.

It is important to note that these are measurements of CME apparent properties as seen projected in two dimensions in an optically thin medium. This projection introduces systematic distortions in the appearance of the object and makes the determination of point properties more difficult and generally model-dependent. The distortions are small for structures close to the "plane of the sky" (i.e., the plane containing the solar limb) but can be severe elsewhere. Objects located away from the plane of the solar limb appear at higher apparent latitudes, have larger apparent widths and lower apparent heights than their true values (Hundhausen, 1993; Burkepile *et al.*, 2004). In addition, the lower apparent heights lead to underestimates of CME speeds (Hundhausen *et al.*, 1994). The underestimation of the height also impacts the brightness and, hence, the mass estimate.

3.2. Morphological and Kinematical Properties

3.2.1. *Position Angles*

The apparent latitude of a CME is typically determined from the position angle of its projected angular centroid (Howard *et al.*, 1985). Hundhausen (1993) showed that this depends strongly upon the CME source location. They also found the distribution of apparent latitudes of CMEs to be unimodal and to center at the heliomagnetic equator. There is a systematic variation with the solar cycle.

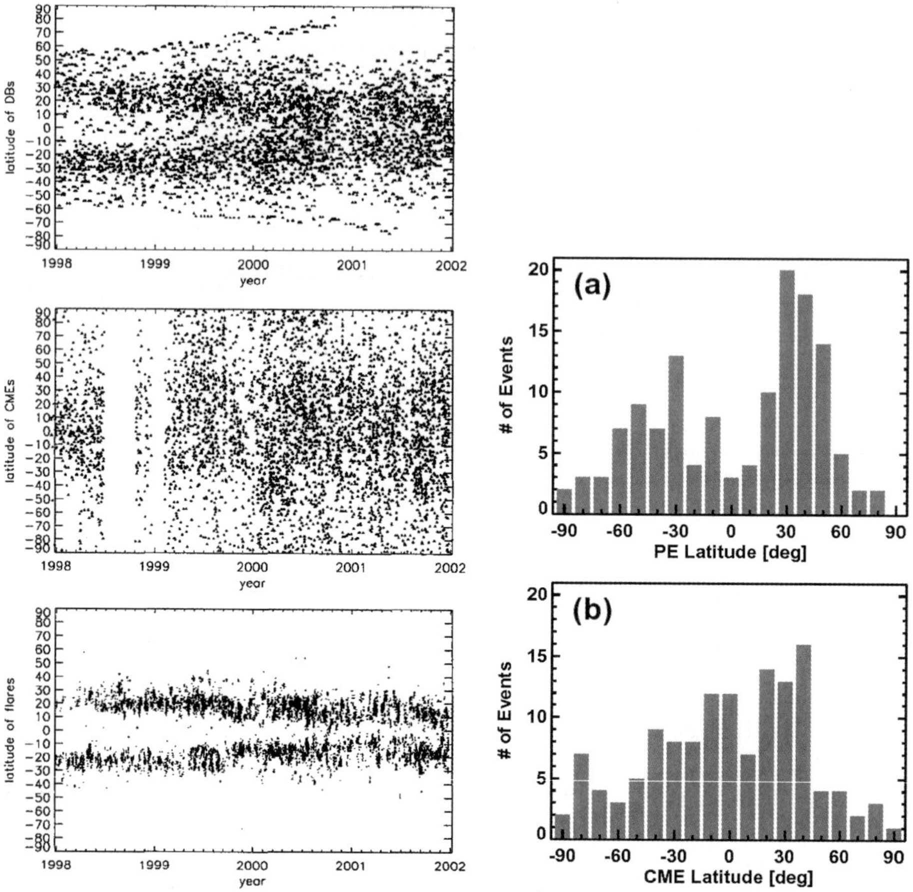

Figure 3. Left: Apparent latitudes (position angles) of CME occurrence, as observed by SOHO (center panel) compared with disappearing filaments (*top*) and flares (*bottom*) (from Pojoga and Huang (2003)). *Right*: Similar comparison between microwave-observed filament locations (*top*) and their corresponding CMEs (Gopalswamy *et al.*, 2003). The statistical views show that CME origins in the low corona (flares or CME eruptions) have a bimodal distribution in latitude, whereas the CMEs have a unimodal distribution concentrated at the equator.

Around solar minimum the CMEs tend to occur at lower latitudes, and as the rise to maximum occurs, the apparent latitudes increase. The CME apparent latitudes are well-correlated with the latitude distribution of the helmet streamers (Hundhausen, 1993) rather than with the "butterfly diagram" latitudes of active regions. The LASCO observations of the current cycle (St. Cyr *et al.*, 2000) confirm this observation (Figure 3).

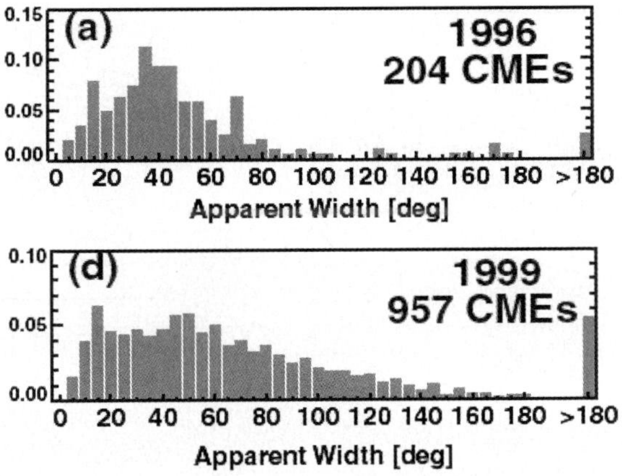

Figure 4. Angular sizes of CMEs vs. phase in the solar cycle, based upon LASCO observations (St. Cyr *et al.*, 2000). The number of wider CMEs increases towards solar maximum (Hundhausen, 1993).

3.2.2. *Angular Sizes*

The smallest average CME angular size in Table I is measured in the low coronal measurements from MK3 (St. Cyr *et al.*, 1999). This suggests that some CMEs may expand in the early stages of their formation and propagation, particularly those events (the majority; see (Subramanian and Dere, 2001) that originate in and near active regions (Dere *et al.*, 1997). The higher average angular sizes determined from the outer coronal observations from LASCO (St. Cyr *et al.*, 2000) probably result from projection, since the LASCO coronagraphs are able to detect many disk-centered CMEs with large apparent widths. Figure 4 compares CME apparent widths between states of low and high solar activity (St. Cyr *et al.*, 2000). The data generally indicate a decrease in the percentage of wide CMEs during the descending or minimum phases of the solar cycle for each of the three datasets.

3.2.3. *Speeds*

The average CME speeds determined from the various datasets do not vary significantly (see Table I). This speed, however, does have a solar-cycle dependence, though not a simple one. Both SMM and Solwind report very low speeds for CMEs in 1984, during the declining phase of activity. However, the average SMM CME speeds are higher in 1985 and 1986, at solar minimum, due to the appearance of new active regions which are associated with a handful of high-speed CMEs. The lowest average LASCO CME speed occurs at solar minimum (1996) and gradually increases through 1998 with the appearance of a high-speed tail in the distribution which may be associated with the occurrence of new-cycle active regions. CMEs associated with active regions have higher average speeds than CMEs associated with eruptive prominences located away from active regions (Gosling *et al.*, 1976).

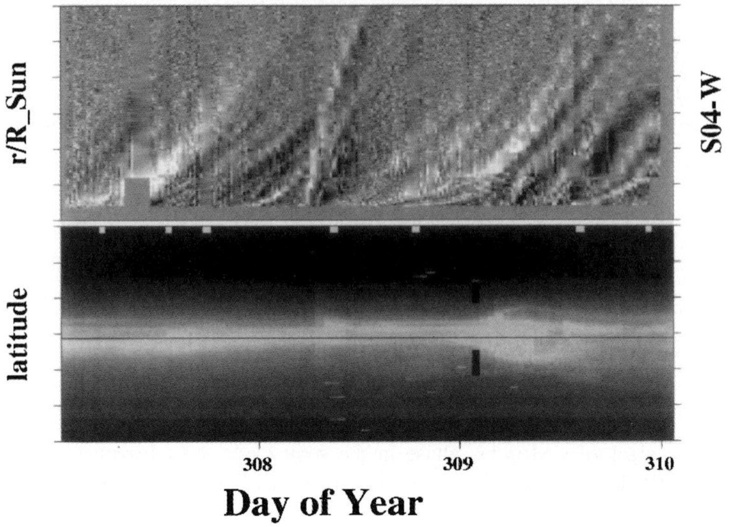

Figure 5. Illustration of the two types of CME motion suggested by Sheeley *et al.* (1999). The upper panel shows brightness distribution along a radial line (in this case 4° N of W, or a position angle of 274°). The decelerating event of Nov. 4, 1997, occurs early on Day 308 and was associated with an X2.1 flare at S14, W33. Many accelerating events can be seen as well.

3.2.4. *Accelerations*

MacQueen and Fisher (1983) found that CMEs associated with flares had more rapid accelerations. Sheeley *et al.* (1999), on this basis, argue for the existence of two types of CMEs: those associated with flares, which tend to appear at full speed and then decelerate, and the filament-eruption CMEs, which slowly accelerate (see Figure 5 for examples).

3.3. Physical Properties

3.3.1. *Masses*

The excess brightness of a given image relative to a pre-event image gives a "snapshot" estimate of CME mass via the plane-of-the-sky assumption. This represents a lower limit, and a snapshot also does not capture the continuing enhanced flow often seen long after the initial eruption. Standard assumptions are (1) that all of the CME material is located in the plane of the sky, and (2) that the corona is a completely ionized plasma consisting of 90% hydrogen and 10% helium (Vourlidas *et al.*, 2000).

3.3.2. *Energies*

The kinetic and potential energies of a CME can be determined from the inferred masses and velocities, subject to the projection biases. The total mechanical energy of a major CME obtained in this manner is of order 10^{31-32} ergs, with the potential energy dominating for flux-rope CMEs (Vourlidas *et al.*, 2000). The magnetic energy of a CME is the dominant factor; it is widely agreed that CMEs result from a conversion of magnetic energy into the other forms, but we have no direct observations and cannot confirm this. The energetics estimates of Vourlidas *et al.* (2000) suggest that the magnetic energy does in fact diminish as the kinetic and potential energies increase.

There are inherent inaccuracies in the estimates of CME energetics. CME masses are underestimated, due to the assumption that all of the material lies in the plane containing the solar limb. CME mass and speed underestimations become significant for CME components more than $\sim$30 degrees from the plane of the solar limb (see Hundhausen, 1993, Appendix A and Hundhausen *et al.*, 1994).

3.3.3. *Energy or Mass Distribution*

Because of the lack of direct estimates of the dominant component, the magnetic energy, it is doubtless premature to draw conclusions from the distribution of CME total energies; but the masses and kinetic energies are available. Vourlidas *et al.* (2002) suggest power-law distributions for the mass and kinetic energy, rather than the exponential distribution of Jackson and Howard (1993). The inferred power laws are flatter than those observed for flares (e.g., Hudson, 1991).

3.4. UV and EUV Limb Spectroscopy

The UV and EUV spectrographic observations of CMEs provide diagnostic information but suffer from limited sensitivity. SOHO carries two UV spectrographs (UVCS for coronal observations, and SUMER for disk observations, but operated for most of the mission with its slit positioned above the limb in a coronagraphic mode).

Raymond *et al.* (2003) discuss three well-observed CMEs, each associated with an X-class flare near the limb. The UVCS observing slit was positioned approximately tangent to the limb at a height of 1.64 $R_\odot$ above it, and with an observing cadence of 120 s for spectra of a variety of UV emission lines, including some with high formation temperatures (notably Fe XVIII above 6×10^6 K). This high-temperature emission occurs in narrow structures the authors identify with the current sheets expected to form after the eruption (Ciaravella *et al.*, 2002; Ko *et al.*, 2003).

SUMER has provided observations that may be more directly related to flare energy release in large-scale reconnection. The original observation of downflows in soft X-rays by McKenzie and Hudson (1999) suggested reconnection outflow with a complex structure and clearly sub-Alfvénic velocities. SUMER observations have confirmed that the principal components of these downflows have low densities, being undetectable in any temperature regime (Innes *et al.*, 2003).

4. Non-Coronagraphic Observations

Much of the interesting development of a CME takes place in the lower corona, below the coronagraph's occulting edge. Even if this edge could be placed exactly at the solar limb, a halo CME originating at disk center would be at a large radial distance from the Sun before any part of it became visible. Luckily there are many wavebands, ranging from radio to X-ray, that in principle reveal the CME development from the photosphere outwards. One must be cautious interpreting these non-coronagraphic observations, however, because they show aspects of the CME disturbance that may not be directly identifiable with the mass distribution as seen in a coronagraph. Radio observations, in particular, normally show only non-thermal particles and thus give a picture of the overall structure that is biased towards those parts containing energetic particles, specifically electrons far out in the tail of the velocity distribution function. The "calibration" of these different kinds of observation presents problems to the extent that we may need to rely upon theory and modeling (or even cartoon descriptions) to link one feature with another observed by very different means (Hudson and Cliver, 2001).

4.1. X-Ray and EUV Imaging

We have now had more than a decade of systematic exploration of the solar corona via soft X-ray and EUV imaging from Yohkoh, SOHO, and TRACE. These new data have gone far beyond the pioneering observations from Skylab, especially in terms of sensitivity and of sampling. The essential contributions of these new observations lie in several domains: the direct observation of ejecta (Klimchuk *et al.*, 1994; Nitta and Akiyama, 1999); the detailed observation of coronal dimming (Hudson and Webb, 1997); and the observation of EIT waves (Moses *et al.*, 1997); Thompson *et al.* (1999). Such observations show that the coronal restructuring underlying the CME phenomenon in fact extends throughout the corona, consistent with the simple idea that the CME simply opens the coronal magnetic field into an enhanced solar-wind flow. Spectroscopic observations from SOHO (Harra and Sterling, 2001; Harrison *et al.*, 2003) confirm that the X-ray dimmings do represent material depletions rather than a temperature effect (Hudson *et al.*, 1996).

The X-ray and EUV observations of eruptions should be considered in the context of the behavior of filaments observed in Hα emission. Filaments give a different glimpse at coronal behavior during the CME process. The onset of filament activity, together with a gradual rising motion presumably related to streamer swelling, may precede the actual eruption by tens of minutes. In some cases the erupting filament continues into the outer corona, where it forms the dense core of a classical three-part CME structure; in other cases the filament appears to stop ("confined explosion" or failed eruption"; (see, e.g., Moore *et al.*, 2001; Ji *et al.*, 2003), and in some CMEs there appears to be no filament involvement at all. The

X-ray observations (Kano, 1994; Hanaoka *et al.*, 1994) show that the filament matter may heat rapidly during the eruption, and the EUV observations often show both cool and hot phases of the filament during its eruption.

The direct observations of CME counterparts in the low corona help greatly with understanding the time sequence of the eruption. The X-ray dimmings could be directly interpreted as a part of the coronal depletion required for a CME (Hudson *et al.*, 1996; Sterling and Hudson, 1997; Hudson and Webb, 1997). The dimmings turn out to coincide with the flare brightening, suggesting that the flare energization and CME acceleration can be identified (Zarro *et al.*, 1999). This close timing relationship has also been found with the LASCO C1 observations, which have the lowest occulting edge and hence the least timing ambiguity (Zhang *et al.*, 2001).

Large-scale shock waves in the corona and heliosphere play a major role in any discussion of CMEs (Schwenn, 1986); indeed the CME disturbance itself is describable in terms of MHD waves (e.g., Chen *et al.*, 2002). The type II bursts provided the first evidence for the passage of global waves through the corona and heliosphere, and the Moreton waves in the chromosphere (e.g., Athay and Moreton, 1961) were put into the same context by the Uchida (1968) theory of weak fast-mode MHD shock emission from solar flares. Interplanetary shocks and geomagnetic impulses (e.g., Chapman and Bartels, 1940), on the other hand, have a natural interpretation in terms of bow shocks driven ahead of the CME ejecta.

4.2. Radio Signatures

Radio-frequency observations provided some of the first clues of large-scale restructuring of the solar corona during a CME. The metric wavelength band (30–300 MHz) led to the well-known event classification (the type I–V bursts; see Kundu, 1965). Space-borne receivers extended the observational domain down to $\sim$30 kHz, and at shorter wavelengths ground-based observations have generally improved in resolution and coverage. These bursts tell us about energetic electrons either trapped in large-scale coronal magnetic structures or propagating through them on open field lines. In particular, the type II bursts reveal MHD shock waves propagating away from coronal disturbances such as flares and CMEs. We also now have clear observations of the elements of the classical 3-part CME structure via gyrosynchrotron emission at decimetric wavelengths and via free-free emission at centimetric wavelengths (Bastian *et al.*, 2001).

The radio observations provide key information about the connectivity of the coronal magnetic field. The type III bursts show that open (i.e., heliospheric) magnetic fields can originate in active regions as well as in coronal holes; the exciter (an electron beam) can be traced over at least four decades in frequency or 8 decades of density.

4.3. In-Situ Measurements

CMEs often have observable consequences further out in the heliosphere. There is still no consensus regarding the mapping of features seen in (coronagraphic) CME observations with the interplanetary phenomena (Interplanetary CMEs, or ICMEs; see Schwenn, 1995; Forsyth *et al.*, 2006, this volume; Wimmer *et al.*, 2006 this volume), but there are many specific signatures (e.g., Gosling and Forsyth, 2001; Zurbuchen and Richardson, 2006, this volume). These range from magnetic clouds (Burlaga *et al.*, 1982), with a highly organized flux-rope magnetic pattern, to solar energetic particles (SEPs) (Cane and Lario, 2006, this volume; Klecker *et al.*, 2006, this volume), whose acceleration is directly related to CME dynamics but which are observed on field lines not directly a part of the solar ejecta. The presence or absence of particular signatures varies from event to event, but counterstreaming electrons (i.e., suprathermal electrons with pitch-angle distributions aligned both parallel and antiparallel to the field) are commonly interpreted as indicating that the connectivity of the field is "closed" (here meaning tied to the Sun at both ends, hence the result of an ejection), even though the observations are carried out in the heliosphere (hence "open" from the solar-wind point of view).

The prevalence of flux-rope signatures in ICMEs, which are especially clear in the Ulysses high-latitude events (Gosling *et al.*, 1995), strongly suggests several aspects of the solar imaging observations. It is now clear that the "disconnection" events long-sought in coronagraphic signatures are rare, but the common occurrence of concave-up structures points instead to flux ropes formed in the corona.

The ICME magnetic properties can in principle be used to learn about the source regions of CMEs (Bothmer and Schwenn, 1994; Rust and Kumar, 1996; Cremades and Bothmer, 2004; Crooker and Horbury, 2006, this volume). Although we do not yet have a complete understanding of the mapping of ICME components to structures in the low corona, filament channels often play an important role. The prevalence of forward-reverse shock pairs in high-latitude ICMEs, an indication of overexpansion (Gosling *et al.*, 1994) may reflect the non-radial expansion observed in the low corona by many techniques e.g., (Cremades and Bothmer, 2004). The particles observed within a CME can also be used as tracers of the magnetic connectivity (Kahler and Reames, 1991; Larson *et al.*, 1997); the nearly relativistic particles at higher energies are especially interesting because of their short propagation times. The observations of impulsive particle events closely associated with flares (e.g., Kahler *et al.*, 2001) confirms the knowledge from radio type III bursts that open (i.e., connected into the solar wind) field lines commonly occur in active regions near sunspots.

5. Remarks on Theory

The theory of coronal mass ejections involves a complicated system with large parameter ranges and an ill-understood coupling between large and small scales during the process of eruption. Accordingly the existing theories (e.g., Forbes, 2000) are more descriptive than predictive in nature (see Figure 6 for a cartoon representation given by Forbes, adding swept-up mass and a bow shock to the eruptive-flare cartoon of Hirayama (1974) and Anzer and Pneuman (1982). Much of the theoretical work must be carried out in large-scale numerical simulations, and the scale of the problem unfortunately limits them to the (resistive) MHD approximation and to scales far larger than those thought necessary to capture the microscopic physics.

Most modern models invoke magnetic energy storage in the corona, which is released either by a dissipative process or by an ideal MHD loss of equilibrium in a low-β environment. Other models invoke direct driving by injection of twist from below the photosphere during the eruption; still others make use of gravitational energy stored in or above the erupting medium. In the dissipative models one describes the restructuring in terms of magnetic reconnection, either below the erupting structure ("tether cutting" or "emerging flux") or above it ("breakout"; Antiochos *et al.*, 1999). These models would share the geometry of Figure 6 but would differ in the

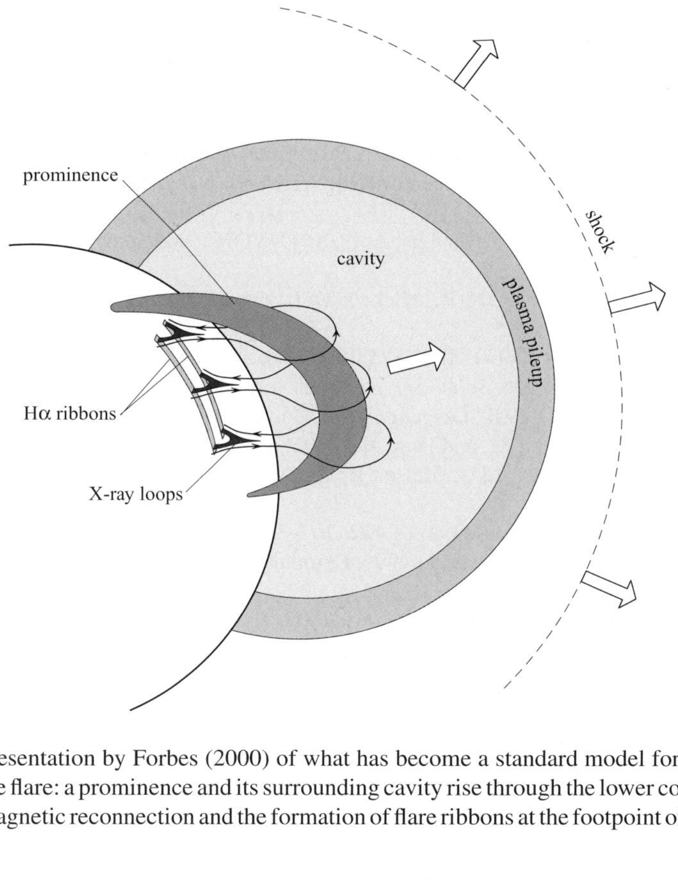

Figure 6. Representation by Forbes (2000) of what has become a standard model for a "three-part" CME or eruptive flare: a prominence and its surrounding cavity rise through the lower corona, followed by sequential magnetic reconnection and the formation of flare ribbons at the footpoint of a loop arcade.

initiation mechanism. Many observational papers strive to identify these processes, but it would be fair to say that the current results are ambiguous. One grave problem with essentially all of the models is that they remain in the MHD framework and thus cannot deal self-consistently with energetic particles.

Finally, the fact of CME existence leads to several further interesting theoretical problems relating to their propagation into the heliosphere. First among these would be the problem of solar open flux (Gold, 1962; Crooker *et al.*, 2002; Crooker and Horbury, 2006, this volume); CMEs regularly increase the fraction of solar open field, and have a strong solar-cycle occurrence pattern, so why doesn't the magnetic intensity in the heliosphere steadily increase? Second, the ejected magnetic flux is often twisted to form flux ropes, and these will transport magnetic helicity away from the Sun (Low, 1994; Kumar and Rust, 1996) – ultimately, from the interior dynamo itself?

References

Alexander, D., Richardson, I. G., and Zurbuchen, T. H.: 2006, *Space Science Rev.* (this volume) doi: 10.1007/s11214-006-9008-y.

Altschuler, M. D., and Newkirk, G.: 1969, *Solar Phys.*, **9**, 131.

Antiochos, S. K., Devore, C. R., and Klimchuk, J. A.: 1999, *ApJ* **510**, 485–493.

Anzer, U., and Pneuman, G. W.: 1982, *Solar Phys.*, **79**, 129–147.

Athay, R. G., and Moreton, G. E.: 1961, *ApJ* **133**, 935.

Bastian, T. S., Pick, M., Kerdraon, A., Maia, D., and Vourlidas, A.: 2001, *ApJL* **558**, L65–L69.

Bothmer, V., and Schwenn, R.: 1994, *Space Science Reviews* **70**, 215.

Buffington, A., Jackson, B. V., and Hick, P. P.: 2003, *Innovative Telescopes and Instrumentation for Solar Astrophysics.* Edited by Stephen L. Keil, Sergey V. Avakyan . Proceedings of the SPIE, pp. 490–503.

Burkepile, J. T., Hundhausen, A. J., Stanger, A. L., St. Cyr, O. C., and Seiden, J. A.: 2004, *J. Geophys. Res. (Space Physics)* pp. 3103.

Burlaga, L. F., Klein, L., Sheeley, N. R., Michels, D. J., Howard, R. A., Koomen, M. J., *et al.*,: 1982, *GRL* **9**, 1317–1320.

Cameron, R., and Sammis, I.: 1999, *ApJL* **525**, L61–L64.

Cane, H. V., and Lario, D.: 2006, *Space Sci. Rev.* (this volume) doi: 10.1007/s11214-006-9011-3.

Chapman, S., and Bartels, J.: 1940, *Geomagnetism*, University Press, Oxford.

Chen, P. F., Wu, S. T., Shibata, K., and Fang, C.: 2002, *ApJL* **572**, L99–L102.

Ciaravella, A., Raymond, J. C., Li, J., Reiser, P., Gardner, L. D., Ko, Y.-K., and Fineschi, S.: 2002, *ApJ* **575**, 1116–1130.

Cremades, H., and Bothmer, V.: 2004, *A&A* **422**, 307–322.

Crooker, N., Joselyn, J., and Feynman, J.: 1997, *Coronal Mass Ejections: Causes and Consequences.* Geophysical Monographs #99.

Crooker, N. U., Gosling, J. T., and Kahler, S. W.: 2002, *J. Geophys. Res. (Space Phys.)* **107**, 3–1.

Crooker, N. U., and Horbury, T. S.: 2006, *Space Sci. Rev.* (this volume) doi: 10.1007/s11214-006-9014-0.

Dere, K. P. *et al.*: 1997, *Solar Phys.*, **175**, 601–612.

Forbes, T. G.: 2000, *JGR* **105**(14), 23153–23166.

Forsyth, R. J., Bothmer, V., *et al.*: 2006, *Space Sci. Rev.* (this volume) doi: 10.1007/s11214-006-9022-0.

Gary, G. A.: 2001, *Solar Phys.*, **203**, 71–86.

Gazis, P. R., and Balogh, A.: 2006, *Space Sci. Rev.* (this volume) doi: 10.1007/s11214-006-9023-z.
Gold, T.: 1962, *Space Sci. Rev.* **1**, 100.
Gopalswamy, N., Shimojo, M., Lu, W., Yashiro, S., Shibasaki, K., and Howard, R. A.: 2003, *ApJ* **586**, 562–578.
Gosling, J. T.: 1991, *NASA STI/Recon Technical Report N* **92**, 17725.
Gosling, J. T., Bame, S. J., McComas, D. J., Phillips, J. L., Balogh, A., and Strong, K. T.: 1995, *Space Sci. Rev.* **72**, 133.
Gosling, J. T., and Forsyth, R. J.: 2001, *Space Sci. Rev.* **97**, 87–98.
Gosling, J. T., Hildner, E., MacQueen, R. M., Munro, R. H., Poland, A. I., and Ross, C. L.: 1974, *JGR* **79**(18), 4581–4587.
Gosling, J. T., Hildner, E., MacQueen, R. M., Munro, R. H., Poland, A. I., and Ross, C. L.: 1976, *Solar Phys.*, **48**, 389–397.
Gosling, J. T., McComas, D. J., Phillips, J. L., Weiss, L. A., Pizzo, V. J., Goldstein, B. E., *et al.*: 1994, *GRL* **21**, 2271–2274.
Hanaoka, Y., Kurokawa, H., Enome, S., Nakajima, H., Shibasaki, K., Nishio, M., *et al.*: 1994, *PASJ* **46**, 205–216.
Handy, B. N., *et al.*: 1999, *Solar Phys.* **187**, 229–260.
Harra, L. K., and Sterling, A. C.: 2001, *ApJL* **561**, L215–L218.
Harrison, R. A., Bryans, P., Simnett, G. M., and Lyons, M.: 2003, *A&A* **400**, 1071–1083.
Harvey, K., Sheeley, N., and Harvey, J.: 1986, *Solar-Terrestrial Predictions* p. 198.
Hildner, E., Gosling, J. T., Hansen, R. T., and Bohlin, J. D.: 1975, *Solar Phys.* **45**, 363–376.
Hirayama, T.: 1974, *Solar Phys.* **34**, 323.
Howard, R. A., Michels, D. J., Sheeley, N. R., and Koomen, M. J.: 1982, *ApJL* **263**, L101–L104.
Howard, R. A., Sheeley, N. R., Michels, D. J., and Koomen, M. J.: 1985, *JGR* **90**(9), 8173–8191.
Howard, R. A., Sheeley, N. R., Michels, D. J., and Koomen, M. J.: 1986, *ASSL Vol. 123: The Sun and the Heliosphere in Three Dimensions*, pp. 107–111.
Hudson, H., Haisch, B., and Strong, K. T.: 1995, *JGR* **100**(9), 3473–3477.
Hudson, H. S.: 1991, *Solar Phys.* **133**, 357–369.
Hudson, H. S., Acton, L. W., and Freeland, S. L.: 1996, *ApJ* **470**, 629.
Hudson, H. S., and Cliver, E. W.: 2001, *JGR* **106**, 25199–25214.
Hudson, H. S., and Webb, D. F.: 1997, *Geophys. Monogr. #99* p. 27.
Hundhausen, A. J.: 1993, *JGR* **98**(17), 13177.
Hundhausen, A. J., Burkepile, J. T., and St. Cyr, O. C.: 1994, *JGR* **99**(18), 6543–6552.
Hundhausen, A. J., Sawyer, C. B., House, L., Illing, R. M. E., and Wagner, W. J.: 1984, *JGR* **89**(18), 2639–2646.
Hundhausen, A. J.: 1987, *Sixth International Solar Wind Conference*, 181.
Illing, R. M. E., and Hundhausen, A. J.: 1985, *JGR* **90**(9), 275–282.
Innes, D. E., McKenzie, D. E., and Wang, T.: 2003, *Solar Phys.* **217**, 247–265.
Jackson, B. V., and Howard, R. A.: 1993, *Solar Phys.* **148**, 359.
Ji, H., Wang, H., Schmahl, E. J., Moon, Y.-J., and Jiang, Y.: 2003, *ApJL* **595**, L135–L138.
Kahler, S. W., and Reames, D. V.: 1991, *JGR* **96**(15), 9419–9424.
Kahler, S. W., Reames, D. V., and Sheeley, N. R.: 2001, *ApJ* **562**, 558–565.
Kano, R.: 1994, *X-ray Solar Physics from Yohkoh* pp. 273.
Klecker, H., Kunow, H., *et al.*: 2006, *Space Sci. Rev.* (this volume) doi: 10.1007/s11214-006-9018-9.
Klimchuk, J. A., Acton, L. W., Harvey, K. L., Hudson, H. S., Kluge, K. L., Sime, D. G., *et al.*: 1994, *X-ray Solar Physics from Yohkoh* pp. 181.
Ko, Y., Raymond, J. C., Lin, J., Lawrence, G., Li, J., and Fludra, A.: 2003, *ApJ* **594**, 1068–1084.
Kumar, A., and Rust, D. M.: 1996, *JGR* **101**(10), 15667–15684.
Kundu, M. R.: 1965, *Solar Radio Astronomy*, Interscience Publication, New York.
Larson, D. E., Lin, R. P., McTiernan, J. M., McFadden, J. P., Ergun, R. E., McCarthy, M., *et al.*: 1997, *GRL* **24**, 1911.

Low, B. C.: 1994, *Phys. Plasmas*, **1**, 1684–1690.
MacQueen, R. M., and Fisher, R. R.: 1983, *Solar Phys.* **89**, 89–102.
McKenzie, D. E., and Hudson, H. S.: 1999, *ApJL* **519**, L93–L96.
Melrose, D. B.: 1995, *ApJ* **451**, 391.
Mikić, Z., and Lee, M. A.: 2006, *Space Sci. Rev.* (this volume) doi: 10.1007/s11214-006-9012-2.
Moore, R. L., Sterling, A. C., Hudson, H. S., and Lemen, J. R.: 2001, *ApJ* **552**, 833–848.
Moses, D. F. *et al.*: 1997, *Solar Phys.* **175**, 571–599.
Munro, R. H., Gosling, J. T., Hildner, E., MacQueen, R. M., Poland, A. I., and Ross, C. L.: 1979, *Solar Phys.* **61**, 201–215.
Nitta, N., and Akiyama, S.: 1999, *ApJL* **525**, L57–L60.
Pojoga, S., and Huang, T. S.: 2003, *Adv. Space Res.* **32**, 2641–2646.
Raymond, J. C., Ciaravella, A., Dobrzycka, D., Strachan, L., Ko, Y.-K., Uzzo, M., *et al.*: 2003, *ApJ* **597**, 1106–1117.
Reames, D. V.: 1999, *Space Sci. Rev.* **90**, 413–491.
Rust, D. M.: 1979, *IAU Colloq. 44: Physics of Solar Prominences*, pp. 252–266.
Rust, D. M., and Kumar, A.: 1996, *ApJL* **464**, L199.
Schatten, K. H., Wilcox, J. M., and Ness, N. F.: 1969, *Solar Phys.* **6**, 442–455.
Schmieder, B., van Driel-Gesztelyi, L., Aulanier, G., Démoulin, P., Thompson, B., de Forest, C., *et al.*: 2002, *Adv. Space Res.* **29**, 1451–1460.
Schwenn, R.: 1986, *Space Sci. Rev.* **44**, 139–168.
Schwenn, R.: 1995, *Solar Wind Conference* pp. 45.
Schwenn, R.: 1996, *APSS* **243**, 187.
Sheeley, N. R., Bohlin, J. D., Brueckner, G. E., Purcell, J. D., Scherrer, V. E., Tousey, R., *et al.*: 1975, *Solar Phys.* **45**, 377–392.
Sheeley, N. R., Howard, R. A., Michels, D. J., Koomen, M. J., Schwenn, R., Muehlhaeuser, K. H., *et al.*: 1985, *JGR* **90**(9), 163–175.
Sheeley, N. R., Walters, J. H., Wang, Y.-M., and Howard, R. A.: 1999, *JGR* **104**, 452–462.
St. Cyr, O. C., Burkepile, J. T., Hundhausen, A. J., and Lecinski, A. R.: 1999, *JGR* **104**(13), 12493–12506.
St. Cyr, O. C., Plunkett, S. P., Michels, D. J., Paswaters, S. E., Koomen, M. J., Simnett, G. M., *et al.*: 2000, *JGR* **105**(14), 18169–18186.
Sterling, A. C., and Hudson, H. S.: 1997, *ApJL* **491**, L55.
Subramanian, P., and Dere, K. P.: 2001, *ApJ* **561**, 372–395.
Thompson, B. J., Gurman, J. B., Neupert, W. M., Newmark, J. S., Delaboudinière, J.-P., St. Cyr, O. C., *et al.*: 1999, *ApJL* **517**, L151–L154.
Tousey, R.: 1973, Rycroft, M. J. and Runcorn, S. K. (eds.), *Space Research*, pp. 713.
Uchida, Y.: 1968, *Solar Phys.* **4**, 30.
Vourlidas, A., Buzasi, D., Howard, R. A., and Esfandiari, E.: 2002, *ESA SP-506: Solar Variability: From Core to Outer Frontiers*, pp. 91.
Vourlidas, A., Subramanian, P., Dere, K. P., and Howard, R. A.: 2000, *ApJ* **534**, 456–467.
Wang, Y., Hawley, S. H., and Sheeley, N. R.: 1996, *Science* **271**, 464–469.
Webb, D. F. and Howard, R. A.: 1994, *JGR* **99**(18), 4201–4220.
Wimmer-Schweingruber, R. F., Crooker, N. U., *et al.*: 2006, *Space Sci. Rev.* (this volume) doi: 10.1007/s11214-006-9017-x.
Withbroe, G. L.: 1988, *ApJ* **325**, 442–467.
Zarro, D. M., Sterling, A. C., Thompson, B. J., Hudson, H. S., and Nitta, N.: 1999, *ApJL* **520**, L139–L142.
Zhang, J., Dere, K. P., Howard, R. A., Kundu, M. R., and White, S. M.: 2001, *ApJ* **559**, 452–462.
Zurbuchen, T. H., and Richardson, I. G.: 2006, *Space Sci. Rev.* (this volume) doi: 10.1007/s11214-006-9010-4.

IN-SITU SOLAR WIND AND MAGNETIC FIELD SIGNATURES OF INTERPLANETARY CORONAL MASS EJECTIONS

THOMAS H. ZURBUCHEN[1] and IAN G. RICHARDSON[2]
[1]*Department of Atmospheric, Oceanic and Space Sciences, University of Michigan, Ann Arbor, MI 48109, U.S.A.*
[2]*The Astroparticle Physics Laboratory, NASA Goddard Space Flight Center, Greenbelt, MD 20771, U.S.A. and The Department of Astronomy, University of Maryland, College Park, MD 20742, U.S.A.*
*(*Author for correspondence, E-mail: thomasz@umich.edu)*

(Received 16 March 2004; Accepted in final form 4 May 2005)

Abstract. The heliospheric counterparts of coronal mass ejections (CMEs) at the Sun, interplanetary coronal mass ejections (ICMEs), can be identified in situ based on a number of magnetic field, plasma, compositional and energetic particle signatures as well as combinations thereof. We summarize these signatures and their implications for understanding the nature of these structures and the physical properties of coronal mass ejections. We conclude that our understanding of ICMEs is far from complete and formulate several challenges that, if addressed, would substantially improve our knowledge of the relationship between CMEs at the Sun and in the heliosphere.

Keywords: interplanetary coronal mass ejections, solar wind plasma, interplanetary magnetic field

1. Introduction

We review the signatures observed by spacecraft that are currently used for the in-situ identification of interplanetary coronal mass ejections (ICMEs), the interplanetary manifestations of coronal mass ejections (CMEs) at the Sun. The emphasis is on near-Earth phenomena. These signatures are summarized in Table I together with a few key references that further define and/or use a specific signature. We separate the ICME identifiers into magnetic field, plasma dynamics, plasma composition, plasma wave and suprathermal particle signatures. See, also, the reviews by Gosling (1990, 2000) and Neugebauer and Goldstein (1997).

2. ICME Signatures

2.1. Magnetic Field Signatures, Magnetic Clouds

Magnetic field signatures are perhaps the most studied because, if a particular model is assumed, the three-dimensional magnetic field structure may be inferred from a single pass through an ICME. An interesting subset of ICMEs (Klein and Burlaga, 1982) have enhanced magnetic fields (>10 nT) that rotate slowly through a large

Space Science Reviews (2006) 123: 31–43
DOI: 10.1007/s11214-006-9010-4

TABLE I

In-situ signatures of ICMEs (description applies to ~1 AU heliospheric distance) in the magnetic field (B), plasma dynamics (P), plasma composition (C), plasma waves (W), and suprathermal particles (S)

Signature	Description	Selected references
B1: *B* Rotation	$\gg 30°$, smooth	Klein and Burlaga (1982)
B2: *B* Enhancement	>10 nT	Hirshberg and Colburn (1969); Klein and Burlaga (1982)
B3: *B* Variance decrease		Pudovkin *et al.* (1979); Klein and Burlaga (1982)
B4: Discontinuity at ICME boundaries		Janoo *et al.* (1998)
B5: Field line draping around ICME		Gosling and McComas (1987); McComas *et al.* (1989)
B6: Magnetic clouds	(B1, B2 and $\beta = \frac{\sum nkT}{B^2/(2\mu_0)} < 1$)	Klein and Burlaga (1982); Lepping *et al.* (1990)
P1: Declining velocity profile/expansion	Monotonic decrease	Klein and Burlaga (1982); Russell and Shinde (2003)
P2: Extreme density decrease	≤ 1 cm^{-3}	Richardson *et al.* (2000a)
P3: Proton temperature decrease	$T_p < 0.5 T_{\mathrm{exp}}$	Gosling *et al.* (1973); Richardson and Cane (1995)
P4: Electron temperature decrease	$T_e < 6 \times 10^4$ K	Montgomery *et al.* (1974)
P5: Electron Temperature increase	$T_e \gg T_p$	Sittler and Burlaga (1998); Richardson *et al.* (1997)
P6: Upstream forward shock/"Bow Wave"	Rankine-Hugoniot relations	Parker (1961)
C1: Enhanced α/proton ratio	$He^{2+}/H^+ > 8\%$	Hirshberg *et al.* (1972); Borrini *et al.* (1982a)
C2: Elevated oxygen charge states	$O^{7+}/O^{6+} > 1$	Henke *et al.* (2001); Zurbuchen *et al.* (2003)
C3: Unusually high Fe charge states	$\langle Q \rangle_{\mathrm{Fe}} > 12$; $Q_{\mathrm{Fe}}^{\geq 15+} > 0.01$	Bame *et al.* (1979); Lepri *et al.* (2001); Lepri and Zurbuchen (2004)
C4: Occurrence of He^+	$He^+/He^{2+} > 0.01$	Schwenn *et al.* (1980); Gosling *et al.* (1980); Gloeckler *et al.* (1999)
C5: Enhancements of Fe/O	$\frac{(\mathrm{Fe/O})_{\mathrm{CME}}}{(\mathrm{Fe/O})_{\mathrm{photosphere}}} > 5$	Ipavich *et al.* (1986)
C6: Unusually high $^3He/^4He$	$\frac{(^3\mathrm{He}/^4\mathrm{He})_{\mathrm{CME}}}{(^3\mathrm{He}/^4\mathrm{He})_{\mathrm{photosphere}}} > 2$	Ho *et al.* (2000)
W1: Ion acoustic waves		Fainberg *et al.* (1996); Lin *et al.* (1999)
S1: Bidirectional strahl electrons		Gosling *et al.* (1987)
S2: Bidirectional ~MeV ions	2nd harmonic >1st harmonic	Palmer *et al.* (1978); Marsden *et al.* (1987)
S3: Cosmic ray depletions	Few % at ~ 1GeV	Forbush (1937); Cane (2000)
S4: Bidirectional cosmic rays	2nd harmonic >1st harmonic	Richardson *et al.* (2000b)

angle, low proton temperatures and low plasma β (ratio of the thermal and magnetic field energies), features that are evident in the event in Figure 1(a). Such ICMEs are termed "magnetic clouds" (MCs). Although spheromak-like plasmoid models have been proposed for magnetic clouds (Vandas *et al.*, 1993), work has focused on flux ropes (Lepping *et al.*, 1990; Osherovich and Burlaga, 1997; Cid *et al.*, 2002; Mulligan and Russell, 2001; Lynch *et al.*, 2003, and references therein). Figure 3 shows a schematic of an ICME with a magnetic flux-rope structure.

It should be emphasized that MC-like features are only present in a subset of all ICMEs. The magnetic field configurations of non-cloud-like ICMEs may be more complicated, leading Burlaga *et al.* (2002) to name them "complex ejecta". Two non-cloud ICMEs are shown in Figures 1(b) and (c). Signatures B1 and B2 (Table I) are not observed even though each ICME includes a number of the other characteristic signatures to be discussed below. Gosling (1990) concluded that $\sim$30% of ICMEs in 1978–1982 were MCs. Other estimates (Bothmer and Schwenn, 1996; Richardson *et al.*, 1997; Cane *et al.*, 1997; Mulligan *et al.*, 1999) range from $\sim$15 to 60%, while Marubashi (2000) has claimed that up to $\sim$80% of the set of ICMEs studied were flux rope encounters, arguing that the absence of MC signatures frequently occurs when the observing spacecraft makes only a glancing encounter with the MC. There is also evidence of a solar cycle effect, ranging from $\gtrsim$60% MCs for the few ICMEs near solar minimum to $\sim$15% around solar maximum (Cane and Richardson, 2003). Non-MC-like configurations may arise if an ICME is a conglomerate of several individual ICMEs (cf. Figure 2(c)), or if the magnetic field configuration of the original CME was more complex than a simple flux rope (Figure 2(b) may be an example). Even an apparently simple MC may consist of several flux tubes (Fainberg *et al.*, 1996).

Magnetic field observations can help identify the boundaries of the ICME. In principle, the boundary between the ICME and ambient solar wind should be a tangential discontinuity, which magnetic field lines do not cross. While in some cases such discontinuities can be identified with little ambiguity, in other cases the boundaries are less distinct and may include complex structures perhaps indicative of waves or field-line reconnection (Vasquez *et al.*, 2001).

Another common feature within ICMEs is a reduction in the magnetic field variability. This is most evident from inspection of field observations with time resolutions of $\sim$5 minutes or less (Figure 1). The relatively smooth magnetic fields within ICMEs are in marked contrast to those in the turbulent "sheaths" found ahead of fast ICMEs.

The southward interplanetary magnetic field component is a dominant parameter governing the intensity of geomagnetic activity (Tsurutani and Gonzalez, 1997). Because this is strongly enhanced within some ICMEs or the associated sheaths, the majority of major geomagnetic storms are ICME-related (Richardson *et al.*, 2001). In Figure 2, the Dst index (increasingly negative values indicate increased activity) illustrates the geomagnetic response to variations in the southward magnetic field intensity during each event (cf. B, and $\theta_B \ll 0^\circ$).

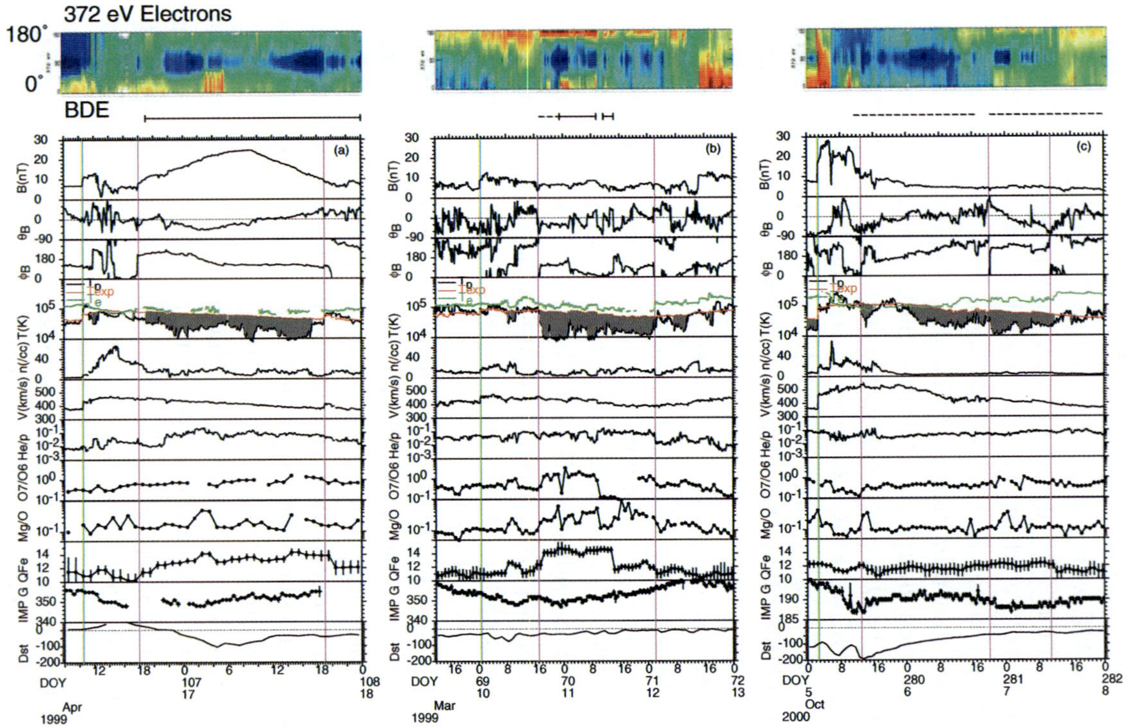

Figure 1. Examples of ICMEs observed by ACE following interplanetary shocks (green vertical lines). ICME boundaries are based on a consensus of plasma/field signatures (Cane and Richardson, 2003). Event (a) is a "classic" magnetic cloud, showing an enhanced, smoothly-rotating magnetic field; (b) has a more irregular and weaker field; (c) may be divided into two regions, possibly separate ICMEs, each with weak but rotating fields. Other ICME characteristics which may be evident and may not necessarily occupy the same regions include: depressed proton temperatures (grey shading indicates $T_p \leq 0.5T_{\rm exp}$); electron temperatures $>T_p$; declining solar wind speed profiles; He/proton abundance enhancements; enhanced oxygen and iron charge states and Mg/O ratio; cosmic ray depressions (IMP 8 GME guard $\gtrsim$60 MeV particle count rate) commencing in the vicinity of shock passage; and geomagnetic storms (indicated by the Dst index). The top panels show 0–180° 372 eV electron pitch-angle distributions, with BDEs at the times indicated (dashed = weak/questionable).

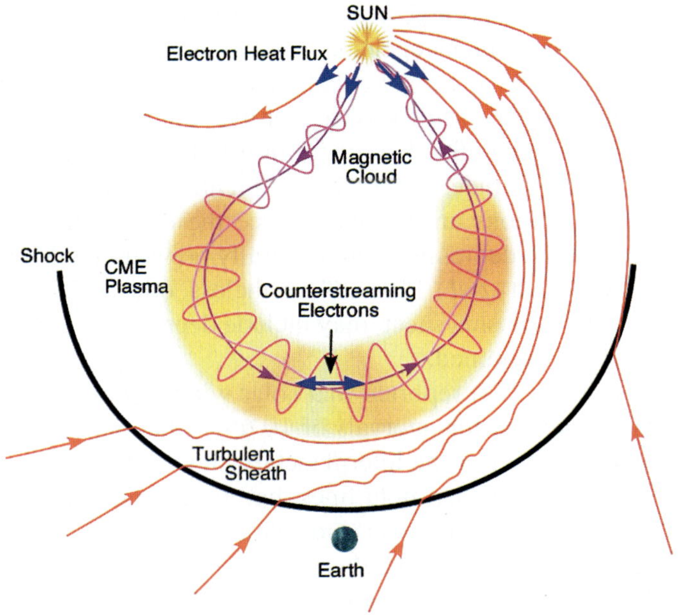

Figure 2. Schematic of the three-dimensional structure of an ICME and upstream shock, relating magnetic field, plasma, and BDE signatures.

2.2. Plasma Dynamics

The solar wind velocity signatures of some ICMEs indicate expansion in the solar wind rest frame (cf. Figure 2). The ICME leading edge moves at a speed $V_{\rm ICME} + V_{\rm EXP}$, with a smooth transition during passage of the ICME to a speed of $V_{\rm ICME} - V_{\rm EXP}$ at the trailing edge. The expansion speed $V_{\rm EXP}$ is typically around half the Alfvén speed in the ICME (Klein and Burlaga, 1982). Not all ICMEs exhibit expansion signatures, however, and similar speed variations in coronal-hole-associated solar wind may lead to false identifications.

In the ambient (non-ICME) solar wind, there is an empirical correlation between the solar wind speed (V_{sw}) and plasma proton temperature (T_p) (Lopez, 1987, and references therein). Gosling *et al.* (1973), however, pointed out occasional intervals of unusually low T_p that do not follow this correlation. These intervals were attributed to magnetically isolated, ejected material expanding at a higher rate than the ambient solar wind. They also tended to follow interplanetary shocks by a few hours, suggesting that they were related to the drivers of these shocks that we now associate with ICMEs. Richardson and Cane (1995) found that ICMEs typically have $T_p < 0.5T_{ex}$, where T_{ex} is the "expected T_p" determined from the empirical $V_{sw} - T_p$ correlation and the simultaneously observed solar wind speed. Grey shading in Figure 1 denotes intervals when this criterion is met. They also noted that the fraction of the solar wind having $T_p < 0.5T_{ex}$ increases from ~4%

at solar minimum to ~12% around solar maximum, consistent with an association with ICMEs. In a similar vein, Neugebauer and Goldstein (1997) defined a thermal index $I_{th} = (500V_p + 1.75 \times 10^5)/T_p$ such that if $I_{th} > 1$, the plasma is likely to be associated with an ICME, while this may or may not be the case when $I_{th} < 1$. Other authors have simply defined an upper threshold for T_p (e.g., thermal speed ≤20 km/s; Russell and Shinde, 2003). ICME identification based on T_p has the advantages that observations are available since the beginning of the space era (with some gaps), and T_p depressions are generally present in ICMEs (Richardson and Cane, 1995; Mulligan *et al.*, 1999). Nevertheless, other solar wind structures, such as the heliospheric plasma sheet, may include depressed T_p, so the solar wind context should also be examined.

Montgomery *et al.* (1974) reported that solar wind electron temperatures (T_e) were temporarily depressed for intervals of 10 to >40 hours commencing 10–20 hours following around half of the interplanetary shocks they studied, concluding that these were regions of closed field lines that were magnetically isolated from the hot corona. Other studies, however, indicate that T_e tends to be enhanced relative to T_p in some magnetic clouds (Osherovich *et al.*, 1993; Fainberg *et al.*, 1996; Sittler and Burlaga, 1998) and non-cloud ICMEs (Richardson *et al.*, 1997), suggesting efficient transport of electron thermal energy along field lines connected to the corona. Richardson *et al.* (1997) proposed $T_e/T_p > 2$ as a more appropriate indicator of an ICME than one based on T_e alone. When $T_e/T_p > 1$, the Landau damping constraint on the excitation of ion acoustic waves is removed, so these waves may accompany ICMEs (Lin *et al.*, 1999). We note that, when this criterion holds, the plasma pressure is dominated by the electron component.

2.3. Plasma Composition Signatures

Observations since the 1970s have identified regions following some interplanetary shocks with helium (He^{2+}) abundances (e.g., He^{2+}/protons >6%) that exceed normal solar wind values, leading to the suggestion that this unusual composition is indicative of ejected solar material (Hirshberg *et al.*, 1971). Helium enhancements are not detected following every shock because they are only present in a subset of the ICMEs identified by other signatures (Zwickl *et al.*, 1983; Mulligan *et al.*, 1999; Richardson and Cane, 2004), and ICMEs are typically less extended than the shocks they generate (Figure 3). Figures 1a and 1b show ICMEs with enhanced He/p. Neugebauer and Goldstein (1997) ascribe the enhanced helium abundances to "a sludge removal phenomenon," whereby helium that has settled at the footpoints of solar wind flow tubes is cleared out by the CME. The predictions from such chromospheric evaporation models with collisional transport, however, have not been tested in the context of the complete set of compositional data now available.

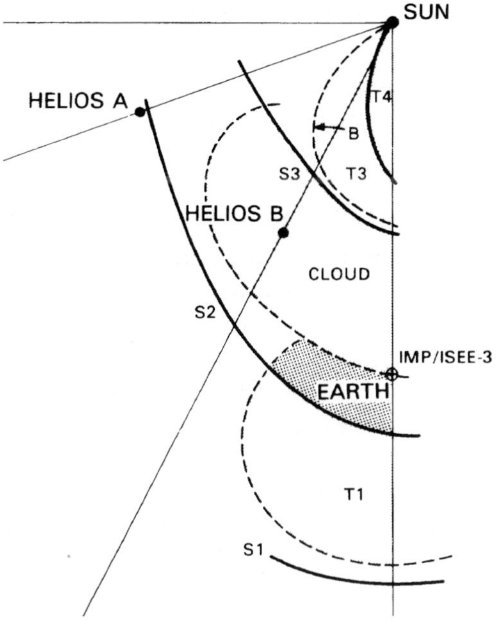

Figure 3. The configuration of interplanetary shocks (S1–S3) and ICMEs (T1–T4) at 2200 UT on April 3, 1979, inferred from multi-spacecraft observations (Helios and IMP 8/ISEE-3) (Burlaga *et al.*, 1987).

Although some observations were available from earlier spacecraft, detailed measurements of solar wind composition other than He/p have only been routinely available since the launch of Ulysses in 1990, and more recently from the ACE spacecraft (Galvin, 1997; Zurbuchen *et al.*, 2003; Richardson and Cane, 2004, and references therein). The relationship of compositional anomalies to ICMEs is an active area of current research (Wimmer *et al.*, 2006, this volume). ICMEs generally show elemental abundances that are fractionated relative to the First Ionization Potential (FIP) in a similar manner to those in slow solar wind associated with streamers (Neukomm, 1998). There are also reports that some ICMEs exhibit substantial mass fractionation, as opposed to FIP fractionation (Gloeckler *et al.*, 1999; Wurz *et al.*, 2000; Zurbuchen *et al.*, 2004). Isotopic fractionation has been observed in ICMEs in the case of $^3He/^4He$ (Ho *et al.*, 2000) but not conclusively for other elements, probably because of the limited precision of current experiments (Wimmer *et al.*, 1999; Wimmer *et al.*, 2006, this volume). Relative to the ambient solar wind, ICMEs may include enhancements in heavy ion abundances (in particular iron) (Mitchell *et al.*, 1983; Ipavich *et al.*, 1986) and enhanced ion charge states. The ionic charge state of heavy ions is a sensitive measure of the thermal environment of CMEs and their interplanetary counterparts (Hundhausen *et al.*, 1968; Buergi and Geiss, 1986). Generally, ICME-associated plasma charge states suggest a CME source that is "hot" relative to the ambient solar wind. Examples were reported by Bame *et al.* (1979) and Fenimore (1980), and more complete surveys have been made by Neukomm (1998), Henke *et al.* (2001) and Rodriguez *et al.* (2004) based on O and C charge states, which freeze in relatively close to the Sun (within $\sim 1 R_s$ above the solar surface). Lepri *et al.* (2001) and Lepri and Zurbuchen (2004) have discussed Fe charge states, which become frozen in during the CME expansion in the outer corona where ICME plasma seems to be well-differentiated from plasma of the ambient solar wind. Roughly 50% to 70% of all ICMEs have enhanced Fe charge states as defined by the criteria in Table I. This fraction is much smaller for $O^{7+}/O^{6+} > 1$, though a relative enhancement of O^{7+}/O^{6+} might be a more reliable ICME indicator (Richardson and Cane, 2004). Compositional signatures relying on "hot" ionic charge states appear to be some of the best indicators of ICMEs currently available, with remarkably few false identifications (Lepri *et al.*, 2001). In particular, they are more generally present in ICMEs than, for example, magnetic cloud signatures. The ICMEs in Figures 1(a) and (b) show enhancements in the helium/proton, O^{7+}/O^{6+}, and Mg/O ratios, and Fe charge states, while these features are essentially absent in the ICME in Figure 1(c). Richardson and Cane (2004) have made a comprehensive survey of enhancements in these compositional signatures during 1996–2002 and demonstrate their close association with ICMEs (see Figure 2 in Wimmer *et al.*, 2006, this volume).

There is also a very small subset of unusually "cold" events with low ion charge states and unusual fractionation patterns that are uncharacteristic of the majority of ICMEs. These were first identified by the presence of singly-charged helium abundances well above solar wind values (Schwenn *et al.*, 1980; Gosling *et al.*, 1980). Zwickl *et al.* (1982) reported only three cases in eight years of observations. Additional cold ICMEs have been reported (Yermolaev *et al.*, 1989; Burlaga *et al.*, 1998; Gloeckler *et al.*, 1999; Skoug *et al.*, 1999). Singly-charged He and other low charge states suggest that the plasma originated in low temperature material at the Sun, possibly the cool, dense prominence material which is observed rising above the solar surface following some CMEs. None of the events in Figure 1 have this signature. Under special circumstances, both unusually "hot" and unusually "cold" ion charge states have been observed within the same ICME, even with simple electrostatic analyzers (see Bame, 1983, and references therein).

2.4. Energetic Particle Signatures

Bidirectional beams of suprathermal ($\gtrsim$100 eV) electrons (BDEs), which normally focus into a single field-aligned "strahl" directed away from the Sun, are typically associated with other ICME signatures (Zwickl *et al.*, 1983; Gosling *et al.*, 1987). The physical interpretation is that the electrons are flowing in opposite directions along magnetic field loops within ICMEs that are rooted at the Sun (Figure 2). BDEs are one of the more widely-used signatures for identifying ICMEs, and the primary signature in some studies. Some care, however, is required in interpreting the electron distributions (Gosling *et al.*, 2001; Wimmer *et al.*, 2006, this volume). Furthermore, BDEs may occur intermittently, or even be absent, within an ICME (Shodhan *et al.*, 2000). Their absence may indicate ICME field lines that have reconnected in the legs of the loops with open interplanetary magnetic field lines (Gosling *et al.*, 1995). Electron flows are also usually stronger in one direction, possibly corresponding to flow away from the field line footpoint that is closer to the observer. Intervals of bidirectional electron flows observed by ACE/SWEPAM are indicated in Figure 1 together with angular distributions of 372 eV electrons.

Other particle signatures of ICMEs include short-term (few day duration) depressions in the galactic cosmic ray intensity, bidirectional energetic particle flows, and unusual flow directions during solar energetic particle onsets. See Cane and Lario (2006, this volume) for an overview of energetic particle phenomena associated with ICMEs.

2.5. Association with Interplanetary Shocks

Fast mass ejections, exceeding the magnetosonic speed in the solar wind, generate fast forward shocks ahead of them. Studies suggest that shocks can be observed over $\sim 90^\circ$ in longitude from the location of energetic solar events, compared with up to $\sim 50^\circ$ (i.e., a total extent of $\sim 100^\circ$) for the related ICMEs (Borrini *et al.*, 1982b; Cane, 1988; Richardson and Cane, 1993). ICMEs from less energetic events may be narrower. For example, remarkably few ICMEs were observed at both the Helios 1 and 2 spacecraft even when separated by only $\sim 40^\circ$ in longitude (Cane *et al.*, 1997).

Relating shocks, ICMEs and solar events can be particularly complicated at times when several ejections are moving away from the Sun. For example, Figure 3 shows the configuration of shocks (S1–S3) and ICMEs (T1–T4) inferred from Helios and IMP 8 observations in early April 1979 (Burlaga *et al.*, 1987). Observations from multiple, well-separated spacecraft are of immense value when studying such structures. Reliable associations between shocks/ICMEs and the related solar event are also important. For energetic events, energetic particle intensity-time profiles or interplanetary type II radio emissions can be helpful. For less energetic events, it can be difficult to make an unambiguous association, in particular if there are several candidate solar events. Based on the estimates of typical ICME longitudinal extents referred to above, it is probably reasonable to treat claimed ICME associations with solar phenomena much beyond $\sim 50^\circ$ longitude from the observer with a degree of skepticism.

3. Summary and Discussion

Despite the plethora of signatures associated with ICMEs and improvements in spacecraft instrumentation, ICME identification "is still something of an art" (Gosling, 1997). The main reasons are that the various signatures do not necessarily occur simultaneously and define precisely the same regions of the solar wind, and they show little event-to-event organization (Zwickl *et al.*, 1983; Crooker *et al.*, 1990; Richardson and Cane, 1995; Neugebauer and Goldstein, 1997; Mulligan *et al.*, 1999). This is not too surprising since they arise from different physical circumstances. For example, plasma composition reflects abundances and electron temperatures near the Sun, depressed proton temperatures result from expansion of the ICME in the solar wind, and suprathermal electrons indicate field line connectivity to the Sun. The most practical approach is to examine as many signatures as possible and reach a consensus based on the grouping of several signatures within a certain region of the solar wind. This region may have distinct boundaries in plasma, magnetic field and other signatures, while in other cases, the boundaries may be more ambiguous. Differences in instrumentation, data analysis and selection criteria will also influence when certain ICME signatures are reported by different researchers. There are also ICMEs that lack some of the characteristic signatures, even those that are relatively ubiquitous, such as a depressed proton temperature. Hence, the most important conclusion of this paper is that a necessary and sufficient condition that defines the presence of an ICME or provides a crisp definition of an ICME remains elusive and is most likely unattainable.

Further progress is necessary in relating the properties of ICMEs to coronal phenomena. One limitation is that most data analysis has been limited to single-point observations whereas ICMEs are three-dimensional structures that can only be disentangled through multi-point observations. Recent three-dimensional simulations of CME propagation into the heliosphere (Riley *et al.*, 2003; Manchester *et al.*, 2004) can provide a context for interpreting observations, but their physical realism is still insufficient to answer many of the questions posed by observers. Second, our limited understanding of the underlying physical processes governing ICME signatures makes it difficult to know how to interpret observations or combine signatures that are intrinsically related. We therefore suggest four challenges, which, if addressed, may provide breakthroughs in our understanding of ICMEs and their signatures:

- Investigate the thermodynamic state of CMEs and ICMEs, based on a combination of theoretical and observational studies, and hence advance our understanding of the physical interpretation of the various ICME signatures and the relationship of compositional signatures to other in-situ observables.
- Develop a theoretical framework for the interpretation of compositional data from ICMEs that can address elemental, isotopic, and charge composition in concert, and relate them to the plasma properties observed in situ. Although compositional data teach us something about the source of ICME material, currently we do not know how to interpret that information.
- Using models and multi-point observations of critical signatures, such as BDEs, magnetic fields, and energetic particles, investigate the three-dimensional topology of ICMEs and their effects on the space environment.

– Provide wider access to ICME models, allowing observers to address questions of specific interest, such as the effect of changing intersection geometries.

Acknowledgements

The authors acknowledge ISSI and the workshop organizers for making this workshop possible. THZ was supported, in part, by NASA Grants NAG–12893 and NAG5–11000, while IGR was supported by NASA Grant NCC5–180. We thank Ruth Skoug and Jack Gosling for providing and interpreting the BDE data in Figure 1.

References

Bame, S. J., Asbridge, J. R., Feldman, W. C., Fenimore, E. E., and Gosling, J. T.: 1979, *Sol. Phys.* **62**, 179.
Bame, S. J.: 1983, in Neugebauer, M. (ed.), *Solar Wind 5 Proceedings*, NASA Conf. Proc., p. 2280.
Borrini, G., Gosling, J. T., Bame, S. J., and Feldman, W. C.: 1982a, *J. Geophys. Res.* **87**, 7370.
Borrini, G., Gosling, J. T., Bame, S. J., and Feldman, W. C.: 1982b, *J. Geophys. Res.* **87**, 4365.
Bothmer, V., and Schwenn, R.: 1996, *Adv. Space Res.* **17**(4/5), 319.
Buergi, A., and Geiss, J.: 1986, *Solar Phys.* **103**, 347.
Burlaga, L. F., Behannon, K. W., and Klein, L. W.: 1987, *J. Geophys. Res.* **92**, 5725.
Burlaga, L. F., *et al.*: 1998, *J. Geophys. Res.* **103**, 227.
Burlaga, L. F., Plunkett, S. P., and St. Cyr, O. C.: 2002, *J. Geophys. Res.* **107**, No. A10, 1266, doi:10.1029/2001JA000255.
Cane, H. V.: 1988, *J. Geophys. Res.* **93**, 1.
Cane, H. V.: 2000, *Space Sci. Rev.* **93**, 55.
Cane, H. V., and Richardson, I. G.: 2003, *J. Geophys. Res.* **108**(4), 1156, doi:10.1029/2002JA009817.
Cane, H. V., and Lario, D.: 2006, *Space Sci. Rev.* this volume, doi: 10.1007/s11214-006-9011-3.
Cane, H. V., Wibberenz, G., and Richardson, I. G.: 1997, *J. Geophys. Res.* **102**, 7075.
Cid, C., Hidalgo, M. A., Nieves-Chinchilla, T., Sequeiros, J., and Viñas, A. F.: 2002, *Solar Phys.* **207**, 187.
Crooker, N. U., Gosling, J. T., Smith, E. J., and Russell, C. T.: 1990, in Russell, C. T., Priest, E. R. and Lee, L. C. (eds.), *Physics of Magnetic Flux Ropes*, Geophys. Monogr. Ser., Vol. 58, AGU, Washington D. C., p. 365.
Fainberg, J., Osherovich, V. A., Stone, R. G., MacDowall, R. J., and Balogh, A.: 1996, in Winterhalter, D., Gosling, J. T., Habbal, S. R., Kurth, W. S., Neugebauer, M. (eds.), *Solar Wind Eight*, AIP Conf. Proc. 382, Woodbury, N.Y., p. 554.
Fenimore, E. E.: 1980, *Astrophys. J.* **235**, 245.
Forbush, S. E.: 1937, *Phys. Rev.* **51**, 1108–1109.
Galvin, A. B.: 1997, in Crooker, N., Joselyn, J. A., Feynman, J. (eds.), *Coronal Mass Ejections*, AGU, Washington D. C., p. 253.
Gloeckler, G., *et al.*: 1999, *Geophys. Res. Lett.* **26**(2), 157.
Gosling, J. T.: 1990, in Russell, C. T., Priest, E. R., Lee, L. C. (eds.), *Physics of Magnetic Flux Ropes*, Geophys. Monogr. Ser., Vol. 58, AGU, Washington D. C., p. 343.

Gosling, J. T.: 1997, in Crooker, N., Joselyn, J. A., Feynman, J. (eds.), *Coronal Mass Ejections*, AGU, Washington D. C., p. 9.
Gosling, J. T.: 2000, in Dingus, B. L., Kieda, D., and Salamon, M. (eds.), *26th Int. Cosmic Ray Conf. Invited, Rapporteur, and Highlight Papers*, AIP Conf. Proc. 516, Melville, N. Y., p. 59.
Gosling, J. T., and McComas, D. J.: 1987, *Geophys. Res. Lett.* **14**, 355.
Gosling, J. T., Pizzo, V., and Bame, S. J.: 1973, *J. Geophys. Res.* **78**, 2001.
Gosling, J. T., Asbridge, J. R., Bame, S. J., Feldman, W. C., and Zwickl, R. D.: 1980, *J. Geophys. Res.* **85**, 3431.
Gosling, J. T., Baker, D. N., Bame, S. J., Feldman, W. C., Zwickl, R. D., and Smith, E. J.: 1987, *J. Geophys. Res.* **92**, 8519.
Gosling, J. T., Birn, J., and Hesse, M.: 1995, *Geophys. Res. Lett.* 22, 869.
Gosling, J. T., Skoug, R. M., and Feldman, W. C.: 2001, *Geophys. Res. Lett.* **28**, 4155.
Henke, T., *et al.*: 2001, *J. Geophys. Res.* **106**, 10,597.
Hirshberg, J., and Colburn, D. S.: 1969, *Planet. Space Sci.* **17**, 1183.
Hirshberg, J., Asbridge, J. R., and Robbins, D. E.: 1971, *Sol. Phys.* **18**, 313.
Hirshberg, J., Bame, S. J., and Robbins, D. E.: 1972, *Sol. Phys.* **23**, 467.
Ho, G. C., Hamilton, D. C., Gloeckler, G., and Bochsler, P.: 2000, *Geophys. Res. Lett.* **27**, 309.
Hundhausen, A. J., Gilbert, H. E., and Bame, S. J.: 1968, *Astrophys. J.* **157**, L3.
Ipavich, F. M., *et al.*: 1986, *J. Geophys. Res.* **91**, 4133.
Janoo, L., *et al.*: 1998, *J. Geophys. Res.* **103**, 17,249.
Klein, L. W., and Burlaga, L. F.: 1982, *J. Geophys. Res.* **87**, 613.
Lepping, R. P., Jones, J. A., and Burlaga, L. F.: 1990, *J. Geophys. Res.* **95**, 11,957.
Lepri, S. T., Zurbuchen, T. H., Fisk, L. A., Richardson, I. G., Cane, H. V., and Gloeckler, G.: 2001, *J. Geophys. Res.* **106**, 29,231.
Lepri, S. T., and Zurbuchen, T. H.: 2004, *J. Geophys. Res.* **109**, doi:10.1029/2003JA009954.
Lin, F. A., *et al.*: 1999, in Habbal, S. (ed.), *Solar Wind Nine*, AIP Conf. Proc., 471, pp. 673–676.
Lopez, R. E.: 1987, *J. Geophys. Res.* **92**, 11,189.
Lynch, B. J., Zurbuchen, T. H., Fisk, L. A., and Antiochos, S. K.: 2003, *J. Geophys. Res.* **108**, CiteID 1239, doi10.1029/2002JA009591.
Manchester, W. B. M., IV, *et al.*: 2004, *J. Geophys. Res* **109**, A01102.
Marubashi, K.: 2000, *Adv. Space Res.* **26**(1), 55.
Marsden, R. G., Sanderson, T. R., Tranquille, C., Wenzel, K.-P., and Smith, E. J.: 1987, *J. Geophys. Res.* **92**, 11,009.
McComas, D. J., Gosling, J. T., Bame, S. J., Smith, E. J., and Cane, H. V.: 1989, *J. Geophys. Res.* **94**, 1465.
Mitchell, D. G., Roelof, E. C., and Bame, S. J.: 1983, *J. Geophys. Res.* **88**, 9059.
Montgomery, M. D., Asbridge, J. R., Bame, S. J., and Feldman, W. C.: 1974, *J. Geophys. Res.* **79**, 3103.
Mulligan, T., Russell, C. T., and Gosling, J. T.: 1999, in Habbal, S. R., Esser, R., Hollweg, J. V., Isenberg, P. A. (eds.), *Solar Wind Nine*, AIP Conf. Proc. 471, Woodbury, N. Y., p. 693.
Mulligan, T., and Russell, C. T.: 2001, *J. Geophys. Res.* **106**, 10,581.
Neugebauer, M., and Goldstein, R.: 1997, in Crooker, N., Joselyn, J. A., Feynman, J. (eds.), *Coronal Mass Ejections*, AGU, Washington D.C., p. 245.
Neukomm, R. O.: 1998, Ph.D. Thesis, University of Bern, 1998.
Osherovich, V., and Burlaga, L. F.: 1997, in Crooker, N., Joselyn, J. A., Feynman, J. (eds.), *Coronal Mass Ejections*, AGU, Washington D.C., p. 157.
Osherovich, V. A., *et al.*: 1993, *J. Geophys. Res.* **98**, 15,331.
Palmer, I. D., Allum, F. R., and Singer, S.: 1978, *J. Geophys. Res.* **83**, 75.
Parker, E.: 1961, *Astrophys. J.* **133**, 1014.
Pudovkin, M. I., Zaitseva, S. A., and Benevslenska, E. E.: 1979, *J. Geophys. Res.* **84**, 6649.

Richardson, I. G., and Cane, H. V.: 1993, *J. Geophys. Res.* **98**, 15,295.
Richardson, I. G., and Cane, H. V.: 1995, *J. Geophys. Res.* **100**, 23,397.
Richardson, I. G., and Cane, H. V.: 2004, *J. Geophys. Res.* **109**, A09104, doi:10.1029/2004JA010598.
Richardson, I. G., Farrugia, C. J., and Cane, H. V.: *J. Geophys. Res.* **102**, 4691.
Richardson, I. G., Berdichevsky, D., Desch, M. D., and Farrugia, C. J.: 2000a, *Geophys. Res. Lett.* **27**, 3761.
Richardson, I. G., Dvornikov, V. M., Sdobnov, V. E., and Cane, H. V.: 2000b, *J. Geophys. Res.* **105**, 12,597.
Richardson, I. G., Cliver, E. W., and Cane, H. V.: 2001, *Geophys. Res. Lett.* **28**, 2569.
Riley, P., *et al.*: 2003, *J. Geophys. Res.* **108**, CiteID 1272, doi:10.1029/2002JA009760.
Rodriguez, L., *et al.*: 2004, *J. Geophys. Res.* **109**, A01108, doi:10.1029/2003JA010156.
Russell, C. T., and Shinde, A. A.: 2003, *Solar Phys.* **216**, 285.
Schwenn, R., Rosenbauer, H., and Mühlhäuser, K.-H.: 1980, *Geophys. Res. Lett.* **7**, 201.
Shodhan, S., *et al.*: 2000, *J. Geophys. Res.* **105**, 27,261.
Sittler, E. C., Jr., and Burlaga, L. F.: 1998, *J. Geophys. Res.* **103**, 17,447.
Skoug, R. M., *et al*: 1999, *Geophys. Res. Lett.*, **26**, 161.
Tsurutani, B. T., and Gonzalez, W. D.: 1990, in Tsurutani, B. T., Gonzalez, W. D., Kamide, Y., Arballo, J. K. (eds.), *Magnetic Storms*, Geophys. Monogr. Ser., Vol. 98, AGU, Washington D. C., p. 77.
Vandas, M., Fischer, S., Pelant, P., and Geranios, A.: 1993, *J. Geophys. Res.* **98**, 21,061.
Vasquez, B. J., *et al.*: 2001; *J. Geophys. Res.* **106**, 29,283.
Wimmer-Schweingruber, R. F., Bochsler, P., and Wurz, P.: 1999, in Habbal, S. R., *et al.* (eds.), *Solar Wind Nine*, AIP Conf. Proc. 471, AIP Press, Woodbury, N. Y, pp. 147–152.
Wimmer-Schweingruber, R. F., Crooker, N. U., *et al.*: 2006, *Space Sci. Rev.* this volume, doi: 10.1007/s11214-006-9017-x.
Wurz, P., *et al.*: 2000, *J. Geophys. Res.* **105**, 27,239.
Yermolaev, Y. I., *et al.*: 1989, *Cosmic. Res.* **27**, 614.
Zurbuchen, T., Fisk, L. A., Lepri, S. T., and von Steiger, R.: 2003, in Velli, M., Bruno, R., Malara, F. (eds.), *Solar Wind Ten*, AIP Conf. Proc. 679, Mellville, N. Y., p. 604.
Zurbuchen, T. H., Gloeckler, G., Ipavich, F. M., Raines, J., Smith, C. W., and Fisk, L. A.: 2004, *Geophys. Res. Lett.*, **31**, L11805, doi:10.1029/2004GL019461.
Zwickl, R. D., Asbridge, J. R., Bame, S. J., Feldman, W. C., and Gosling, J. T.: 1982, *J. Geophys. Res.* **87**, 7379.
Zwickl, R. D., Asbridge, J. R., Bame, S. J., Feldman, W. C., Gosling, J. T., and Smith, E. J.: 1983, in Neugebauer, M. (ed.), *Solar Wind Five; NASA Conference Proceedings 2280*, NASA, Washington, D. C., p. 711.

AN INTRODUCTION TO CMES AND ENERGETIC PARTICLES

H. V. CANE[1,*] and D. LARIO[2]
[1] *School of Mathematics and Physics, University of Tasmania, Tasmania, Australia*
[2] *Applied Physics Laboratory, Johns Hopkins University, Laurel, MD, USA*
*(*Author for correspondence: E-mail: hilary.cane@utas.edu.au)*

(Received 11 May 2004; Accepted in final form 22 March 2006)

Abstract. Energetic particle observations in the interplanetary medium provide fundamental information about the origin, development and structure of coronal mass ejections. This paper reviews the status of our understanding of the ways in which particles are energised at the Sun in association with CMEs. This understanding will remain incomplete until the relationship between CMEs and flares is determined and we know the topology of the associated magnetic fields. The paper also discusses the characteristics of interplanetary CMEs that may be probed using particle observations.

1. Introduction

From the occurrence of a coronal mass ejection (CME) on the Sun until even after its passage over a spacecraft, energetic particle observations in the interplanetary medium help us to discern the development and structure of CMEs both close to the Sun and in the interplanetary (IP) medium. Solar energetic particles (SEPs) originate in at least two different ways both of which are likely related to CMEs. The shocks that CMEs create are accelerators of energetic particles as are the reconnection processes that must occur because of the CME–associated solar magnetic field topology changes. Crucial questions remain about both processes. With respect to shock acceleration the major question concerns the distance from the solar surface that CME shocks form. The major question in the case of reconnection regions is the connectivity of such regions to the IP medium, that is the accessibility to, and extent of, open field lines. Composition and charge state measurements indicate that some solar particles have their origin in heated and/or dense plasma. These observations place limits on the height in the solar corona where particles are accelerated and injected into the IP medium. Once the source regions of particles are understood the particles themselves may provide answers to other questions about CMEs. Because particles tend to follow field lines they can be used to trace field line topologies. Indeed, decreases in the intensity of galactic cosmic rays can indicate the presence of a CME in the IP medium (known as an ICME). Particle flows and intensity changes track magnetic structures within ICMEs. Also, shock accelerated populations provide information about the sizes of CME shocks as they travel from the Sun to the observer.

Space Science Reviews (2006) 123: 45–56
DOI: 10.1007/s11214-006-9011-3

2. CMEs at the Sun

Solar activity associated with the onset of an SEP event involves many related phenomena of which the most prominent, for all but the weakest events, is a CME. However, prompt events originating on the disk are also associated with flares. In almost all of the largest events the flare emissions are intense and long-lasting, suggesting that there is possibly a relationship between these emissions and the earliest energetic particles.

2.1. CMEs and Flares: Classes of SEP Events

The division of SEP events into two classes goes back to the early work of Lin (1970) in which he found that some electron increases were accompanied by proton increases and some were not. The 'pure' electron events were found to be associated with small flares that produced type III bursts and impulsive microwave and hard X-ray bursts. It was suggested that the 10–100 keV electrons were responsible for the electromagnetic emissions and were an integral part of the initial rapid, bright expansion phase of flares. Observations in the 1980's and 1990's showed that the acceleration mechanism could also produce high energy electrons (up to ~100 MeV) and ions to about 1 GeV as evidenced primarily by gamma ray observations but also by more sensitive in-situ observations.

Proton events were found by Lin (1970) to be associated with large flares and with type II and type IV radio bursts. The associated microwave bursts had complex structure. Previously it had been suggested that, since proton events were associated with type II bursts (taken as evidence of a coronal shock) protons are likely to be shock accelerated. In the late 1970's it was determined that large proton events also occurred at the times of CMEs (Kahler *et al.*, 1978) and it was assumed that type II bursts are a signature of the bow shocks of CMEs. But the picture is more complicated because proton events are actually best associated with long lasting type III emissions (Cane *et al.*, 2002). The importance of late low frequency radio emissions was previously stressed by Klein *et al.* (1999). Also it is unlikely that type II bursts observed from the ground are the high frequency component of the CME shock (Wagner and MacQueen, 1983; Cane, 1983). Nevertheless proton events are well associated with CMEs that do drive shocks, but it is not clear at what coronal height these shocks form and at what height accelerated energetic particles escape the shock. Furthermore, it seems likely that particles accelerated during the flare process contribute in large SEP events (Klein *et al.*, 1999; Torsti *et al.*, 2001; Cane *et al.*, 2002). This possibility is supported by charge state and abundance measurements.

Another complication is that the more intense electron events are also associated with CMEs, albeit ones that affect a smaller region of the corona. However, in the paradigm espoused by Reames (1999) the presence of a CME is what distinguishes two classes of particle events. Taken together the associations suggest that there are two ways in which particles are accelerated and, in the largest events, both occur to some extent dependent on energy. Thus it is unlikely that there is a sharp division separating SEP events into two classes.

2.2. Charge States and Composition

SEP charge states provide crucial clues as to the particle acceleration site, the acceleration process and their transportation out of the low corona. Whereas prior work, in cycle 21, was limited to long averaging and determinations of mean charge states in a small, low energy range, the Solar Anomalous and Magnetospheric Particle Explorer (SAMPEX) instrumentation can examine the average charge state over a wide energy range from 0.3 to 70 MeV/nuc (Oetliker *et al.*, 1997; and references therein). Furthermore, the Solar Energetic Particle Ionic Charge Analyzer (SEPICA) instrument on the Advanced Composition Explorer (ACE) can measure charge state distributions and their energy dependence. These new data show that the idea based on the work of Luhn *et al.* (1985) that there are two classes of SEP events distinguished by very different charge states is no longer tenable. Of particular interest are the high energy measurements (Leske *et al.*, 2001) that show that many large western events have high charge states ($\sim+20$ for Fe), just like in the smaller events. The origin of these high charge states is not yet clear. It has also been found that the charge states in all events are strongly dependent on energy (Möbius *et al.*, 2003; Popecki *et al.*, 2000). This means that acceleration at heights above 2 solar radii, as is thought to be the case in large events (Reames, 1999), is unlikely (Kochorov *et al.*, 2000). The new results imply that for small events the temperature of the ambient plasma is lower than previously deduced. The average $\sim$0.5 MeV/nuc Fe charge state of near $+21$ found in cycle 21 must reflect additional stripping during and after acceleration (Möbius *et al.*, 2003).

Abundance variations are another important diagnostic tool. As noted above it was a comparison of electron and protons that first indicated that there were possibly two classes of SEP events. Later two classes were also indicated by measurements on the isotopes of He (viz ^{3}He/^{4}He) and of ratios of heavy ions. Most of these earlier measurements were made at low energies ($<\sim$25 MeV) where the largest events are those in which there is strong IP shock acceleration. Thus in cycle 21 it was determined that large proton events had abundances comparable to that of the solar wind and very different from that seen in small, Fe–rich, ^{3}He–rich events (Reames, 1999). In cycle 23 with the large geometry factor instruments on ACE and on the Solar Heliospheric Observatory (SOHO) it has been found that smaller 'proton' events and larger ones at high energies, are also Fe–rich (Cohen *et al.*, 1999) and ^{3}He–rich (Torsti *et al.*, 2003) although not to the extent of the small 'electron' events. Thus abundance variations no longer indicate a clear separation into two classes. The observations suggest that abundance variations may be related to flare duration (Kocharov *et al.*, 1986; Cane *et al.*, 1986).

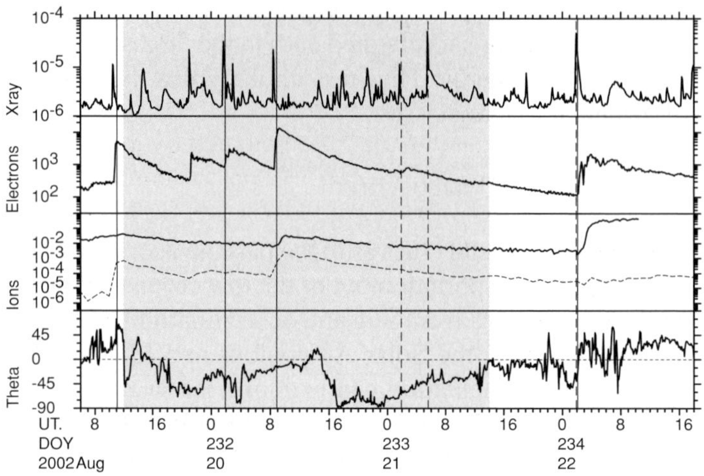

Figure 1. Observations during ~3 days in August 2002. There are (at least) four electron–rich events and one proton–rich event. Several electron events, including the largest at ~0800 UT on August 20, occurred inside an ICME (indicated by the gray shading). The proton event on August 22 occurred in association with the largest, but not the fastest, CME. Large CMEs are indicated by vertical solid lines and type II bursts by dashed lines.

The difficulty in untangling the relevant physics of solar particle acceleration is illustrated by the time period shown in Figure 1. The figure shows (from top to bottom) 1–8 Å soft X-ray intensity, ~200 keV electron, ~25 MeV proton and ~10 MeV/nuc Fe intensities. The bottom panel shows the elevation angle of the ambient interplanetary magnetic field. The particle events were some of the many events arising from the same active region in August 2002. All of them were associated with CMEs. The solid lines in the figure indicate the four CMEs with angular widths >100°. These CMEs had sky–plane speeds of 549, 961, 1099 and 1005 km/s. One would expect the second, third and fourth of these to drive shocks yet only for the last was a type II burst reported. The last SEP event has high proton to electron ratio and a relatively low Fe intensity making it a 'proton' event. The other SEP events are 'electron' events. Although only the proton event has a type II burst it was not observed beyond a few solar radii from the Sun. The presence of a fast CME does not differentiate the proton event from the electron events. The clearest difference is in the behavior of the metric type III radio emissions. These lasted much longer and started at lower frequencies for the proton event indicating the occurrence of extended particle acceleration in the middle corona. The magnetic field angle in the bottom panel shows that the electron event near 0800 UT on August 20 occurred inside an ICME as indicated by the field rotation and relatively smooth field. The extremely short flare suggests a short solar injection that was only seen at 1 AU because of interplanetary conditions appropriate for weak particle scattering.

2.3. Timing

Deductions about when particles left the Sun are crucially dependent on what assumptions are made about particle propagation. The term "scatter–free" is often applied to events where the injection time is calculated by determining how long it takes particles of a particular energy to travel along an Archimedean spiral of length 1.2 AU (corresponding to a solar wind speed of 400 km/s). A more detailed analysis involves plotting the arrival time of particles as a function of their inverse speed. Such a plot is expected to produce a straight line having a slope corresponding to the path length and the intercept on the time axis indicating the injection time. Indeed, it is commonly assumed that the straight line that is usually found with a slope implying a path length of $\approx$1.2 AU is proof that the particles travel scatter free. Recent analyses using the new experiments with good counting statistics have indicated that the onset times are nearly always delayed relative to the flare emissions. This is true even in small electron events in which the electrons are believed to have caused the flare emissions. This delay has been interpreted by some reserachers to indicate that the $>$25 keV electrons that they observe are not related to the flare emissions but rather are probably shock accelerated at the leading edge of a CME (Haggerty and Roelof, 2002). Alternatively, Cane (2003) has proposed that interplanetary scattering must be occurring based on an analysis of the radio emissions. Delayed injections are also deduced for ions (see Posner and Kunow 2003). Clearly propagation effects require further investigation. It is possible that mean free paths vary with rigidity in such a way that the straight lines obtained in "1/beta plots" are fortuitous and not indicative of a lack of transport effects. Also suggestive that scattering is likely occurring is the fact that the particle events with the shortest inferred delays also have intensity–time profiles indicative of little scattering.

3. Propagation of CMEs

A fast CME driven shock will accelerate particles out of the bulk solar wind and its suprathermal tail, and/or out of the suprathermal remnants left over from prior SEP events or injected during the flare process (Desai *et al.*, 2003). The way in which these energetic particles are observed depends on (1) how they are accelerated and injected into the IP medium by the traveling CME-driven shock, and (2) how they are transported along the interplanetary magnetic field (IMF). These two factors implicitly depend on the energy of the particles, the particle species and mass per charge, the characteristics of the CME-driven shock (i.e., its speed, size, shape, strength and efficiency in particle acceleration), and the IMF topology that determines the magnetic connection between the observer and the expanding CME-driven shock. A great effort to model all these processes has been undertaken (e.g., Lario *et al.*, 1998; Kallenrode, 2001; Ng *et al.*, 2003; Rice *et al.*, 2003 and references therein). These efforts, however, are still in their infancy since none of them treats particle acceleration and transport, as well as CME-driven shock propagation, in their entirety. Models simplify the variety of processes involved in the particle shock-acceleration by assuming either arbitrary injection functions or quasi-steady diffusive shock-acceleration mechanisms. None of the models has yet treated the injection process self-consistently, or the complete evolution of the shocks from their formation close to the Sun to their propagation towards the outer heliosphere (see review of these models in Lario, 2005).

Energetic particle observations from IP spacecraft can be used to infer the properties of the traveling CME. For example, the time-intensity profiles of the SEP events observed in the ecliptic plane at 1 AU are organized in terms of the longitude of the observer with respect to the traveling CME-driven shock (Cane *et al.*, 1988). Figure 2 shows proton intensity profiles of several SEP events observed by the IMP-8 spacecraft as a function of the longitude of the parent solar event. (Note that

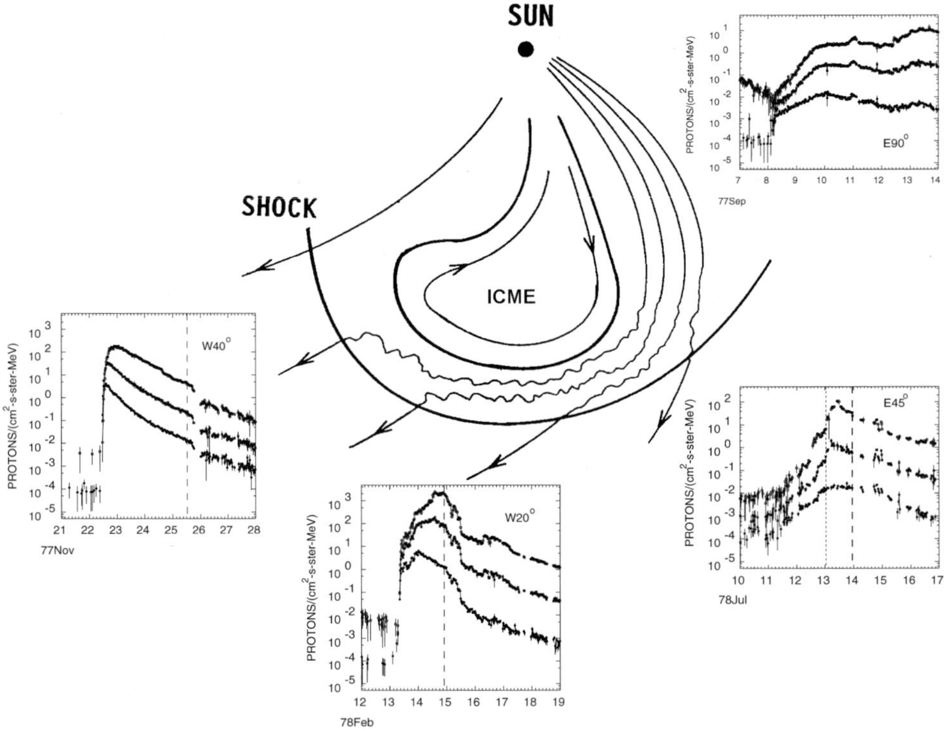

Figure 2. Cartoon showing the shape of an ICME and surrounding IP field structure including the presence of a shock. A strong shock will accelerate particles to an extent dependent on energy and the location of the observer. Thus particle intensity profiles are organised by the longitude of the associated solar event. Proton intensities in three energy ranges (~5, ~15 and ~30 MeV) are shown. Dashed lines indicate the passage of shocks. Figure adapted from Cane *et al.* (1988).

the events illustrated are typical, but event to event variations can be quite large in particular when there are additional CMEs either at the Sun or in the IP medium). Whereas events generated from the western longitudes have rapid rises followed by gradual decreasing intensities, events generated from eastern longitudes show slowly rising intensity enhancements structured around the arrival of the CME-driven shocks. This longitudinal dependence of the time-intensity profiles together with the rate at which the particle intensities increase or decrease have been used to predict the arrival of CME-driven shocks at 1 AU (Smith *et al.*, 2004; Vandegriff *et al.*, 2005). At large heliocentric distances and at high heliolatitudes, however, the relation between the origin of the event and the time-intensity profiles is less clear (e.g., McKibben *et al.*, 2001).

Simultaneous observations by widely separated spacecraft show that in large events particles reach widespread regions of the heliosphere, up to 300° in longitude (Cliver *et al.*, 1995); and up to at least 80° in latitude (Lario *et al.*, 2003). This widespread observation of SEP events suggests that there are magnetic connections to broad sources of particles that are able to both accelerate and inject particles into wide regions of the heliosphere. Alternatively, or in addition, the transport of particles across magnetic field lines might be very efficient as suggested by a number of authors (McKibben *et al.*, 2001; Cane and Erickson, 2003; Dalla *et al.*, 2003). However, particle anisotropies observed at the onset of large SEP events (at both low and high latitudes) are field-aligned with small or zero flow transverse to the magnetic field (Sanderson *et al.*, 2003). This suggests that perpendicular transport is inefficient. If there is widespread shock acceleration of particles close to the Sun then the shocks must decrease in latitudinal and longitudinal extent as they move away from the Sun since the necessary low coronal sizes are much more extended than implied from in situ observations (Cane, 1996). Only when the CME-driven shock arrives at the observer, is it possible to study, in situ, both the shock properties and the mechanisms working on particle acceleration (e.g., Tsurutani and Lin, 1985; and references therein). Whereas the study of specific events helps us to understand the underlying physics of the mechanisms involved in the generation of particular events, it is necessary to extend these studies to a comprehensive analysis of diverse events and thus, determine the multitude of processes involved in the generation of energetic particles by traveling CME-driven shocks.

4. Structure of ICMEs

Observations of energetic particles during the passage of an ICME over the observer provide valuable information about the structure of the ICME and its magnetic field topology (Richardson, 1997 and references therein). Energetic particle signatures associated with the passage of ICMEs in the ecliptic plane at 1 AU include (1) energetic particle intensity depressions (Forbush decreases) (Cane, 2000 and references

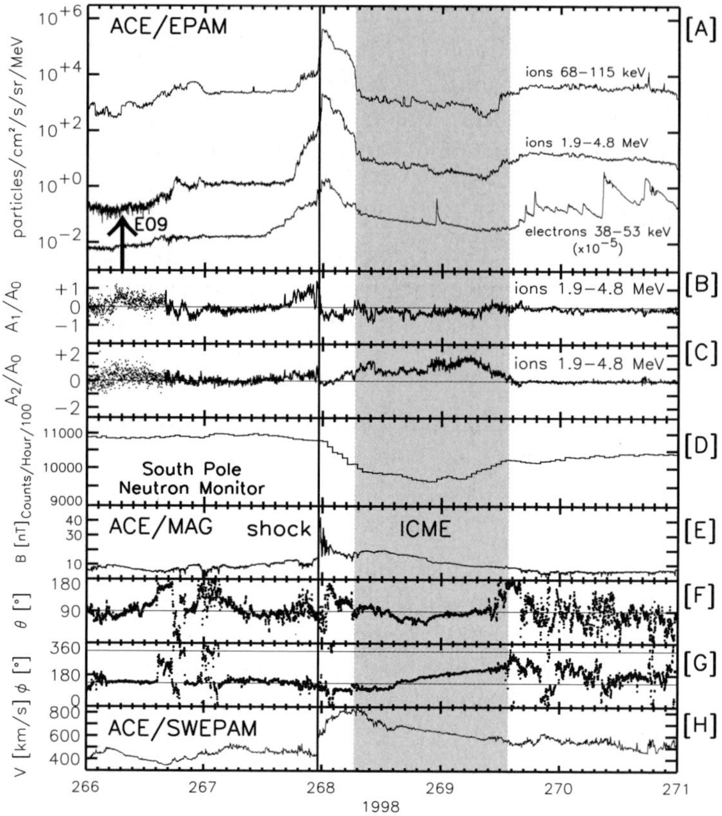

Figure 3. From top to bottom. [A] 96-s averages of the ion and electron intensities as measured by the ACE spacecraft. [B] 1.9–4.8 MeV ion first-order parallel anisotropy coefficient in the solar wind frame. [C] 1.9–4.8 MeV ion second-order anisotropy coefficient in the solar wind frame. [D] Count rates measured by the South Pole cosmic ray monitor. [E-G] Magnetic field magnitude and directions (in the GSE coordinate system) as measured by the ACE spacecraft. [H] Solar wind speed as measured by the ACE spacecraft.

therein); (2) bidirectional ~1 MeV ion flows (Marsden *et al.*, 1987); (3) bidirectional cosmic ray flows (Richardson *et al.*, 2000); and (4) occasionally, unusual SEP flow directions due to the fresh injection of SEPs by unrelated solar events (Richardson and Cane, 1996). In contrast to in the ecliptic plane, observations of ICMEs in high-speed streams at high heliographic latitudes show enhanced particle intensities instead of depressions (Bothmer *et al.*, 1995; Lario *et al.*, 2004).

Figure 3 shows the energetic particle response to the passage of an ICME through the near-earth solar wind observed by the ACE spacecraft in September 1998. (Note that this ICME was atypical in having well-defined boundaries; Cane and Richardson, 2003). This fast ICME was able to drive a strong IP shock (solid vertical line) that locally accelerated ions to at least ~60 MeV and electrons to at least ~50 keV at its arrival at 1 AU. The panels [B] and [C] show the 1.9–4.8 MeV first-order parallel (A_1) and second-order (A_2) anisotropy coefficients, respectively, computed in the solar wind frame following the method described in Lario *et al.* (2004). Note that A_1 changed its sign at the passage of the shock indicating that these particles were flowing away from the shock in the solar wind frame of reference. The entry of the spacecraft into the ICME showed an abrupt decrease in the low-energy ion intensities. Panels [B] and [C] show that $A_2 > A_1$ throughout the passage of the ICME indicating the presence of bidirectional ion flows (BIFs). A small impulsive electron event was observed at the end of day 268 when ACE was within the ICME, showing that the spacecraft was still magnetically connected to the Sun. Panel [D] shows that the particle intensity depressions extended to high energies indicating that the access of galactic cosmic rays into the ICME was limited. After exit from the ICME, low-energy ion intensities recovered to values similar to those observed prior to the passage of the ICME (after allowing for their gradual fall off with distance from the shock). The recovery of cosmic ray intensities, however, was more gradual and extended for several days after the ICME passage. This is because at these energies the post-shock turbulence causes an additional longer lasting decrease.

The presence of BIFs and the rapid onset of the electron event inside the ICME, are usually interpreted as evidence for the presence of looped magnetic field lines with the legs rooted at the Sun. The sharp decrease of the low-energy ion intensities observed upon entry into ICMEs at 1 AU show that the penetration of shock-accelerated particles into the ICME is restricted. Other particles inside ICMEs could come from particles accelerated at the time when the CME leaves the Sun (which implies the existence of a particle acceleration mechanism different from the CME-driven shock), and/or particles injected into the ICME by unrelated solar events. Although Figure 3 shows only a particular event, the study of energetic particle observations around and within ICMEs can be used not only to determine the magnetic topology of ICMEs but also the origin of intra-ICME particles and the transport conditions of these particles within and around the ICME (Lario *et al.*, 2004; and references therein).

5. Summary

Although energetic particle observations help us to study CMEs from their origin close to the Sun up to their arrival at the spacecraft, there are still many unknowns. Theoretical models of CME initiation at the Sun, three-dimensional simulations of the interplanetary transport of the CMEs and energetic particles, combined with multi-spacecraft observations of both ICMEs and SEPs (including composition measurements, ionic charge-state distributions and anisotropy analyses) will help us to understand the underlying physical mechanisms involved in the origin, acceleration and transport of energetic particles.

In order to discern the origin of the SEPs it is essential to determine both the relationship between flares and CMEs as well as the coronal magnetic topology during the eruption of the CMEs. In particular, challenges for future theoretical models of CME initiation include the following questions:

- Where are the flaring regions relative to the CME?
- Are there open field lines in the reconnection region behind a CME along which the energetic particles may propagate and escape to the IP medium? Is there evidence for progressive field line opening away from CMEs?
- Where, when and how do the CME-driven shocks form?
- What is the relationship between coronal shock waves and interplanetary shocks?
- When, where and how does particle injection start?
- What are the values of the physical parameters that are required to reproduce the abundance measurements, the energy dependence of the ionic charge states, and the maximum achievable energy of the particles?

Energetic particle observations by spacecraft are modulated by transport effects. Future and present models of energetic particle propagation and acceleration in the IP medium should include:

- Three-dimensional simulation of shock propagation from their formation to beyond the spacecraft location.
- Realistic seed particle populations for the time-dependent mechanisms of shock-acceleration including possible contributions from suprathermal remnants and particles accelerated during the flare processes.
- Evolution of the shock characteristics and its efficiency in accelerating and injecting particles into the IP medium.
- The influence of the IMF structure on the particle transport, i.e., on determining the onset times, spectra, anisotropy flows and time-intensity profiles of the SEP events at different regions of the heliosphere.

Finally, energetic particle observations within and around ICMEs should help us to determine both the origin of the intra-CME particle populations and the magnetic topology of the ICMEs. Energetic particle measurements should be used to improve both the methods of ICME identification and the models used to infer the three-dimensional structure of the ICMEs. Multi-spacecraft observations are essential to achieve these purposes. Most of the topics mentioned above are discussed in more detail in Klecker *et al.* (2006, this volume) or in Forbes *et al.* (2006, this volume).

References

Bothmer, V., Marsden, R. G., Sanderson, T. R., Trattner, K. J., Wenzel, K.-P., Balogh, A., *et al.*: 1995, *Geophys. Res. Lett.* **22**, 3369–3372.

Cane, H.: 1982, in *Solar Wind Five*, JPL, Pasadena, pp. 703–709.

Cane, H. V.: 1996, in: Ramaty, R., Mandzhavidze, N., and Hua, X. M. (eds.), *High Energy Solar Physics*, AIP, Woodberry, N.Y., pp. 124–130.
Cane, H. V.: 2000, *Space Sci. Rev.* **93**, 55–77.
Cane, H. V.: 2003, *Astrophys. J.* **598**, 1403–1408.
Cane, H. V. and Erickson, W. C.: 2003, *J. Geophys. Res.* **108**(A5), SSH 8–1, doi: 10.1029/2002JA009488.
Cane, H. V., Erickson, W. C., and Prestage, N. P.: 2002, *J. Geophys. Res.* **107**(A10), SSH 14–1, doi: 10.1029/2001JA000320.
Cane, H. V., McGuire, R. E., and von Rosenvinge, T. T.: 1986, *Astrophys. J.* **301**, 448–459.
Cane, H. V., Reames, D. V., and von Rosenvinge, T. T.: 1988, *J. Geophys. Res.* **93**, 9555–9567.
Cane, H. V. and Richardson, I. G.: 2003, *J. Geophys. Res.* **108**(A4), SSH 6–1, doi: 10.1029/2002JA009817.
Cliver, E. W., Kahler, S. W., Neidig, D. F., Cane, H. V., Richardson, I. G., Kallenrode, M.-B., *et al.*: 1995, Vol. 4 in: *Proc. 24th Int. Cosmic Ray Conf.*, pp. 257–260.
Cohen, C. M. S., Mewaldt, R. A., Leske, R. A., Cummings, A. C., Stone, E. C., Wiedenbeck, M. E., *et al.*: 1999, *Geophys. Res. Lett.* **26**, 2697–3000.
Dalla, S., Balogh, A., Krucker, S., Posner, A., Müller-Mellin, R., Anglin, J. D., *et al.*: 2003, *Geophys. Res. Lett.* **30**, ULY9–1, doi: 10.1029/2003GL017139.
Desai, M. I., Mason, G. M., Dwyer, J. R., Mazur, J. E., Gold, S. M., Krimigis, R. E., *et al.*: 2003, *Astrophys. J.* **588**, 1149–1162.
Forbes, T. G., Linker, J. A., *et al.*: 2006, *Space Sci. Rev.*, this volume, 10.1007/s11214-006-9019-8.
Haggerty, D. K. and Roelof, E. C.: 2002, *Astrophys. J.* **579**, 841–853.
Kahler, S. W., Hildner, E., and van Hollebeke, M. A. I.: 1978, *Solar Phys.* **57**, 429–443.
Kallenrode, M.-B.: 2001, *J. Geophys. Res.* **106**, 24989–25004.
Klecker, H., Kunow, H., *et al.*: 2006, *Space Sci. Rev.*, this volume, 10.1007/s11214-006-9018-9.
Klein, K.-L., Chupp, E. L., Trottet, G., Magun, A., Dunphy, P. P., Rieger, E., *et al.*: 1999, *Astron. Astrophys.* **348**, 271–285.
Kocharov, G. E., Kovaltsov, G. A., and Kocharov, L. G.: 1983, *Proc. 18th Intern. Cosmic Ray Conference* **4**, 105–108.
Kochorov, L., Kovaltsov, G. A., Torsti, J., and Ostryakov, V. M.: 2000, *Astron. Astrophys.* **357**, 716–724.
Lario, D.: 2005, *Adv. Space Res.* **36**, 2279–2288.
Lario, D., Decker, R. B., Roelof, E. C., Reisenfeld, D. B., and Sanderson, T. R.: 2004, *J. Geophys. Res.* **109**, A01107, doi:10.1029/2003JA010071.
Lario, D., Roelof, E. C., Decker, R. B., and Reisenfled, D. B.: 2003, *Adv. Space Res.* **32**, 579–584.
Lario, D., Sanahuja, B., and Heras, A. M.: 1998, *Astrophys. J.* **509**, 415–434.
Leske, R. A., Mewaldt, R. A., Cummings, A. C., Stone, E. C., and von Rosenvinge, T. T.: 2001, *AIP Conf. Proc. 598: Joint SOHO/ACE workshop "Solar and Galactic Composition"*. pp. 171–176.
Lin, R. P.: 1970, *Solar Phys.* **12**, 266–303.
Luhn, A., Hovestadt, D., Klecker, B., Scholer, M., Gloeckler, G., Ipavich, F. M., *et al.*: 1985, *Proc. 19th Intern. Cosmic Ray Conf.* **4**, 241–244.
Marsden, R. G., Sanderson, T. R., Tranquille, K.-P., Wenzel, C., and Smith, E. J.: 1987, *J. Geophys. Res.* **92**, 11009–11019.
McKibben, R. B., Lopate, C., and Zhang, M.: 2001, *Space Sci. Rev.* **97**, 257–262.
Möbius, E., Cao, Y., Popecki, M. A., Kistler, L. M., Kucharek, H., Morris, D., *et al.*: 2003, *Proc. 28th Intern. Cosmic Ray Conf.* **4**, 3273–3276.
Ng, C. K., Reames, D. V., and Tylka, A. J.: 2003, *Astrophys. J.* **591**, 461–485.
Oetliker, M., Klecker, B., Hovestadt, D., Mason, G. M., Mazur, J. E., Leske, R. A., *et al.*: 1997, *Astrophys. J.* **477**, 495–501.

Popecki, M. A., Mazur, J., Möbius, E., Klecker, B., Bogdanov, A., Mason, M. A., *et al.*: 2003, *Int. Cosmic Ray Conf.* **4**, 3283–3286.

Posner, A. and Kunow, H.: 2003, *Proc. 28th Intern. Cosmic Ray Conf.* **4**, 3309–3312.

Reames, D. V.: 1999, *Space Sci. Rev.* **90**, 413–491.

Rice, W. K. M., Zank, G. P., and Li, G.: 2003, *J. Geophys. Res.* **108**(A10), 1369, DOI:10.1029/2002JA009756.

Richardson, I. G.: 1997, in: Crooker, N., Joselyn, J. A., and Feynman, J. (eds.), *Coronal Mass Ejections*, AGU Geophysical Monograph, 99, pp. 189–198.

Richardson, I. G. and Cane, H. V.: 1996, *J. Geophys. Res.* **101**, 27521–27532.

Richardson, I. G., Dvornikov, V. M., Sdobnov, V. E., and Cane, H. V.: 2000, *J. Geophys. Res.* **105**, 12579–12592.

Sanderson, T. R., Marsden, R. G., Tranquille, C., Dalla, S., Forsyth, R. J., Gosling, J. T., *et al.*: 2003, *Geophys. Res. Lett.* **30**(19), 8036, doi:10.1029/2003GL017306.

Smith, Z., Murtagh, W., and Smithtro, C.: 2004, *J. Geophys. Res.* **109**(A1), A01110, doi:10.1029/2004JA010044.

Torsti, J., Kocharov, L., Innes, D. E., Laivola, J., and Sahla, T.: 2001, *Astron. Astrophys.* **365**, 198–203.

Torsti, J., Laivola, J., and Kocharov, L.: 2003, *Astron. Astrophys.* **408**, L1–L4.

Tsurutani, B. T. and Lin, R.: 1985, *J. Geophys. Res.* **90**, 1–11.

Vandegriff, J., Wagstaff, K., Ho, G., and Plauger, J.: 2005, *Adv. Space Res.* **36**(12), 2323–2327.

Wagner, W. J. and MacQueen, R. M.: 1983, *Astron. Astrophys.* **120**, 136–138.

AN INTRODUCTION TO THEORY AND MODELS OF CMES, SHOCKS, AND SOLAR ENERGETIC PARTICLES

Z. MIKIĆ[1,*] and M. A. LEE[2]
[1]*Science Applications International Corporation, 10260 Campus Point Drive, San Diego, CA 92121, USA*
[2]*Institute for Earth, Oceans, and Space, 39 College Rd., University of New Hampshire, Durham, NH 03824, USA*
*(*Author for correspondence: E-mail: mikicz@saic.com)*

(Received 1 February 2006; Accepted in final form 12 June 2006)

Abstract. We present a brief introduction to the essential physics of coronal mass ejections as well as a review of theory and models of CME initiation, solar energetic particle (SEP) acceleration, and shock propagation. A brief review of the history of CME models demonstrates steady progress toward an understanding of CME initiation, but it is clear that the question of what initiates CMEs has still not been solved. For illustration, we focus on the flux cancellation model and the breakout model. We contrast the similarities and differences between these models, and we examine how their essential features compare with observations. We review the generation of shocks by CMEs. We also outline the theoretical ideas behind the origin of a gradual SEP event at the evolving CME-driven coronal/interplanetary shock and the origin of "impulsive" SEP events at flare sites of magnetic reconnection below CMEs. We argue that future developments in models require focused study of "campaign events" to best utilize the wealth of available CME and SEP observations.

1. Introduction

One of the primary focuses of present theoretical coronal mass ejection (CME) research is the initiation problem. Many of the theoretical interpretations of observations in the lower corona and inner heliosphere, including radio emission, shock acceleration of particles, and the structure and properties of interplanetary CMEs (ICMEs) and flux ropes, and their relationship to their solar source regions, hinge on the details of the CME initiation mechanism.

Modern observations, starting with Skylab in the 1970s, the Solar Maximum Mission (SMM) in the 1980s, and with Yohkoh, SOHO, and TRACE in the 1990s and present decade, have provided a rich source of observations to classify the morphology and characteristics of CMEs. Why, then, has the solution to the CME initiation problem remained elusive, in light of this wealth of observations? It is fair to say that we strongly suspect we know the key phenomena involved in CME initiation, and several candidate models, but no confirmation yet. Alexander *et al.* (2006, this volume) have provided a brief historical review of CME observations in the last century and a half. During this time period we have gradually come to the realization, which is universally held today, that CMEs are magnetically driven

Space Science Reviews (2006) 123: 57–80
DOI: 10.1007/s11214-006-9012-2

phenomena. That is not to say that pressure and gravity forces do not play a role in the destabilization of CMEs (they very well may). Indeed, the solar wind itself is a phenomenon driven by pressure and gravity forces (Parker, 1963). The role of non-magnetic forces will most likely grow in importance as our understanding of CMEs improves.

One of the principal reasons why the CME initiation problem has not been solved is because it is not possible (in general) to measure coronal magnetic fields in detail. We therefore have to rely on photospheric (and sometimes chromospheric) measurements, extrapolated using models, to infer the magnetic field in the corona. Furthermore, we routinely only measure the line-of-sight (longitudinal) component of the magnetic field, and not the transverse component, even though the energization of the coronal field (the very energy that drives the CME) can only be quantified with vector magnetic field measurements. Because measured coronal emission (e.g., in white light, EUV, and X-rays) is optically thin, it is necessary to deconvolve the effect of line-of-sight integration to interpret the emission, complicating the situation further. Radio emission measurements also need to be deconvolved in a non-trivial way to infer the coronal magnetic field (e.g., Lee *et al.*, 1998a,b, 1999). In-situ measurements of ICMEs afford limited ability to diagnose the 3D structure of CME ejecta but nevertheless provide important constraints. In Section 3 we list observations that are useful in characterizing the properties of CMEs.

In addition to these observational difficulties, there are considerable theoretical difficulties: the models are too idealized; they cannot address realistic geometry; they don't include fine-scale structure; they are too dissipative; they are not fully self-consistent (e.g., energy transport is neglected, prominences are not included, parallel flows are not modeled); and they don't produce the quantities that are measured (e.g., EUV, X-ray, and H-α images).

We are therefore forced to deduce the structure and topology of the pre-CME and post-CME plasma from indirect measurements and interpret them with incomplete models, which explains why it has been difficult to unravel the mystery of CME initiation.

1.1. A Brief History of CME Models

CME models have evolved from the early "cartoon" models, in which the description was qualitative and imprecise, to simple analytic and semi-analytic models, to idealized 2D and 3D numerical models. The next generation of 3D numerical models that are being implemented on massively parallel computers will be able to directly address observations, as discussed in Section 9. It is evident that CMEs are driven by the energy in the magnetic field. The main question that remains is: how is this energy released, and, most importantly, how is it released rapidly enough to explain fast CMEs that are observed to travel at speeds exceeding 1,000 km/s? Explaining fast CMEs has remained a difficulty of present models.

Barnes and Sturrock (1972) examined the energy stored in a twisted force-free field in order to explain how a magnetic field configuration could release energy while the field is opening (and presumably leading to an eruption). In subsequent work, Yang *et al.* (1986) found that the energy in the open field appears to be an upper limit to the magnetic energy, a result that has been formalized into a conjecture (Aly, 1991; Sturrock, 1991). This key development in the theory of CME initiation is worth restating: a fully open magnetic field provides an upper limit to the energy in a force-free field with the same normal magnetic field distribution on the solar surface (Aly, 1984, 1991; Sturrock, 1991). Therefore, in a model in which all the energy is stored in the magnetic field, it does not appear to be energetically possible for a closed magnetic field configuration to spontaneously make a transition to an open field. This seemingly implies that magnetic fields cannot open dynamically. However, there are many ways in which CMEs can open the magnetic field, including *partial* opening of the field (Wolfson and Low, 1992; Wolfson, 1993; Mikić and Linker, 1994), and by including the effect of pressure and gravity forces (Low, 1993; Low and Smith, 1993; Wolfson and Dlamini, 1997, 1999; Wolfson and Saran, 1998). See the review by Forbes (2000) for a discussion.

Early theory (e.g., Low, 1977, 1981; Birn and Schindler, 1981) examined the properties of equilibria using "generating function" solutions of the Grad-Shafranov equation in which the variation of a parameter was taken to represent evolution through a sequence of equilibria. In these models, it was presumed that "loss of equilibrium" would occur when solutions ceased to exist. However, Klimchuk and Sturrock (1989) cautioned against interpreting loss of equilibrium due to an artificial parametric specification as evidence of dynamical evolution.

Aly (1984, 1985, 1988, 1990) has investigated the mathematical properties of magnetic field configurations to deduce limits on their energy, stability, and the existence of solutions. In the highly conducting solar corona, the footpoints of the magnetic field lines are dragged by motions in the dense photosphere, a situation that is referred to as "line tying." Although line tying provides a stabilizing effect, it also allows the convective motions on the Sun to deform the coronal magnetic field, leading to the possibility of eruptive behavior.

The evolution of line-tied 2D magnetic arcades deformed by shearing photospheric motions has been studied by Mikić *et al.* (1988), Biskamp and Welter (1989), Finn and Chen (1990), Finn and Guzdar (1993), Choe and Lee (1996a,b) and Amari *et al.* (1996). When converging motions are applied at the neutral line, the arcade ejects a plasmoid (Inhester *et al.*, 1992) due to reconnection. Manchester (2003) studied the disruption of buoyant 2D arcades. Arcade models have also been extended to spherical geometry (Mikić and Linker, 1994; Antiochos *et al.*, 1999), including the effect of the solar wind (Linker and Mikić, 1995; Wu *et al.*, 2001). Recently, models have been extended to study idealized 3D magnetic configurations (e.g., Amari *et al.*, 2003a,b; Linker *et al.*, 2003a,b; Roussev *et al.*, 2003, 2004; Manchester *et al.*, 2004a b).

In another class of semi-analytic models, the CME problem is formulated as loss of equilibrium due to a catastrophe (Forbes and Isenberg, 1991; Isenberg *et al.*, 1993; Forbes *et al.*, 1994; Forbes and Priest, 1995). Recent improvements to this "catastrophe model" (Lin *et al.*, 1998, 2001, 2002; Forbes and Lin, 2000; Lin and Forbes, 2000; Lin and van Ballegooijen, 2002) have significantly extended its applicability to the CME initiation problem. There is a close relationship between the catastrophe model and the flux cancellation model discussed in Section 5.1.

2. The MHD Model

Most large-scale theories of CMEs in the corona are based on the resistive magnetohydrodynamic (MHD) equations and their simplifications (e.g., zero-β, force-free, etc.), although kinetic extensions are needed to study the evolution in the inner heliosphere and especially to model the acceleration of solar energetic particles (SEPs), as described in Sections 7 and 8. In its most comprehensive form, the MHD model includes equations for mass, momentum, and energy conservation as well as the resistive Ohm's law. Energy transport includes parallel thermal conduction (along the magnetic field lines), radiation loss, coronal heating, and acceleration by Alfvén waves, usually treated according to a WKB formalism (Jacques, 1977; Usmanov *et al.*, 2000). For a description of these equations and their application to coronal modeling, see, for example, Mikić *et al.* (1999). Early models considered a simple polytropic energy equation (e.g., Linker and Mikić, 1995) with a reduced polytropic index (Parker, 1963) to model the solar wind. The realism needed to model campaign events (as discussed in Section 9) is pushing the models toward an improved description of energy transport.

The central role of the magnetic field, combined with a desire to simplify the problem, has led theorists to focus on force-free models of the corona, in which all forces other than magnetic forces are neglected. In this model, the equilibrium force-balance condition simplifies to

$$\mathbf{J} \times \mathbf{B} = 0 \tag{1}$$

where $\mathbf{J} = c\nabla \times \mathbf{B}/4\pi$ is the electric current density and $\mathbf{B}$ is the magnetic field intensity, which implies that $\mathbf{J} = \alpha\mathbf{B}$, with α, the torsion, an (unknown) function of position. Much theoretical research has focused on the study of equilibria satisfying Equation (1), which in itself is a difficult nonlinear problem.

3. Relevant Observations

Observations that help to determine the magnetic field topology in the pre-eruptive state (Gopalswamy, 2003) include the orientation of flows along filaments in H-α,

X-ray and EUV loops (as beautifully seen in TRACE images), radio emission, longitudinal and vector magnetograms, the history of **B** in the photosphere as determined from sequences of magnetograms, and limb measurements of prominences. Magnetograph measurements can help to estimate the magnetic energy (Klimchuk *et al.*, 1992; Metcalf *et al.*, 2005). For details on how these observations can be used to determine the properties of the pre-CME corona see Gopalswamy *et al.* (2006) in this volume.

In the post-eruptive state, clues to the topology of the magnetic field can be obtained from H-α flare ribbons, EUV and soft-X-ray emission in post-flare loops and post-CME coronal dimmings, hard X-ray footpoint emission, measurements of bidirectional electrons and heat flux dropouts (Gosling *et al.*, 1987), and estimates of field line length (Larson *et al.*, 1997). Coronagraph measurements can help to estimate CME velocities and masses, and hence kinetic energies (Vourlidas *et al.*, 2000, 2002), as well as their morphology. Radio signatures (Reiner *et al.*, 2001; Bastian *et al.*, 2001; Cliver *et al.*, 2004) and solar energetic particle (SEP) measurements can probe the properties of shocks in the corona. Composition and charge state measurements can help to relate *in situ* plasma to its solar sources. Interplanetary measurements can be used to estimate flux rope helicity and twist. For details on how these observations can be used to determine the properties of CMEs see Schwenn *et al.* (2006, this volume) and Pick *et al.* (2006, this volume).

4. Classification of Models

There have been several reviews of the theory of CME initiation (Low, 1994, 1996, 1999, 2001; Forbes, 2000; Klimchuk, 2001; Linker *et al.*, 2003b), including a recent comprehensive review (Lin *et al.*, 2003) to which the reader is referred for more detailed discussions; see also Forbes *et al.* (2006, this volume) for a detailed presentation of the various models. Klimchuk (2001) has presented a classification of models into two broad classes: "storage and release" models and "directly driven" models. The class of directly driven models, in which the energy released during CME eruption is injected into the corona during the eruption, includes dynamo models (Chen, 1989; Chen *et al.*, 1997, 2000; Krall *et al.*, 2000) and thermal blast models, in which a pressure pulse is used to initiate the eruption (Dryer, 1982; Wu *et al.*, 1982). These are presently not considered as viable CME initiation models since they are not supported by observations (Forbes, 2000; Lin *et al.*, 2003). We therefore restrict our attention to storage and release models.

5. Examples of "Storage and Release" Models

In storage and release models, the CME is driven by the energy stored in the magnetic field, which is built up over a long period of time (days to weeks) and is released in a short time (minutes to hours). Rough estimates indicate that the coronal magnetic field can store sufficient energy to power even the largest flares and CMEs (Forbes, 2000). Here we will only touch on the essential principles and broadly survey the relevant phenomena, with specific references to two particular models with which we are familiar, the flux cancellation model and the breakout model. The catastrophe models mentioned in Section 1.1 are closely related to the flux cancellation model. See Forbes *et al.* (2006, this volume) for a more detailed discussion of these models.

In equilibrium, if gravity can be neglected (when very strong magnetic fields are present), the momentum equation expresses a balance between the tension in magnetic field lines that are line-tied in the photosphere and magnetic and thermal pressure: $\mathbf{B} \cdot \nabla \mathbf{B} = \nabla(4\pi p + B^2/2)$. Eruption involves forcing the system to evolve into a state in which this delicate balance can no longer be maintained. The two models differ in the way that this balance is upset, as described below.

5.1. The Flux Cancellation Model

Detailed observations of magnetic fields in and around active regions indicate that the emergence of new magnetic flux, especially in the vicinity of pre-existing magnetic fields (Gaizauskas *et al.*, 1983; Zwaan, 1985; Feynman and Martin, 1995), and flux cancellation (Martin *et al.*, 1985; Livi *et al.*, 1985, 1989; Zwaan, 1987, 1996; Gaizauskas, 1993), are connected with solar eruptions (flares and CMEs). This led to the development of models that incorporate flux cancellation and magnetic field diffusion in the neighborhood of polarity inversion lines as a key ingredient in prominence formation and eruption (Pneuman, 1983; van Ballegooijen and Martens, 1989, 1990; Mackay *et al.*, 1998; Litvinenko and Martin, 1999; Amari *et al.*, 1999; Mackay and van Ballegooijen, 2001, 2005; Lionello *et al.*, 2002; Mackay and Gaizauskas, 2003). Flux cancellation has been identified as a key element in the formation of prominences, which are also known as filaments (Gaizauskas *et al.*, 1997; Martin, 1998; van Ballegooijen *et al.*, 2000; Martens and Zwaan, 2001). Démoulin (1998) has reviewed the structure of magnetic fields in filaments.

Flux cancellation has been studied in prominence formation, eruption, and CME initiation with simulations of the large-scale corona (Linker *et al.*, 2001; Linker *et al.*, 2003a,b; Roussev *et al.*, 2004) and also on active-region scales (Amari *et al.*, 1999, 2000, 2003a,b; Lionello *et al.*, 2002; Welsch *et al.*, 2005). When the amount of cancelled flux does not exceed a threshold value, a magnetic flux rope forms above the neutral line in 3D arcades. This structure is stable and can support prominence material (Linker *et al.*, 2001; Lionello *et al.*, 2002). If flux cancellation is continued beyond this threshold, the configuration erupts (Amari *et al.*, 2000, 2003a,b). The eruption converts a significant fraction of the magnetic energy into kinetic energy. When the configuration is close to the critical state, even a small amount of flux cancellation can trigger a violent eruption. The crossing of the threshold is unremarkable as far as the photospheric field is concerned, and it is likely to be missed in magnetograph observations. This critical behavior resembles that seen in the catastrophe models (e.g., Forbes and Isenberg, 1991).

5.2. The Breakout Model

Syrovatskii (1982) first noted the possible importance of multipolar configurations in explaining eruptive behavior. The breakout model, which describes the eruption of multipolar configurations, was developed by Antiochos (1998), Antiochos *et al.* (1999), MacNeice *et al.* (2004), Lynch *et al.* (2004), and DeVore and Antiochos (2005). Like the flux cancellation model, breakout requires strongly sheared fields near the neutral line, as observed in filament channels. A key feature of the model is that it requires a multipolar flux distribution and a magnetic null to be present. When the central arcade is sheared, causing the field lines to rise, slow reconnection at the null point transfers overlying flux in the central arcade to the neighboring arcades, eventually destabilizing the central arcade. An application of the breakout model to flare observations is given by Gary and Moore (2004).

5.3. Similarities and Differences

In our opinion, the flux cancellation and breakout models have more fundamental similarities than differences. Both models require build-up of significant parallel electric current (shear, twist) prior to eruption; the magnetic field is highly aligned along the neutral line; eruption requires build up of magnetic pressure within the central arcade and/or release of tension in the overlying field lines. The exact mechanism by which this occurs is different in the two models. In the flux cancellation model, magnetic pressure is built up in the flux rope by the slow reconnection of magnetic field lines at the neutral line, which at the same time relieves the tension in the overlying field lines by severing connections to the photosphere. In the breakout model, the reconnection of the high field lines in the arcade with the overlying field lines in the surrounding flux systems releases the magnetic tension that holds down the central arcade. In both cases, reconnection in the lower corona within the arcade completes the eruption process and ejects a flux rope. Any prominence material that happens to be trapped in the dips of magnetic field lines is also ejected.

The models also have differences. For example, the photospheric magnetic field has a different character before eruption (the flux cancellation model has a flux rope whereas the breakout model does not). Flux cancellation requires converging flows and cancellation of flux at the neutral line, combined with (slow) magnetic reconnection there. Flux cancellation can occur in a simple dipolar configuration, whereas breakout requires a more complex topology. In the flux cancellation mechanism, prominence material can become trapped on the flux-rope magnetic field line dips as they form and rise into the corona, leading to the natural formation of a prominence within the arcade. This material is observed to be ejected together with the streamer. In contrast, in the breakout model the prominence material needs to condense in the corona or be siphoned from the chromosphere.

It is important to note that the idealizations inherent in axisymmetric (2D) geometry tend to emphasize the differences between the two models. In a fully 3D geometry it is more difficult to distinguish between these two models; there is a continuum between dipolar and multipolar configurations as the relative orientation between the overlying field and a bipolar active region is changed. Furthermore, the distinctions between the two topologies may be blurred by the similarities in behavior of magnetic field configurations that have true separatrices versus quasi-separatrix layers (Démoulin *et al.*, 1996, 1997; Titov *et al.*, 2002). In addition, in 3D it is difficult to distinguish between a "flux rope" with a fraction of a turn of twist, whose legs are attached to the photosphere, and a highly-sheared, dipped field line (e.g., as depicted in the prominence support model of Antiochos *et al.*, 1994).

5.4. Comparison with Observations

Several features of these models agree with observations. A magnetic field topology that can support a filament (i.e., dipped field lines) is formed naturally and erupts together with the streamer in the flux cancellation mechanism, as seen in many CME observations. Also, CMEs tend to occur in decaying active regions with dispersing magnetic flux, in accordance with the flux cancellation scenario. Converging flows and the disappearance of magnetic elements of opposite polarity are also observed at neutral lines. Breakout requires a complex topology, a feature that is consistent with flare-productive regions.

Other features do not agree with observations. In particular, it has been difficult to show that fast CMEs can be produced with the models, although this may be related to the geometrical simplicity of the models and because they have not been able to simulate the strong localized magnetic fields in active regions. Additionally, the large-scale shear flows that have been used to energize the models are typically not observed, although a large part of the twist in the active regions is present when they emerge from below the photosphere (Leka *et al.*, 1996). In Section 9 we discuss future improvements to these models that will greatly enhance their ability to address observations.

6. Connecting the Corona to the Heliosphere: The CME–ICME Connection

CMEs that are observed in the corona produce signatures in interplanetary space which can be measured by *in-situ* spacecraft (Wimmer *et al.*, 2006, this volume). These signatures often reveal a great deal about their properties and origin. Many studies have attempted to link observations of magnetic clouds to their inferred active-region sources (e.g., Webb *et al.*, 2000; Leamon *et al.*, 2004). Simulations have shown that CMEs can undergo a significant amount of distortion as they expand and encounter solar wind streams with different velocities (Riley *et al.*, 1997, 1998; Odstrcil and Pizzo, 1999a,b,c).

Odstrcil *et al.* (2002) and Riley *et al.* (2003) have followed a CME from its eruption in the corona to 5 AU. A 2D CME was initiated in the corona by the flux cancellation mechanism (Linker *et al.*, 2003a). Significant distortion and "pancaking" of the ICME is observed as it propagates away from the Sun. This simulation was used to test several flux-rope fitting techniques (Riley *et al.*, 2004) and to interpret the global context of a CME that was observed by ACE and Ulysses (Riley *et al.*, 2003). Simulations of an idealized 3D eruption have also been performed (Odstrcil, 2003; Luhmann *et al.*, 2004). Although these are promising first steps, we are just beginning to explore the detailed relationship between CMEs and ICMEs. Little is known about the relationship between CME initiation mechanisms and ICME signatures, so that it is difficult to use these signatures to discriminate between the models. Furthermore, the topology of the magnetic field lines that connect the magnetic cloud with the Sun and the heliosphere is not well known (e.g., Gosling *et al.*, 1995; Crooker *et al.*, 2002). Adressing these issues will have to await improvements to the models.

7. CME-Driven Shock Propagation

An immediate consequence of a fast CME is a magnetic field and pressure enhancement ahead of it. If the ejected mass is or becomes superalfvénic, then the enhancement forms a bow shock that drapes around the CME and propagates ahead of it into the heliosphere. The flanks of the bow shock/wave may extend to the base of the corona (Sheeley *et al.*, 2000) and be observed as a Moreton wave (Moreton, 1960) or an EIT-wave (Brueckner *et al.*, 1998). However, this picture is probably oversimplified. Multiple shock waves may be produced low in the corona, where the Alfvén speed is small, due to a complex release of magnetic energy during the eruption process. Thus, the interpretation of the observed disturbances and type II radio bursts indicating shock formation in a given event may be difficult.

The governing equations for the macroscopic behavior of waves and shocks in the corona and solar wind are generally taken to be the 1-fluid ideal MHD equations supplemented by an adiabatic equation of state with $\gamma = 5/3$. They specify the time evolution of the fluid velocity $\mathbf{V}$, magnetic field $\mathbf{B}$, pressure P, and mass density ρ. These equations are not valid at shocks since non-ideal terms involving viscosity, heat flux, and electrical resistivity are important there. However, for the macroscopic behavior of the fluid, it is sufficient to impose the Rankine-Hugoniot jump conditions at the discontinuity that forms in the flow. There are several algorithms available for this purpose (Hundhausen, 1985; Pizzo, 1985; Powell *et al.*, 2003). The linearized equations describe the "fast," "slow" and Alfvén modes of homogeneous MHD and their generalizations in a solar wind with inhomogeneous velocity $\mathbf{V}_0$ and magnetic field $\mathbf{B}_0$. The "fast" and "slow" modes are compressive and form shocks. However, the "slow" shock is indeed slow with subalfvénic flows both upstream and downstream of the shock in the shock frame (Hundhausen, 1985). The shocks observed in the solar wind and predicted in the corona are generally "fast" shocks (but see Whang *et al.*, 1996).

The two simplest solutions of these equations, assuming spherically symmetric, hydrodynamic flows ($\mathbf{B} = 0$), and neglecting gravity, are the blast wave with constant total energy (created, say, by an impulsive enhancement in solar wind speed) and the driven shock with increasing energy content (created, say, by a sudden and persistent increase in wind speed). Under further simplifying assumptions, these two cases may be described by analytical "self-similar" solutions (Rogers, 1957; Sedov, 1959; Parker, 1961, 1963; Chevalier, 1982). The blast wave is a single forward shock, whereas the driven configuration involves a forward shock propagating into the solar wind ahead and a reverse shock propagating backwards into the ambient wind behind, but swept outwards by the flow. These two simple cases provide a framework for interpreting shock propagation in the corona and solar wind in more general cases. The CME-associated shock is initially driven, since CMEs appear to retain their high speeds for tens of R_s. However, as CME speeds decay with a spatial scalelength of $r \sim 50R_s$ (Reiner *et al.*, 2003) and assimilate into the solar wind as interplanetary CMEs (ICMEs), the shocks probably transform into blast waves. The simple models omit many features which are important in the solar wind and corona. Inhomogeneity, particularly in coronal active regions and the streamer belt, causes refraction of the shock waves (Vainio and Khan, 2004). Beyond Earth orbit, shocks may interact and coalesce (Pyle *et al.*, 1984). Numerous simulations over three decades have revealed various aspects of interplanetary shock propagation (e.g., Dryer, 1974; Steinolfson, 1985; Whang and Burlaga, 1985; Tsurutani *et al.*, 2003).

The US National Space Weather Program has revitalized studies of CME-driven shock propagation, since these shocks contribute to geomagnetic disturbances. Odstrcil and Pizzo (1999c,a,b) developed a 3-D hydrodynamic and an MHD code to investigate how a CME, simulated by a localized pressure and velocity enhancement at the inner boundary, interacts with a solar wind that includes a tilted magnetic dipole configuration with streamer belt and stream structure. Odstrcil *et al.* (2002) combined the heliospheric MHD code of Odstrcil and Pizzo (1999a) with the coronal MHD code of Mikić and Linker (1994) to treat a 2-D axisymmetric eruption of a CME into the heliosphere beyond Earth orbit. A competing 3-D MHD code with adaptive mesh refinement has been developed by the U. Michigan group (Powell *et al.*, 2003). Groth *et al.* (2000) and Manchester *et al.* (2004c) presented results of the Michigan code, which described the eruption of a CME into a structured solar wind. The CME was modeled as a 3-D flux rope with initial force imbalance, resulting in rapid outward acceleration.

A crucial issue for particle acceleration at a CME-driven shock (see Section 8.1(5)) is the formation time and strength of the shock close to the Sun. These features depend primarily on CME speed and the spatial distribution of the Alfvén speed, V_A. Although V_A clearly depends on local coronal structure, generally, away from active regions V_A is low in the chromosphere and low corona where the density is high, increases with height as the density decreases, and finally decreases in the solar wind as $V_A \propto r^{-1}$ inside Earth orbit. We expect a maximum of $V_A \sim$500 km/s at a heliocentric radial distance of $r \sim 3 - 5R_s$ (Gopalswamy *et al.*, 2001; Mann *et al.*, 2003). Thus, we expect CMEs that accelerate to speeds $\sim$500 km/s to form bow shocks at $r \sim 3 - 5R_s$. This expectation is consistent with the onset time of energetic ion acceleration to GeV energies (Kahler, 1994) and the strong correlation of energetic ion intensity and CME speed (Reames *et al.*, 1997).

A recent simulation by Roussev *et al.* (2004) initiates a CME based on the evolution of the observed photospheric and coronal magnetic fields for the event of 2 May 1998. They find that the nose of the driven shock reaches a speed of $\sim$1200 km/s at $r \sim 4R_s$, with a compression ratio $\sim$3. Assuming that the scattering mean free path of protons is approximately their gyroradius, this shock is predicted to have an energetic particle cutoff energy $\sim$10 GeV, consistent with the observation that this event was a "ground-level event" (Lopate, 2001, but see Section 8.1(5)).

8. Acceleration of Energetic Particles

An important product of CMEs are energetic particles, which are detected either directly in space or by secondary electromagnetic and neutron emissions. Energetic particles generally contain a small fraction of the energy released by a CME. Nevertheless, by virtue of their high speed and energy, they can have deleterious effects on humans and assets in space and may be utilized in space weather forecasting (Reames, 1999; Feynman and Gabriel, 2000; Tóth *et al.*, 2005).

Charged particles are accelerated by **E**, the electric field. The fact that the accelerated electrons are very effective in canceling electric field enhancements generally ensures that $\mathbf{E} \approx 0$ in the plasma frame of reference. Thus, acceleration in space plasmas generally depends on relatively subtle effects. These involve variations of the plasma velocity $\delta\mathbf{V}$ in a particular configuration so that both $\mathbf{E} \approx -c^{-1}\delta\mathbf{V} \times \mathbf{B}_0$ and the motion of the particles allows a nonzero average value of $\mathbf{v} \cdot \mathbf{E}$. Here $\mathbf{v}$ is particle velocity, and $|\mathbf{E}|/|\mathbf{B}_0| \sim |\delta\mathbf{V}|/c \ll 1$.

Observations of solar energetic particles (SEPs) in space provide some guidance in determining the relevant configuration for acceleration. SEPs appear in two fairly distinct classes of events: "impulsive" and "gradual." Impulsive events are small, last for hours, occur at a rate of $\sim 10^3$/y during solar maximum, are rich in electrons, ^{3}He and heavy ions, and have relatively high charge states. In contrast, gradual events are large, last for days, occur at a rate of $\sim$10/y during solar maximum, have approximately solar wind or coronal ion composition, are electron poor, and have relatively low charge states (Lee, 1991). Although this classification has become blurred by recent measurements of elemental and ionic charge composition as described in detail by (Cane and Lario, 2006, this volume) and (Klecker *et al.*, 2006, this volume), nevertheless it implies distinct acceleration mechanisms for impulsive and gradual events.

8.1. Gradual Events and Shock Acceleration

The characteristics of gradual events are generally consistent with their origin at a coronal/interplanetary shock driven by a CME. In addition to those characteristics listed above, these events are strongly correlated with fast CMEs and type II radio bursts and have a broad extent in heliographic solid angle. All of these features are expected for a shock origin.

Figure 1 is a schematic diagram of a CME-driven coronal/interplanetary shock as it propagates toward Earth and into the solar wind across the Archimedes-spiral magnetic field. The dots indicate the SEPs. They are accelerated by criss-crossing the shock and, in the process, both drifting in the inhomogeneous shock magnetic field parallel to **E** in the shock frame and scattering between the convergent

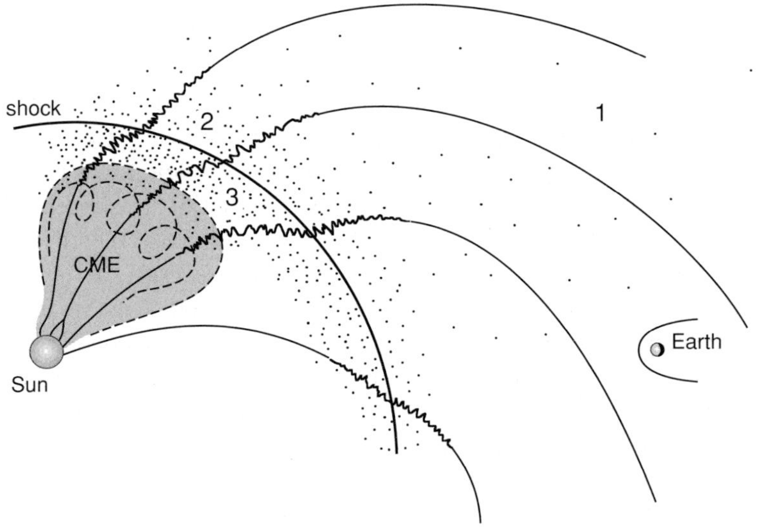

Figure 1. Schematic snapshot of an evolving coronal/interplanetary shock driven by a CME. Accelerated ions are denoted by dots. Magnetic field lines are shown, with wiggles denoting magnetic fluctuations. The spatial domain accessible to the ions is divided into solar wind (1), a proton-excited turbulent sheath upstream of the shock (2), and the turbulent shock-heated solar wind downstream of the shock (3).

electromagnetic irregularities on either side of the shock. These two aspects of shock acceleration are known as "shock drift" acceleration (Hudson, 1965; Sarris and VanAllen, 1974) and "first-order Fermi" acceleration (Fermi, 1954). Their relative contributions are dependent on the chosen frame of reference. Both aspects are combined in the theory of diffusive shock acceleration (Jokipii, 1982).

Figure 1 draws attention to the possible temporal and spatial complexity of the shock acceleration process. The solar wind is inhomogeneous, the shock evolves in strength and shape, and the magnetic connection between observer and shock can be complicated. Another complication is that the ion scattering mean free path λ in the solar wind (Region 1 in Figure 1) is too large ($\sim$0.1–1 AU) to yield the rapid multiple traversals of the shock required for particles to attain the observed SEP energies. However, the accelerating ions excite hydromagnetic waves to form a turbulent sheath upstream of the shock. Within this sheath, denoted by Region 2 in Figure 1 with its fluctuating magnetic field components, λ is small and acceleration is rapid. Region 3, adjacent to and downstream of the shock, is also turbulent. There the upstream turbulence is compressed and enhanced by the shock. These three regions are distinct and require different ion transport equations for the ion distribution function.

The basic particle transport equation for application to SEPs is the focused transport equation (Roelof, 1969; Skilling, 1971; Earl, 1976, 1981; Isenberg, 1997; Forbes *et al.*, 2006, this volume), which describes the convection, adiabatic deceleration, magnetic focusing and pitch-angle diffusion (with diffusion coefficient D_μ) of particles confined to a magnetic flux tube. Although the equation treats particles with $v \sim |\mathbf{V}|$ and accommodates large anisotropy, it neglects drift transport, which is generally negligible for SEPs. If scattering is efficient ($D_\mu \gg |\mathbf{V}|/r$), the particle distribution is nearly isotropic, and $v \gg |\mathbf{V}|$, then the focused transport equation may be integrated over pitch-angle to yield the spatial diffusion equation with diffusion coefficient $\kappa_\parallel$ (Parker, 1965; Gleeson and Axford, 1967; Forbes *et al.*, 2006, this volume). The spatial diffusion equation may be readily generalized to include drift transport and diffusion perpendicular to $\mathbf{B}$ (Jokipii and Levy, 1977) and is the basis for the theory of diffusive shock acceleration. The perpendicular diffusion $\boldsymbol{\kappa}_\perp$ is generally small and negligible for SEP transport. A possible exception is the region close to a quasi-perpendicular shock with large magnetic fluctuation intensities (Dwyer *et al.*, 1997). The parallel spatial diffusion coefficient $\kappa_\parallel$ may be expressed in terms of D_μ, and D_μ in terms of the fluctuation intensity, I (Lee, 1983; Gordon *et al.*, 1999). In principle, then, a wave kinetic equation for I is required, which describes wave excitation by the accelerated protons and closes the nonlinear system of equations.

Early theoretical work on shock acceleration proceeded in two directions. Firstly, following the development of the theory of diffusive shock acceleration (Axford *et al.*, 1978; Krymsky, 1977; Blandford and Ostriker, 1978; Bell, 1978), there were applications of the theory to SEPs by Achterberg and Norman (1980), Lee and Fisk (1982) and Lee and Ryan (1986). These were simplified in both geometry and the form of the diffusion coefficient. There were also applications of the theory to energetic storm particle (ESP) enhancements observed at Earth orbit (Forman, 1981; Lee, 1983; Gordon *et al.*, 1999). An ESP event is actually one phase of a gradual event, which occurs if the shock still accelerates ions when it passes Earth. With the planar geometry appropriate for ESP events, Lee (1983) was able to include wave excitation in the theory. These models provided a reasonable description of the ion (and wave) enhancements near the shock.

Secondly, there were many applications of the focused transport equation to the nearly scatter-free transport of SEPs in interplanetary space early in an event when ion anisotropy may be large (Heras *et al.*, 1992, 1995; Kallenrode, 1993; Ruffolo, 1995; Kallenrode and Wibberenz, 1997; Lario *et al.*, 1998). These particles constitute an important phase of the event for space weather forecasting and usually include the most energetic particles. These models include the shock acceleration heuristically as a source term, which is chosen with a power-law energy spectrum appropriate to a moving shock that is weaker on the flanks. They provide a reasonable description of the early phase of gradual events.

Other studies have attempted to combine the advantages of these two research directions in order to accommodate more realistic geometry, wave excitation, and the transition from scatter-dominated to nearly scatter-free ion transport with increasing distance upstream of the shock. Ng *et al.* (1999) have combined the focused transport and wave kinetic equations describing the upstream propagation of all ion species, including the wave excitation essential to the turbulent sheath adjacent to the shock. Although this approach cannot describe the acceleration process, it does predict the upstream fractionation of different ion species. For the event of 20 April 1998 they find excellent agreement with observed abundance ratios (Tylka *et al.*, 1999). Zank *et al.* (2000) used a "hybrid" approach to calculate the proton time profiles expected in gradual events. They combined the shocked plasma flow from hydrodynamic numerical simulations, the upstream ion/wave configuration from Gordon *et al.* (1999) assuming a free-escape boundary at a prescribed position upstream of the shock, and a numerical calculation of the ion distribution downstream of the shock. In spite of this patchwork approach, the predicted time profiles provide a good match to observations. Lee (1999, 2005) combined the wave kinetic equation with the two-stream moments of the focused transport equation to accommodate large streaming anisotropy in the theory of diffusive shock acceleration combined with wave excitation. Although this model is effectively stationary, assumes a simple geometry, and neglects adiabatic deceleration and drift of ions, it does describe analytically the extraction of ions from the turbulent sheath adjacent to the shock by magnetic focusing and the resulting cutoff in the power-law energy spectrum.

In attempting to develop a theory that can account for the characteristics of any given event, several challenging issues must be recognized:

(1) The geometry of the magnetic field, the shock, and the connection to the observer are crucial to a quantitative prediction of the event characteristics, yet it

is generally unknown, particularly near the Sun, and most models are restricted to spherical or simplified geometry. Ironically it was the dependence of gradual event morphology on the solar longitude of the flare/CME which was one of the strongest arguments in favor of a shock origin of these events (Cane *et al.*, 1988). Also, the obliquity of the shock (the angle θ_{bn} between the upstream magnetic field and the shock normal) is crucial in determining the appropriate injection rate Q into the acceleration process (see Point 2 below) and the acceleration rate, yet the appropriate value of θ_{bn} for the observer's field line is variable and difficult to determine. These aspects of the problem present severe challenges to a predictive theory.

(2) The traditional transport equations cannot describe the extraction of particles from the ambient plasma at the shock front to create a population of energetic particles satisfying $v \gg |\mathbf{V}|$ or gyrotropy for which the equations are valid. Accordingly, there is no rigorous way to determine Q as a function of v and species. This is a particularly challenging issue for SEPs since the composition of different events may be largely determined by the relative injection rates of solar wind, the suprathermal tail of the solar wind, the "inner source" pickup ions, ambient gradual event material, and ambient impulsive event material (Gloeckler *et al.*, 2000; Mason *et al.*, 1999; Desai *et al.*, 2003). Not only is Q expected to be different for these populations, it also is expected to depend on the shock Mach number and very sensitively on θ_{bn}. The unknown nature of Q makes compositional predictions difficult.

(3) The traditional transport equations depend on quasilinear theory, whose accuracy in general is difficult to assess. Certainly at Earth's bow shock, for example, "shocklets" are observed to form as large-amplitude upstream waves steepen; this nonlinear process modifies the power spectrum markedly (Hoppe *et al.*, 1981; Hada *et al.*, 1987). In addition, the models employ approximate solutions of the wave kinetic equation even though the coupled configuration of ions and waves is expected to be very sensitive to I.

(4) Gradual events, which are magnetically well-connected to the observer when the CME/flare is near the Sun, may also contain energetic particles that originate at the flare (see Section 8.2). The resulting admixture of impulsive event material should have important spectral and compositional signatures. There is a tendency for Fe/O to be enhanced, as in typical impulsive events, early in well-connected gradual events (Cane *et al.*, 1991). This enhancement, however, can also be explained by rigidity-dependent propagation of ions from the shock to the observer (Tylka *et al.*, 1999) or by a combination of shock geometry and accessible seed particles (Tylka *et al.*, 2005). The possible admixture of flare-accelerated ions in gradual events remains a controversial topic.

(5) Although the formation of a CME-driven shock at $r \sim 3$–$5\ R_s$ is in principle consistent with the onset times of GeV protons (Kahler, 1994), these onset times require scattering mean free paths on the order of the proton gyroradius to achieve the required acceleration rate (Roussev *et al.*, 2004). It is unclear whether such

small scattering mean free paths exist parallel to the shock normal upstream of the shock in this region of the corona.

8.2. Impulsive Events and Magnetic Reconnection

The characteristics of impulsive events imply that they originate in solar flares low in the corona. Flare electromagnetic emissions indicate that most of the energetic particles accelerated in the flare remain in the low corona; only a small fraction find their way to magnetic field lines open to interplanetary space. Many or most impulsive events are not associated with CMEs. However, since most CMEs are associated with large flares, it is reasonable to suppose that the energetic particles in gradual events have an "impulsive" component (see Section 8.1(4)).

The acceleration mechanism of impulsive events is unknown. The two leading candidates are direct acceleration by **E** at the site of magnetic reconnection, where the reconnecting component of **B** normal to **E** is small (Litvinenko, 1996). This field configuration is able to accelerate both electrons and ions. However, it is unclear how it can lead to the remarkable enhancements and variability in ^{3}He. These enhancements appear to require stochastic acceleration by plasma waves which can resonate selectively with different ion species. Fisk (1978) suggested electrostatic ion-cyclotron waves because they selectively resonate with ^{3}He if the ^{4}He/H ratio is enhanced. Others have suggested that the waves are excited by the electric-field-accelerated electrons (Temerin and Roth, 1992; Miller and Viñas, 1993). Several authors (e.g., Ramaty, 1979; Möbius *et al.*, 1982; Miller *et al.*, 1990; Ryan and Lee, 1991) have constructed models for the stochastic acceleration of impulsive event ions based on the diffusion equation including an energy diffusion term. However, these models are limited by unknown geometry, origin of the turbulence, and particle escape rates.

Although these two mechanisms are the leading candidates for particle acceleration in impulsive events, shock acceleration is also possible, either at the shock produced by the downward reconnection jet (Tsuneta and Naito, 1998) or a shock generated by impulsive plasma heating at the reconnection site. Clearly, the current theoretical framework for impulsive events is more rudimentary and challenging than for gradual events.

9. Future Directions: Confronting Models with Observations

Presently, there exist a broad spectrum of models for CME initiation that address selected aspects of the observations, though not in a consistent and complete manner. Progress in CME theory will most likely be achieved by confronting models with observations. Although the present complement of CME observations is rich and abundant, the models are too idealized to address them in detail. Modeling an actual

CME event and producing quantities that are observed is presently not possible. Once the models improve sufficiently, these observations will serve to distinguish the various models. In this sense, the observations are ahead of the models. Impending advances in CME and active region observations, especially those from the upcoming Solar-B and STEREO missions, will not only present further challenges to the models but will also undoubtedly provide additional insight into the CME initiation problem.

In our opinion, an effective way to resolve which physical mechanism initiates CMEs from among the many proposed possibilities will require the models to be refined until they can directly address observations, i.e., by producing as output measured observables, such as white light coronagraph images, radio, EUV and X-ray emission images, shock and particle signatures, and predicted in-situ ICME properties. The following improvements are presently being considered:

1. Extension of the models to 3D, including the effect of the strong magnetic fields that characterize active regions;
2. Modeling active-region length scales while at the same time including the coupling to large-scale (global) fields;
3. Models that are driven by observed boundary conditions, such as photospheric line-of-sight magnetic fields (e.g., from SOHO/MDI magnetograms) and transverse magnetic fields from vector magnetograms;
4. Use of time-dependent photospheric magnetic field boundary conditions to follow the evolution of the coronal magnetic field and the triggering of eruptions;
5. A more sophisticated treatment of energy transport in the corona;
6. Improved coupling of coronal and heliospheric models to follow the propagation of a CME into the heliosphere;
7. Improved modeling of the quasi-steady solar wind structure to better track the trajectory of a CME and to describe its evolution;
8. Direct comparison of model outputs with X-ray and EUV emission images;
9. Relating observed ICME characteristics to their solar source regions;
10. Focused study of specific CME events.

The requirement that models directly address observations in order to make progress also holds for the configuration of the ICME in interplanetary space, the behavior of the CME-driven shock, and the distribution of energetic ions and excited waves throughout the inner heliosphere. This challenge is being faced in part by global heliospheric MHD codes (Odstrcil *et al.*, 2002; Manchester *et al.*, 2004c). The SEP models are not yet ready for the severe challenges posed by energy spectra, anisotropies and time profiles for electrons and multiple ion species, and charge states for a complex variety of events with a variety of magnetic connection geometries. However, this bewildering array of particle data is slowly achieving some level of organization through consideration of multiple seed populations (Mason *et al.*, 1999; Desai *et al.*, 2003) and shock geometry (Tylka *et al.*, 2005). Insights gained through these considerations should lead to a predictive class of models.

This prospect is particularly exciting since SEPs are in principle a very effective probe of the CME/shock configuration in the inner heliosphere and even close to the Sun for magnetically well-connected events.

The realization that detailed simulations of specific events is needed to advance our understanding further has been espoused by the CME community. For example, the "Solar, Heliospheric, and INterplanetary Environment" group (SHINE) has selected a set of "Campaign Events" for detailed coordinated study (Gopalswamy, 2005). The list of events can be accessed at the group website (`http://www.shinegroup.org`). One of these, perhaps the simplest for modeling purposes, is the CME that occurred on May 12, 1997. We expect that such detailed studies will solve the CME initiation problem and establish the behavior of the ICME and its driven shock in interplanetary space.

Acknowledgements

The authors are grateful for the patience and hospitality of the Workshop organizers at Schloss Elmau and ISSI. ZM acknowledges support from NASA's Sun-Earth Connection Theory, Supporting Research and Technology, and Living With a Star Programs, and NSF's Center for Integrated Space Weather Modeling. The computations were performed at NSF's San Diego Supercomputer Center. ML acknowledges support from NASA grant NNG05GL40G, NSF grant ATM-0091527, and the DoD MURI grants to the University of Michigan and the University of California at Berkeley (subcontracts to the University of New Hampshire).

References

Achterberg, A., and Norman, C. A.: 1980, *Astron. Astrophys.* **89**, 353.
Alexander, D., Richardson, I., and Zurbuchen, T. H.: 2006, *Space Sci. Rev.*, (this volume), doi: 10.1007/s11214-006-9008-y.
Aly, J. J.: 1984, *ApJ* **283**, 349–362.
Aly, J. J.: 1985, *A&A* **143**, 19–22.
Aly, J. J.: 1988, *A&A* **203**, 183–188.
Aly, J. J.: 1990, *Physics of Fluids B* **2**, 1928–1935.
Aly, J. J.: 1991, *ApJ* **375**, L61–L64.
Amari, T., Luciani, J. F., Aly, J. J., Mikić, Z., and Linker, J.: 2003a, *ApJ* **585**, 1073–1086.
Amari, T., Luciani, J. F., Aly, J. J., Mikić, Z., and Linker, J.: 2003b, *ApJ* **595**, 1231–1250.
Amari, T., Luciani, J. F., Aly, J. J., and Tagger, M.: 1996, *A&A* **306**, 913–923.
Amari, T., Luciani, J. F., Mikić, Z., and Linker, J.: 1999, *ApJ* **518**, L57–L60.
Amari, T., Luciani, J. F., Mikić, Z., and Linker, J.: 2000, *ApJ* **529**, L49–L52.
Antiochos, S. K., Dahlburg, R. B., and Klimchuk, J. A.: 1994, *ApJ* **420**, L41–L44.
Antiochos, S. K.: 1998, *ApJ* **502**, L181–L184.
Antiochos, S. K., DeVore, C. R., and Klimchuk, J. A.: 1999, *ApJ* **510**, 485–493.

Axford, W. I., Leer, E., and Skadron, G.: 1978, *Proceedings of the International Conference Cosmic Rays 15th*, vol. 11, pp. 132–137.
Barnes, C. W., and Sturrock, P. A.: 1972, *ApJ* **174**, 659.
Bastian, T. S., Pick, M., Kerdraon, A., Maia, D., and Vourlidas, A.: 2001, *ApJ* **558**, L65–L69.
Bell, A. R.: 1978, *Mon. Not. R. Astr. Soc.* **182**, 147.
Birn, J., and Schindler, K.: 1981, in Priest, E. R. (ed.), *Solar Flare Magnetohydrodynamics*, Gordon and Breach, New York. pp. 337–378.
Biskamp, D., and Welter, H.: 1989, *Sol. Phys.* **120**, 49–77.
Blandford, R. D., and Ostriker, J. P.: 1978, *Astrophys. J.* **221**, L29–L32.
Brueckner, G. E., Delaboudiniere, J.-P., Howard, R. A., Paswaters, S. E., St. Cyr, O. C., Schwenn, R., *et al.*: 1998, *Geophys. Res. Lett.* **25**, 3019–3022.
Cane, H. V., and Lario, D.: 2006, *Space Sci. Rev.*, (this volume), doi: 10.1007/s11214-006-9011-3.
Cane, H. V., Reames, D. V., and von Rosenvinge, T. T.: 1988, *J. Geophys. Res.* **93**, 9555–9567.
Cane, H. V., Reames, D. V., and von Rosenvinge, T. T.: 1991, *Astrophys. J.* **373**, 675–682.
Chen, J.: 1989, *ApJ* **338**, 453–470.
Chen, J., Howard, R. A., Brueckner, G. E., Santoro, R., Krall, J., Paswaters, S. E., *et al.*: 1997, *ApJ* **490**, L191+.
Chen, J., Santoro, R. A., Krall, J., Howard, R. A., Duffin, R., Moses, J. D., *et al.*: 2000, *ApJ* **533**, 481–500.
Chevalier, R. A.: 1982, *Astrophys. J.* **258**, 790.
Choe, G. S., and Lee, L. C.: 1996a, *ApJ* **472**, 360–371.
Choe, G. S., and Lee, L. C.: 1996b, *ApJ* **472**, 372–388.
Cliver, E. W., Nitta, N. V., Thompson, B. J., and Zhang, J.: 2004, *Sol. Phys.* **225**, 105–139.
Crooker, N. U., Gosling, J. T., and Kahler, S. W.: 2002, *J. Geophys. Res.* **107**, 1028.
Démoulin, P.: 1998, Webb, D., Rust, D., and Schmieder, B. (eds.), *ASP Conference Series: New Perspectives on Solar Prominences*, Vol. 150, pp. 78–85.
Démoulin, P., Bagala, L. G., Mandrini, C. H., Henoux, J. C., and Rovira, M. G.: 1997, *A&A* **325**, 305–317.
Démoulin, P., Henoux, J. C., Priest, E. R., and Mandrini, C. H.: 1996, *A&A* **308**, 643–655.
Desai, M. I., Mason, G. M., Dwyer, J. R., Mazur, J. E., Gold, R. E., Krimigis, S. M., *et al.*: 2003, *Astrophys. J.* **588**, 1149–1162.
DeVore, C. R., and Antiochos, S. K.: 2005, *ApJ* **628**, 1031–1045.
Dryer, M.: 1974, *Space Sci. Rev.* **15**, 403.
Dryer, M.: 1982, *Space Sci. Rev.* **33**, 233–275.
Dwyer, J. R., Mason, G. M., Mazur, J. E., Jokipii, J. R., von Rosenvinge, T. T., and Lepping, R. P.: 1997, *Astrophys. J.* **490**, L115.
Earl, J. A.: 1976, *Astrophys. J.* **205**, 900.
Earl, J. A.: 1981, *Astrophys. J.* **251**, 739–755.
Fermi, E.: 1954, *Astrophys. J.* **119**, 1–6.
Feynman, J., and Gabriel, S. B.: 2000, *J. Geophys. Res.* **105**, 10543–10564.
Feynman, J., and Martin, S. F.: 1995, *J. Geophys. Res.* **100**, 3355–3367.
Finn, J. M., and Chen, J.: 1990, *ApJ* **349**, 345–361.
Finn, J. M., and Guzdar, P. N.: 1993, *Physics of Fluids B* **5**, 2870–2876.
Fisk, L. A.: 1978, *Astrophys. J.* **224**, 1048–1055.
Forbes, T. G.: 2000, *J. Geophys. Res.* **105**, 23153–23166.
Forbes, T. G., Linker, J. A., *et al.*: 2006, *Space Science Reviews*, (this volume), doi: 10.1007/s11214-006-9019-8.
Forbes, T. G., and Isenberg, P. A.: 1991, *ApJ* **373**, 294–307.
Forbes, T. G., and Lin, J.: 2000, *J. Geophys. Res.* **62**, 1499–1507.
Forbes, T. G., and Priest, E. R.: 1995, *ApJ* **446**, 377–389.

Forbes, T. G., Priest, E. R., and Isenberg, P. A.: 1994, *Sol. Phys.* **150**, 245–266.
Forman, M. A.: 1981, *Adv. Space Res.* **1**(3), 97.
Gaizauskas, V.: 1993, *Adv. Space Res.* **13**, 5–14.
Gaizauskas, V., Harvey, K. L., Harvey, J. W., and Zwaan, C.: 1983, *ApJ* **265**, 1056–1065.
Gaizauskas, V., Zirker, J. B., Sweetland, C., and Kovacs, A.: 1997, *ApJ* **479**, 448–457.
Gary, G. A., and Moore, R. L.: 2004, *ApJ* **611**, 545–556.
Gleeson, L. J., and Axford, W. I.: 1967, *Astrophys. J.* **149**, L115.
Gloeckler, G., Fisk, L. A., Geiss, J., Schwadron, N. A., and Zurbuchen, T. H.: 2000, *J. Geophys. Res.* **105**, 7459–7464.
Gopalswamy, N.: 2003, *Adv. Space Res.* **31**, 869–881.
Gopalswamy, N.: 2005, *EOS Trans. AGU* **86**(50), 525.
Gopalswamy, N. *et al.*: 2006, *Space Sci. Rev.*, (this volume), doi: 10.1007/s11214-006-9020-2.
Gopalswamy, N., Lara, A., Kaiser, M. L., and Bougeret, J.-L.: 2001, *J. Geophys. Res.* **106**, 25,261–25,277.
Gordon, B. E., Lee, M. A., Moebius, E., and Trattner, K. H.: 1999, *J. Geophys. Res.* **104**, 28263–28277.
Gosling, J. T., Baker, D. N., Bame, S. J., Feldman, W. C., Zwickl, R. D., and Smith, E. J.: 1987, *J. Geophys. Res.* **92**, 8519–8535.
Gosling, J. T., Birn, J., and Hesse, M.: 1995, *Geophys. Res. Lett.* **22**, 869–872.
Groth, C. P. T., de Zeeuw, D. L., Gombosi, T. I., and Powell, K. G.: 2000, *J. Geophys. Res.* **105**, 25053–25078.
Hada, T., Kennel, C. F., and Terasawa, T.: 1987, *J. Geophys. Res.* **92**, 4423–4435.
Heras, A. M., Sanahuja, B., Lario, D., Smith, Z. K., Detman, T., and Dryer, M.: 1995, *Astrophys. J.* **445**, 497–508.
Heras, A. M., Sanahuja, B., Smith, Z. K., Detman, T., and Dryer, M.: 1992, *Astrophys. J.* **391**, 359–369.
Hoppe, M. M., Russell, C. T., Frank, L. A., Eastman, T. E., and Greenstadt, E. W.: 1981, *J. Geophys. Res.* **86**, 4471–4492.
Hudson, P. D.: 1965, *Mon. Not. R. Astr. Soc.* **131**, 23–49.
Hundhausen, A. J.: 1985, *AGU Geophys. Monogr.* **34**, 37–58.
Inhester, B., Birn, J., and Hesse, M.: 1992, *Sol. Phys.* **138**, 257–281.
Isenberg, P. A.: 1997, *J. Geophys. Res.* **102**, 4719–4724.
Isenberg, P. A., Forbes, T. G., and Démoulin, P.: 1993, *ApJ* **417**, 368–386.
Jacques, S. A.: 1977, *ApJ* **215**, 942–951.
Jokipii, J. R.: 1982, *Astrophys. J.* **255**, 716–720.
Jokipii, J. R., and Levy, E. H.: 1977, *Astrophys. J.* **213**, L85–L88.
Kahler, S.: 1994, *Astrophys. J.* **428**, 837–842.
Kallenrode, M. B.: 1993, *J. Geophys. Res.* **98**, 19037–19047.
Kallenrode, M. B., and Wibberenz, G.: 1997, *J. Geophys. Res.* **102**, 22311–22334.
Klecker, H., Kunow, H., *et al.*: 2006, *Space Science Reviews*, (this volume), doi: 10.1007/s11214-006-9018-9.
Klimchuk, J. A.: 2001, *Space Weather*, Song, P., Singer, H., and Siscoe, G. (eds.), AGU, Washington DC, Geophys. Monograph, Vol. 125, pp. 143–158.
Klimchuk, J. A., Canfield, R. C., and Rhoads, J. E.: 1992, *ApJ* **385**, 327–343.
Klimchuk, J. A., and Sturrock, P. A.: 1989, *ApJ* **345**, 1034–1041.
Krall, J., Chen, J., and Santoro, R.: 2000, *ApJ* **539**, 964–982.
Krymsky, G. F.: 1977, *Dokl. Akad. Nauk SSSR* **234**, 1306.
Lario, D., Sanahuja, B., and Heras, A. M.: 1998, *Astrophys. J.* **509**, 415–434.
Larson, D. E., Lin, R. P., Ergun, R. E., McTiernan, J. M., McFadden, J. P., Carlson, C. W., *et al.*: 1997, *Adv. Space Res.* **20**, 655–658.
Leamon, R. J., Canfield, R. C., Jones, S. L., Lambkin, K., Lundberg, B. J., and Pevtsov, A. A.: 2004, *J. Geophys. Res.* **109**, 5106.

Lee, J., McClymont, A. N., Mikić, Z., White, S. M., and Kundu, M. R.: 1998a, *ApJ* **501**, 853–865.
Lee, J., White, S. M., Kundu, M. R., Mikić, Z., and McClymont, A. N.: 1998b, *Sol. Phys.* **180**, 193–211.
Lee, J., White, S. M., Kundu, M. R., Mikić, Z., and McClymont, A. N.: 1999, *ApJ* **510**, 413–421.
Lee, M. A.: 1983, *J. Geophys. Res.* **88**, 6109–6119.
Lee, M. A.: 1991, *22nd International Cosmic Ray Conference*, Vol. 5, pp. 293–312.
Lee, M. A.: 1999, *EOS* **80**, S255. abstract.
Lee, M. A.: 2005, *Astrophys. J. Suppl.* **158**, 38–67.
Lee, M. A., and Fisk, L. A.: 1982, *Space Sci. Rev.* **32**, 205–228.
Lee, M. A., and Ryan, J. M.: 1986, *Astrophys. J.* **303**, 829–842.
Leka, K. D., Canfield, R. C., McClymont, A. N., and van Driel-Gesztelyi, L.: 1996, *ApJ* **462**, 547–560.
Lin, J. and Forbes, T. G.: 2000, *J. Geophys. Res.* **105**, 2375–2392.
Lin, J., Forbes, T. G., and Isenberg, P. A.: 2001, *J. Geophys. Res.* **106**, 25053–25074.
Lin, J., Forbes, T. G., Isenberg, P. A., and Démoulin, P.: 1998, *ApJ* **504**, 1006–1019.
Lin, J., Soon, W., and Baliunas, S. L.: 2003, *New Astron. Rev.* **47**, 53–84.
Lin, J., and van Ballegooijen, A. A.: 2002, *ApJ* **576**, 485–492.
Lin, J., van Ballegooijen, A. A., and Forbes, T. G.: 2002, *J. Geophys. Res.* **107**, 3–1.
Linker, J. A., Lionello, R., Mikić, Z., and Amari, T.: 2001, *J. Geophys. Res.* **106**, 25165–25176.
Linker, J. A., and Mikić, Z.: 1995, *ApJ* **438**, L45–L48.
Linker, J. A., Mikić, Z., Lionello, R., Riley, P., Amari, T., and Odstrcil, D.: 2003a, *Phys. Plasmas* **10**, 1971–1978.
Linker, J. A., Mikić, Z., Riley, P., Lionello, R., and Odstrcil, D.: 2003b, *AIP Conference Proceedings: Solar Wind Ten*, Velli, M., Bruno, R., and Malara, F. (eds.), American Institute of Physics, New York, Vol. 679, pp. 703–710.
Lionello, R., Mikić, Z., Linker, J. A., and Amari, T.: 2002, *ApJ* **581**, 718–725.
Litvinenko, Y. E.: 1996, *Astrophys. J.* **462**, 997–1004.
Litvinenko, Y. E., and Martin, S. F.: 1999, *Sol. Phys.* **190**, 45–58.
Livi, S. H. B., Martin, S., Wang, H., and Ai, G.: 1989, *Sol. Phys.* **121**, 197–214.
Livi, S. H. B., Wang, J., and Martin, S. F.: 1985, *Australian J. Phys.* **38**, 855–873.
Lopate, C.: 2001, *Proceedings of the International Conference Cosmic Rays 27th*, Vol. 8, pp. 3398–3400.
Low, B. C.: 1977, *ApJ* **212**, 234–242.
Low, B. C.: 1981, *ApJ* **251**, 352–363.
Low, B. C.: 1993, *Adv. Space Res.* **13**, 63–69.
Low, B. C.: 1994, *Phys. Plasmas* **1**, 1684–1690.
Low, B. C.: 1996, *Sol. Phys.* **167**, 217–265.
Low, B. C.: 1999, *AIP Conference Proceedings: Solar Wind Nine*, Habbal, S. R., Esser, R., Hollweg, J. V., and Isenberg, P. A. (eds.), American Inst. Phys., New York, Vol. 471, pp. 109–114.
Low, B. C.: 2001, *J. Geophys. Res.* **106**, 25141–25164.
Low, B. C., and Smith, D. F.: 1993, *ApJ* **410**, 412–425.
Luhmann, J. G., Solomon, S. C., Linker, J. A., Lyon, J. G., Mikić, Z., Odstrcil, D., *et al.*: 2004, *J. Atmos. Terrestrial Phys.* **66**, 1243–1256.
Lynch, B. J., Antiochos, S. K., MacNeice, P. J., Zurbuchen, T. H., and Fisk, L. A.: 2004, *ApJ* **617**, 589–599.
Mackay, D. H., and Gaizauskas, V.: 2003, *Sol. Phys.* **216**, 121–142.
Mackay, D. H., Priest, E. R., Gaizauskas, V., and van Ballegooijen, A. A.: 1998, *Sol. Phys.* **180**, 299–312.
Mackay, D. H., and van Ballegooijen, A. A.: 2001, *ApJ* **560**, 445–455.
Mackay, D. H., and van Ballegooijen, A. A.: 2005, *ApJ* **621**, L77–L80.

MacNeice, P., Antiochos, S. K., Phillips, A., Spicer, D. S., DeVore, C. R., and Olson, K.: 2004, *ApJ* **614**, 1028–1041.

Manchester, W.: 2003, *J. Geophys. Res.* **108**, 10–1.

Manchester, W., Gombosi, T., DeZeeuw, D., and Fan, Y.: 2004a, *ApJ* **610**, 588–596.

Manchester, W. B., Gombosi, T. I., Roussev, I., De Zeeuw, D. L., Sokolov, I. V., Powell, K. G., Tóth, G., and Opher, M.: 2004b, *J. Geophys. Res.* **109**, 10.1029/2002JA009672.

Manchester, W. B., Gombosi, T. I., Roussev, I. I., Ridley, A., de Zeeuw, D. L., Sokolov, I. V., *et al.*: 2004c, *J. Geophys. Res.* **109**, CiteID A02107 DOI 10.1029/2003JA010150.

Mann, G., Klassen, A., Aurass, H., and Classen, H.-T.: 2003, *Astron. Astrophys.* **400**, 329–336.

Martens, P. C., and Zwaan, C.: 2001, *ApJ* **558**, 872–887.

Martin, S. F.: 1998, *Sol. Phys.* **182**, 107–137.

Martin, S. F., Livi, S. H. B., and Wang, J.: 1985, *Australian Journal of Physics* **38**, 929–959.

Mason, G. M., Mazur, J. E., and Dwyer, J. R.: 1999, *Astrophys. J.* **525**, L133–L136.

Metcalf, T. R., Leka, K. D., and Mickey, D. L.: 2005, *ApJ* **623**, L53–L56.

Mikić, Z., Barnes, D. C., and Schnack, D. D.: 1988, *ApJ* **328**, 830–847.

Mikić, Z., and Linker, J. A.: 1994, *ApJ* **430**, 898–912.

Mikić, Z., Linker, J. A., Schnack, D. D., Lionello, R., and Tarditi, A.: 1999, *Phys. Plasmas* **6**, 2217–2224.

Miller, J. A., Guessoum, N., and Ramaty, R.: 1990, *Astrophys. J.* **361**, 701–708.

Miller, J. A., and Vinas, A. F.: 1993, *Astrophys. J.* **412**, 386–400.

Moebius, E., Scholer, M., Hovestadt, D., Klecker, B., and Gloeckler, G.: 1982, *Astrophys. J.* **259**, 397–410.

Moreton, G. E.: 1960, *Astron. J.* **65**, 494–495.

Ng, C. K., Reames, D. V., and Tylka, A. J.: 1999, *Geophys. Res. Lett.* **26**, 2145–2148.

Odstrcil, D.: 2003, *Adv. Space Res.* **32**, 497–506.

Odstrcil, D., Linker, J. A., Lionello, R., Mikić, Z., Riley, P., Pizzo, V. J., *et al.*: 2002, *J. Geophys. Res.* **107**, 10.1029/2002JA009334.

Odstrcil, D., and Pizzo, V. J.: 1999a, *J. Geophys. Res.* **104**, 28225–28239.

Odstrcil, D., and Pizzo, V. J.: 1999b, *J. Geophys. Res.* **104**, 483–492.

Odstrcil, D., and Pizzo, V. J.: 1999c, *J. Geophys. Res.* **104**, 493–504.

Parker, E. N.: 1961, *Astrophys. J.* **133**, 1014.

Parker, E. N.: 1963, *Interplanetary Dynamical Processes*, New York, Interscience Publishers.

Parker, E. N.: 1965, *Planet. Space Sci.* **13**, 9.

Pick, M., Forbes, T. G., Mann, G., *et al.*: 2006, *Space Sci. Rev.*, (this volume), doi: 10.1007/s11214-006-9021-1.

Pizzo, V. J.: 1985, *AGU Geophys. Monogr.* **35**, 51–68.

Pneuman, G. W.: 1983, *Sol. Phys.* **88**, 219–239.

Powell, K. G., Gombosi, T. I., deZeeuw, D. L., Ridley, A. J., Sokolov, I. V., Stout, Q. F., *et al.*: 2003, *SOLAR WIND TEN: Proceedings of the Tenth International Solar Wind Conference, AIP Conference Proceedings*, Vol. 679, pp. 807–814.

Pyle, K. R., Simpson, J. A., Barnes, A., and Mihalov, J. D.: 1984, *Astrophys. J.* **282**, L107–L111.

Ramaty, R.: 1979, in *Particle Acceleration Mechanisms in Astrophysics: Proceedings of the Workshop held at La Jolla, Calif., January 3–5, 1979*, AIP, New York, pp. 135–154.

Reames, D. V.: 1999, *Space Sci. Rev.* **90**, 413–491.

Reames, D. V., Kahler, S. W., and Ng, C. K.: 1997, *Astrophys. J.* **491**, 414–420.

Reiner, M. J., Kaiser, M. L., and Bougeret, J.-L.: 2001, *J. Geophys. Res.* **106**, 29989–30000.

Reiner, M. J., Kaiser, M. L., and Bougeret, J.-L.: 2003, *Solar Wind Ten, Proceedings of the Tenth International Solar Wind Conference, AIP Conference Proceedings*, Vol. 679, pp. 152–155.

Riley, P., and Gosling, J. T.: 1998, *Geophys. Res. Lett.* **25**, 1529–1532.

Riley, P., Gosling, J. T., and Pizzo, V. J.: 1997, *J. Geophys. Res.* **102**, 14677–14686.

Riley, P., Linker, J. A., Lionello, R., Mikić, Z., Odstrcil, D., Hidalgo, M. A., *et al.*: 2004, *J. Atmos. Terrestrial Phys.* **66**, 1321–1331.
Riley, P., Linker, J. A., Mikić, Z., Odstrcil, D., Zurbuchen, T. H., Lario, D., *et al.*: 2003, *J. Geophys. Res.* **108**, 1272.
Roelof, E. C.: 1969, *Lectures in High Energy Astrophysics*, Ogelmann, H., and Wayland, J. R. (eds.), NASA Spec. Publ. SP-199, p. 111.
Rogers, M. H.: 1957, *Astrophys. J.* **125**, 478.
Roussev, I. I., Forbes, T. G., Gombosi, T. I., Sokolov, I. V., DeZeeuw, D. L., and Birn, J.: 2003, *ApJ* **588**, L45–L48.
Roussev, I. I., Sokolov, I. V., Forbes, T. G., Gombosi, T. I., Lee, M. A., and Sakai, J. I.: 2004, *ApJ* **605**, L73–L76.
Ruffolo, D.: 1995, *Astrophys. J.* **442**, 861–874.
Ryan, J. M., and Lee, M. A.: 1991, *Astrophys. J.* **368**, 316–324.
Sarris, E. T., and Van Allen, J. A.: 1974, *J. Geophys. Res.* **79**, 4157–4173.
Schwenn, R., Raymond, J. C., *et al.*: 2006, *Space Sci. Rev.*, (this volume), doi: 10.1007/s11214-006-9016-y.
Sedov, L.: 1959, *Similarity and Dimensional Methods in Mechanics*, Academic Press, New York.
Sheeley, N. R., Hakala, W. N., and Wang, Y.-M.: 2000, *J. Geophys. Res.* **105**, 5081–5092.
Skilling, J.: 1971, *Astrophys. J.* **170**, 265.
Steinolfson, R. S.: 1985, *AGU Geophys. Monographs* **35**, 1–12.
Sturrock, P. A.: 1991, *ApJ* **380**, 655–659.
Syrovatskii, S. I.: 1982, *Sol. Phys.* **76**, 3–20.
Temerin, M., and Roth, I.: 1992, *Astrophys. J.* **391**, L105–L108.
Titov, V. S., Hornig, G., and Démoulin, P.: 2002, *J. Geophys. Res.* **107**, 1164.
Tóth, G., Sokolov, I. V., Gombosi, T. I., Chesney, D. R., Clauer, C. R., De Zeeuw, D. L., *et al.*: 2005, *J. Geophys. Res.* **110**, CiteID A12226, DOI 10.1029/2005JA011126.
Tsuneta, S., and Naito, T.: 1998, *Astrophys. J.* **495**, L67–L70.
Tsurutani, B., Wu, S. T., Zhang, T. X., and Dryer, M.: 2003, *Astron. and Astrophys.* **412**, 293–304.
Tylka, A. J., Cohen, C. M. S., Dietrich, W. F., Lee, M. A., Maclennan, C. G., Mewaldt, R. A., *et al.*: 2005, *Astrophys. J.* **625**, 474–495.
Tylka, A. J., Reames, D. V., and Ng, C. K.: 1999, *Geophys. Res. Lett.* **26**, 2141–2144.
Usmanov, A. V., Goldstein, M. L., Besser, B. P., and Fritzer, J. M.: 2000, *J. Geophys. Res.* **105**, 12675–12696.
Vainio, R., and Khan, J. I.: 2004, *Astrophys. J.* **600**, 451–457.
van Ballegooijen, A. A., and Martens, P. C. H.: 1989, *ApJ* **343**, 971–984.
van Ballegooijen, A. A., and Martens, P. C. H.: 1990, *ApJ* **361**, 283–289.
van Ballegooijen, A. A., Priest, E. R., and Mackay, D. H.: 2000, *ApJ* **539**, 983–994.
Vourlidas, A., Buzasi, D., Howard, R. A., and Esfandiari, E.: 2002, *ESA SP-506: Solar Variability: From Core to Outer Frontiers*, pp. 91–94.
Vourlidas, A., Subramanian, P., Dere, K. P., and Howard, R. A.: 2000, *ApJ* **534**, 456–467.
Webb, D. F., Lepping, R. P., Burlaga, L. F., DeForest, C. E., Larson, D. E., Martin, S. F., *et al.*: 2000, *J. Geophys. Res.* **105**, 27251–27260.
Welsch, B. T., DeVore, C. R., and Antiochos, S. K.: 2005, *ApJ* **634**, 1395–1404.
Whang, Y. C., and Burlaga, L. F.: 1985, *J. Geophys. Res.* **90**, 10,765–10,778.
Whang, Y. C., Zhou, J., Lepping, R. P., and Ogilvie, K. W.: 1996, *Geophy. Res. Lett.* **23**, 1239–1242.
Wimmer-Schweingruber, R. F., Crooker, N. U., *et al.*: 2006, *Space Science Reviews*, this volume.
Wolfson, R.: 1993, *ApJ* **419**, 382–387.
Wolfson, R., and Dlamini, B.: 1997, *ApJ* **483**, 961–971.
Wolfson, R., and Dlamini, B.: 1999, *ApJ* **526**, 1046–1051.
Wolfson, R., and Low, B. C.: 1992, *Apj* **391**, 353–358.

Wolfson, R., and Saran, S.: 1998, *ApJ* **499**, 496–503.
Wu, S. T., Andrews, M. D., and Plunkett, S. P.: 2001, *Space Sci. Rev.* **95**, 191–213.
Wu, S. T., Hu, Y. Q., Wang, S., Dryer, M., and Tandberg-Hanssen, E.: 1982, *Ap&SS* **83**, 189–194.
Yang, W. H., Sturrock, P. A., and Antiochos, S. K.: 1986, *ApJ* **309**, 383–391.
Zank, G. P., Rice, W. K. M., and Wu, C. C.: 2000, *J. Geophys. Res.* **105**, 25079–25095.
Zwaan, C.: 1985, *Sol. Phys.* **100**, 397–414.
Zwaan, C.: 1987, *ARA&A* **25**, 83–111.
Zwaan, C.: 1996, *Sol. Phys.* **169**, 265–276.

AN INTRODUCTION TO THE PRE-CME CORONA

DAVID ALEXANDER
Department of Physics and Astronomy, Rice University, 6100 Main St., Houston, TX 77005, USA
(E-mail: dalex@rice.edu)

(Received 17 May 2005; Accepted in final form 11 April 2006)

Abstract. Coronal mass ejections provide a gateway to understanding the physics of energy release and conversion in the solar corona. While it is generally accepted that the energy required to power a CME is contained in the pre-eruption coronal magnetic field, the pre-CME state of that field and the conditions leading up to the release of the magnetic energy are still not entirely clear. Recent studies point to various phenomena which are common to many, if not all, CME events, suggesting that there may be identifiable characteristics of the pre-CME corona which signal the impending eruption. However, determining whether these phenomena are necessary or even sufficient has yet to be achieved. In this paper we attempt to summarize the state of the solar corona and its evolution in the build up to a CME.

Keywords: corona, CMEs, magnetic field

1. Introduction

One of the greatest challenges in understanding the energy release process resulting in a coronal mass ejection (CME) is to separate "the gold from the dross"[1] and to determine which of all of the observable characteristics of a CME source region are key in driving the corona to erupt. Given the sheer number of studies characterizing pre-CME conditions and the limited space available in this paper we adopt a breadth over depth approach to discuss some of the more recent results pertaining to the state of the corona prior to a CME.

To understand the energy build-up, storage and release processes which govern CME initiation one must understand the magnetic field and its variations before, during, and after an eruption. Several advances have been made in recent years in measuring, modeling, interpreting, and understanding the development of the source region magnetic field both as a photospheric boundary condition and as a 3D topological system. How this field manifests itself in the corona and how the corona responds to its evolution provides the main focus for this paper.

2. Energy Requirements for CMEs

CMEs have many characteristics signifying the conversion of the free magnetic energy (the difference between the total energy in the magnetic field and that in

[1]Or "the wheat from the chaff" for a less Scottish version

Space Science Reviews (2006) 123: 81–92
DOI: 10.1007/s11214-006-9013-1

the corresponding potential field) in the pre-CME corona to other forms, the most notable of which is the rapid acceleration of some 10^{16} g of material. Energy is not only required to accelerate the plasma but also to combat solar gravity, open magnetic field, heat in situ plasma to temperatures in excess of 10 MK, and to accelerate particles to GeV energies. These individual components all have comparable energy budgets of around 10^{30-32} ergs.

The energy to power these various CME phenomena comes from the free energy available in the magnetic field, which must, by necessity, contain significant electric currents. These electric currents are generally expected to be field-aligned in order to satisfy the force-free field environment assumed for the solar corona. For the energy requirement of order 10^{32} ergs, the solar corona must convert a 100 G field over a volume of $\sim 10^{29}$ cm^3, which is equivalent to about 100 post-flare loop structures.

The association between current distributions and coronal energy release is further strengthened by the fact that current concentrations, determined from vector magnetic field measurements, are found to be connected by extrapolated coronal field lines that extend along separatrices (e.g. Mandrini *et al.*, 1995). This suggests that the energy released during CMEs is stored in these field-aligned currents and that the energy release takes place when the currents are interrupted by reconnection either at a separator or on separatrix surfaces (see section 5).

Many of these issues are studied in their own right as part of the CME/flare initiation process. However, we are primarily concerned here with the state of the corona which determines the amount of free magnetic energy available and the temporal evolution which serves to release it as a CME.

3. Photospheric and Chromospheric Fields

The solar photospheric magnetic field is routinely measured with constantly improving instrumentation allowing the full magnetic vector to be determined. Recently, Leka and Barnes (2003a,b) have used the photospheric vector magnetic field data from the Mees Imaging Vector Magnetograph (Mees/IVM) in an attempt to identify pre-eruption signatures in parameters derived from the magnetic field. These authors concentrated on solar flares but many of the results apply directly to active region CMEs. While there are many reported correlations between certain field parameters and associated flare phenomena, the correlations are not perfect nor was much attention paid to the diverse array of similar behavior exhibited in active regions which do not produce flares and/or CMEs (e.g. Mandrini *et al.*, 1995; Song *et al.*, 2002).

Leka and Barnes (2003a,b) identify, and quantify, such parameters as horizontal field gradients, vertical current density, measure of field twist, current helicity density and magnetic shear angles, together with their moments, as potential examples of field quantities related to coronal energy storage and release. They concluded that no obvious flare-imminent signatures were evident in the active regions studied and that to ensure a flare-unique signature one must simultaneously consider numerous field parameters since many candidate parameters can be excluded because of similar behavior in flare-productive and flare-quiet regions. *In other words, considering parameters one at a time, as is often done for specific events, is inadequate.*

While photospheric vector magnetic field measurements generally provide the boundary condition for force-free extrapolations into the corona one must consider the fact that the photosphere is demonstrably not force free and hence may be physically disconnected from chromospheric/coronal sites of magnetic reconnection. Moreover, because of the forced nature of the photosphere, the free magnetic energy available for a CME may not be accurately determined. Metcalf *et al.* (1995) have shown, using the Na I D-line, that chromospheric fields become essentially force free some 400 km above the photosphere (see Figure 1). It has been shown that force-free field extrapolations starting with a chromospheric boundary provide better agreement with coronal structures than those using a photospheric boundary (Leka and Metcalf, 2003).

Solar eruptive phenomena such as CMEs are ultimately driven by energy released from the magnetic field. While infrared and radio techniques for determining the magnetic field in the corona are rapidly being developed, the detail to which we understand the coronal field relies entirely on how well we understand the photospheric and chromospheric boundary condition for that field and the validity of the physical assumptions made to extrapolate the observed boundary field into the region of interest. The ability to measure all three components of the magnetic

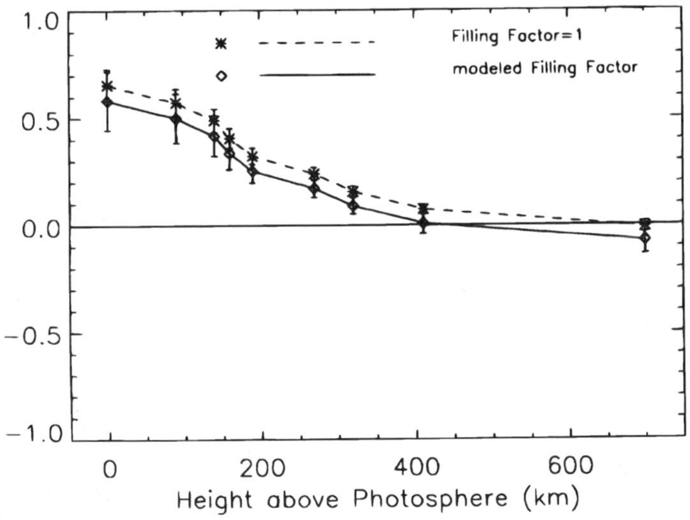

Figure 1. Scaled z-component of net Lorentz force measured in AR7216 as a function of height above the photosphere (from Metcalf *et al.*, 1995, courtesy of T. R. Metcalf).

field in the photosphere and, more interestingly, the chromosphere with increasing resolution (spatial and temporal) and accuracy is making an impact on our understanding of the role of the magnetic field in developing the conditions necessary for a CME to occur.

4. Energy Budgets from Field Measurements

The release of the non-potential magnetic energy required to drive transient activity must be accompanied by a change in the magnetic field topology as it relaxes to a more potential state. One major reconfiguration frequently invoked to describe CMEs is the opening of previously closed field lines. It has been conjectured that, for simple geometries, the energy stored in pre-eruption closed force-free fields can never exceed that of a fully open coronal magnetic field with the same boundary conditions (Aly, 1991; Sturrock, 1991). This has been confirmed by numerical experiments (e.g. Mikic and Linker, 1994). Thus, if a CME was required to open all of the field then the energy source could not be solely magnetic in nature. The Aly-Sturrock conjecture has also been found to apply to more complex magnetic topologies, most notably ones which contain a current-carrying fluxrope of the type often used to model filaments (e.g. Lin *et al.*, 1998). The impact of the Aly-Sturrock conjecture has led many authors to develop schemes with which to maintain the purely magnetic nature of the free-energy released in a CME. Three popular approaches are to assume that

(a) the corona is not, in fact, force-free and that significant energy is stored in cross-field currents (Wolfson and Dlamini, 1997; Gary and Alexander, 1999; Georgoulis and LaBonte, 2004),
(b) the coronal field is only partially opened and that the energy required from the non-potential field need only be sufficient to open part of the closed field (Wolfson, 1993; Antoichos, DeVore and Klimchuk, 1999), or
(c) non-ideal MHD processes, such as magnetic reconnection, are an integral part of the eruption process (e.g. Lin and Forbes, 2000; MacNiece *et al.*, 2004).

For a more detailed discussion on the implications for the Aly-Sturrock conjecture for solar eruptions see Lin *et al.* (2003). One must also note that in addition to the energy required to open the field, the magnetic field must also provide the energy to heat the corona, generate energetic particles, lift the ejected material against the Sun's gravity and accelerate this material into the interplanetary medium.

To fully understand the role played by the magnetic field in powering CMEs, one must be able to determine the available 'free' energy in the magnetic field and to measure how much of this free energy is released during an event. Recently, Metcalf *et al.* (2002) performed an interesting analysis of NOAA Active Region 8299 observed in the Na I 5896 Å spectral line by the Mees/IVM on 1998 August

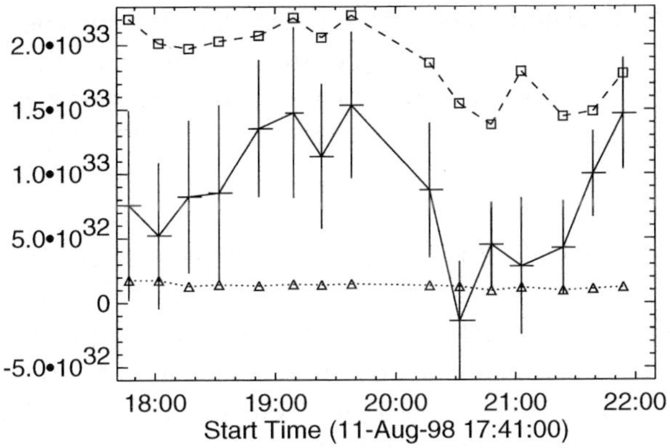

Figure 2. The total magnetic energy in ergs above the chromosphere in AR8299 (*solid line*). The dotted line shows the energy of the equivalent potential field and the dashed line shows the equivalent open field energy. Courtesy of T. R. Metcalf.

11. Using the magnetic Virial theorem, the total (force-free) magnetic energy was calculated as a function of time (Figure 2).

The total magnetic energy shows a rapid decrease beginning around 19:40 UT, falling to the potential field value at ~20:30 UT before rising again more slowly over the remainder of the observation time. This drop in energy corresponds to approximately $\sim 10^{33}$ ergs, more than enough to power a substantial CME. A similar analysis has been performed more recently for the active region 10486 (Metcalf *et al.*, 2005).

5. Role of Multipolar Flux Systems

One of the most vibrant debates over the last few years has been the role of magnetic complexity in the CME process. The magnetic breakout model of Antiochos *et al.* (1999) requires a multi-polar flux configuration as a pre-requisite for a CME eruption. In this scenario, the energy for the eruption builds up in one flux system, evidenced, for example, as shearing of a magnetic arcade, while the presence of a second flux system serves to regulate the coronal response to this build-up in energy by providing a magnetic tension force which restricts the natural expansion of the sheared system. The interaction between these two flux systems then triggers a reconnection in the overlying field allowing the sheared field to erupt.

In recent years, significant advances have been made in understanding the role of the three-dimensional magnetic topology in providing the conditions for the energy release associated with CMEs. In particular, the development of theoretical models of separatrices, separators, and quasi-separatrix layers, coupled to observational studies, have led to the notion that these topological structures, defined by the magnetic field, are the natural locations for current sheets to form and for magnetic reconnection to occur (e.g. Longcope and Silva, 1998).

One clear manifestation of magnetic complexity is that of δ spot active regions which have been found to be directly related to flare and CME productivity (Innes *et al.*, 1999). A δ-configuration active region has two sunspot umbra with a shared penumbra and is frequently observed to have strong localized shear between the two sunspot umbra, providing the conditions for the presence of substantial free magnetic energy. Recently, Tian *et al.* (2005a) performed a statistical study on 104 δ active regions and found that those active regions violating the Hale-Nicholson and Joys Laws but following the hemispherical helicity rule have a much stronger tendency to produce X-class flares, CMEs and strong proton events. There is, therefore, clear observational evidence that increasing magnetic complexity results in more and stronger solar transient activity.

On the theoretical side, the 3D characteristics of magnetic reconnection are highly complex and are only just beginning to be understood. A theoretical understanding of CMEs requires knowledge of the magnetic topology of the parent active region. Given this, CME models must explain not only how and where magnetic energy is released but also the link between the release site and the various CME signatures. Recent developments on the role of separators, separatrices (Mandrini *et al.*, 1995; Longcope and Silva, 1998), and quasi-separatrix layers (Bagalá *et al.*, 2000) in the solar corona, and their application to solar flares and CMEs, have shed new light on the coronal energization story. However, details of how and where the energy storage, release and response occur are still unclear.

6. Role of Filaments

6.1. Filament/CME Association

The relationship between filament/prominence eruptions and CMEs is difficult to fully assess. Many studies typically show a strong but not perfect correlation between the two phenomena with a large spread due to the various data and filament eruption definitions used as well as when in the solar cycle and over what time duration the study was performed. Munro *et al.* (1979) used *Skylab* data to determine that ~55% of CMEs were associated with erupting filaments, while SMM data showed ~45% association (Webb and Hundhausen, 1987; St. Cyr and Webb, 1991). Conversely, Gilbert *et al.* (2000) found from Mauna Loa H_α data that 94% of eruptive prominences had an associated CME. A more recent study by Subramanian and Dere (2001), which concentrated on CMEs emanating from source regions near disk center, found that:

- 44% of CMEs were associated with filament eruptions in active regions
- 15% are associated with filament eruptions outside of active regions

– 41% are associated with active regions with no filament eruptions

giving a total association in the same range as previous studies.

The filament/CME relationship issue is complicated by the fact that filaments only form above the parts of the magnetic polarity inversion line which are also filament channels. Filament channels are chromospheric regions defined by the approximately parallel alignment of fibrils along the magnetic neutral line. In models of CME initiation it is the magnetic configuration of the filament channel which is more important than any mass loading which may serve to define a filament (see, for example, Lin, 2004). The filament/CME relationship studies quoted above do not take into account the possible contribution from filament channel eruptions and so there may be a larger correspondence between the filament-related magnetic configuration and CME initiation. Such as study has yet to be performed.

Recent work by Zhang *et al.* (2001) has looked more closely at the physical connection between the filaments and flares/CMEs with the principal conclusion being that both the magnetic eruption traced by the erupting filament and the impulsive energy release are driven by a destabilization of the overall magnetic field configuration in which the filament and flare are embedded.

6.2. Pre-Eruption Filament Activation

The magnetic field configuration in the solar atmosphere plays a crucial role in the formation and subsequent evolution of filaments. The interaction of a filament/filament channel with the small scale evolution of the nearby magnetic field frequently results in dynamic activation of the filament material, often including counter-streaming bulk flows. While it has often been argued that dips in the magnetic field are required to support the filament material against gravity, recent results (Karpen *et al.*, 2001) have also suggested that the dynamic motions, observed to occur in filaments, can serve to create a high density cool filament in the corona without recourse to dipped field geometries. The importance of this dynamic nature of filaments to the potential for eruption and CME initiation is still being explored but the interaction between the filament magnetic field and the dynamical motions is such that any external disturbance, such as emerging or canceling flux in the filament vicinity, could have dramatic consequences for the filament itself (e.g. Romano, Contarino, and Zuccarello, 2005).

Song *et al.* (2002) found that the observed evolution of the magnetic field in relationship to filament activation implied a continuous transport of magnetic energy and complexity from the lower atmosphere to the corona. In their interpretation, slow magnetic reconnection and helicity re-distribution appeared to play a key role in the energy build-up process resulting in the initiation of a halo CME. Sterling *et al.* (2001) used observations of H_α filament activation in the build-up to a flare and associated CME to demonstrate that while, in this case, the filament itself did not appear to erupt, it underwent significant dynamic motion and morphological changes in the early stages of the CME initiation. The cospatial and cotemporal association with flare-associated brightenings at other wavelengths allowed these authors to conclude that models which allow reconnection high above the core region are more relevant to the CME initiation process. The role played by reconnection in erupting filaments has important consequences for models of CME initiation (see below).

7. Existence of Pre-CME Fluxropes

The existence of fluxropes in the pre-CME corona and the role they play in the CME process is a topic of much debate. There is significant observational and theoretical evidence to support the idea that the coronal cavity surrounding a prominence is an example of a large-scale twisted fluxrope (see Gibson and Low, 2000). In this scenario, the fluxrope geometry is required to support the filament mass against gravity. However, it has been argued from force-free and MHD simulations that dips in the magnetic field form as a result of shearing motions near the neutral line and that such dips can readily support the mass in a filament with no need to resort to the helical structure of a fluxrope (e.g. DeVore and Antiochos, 2000).

The arguments in favor of a fluxrope topology preceding the eruption is based on a combination of modeling and observations. The presence of X-ray sigmoids, the observed three-part structure in CMEs, and observations of twisted fluxropes emerging through the photosphere all point to the presence of a fluxrope configuration in the solar corona prior to any CME eruption with fluxrope models naturally explaining many of the observed phenomena. Lites (2005) concluded, from a study using high angular resolution data with high polarimetric precision from the Advanced Stokes Polarimeter, that low-lying filaments have a profound influence on the photospheric magnetic field and thereby supports the idea of the emrgence of a fluxrope from the solar interior (see also Fan and Gibson, 2004; Tian *et al.*, 2005b).

8. Role of Sigmoids

In recent years the role of helicity injection has been a focal point in the discussion of eruptive events. The attractiveness of magnetic helicity for such studies lies in the fact that it is a globally conserved quantity in ideal MHD and can also be considered to be conserved in resistive MHD on time scales shorter than the global diffusion time scale. This property opens up an array of possibilities for exploring the CME process both theoretically and observationally (see articles in Brown, Canfield, and Pevtsov, 1999).

An observational manifestation of the connection between helicity and CME production is the soft X-ray sigmoid. Sigmoids may indicate the presence of twisted magnetic structures and it has been shown that active regions exhibiting S shapes exhibit a greater tendency to erupt (Canfield, Hudson, and McKenzie, 1999). It is important to understand more about the formation and evolution of sigmoid structures in active regions and to explore the conditions that drive them to eruption if we are to fully understand the conditions leading to solar eruptive events. A key issue here is how the helicity injection is driven: via shearing or direct emergence of twisted flux.

Recent results have been confusing about this issue. On the one hand, Devore (2000) has argued that a significant quantity of magnetic helicity is injected by the action of differential rotation over the lifetime of an active region; enough to explain the total 'ejected' helicity detected in interplanetary magnetic clouds. This assertion has been contested by Démoulin *et al.* (2002) and Green *et al.* (2002) who argue that the helicity injected by differential rotation is 5 to 50 times smaller than that inferred to be carried away in CMEs, leaving these authors to conclude that the bulk of the helicity injection is provided by the twist in the sub-photospheric part of the magnetic fluxtubes forming active regions.

In the debate over the role of differential rotation, the strong local shearing often observed near the magnetic neutral line(s) of flare-productive active regions is frequently neglected. Such strong local shear may contribute significantly to the helicity injection into large but otherwise local structures associated with the active region. Recent studies by Kusano *et al.* (2002) have shown that the shearing motions can contribute as much, if not more, helicity as the flux emergence. Converging motions and the subsequent magnetic reconnection at coronal loop footpoints also contribute to the injection of magnetic helicity into the corona from below (e.g. MacKay and van Ballegooijen, 2005).

9. Rotating Sunspots and Sigmoids

Recent observations of rotating sunspots in TRACE white light images and their apparent association with soft X-ray sigmoids have led to the possibility sunspot rotation is a key component in driving sigmoid formation and evolution. A number of rotating sunspot events have now been observed; many associated with some of the largest solar flares of this solar cycle (Brown *et al.*, 2003; Tian *et al.*, 2005a). Tian and Alexander (2006a) found for NOAA AR 9684 that the whole sunspot-group rotated in the same direction as the main sunspot implying that sunspot rotation is a primary driver of helicity production and injection into the corona (see also Tian and Alexander, 2006b).

The role of helicity injection in driving the corona to eruption has been explored by several authors. Rust and Kumar (1994) calculated that a fluxrope becomes unstable when the injected helicity exceeds a critical value, $H_{\mathrm{crit}} > 1.85\phi^2$, where ϕ is the magnetic flux. These instability conditions are supported by recent numerical simulations of fluxrope emergence by Fan and Gibson (2004).

A recent analysis of a long-lived active region (AR 9632) by Tian *et al.* (2005b) finds that the active region exhibited a prolonged period of clockwise rotation. The best-fit twist parameter observed from vector magnetic fields was found to be positive suggesting that the fluxtube making up the active region had a right-handed twist. Coupled with the clockwise group rotation, it is argued that AR 9632 was comprised of a magnetic configuration with the same-handedness of twist and writhe helicity. This points to an active region formation process involving the emergence of a highly twisted and kinked fluxtube through the photosphere.

The close association between soft X-ray sigmoids and CMEs has been established as a possible driver in understanding the physical connection between active region magnetic topology and the potential for eruption. What remains less clear, however, are the physical processes governing this association and the conditions that determine whether an eruption will occur. The rotating sunspot phenomena allows us insight into the formation of the active regions and the source of the observed dynamics while providing crucial diagnostic information on the energization of the corona in the build-up to an eruption.

10. Models of the Pre-CME Sun

A variety of models exist for exploring the CME formation and initiation. The evolution of magnetic flux from the solar interior to the corona is being addressed by several models (e.g. Abbett, and Fisher, 2003) and the results are being coupled to theoretical developments on helicity injection, atmospheric current distributions and magnetic topology (see Lin, Soon, and Baliunas, 2003, for an excellent review).

Distinguishing between the various models of CME initiation is extremely difficult and, to date, has only been performed for very specific cases. Critical to many of them is the pre-eruption conditions of the ambient magnetic field and the subsequent development of the field through the coronal destabilization. The presence, or lack thereof, of a fluxrope geometry in the pre-eruption corona, the location and drivers for magnetic reconnection, the complexity of the magnetic configuration all play significant roles in the various models and all are difficult to measure quantitatively. As theoretical developments progress in tandem with improved models and observations, we should be able to focus on the key physical conditions in the pre-CME Sun which lead to an eruption and understand how variations in these key conditions influence the subsequent initiation and evolution of the CME.

11. Concluding Remarks

Understanding the pre-CME corona is clearly a crucial step in defining the physics which govern CME initiation. It is important in providing the necessary inputs to theoretical models, to increase the accuracy of event prediction and forecasting, and to better understand the physical interaction between magnetic field and plasma in astrophysical systems. As we have stressed in this brief introductory paper, the pre-CME corona cannot be considered in isolation from the pre-CME photosphere and the pre-CME solar interior. The build-up to a CME involves the dynamic coupling between a wide range of phenomena in a wide range of physical environments. Knowing the 'correct' combination of parameters required to initiate a CME involves many different facets and, at present, remains elusive.

Many studies have pointed to the apparent importance of a number of individual factors related to CME production. However, the detailed analysis by Leka and Barnes (2003a,b) gives a glimpse of the complexity involved in trying to determine which aspects are CME /flare specific and which are the day-to-day behavior of the parent active region.

Techniques for observing chromospheric and coronal magnetic fields are continuously improving (STEREO, Solar-B and the Solar Dynamics Observatory are all due for launch within the next 2–3 years), while computational and data access and handling resources are rapidly being developed. Thus, in the near-term we can expect significant advances in a number of areas which will significantly improve our chances of identifying key characteristics of the pre-CME corona.

Acknowledgements

This work was partially supported by SHINE under NSF Grant ATM-0353345.

References

Abbett, W. P. and Fisher, G. H.: 2003, *ApJ* **582**, 475.
Aly, J. J.: 1991, *ApJ* **375**, 61.
Antiochos, S. K., DeVore, C. R., and Klimchuk, J. A.: 1999, *ApJ* **510**, 485.
Bagalá, L. G., Mandrini, C. H., Rovira, M. G., and Démoulin, P.: 2000, *A&A* **363**, 779.
Brown, D. S., Nightingale, R. W., Alexander, D., Schrijver, C. J., Metcalf, T., R., Shine, R. A., *et al.*: 2003, *Solar Phys.* **216**, 79.
Brown, M. R., Canfield, R. C., and Pevtsov, A. A. (eds.): 1999, Geophysical Monograph, AGU.
Canfield, R. C., Hudson, H. S., and McKenzie, D. E.: 1999, *Geophys. Res. Lett.* **26**, 627.
Démoulin, P., Mandrini, C. H., Van Driel-Gesztelyi, L., Lopez Fuentes, M. C., and Aulanier, G.: 2002, *Solar Phys.* **207**, 87.
Devore, C. R.: 2000, *ApJ* **539**, 944.
DeVore, C. R. and Antiochos, S. K.: 2000, *ApJ* **539**, 954.
Fan, Y. and Gibson, S. E.: 2004, *ApJ* **609**, 1123.
Gary, A. G. and Alexander, D.: 1999, *Solar Phys.* **186**, 123.
Gibson, S. E. and Low, B. C.: 2000, *JGR* **105**, 18187.
Gilbert, H. R., Holzer, T. E., Burkepile, J. T., and Hundhausen, A. J.: 2000, *ApJ* **537**, 503.

Green, L. M., López Fuentes, M. C., Mandrini, C. H., Démoulin, P., Van Driel-Gesztelyi, L., and Culhane, J. L.: 2002, *Solar Phys.* **208**, 43.
Karpen, J. T., Antiochos, S. K., Hohensee, M., Klimchuk, J. A., and MacNeice, P. J.: 2001, *ApJ* **553**, 85.
Kusano, K., Maeshiro, T., Yokoyama, T., and Sakurai, T.: 2002, *ApJ* **577**, 501.
Leka, K. D. and Barnes, G.: 2003a, *ApJ* **595**, 1277.
Leka, K. D. and Barnes, G.: 2003b, *ApJ* **595**, 1296.
Leka, K. D. and Metcalf, T. R.: 2003, *Solar Phys.* **212**, 361.
Lin, J.: 2004, *Solar Phys.* **219**, 169.
Lin, J., Forbes, T. G., Isenberg, P. A., and Démoulin, P.: 1998, *ApJ* **504**, 1006.
Lin, J. and Forbes, T. G.: 2000, *JGR* **105**, 2375.
Lin, J., Soon, W., and Baliunas, S. L.: 2003, *New Astron. Rev.* **47**, 53.
Lites, B. W.: 2005, *ApJ* **622**, 1275.
Longcope, D. W. and Silva, A. V. R.: 1998, *Solar Phys.* **179**, 349.
MacKay, D. H. and van Ballegooijen, A. A.: 2005, *ApJ Lett.* **621**, L77.
MacNeice, P. *et al.*: 2004, *ApJ* **614**, 1028.
Mandrini, C. H., Demoulin, P., Rovira, M. G., de La Beaujardiere, J.-F., and Henoux, J. C.: 1995, *A& A* **303**, 927.
Metcalf, T. R., Jiao, L., McClymont, A. N., Canfield, R. C., and Uitenbroek, H.: 1995, *ApJ* **439**, 474.
Metcalf, T. R., Mickey, D. L., Labonte, B. J., and Ryder, L. A.: 2002, in: P.C.H. Martens, and D. Cauffman, (eds.), *Multi-Wavelength Observations of Coronal Structure and Dynamics*, Elsevier, p. 249.
Metcalf, T. R., Leka, K. D., and Mickey, D. L.: 2005, *ApJ Lett.* **623**, L53.
Mikic, Z. and Linker, J. A.: 1994, *ApJ* **430**, 898.
Munro, R. H., Gosling, J. T., Hildner, E., MacQueen, R. M., Poland, A. I., and Ross, C. L.: 1979, *Solar Phys.* **61**, 201.
Romano, P., Contarino, L., and Zuccarello, F.: 2005, *A&A* **433**, 683.
Rust, D. M. and Kumar, A.: 1994, *Solar Phys.* **155**, 69.
Song, L., Zhang, J., Yang, Z., and Wang, J.: 2002, *Solar Phys.* **211**, 315.
St. Cyr, O. C. and Webb, D. F.: 1991, *Solar Phys.* **136**, 379.
Sterling, A. C., Moore, R. L., Qiu, J., and Wang, H.: 2001, *ApJ* **561**, 1116.
Sturrock, P. A.: 1991, *ApJ* **380**, 655.
Subramanian, P. and Dere, K. P.: 2001, *ApJ* **561**, 372.
Tian, L., Alexander, D., Liu, Y., and Jing, Y.: 2005a, *Solar Phys.* **229**, 63.
Tian, L., Liu, Y., Jing, Y., and Alexander, D.: 2005b, *Solar Phys.* **229**, 237.
Tian, L. and Alexander, D.: 2006a, *Solar Phys.* **233**, 29.
Tian, L. and Alexander, D.: 2006b, *Solar Phys.*, submitted.
Webb, D. and Hundhausen, A. J.: 1987, *Solar Phys.* **108**, 383.
Wolfson, R.: 1993, *ApJ* **419**, 382.
Wolfson, R. and Dlamini, B.: 1997, *ApJ* **483**, 961.
Zhang, J., Dere, K. P., Howard, R. A., Kundu, M. R., and White, S. M.: 2001, *ApJ* **559**, 452.

SOLAR IMPRINT ON ICMES, THEIR MAGNETIC CONNECTIVITY, AND HELIOSPHERIC EVOLUTION

N. U. CROOKER[1,*] and T. S. HORBURY[2]
[1] *Center for Space Physics, Boston University, Boston, Massachusetts, USA*
[2] *The Blackett Laboratory, Imperial College, London, UK*
*(*Author for correspondence: E-mail: crooker@bu.edu)*

(Received 21 April 2004; Accepted in final form 29 August 2005)

Abstract. Interplanetary outflows from coronal mass ejections (ICMEs) are structures shaped by their magnetic fields. Sometimes these fields are highly ordered and reflect properties of the solar magnetic field. Field lines emerging in CMEs are presumably connected to the Sun at both ends, but about half lose their connection at one end by the time they are observed in ICMEs. All must eventually lose one connection in order to prevent a build-up of flux in the heliosphere; but since little change is observed between 1 AU and 5 AU, this process may take months to years to complete. As ICMEs propagate out into the heliosphere, they kinematically elongate in angular extent, expand from higher pressure within, distort owing to inhomogeneous solar wind structure, and can compress the ambient solar wind, depending upon their relative speed. Their magnetic fields may reconnect with solar wind fields or those of other ICMEs with which they interact, creating complicated signatures in spacecraft data.

1. Introduction

How do the properties of interplanetary coronal mass ejections (ICMEs) relate to their origins on the Sun, and how do the kinematics and dynamics of propagation into the heliosphere affect ICMEs and their environment? These two questions structure the content of this paper. The first concerns internal structure and magnetic connection to the Sun and is addressed in Section 2. The second concerns external processes and is addressed in Section 3.

2. Internal Structure and Connectivity

As reviewed by Zurbuchen and Richardson (2006, this volume), ICMEs range in complexity from fairly simple magnetic clouds characterized by smooth field rotations, high magnetic field strength, and low temperature (e.g., Burlaga, 1988) to complicated, compound structures with signatures that have non-matching boundaries. This section focuses on the simple structures, magnetic clouds, whose magnetic parameters, usually calculated from flux rope model fits, can be classified and related to solar parameters. Sections 2.1, 2.2, and 2.3, respectively, address

Space Science Reviews (2006) 123: 93–109
DOI: 10.1007/s11214-006-9014-0

the imprint of solar magnetic fields on clouds, the remote connections of magnetic field lines in clouds, and the relation between cloud properties and solar features observed in coronagraphs.

2.1. Solar Magnetic Field Imprint

Various aspects of solar magnetic structures are reflected in the structure of magnetic clouds. Section 2.1.1 discusses how CME formation under the helmet streamer belt can create ICMEs that blend into the heliospheric sector structure, and Section 2.1.2 discusses how the chirality, leading magnetic field orientation, and axis orientation of magnetic clouds reflects magnetic properties of filaments and the helmet streamer belt.

2.1.1. *ICMEs and Sector Boundaries*

Coronagraphs have long shown that CMEs arise from the predominantly closed field line regions of the Sun under the umbrella of the helmet streamer belt (e.g., Hundhausen, 1993). The helmet streamer belt, in turn, forms the base from which stems the boundary between sectors of oppositely directed magnetic fields in the heliosphere, or the heliospheric current sheet (HCS) (Figure 1a).If field lines from the arcade of loops comprising the streamer belt rise, shear, and reconnect to form a CME flux rope, as pictured in Figure 1a and commonly modeled (e.g., Mikic

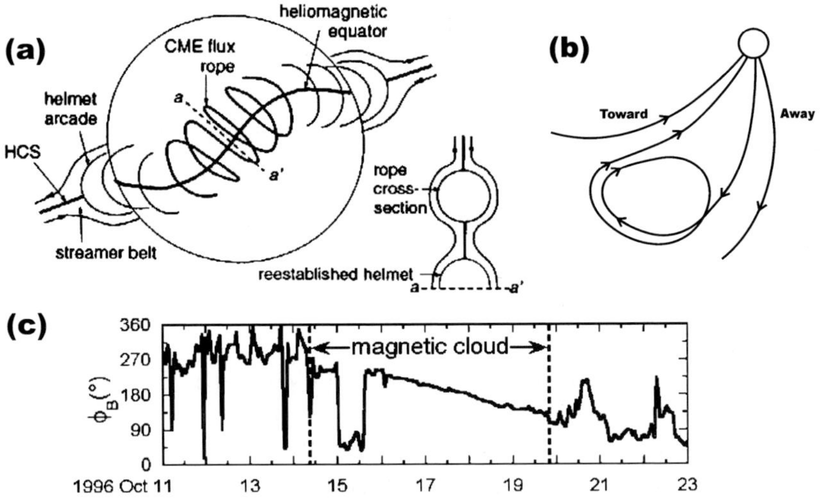

Figure 1. Relationship between magnetic clouds and sector boundaries. (a) A CME flux rope forms from the helmet arcade at the base of the heliospheric current sheet (HCS) separating sectors of opposite magnetic polarity (Crooker *et al.*, 1998). (b) Fields in flux rope legs match away and toward polarity of adjacent sectors. (c) Magnetic azimuth angle measured by Ulysses rotates from away to toward polarity across a magnetic cloud (flux rope) at a sector boundary (Forsyth *et al.*, 1997).

and Lee, 2006, this volume), it follows that the field lines comprising the flux rope will match the surrounding sector structure. Further into the heliosphere, Figure 1b illustrates how the fields in the legs of the flux rope and the sides of its loops will have the same local polarity as the true polarity of the adjacent open field lines on either side. Moreover, the current that creates the flux rope configuration embeds itself in the HCS so that the CME constitutes a bulge of distributed current in what is otherwise a current sheet.

Some observations clearly support the Figures 1a and 1b views (e.g., Crooker *et al.*, 1998). Figure 1c gives an example of the time variation of the magnetic azimuth angle across a magnetic cloud at a sector boundary encountered by Ulysses at 4.4 AU (Forsyth *et al.*, 1997). Instead of a sharp change from 270° marking polarity away from the Sun to 90° marking polarity toward the Sun, as expected for an HCS crossing, the polarity change is accomplished through the days-long field rotation intrinsic to the cloud. As noted by Forsyth *et al.* (1997), "The HCS is neither pushed aside nor draped around the CME but is replaced locally by the CME."

Many ICMEs are not encountered at sector boundaries, presumably because ICMEs are large and orbits through them skim the vicinity of the HCS rather than pass through it. Supporting this view, Kahler *et al.* (1999) found that the "polarity" of ICMEs, assuming passage through one leg rather than the apex of an ICME loop (cf. Figure 1b), is 10 times more likely to match than not to match the surrounding sector polarity. In their study, ICME leg "polarity" was determined not from local magnetic fields, which can turn back on themselves, but from the direction of the strongest counterstreaming suprathermal electron beam relative to the magnetic field direction (see Section 2.2). The fact that one beam is usually stronger supports the assumption that passage is through one leg, since the stronger beam presumably comes from the nearest solar connection point (Pilipp *et al.*, 1987). The Kahler *et al.* (1999) study is the most thorough confirmation to date that ICMEs blend into the sector structure, consistent with the expected solar imprint.

2.1.2. *Magnetic Cloud Flux Rope Parameters, Filaments, and the Heliomagnetic Equator*

A magnetic flux rope expanding into the heliosphere as a loop of nested coils connected to the Sun at both ends (Figure 2 in Zurbuchen and Richardson, 2006, this volume) can be characterized by the directions of its axial and leading fields at the apex of the loop, which together determine the handedness of the twist (Bothmer and Schwenn, 1998). These parameters carry the imprint of both high- and low-altitude solar features (see review by Crooker (2000) and references therein).

From Figure 1a one might expect the direction of the magnetic field at the leading edge of an ICME flux rope or magnetic cloud to reflect the dipole component of the solar magnetic field inherent in the helmet streamer belt, pointing south (north) from the maximum of an even (odd) cycle to the maximum of an odd (even) cycle. Observations show this to be true for 77% of a total of 79 clouds tested

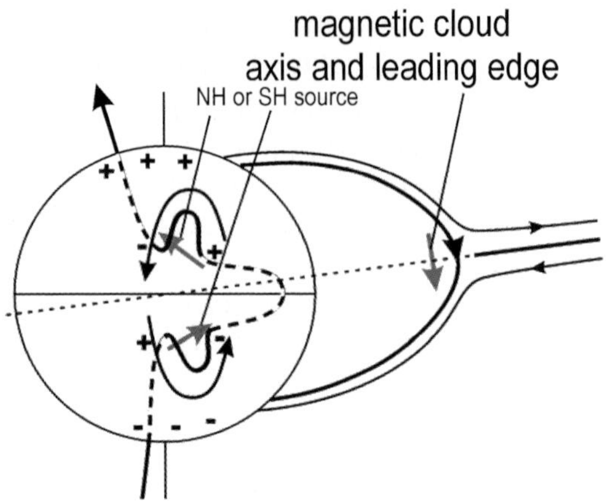

Figure 2. Schematic diagram of solar magnetic features that control magnetic cloud parameters. The direction of the field line distorted by differential rotation gives the direction of the cloud axis, depending upon its hemisphere of origin, and the direction of the dipole component (with a phase lag, see text) gives the direction of the leading field.

in the period spanning 1974 to 1991 (Bothmer and Rust, 1997; Mulligan *et al.*, 1998), with the caveat that the sign change expected at solar maximum shifts to the declining phase. This phase shift may reflect higher-order field components lower in the solar atmosphere, where arcades over filaments retain the old cycle polarity until presumably they are shed as CMEs (cf. Gopalswamy *et al.*, 2003). Although (Leamon *et al.*, 2002) report no correspondence between the solar dipolar component and the leading field direction in magnetic clouds arising from sigmoids in active regions, when the phase shift is taken into account, 65% of their 34 cases fit the pattern.

With the possible exception of the early declining phase, magnetic fields high in the solar atmosphere appear to be systematically related to those in the lower atmosphere (Martin and McAllister, 1997; McAllister *et al.*, 2002), with the result that magnetic cloud parameters reflect filament as well as streamer belt characteristics. Filaments align with neutral lines which are convoluted at low altitudes owing to the influence of higher-order fields but map up to the smoother HCS, which serves as the heliomagnetic equator (Figure 1a). Thus there is some correspondence between the tilts of cloud axes and HCS tilt with respect to the ecliptic plane (Mulligan *et al.*, 1998) as well as the tilts of filament axes (Marubashi, 1997). Zhao and Hoeksema (1997) have shown that on average cloud axes are less tilted than filament axes by a factor of 0.7, consistent with the influence of higher-order fields on filaments (cf. Section 2 of Forsyth *et al.* (2006, this volume)).

In addition to tilt angle, the handedness of twist determined from filament structure is reflected in magnetic clouds. Although filaments may not be flux ropes themselves (Martin and McAllister, 1997), the pattern of magnetic fields surrounding filaments, consisting of barbs and fibrils, displays a skew. Martin *et al.* (1994) found the skew to be dextral in the northern hemisphere and sinistral in the southern hemisphere for 89% of 73 quiescent filaments, independent of solar cycle, although no pattern was found for 31 active-region filaments. At higher altitudes, the coronal arcades overlying quiescent filaments have the opposite skew (Martin and McAllister, 1997). When these arcade fields reconnect to form a CME flux rope, the rope will tend to have left-handed twist if it emerges from the northern hemisphere and right-handed twist if it emerges from the southern hemisphere. Rust (1994) found this to be true for 13 out of 16 magnetic clouds. Somewhat surprisingly, for 36 clouds arising from active regions, (Leamon *et al.*, 2002) found the same hemispheric pattern for 75% of them.

Figure 2 summarizes the solar magnetic imprint patterns on magnetic clouds. The predicted direction of the axial field of a cloud, marked by a short gray arrow in each hemisphere, is the direction of a field line distorted by differential rotation, as in the Babcock model and in the filament pattern low in the solar atmosphere (cf. Bothmer and Schwenn, 1998). At higher altitudes, one can imagine the tilt of the axis lowering toward the dotted line representing the heliomagnetic equator as the neutral line of the filament channel maps up to the HCS. The bipolar field line arched over each filament axis, as in the Babcock view of sunspot formation, represents a low-level arcade. At higher altitudes, the skew of the arcade fields increases until they point in the direction of the solar dipole component, at least until solar maximum. This is the predicted direction of the leading field of a magnetic cloud, as indicated. For the subsequent cycle, when the dipolar fields have the opposite sign, the directions of both the cloud axes and their leading fields will be reversed, which maintains the observed hemispheric pattern of handedness. While the Figure 2 sketch does not capture the lag between filament and polar fields during the declining phase that can account for the phase shift in the sign change of leading fields, it is physically accurate for the ascending phase and serves as a mnemonic device for most of the solar cycle between maxima.

2.2. Magnetic Connectivity to the Sun

Sketches of ICMEs usually show their magnetic field lines connected to the Sun at both ends, as in Figure 1b. The degree to which this is true, our understanding of how connections change, and implications for the heliospheric magnetic flux budget are the respective topics of Sections 2.2.1, 2.2.2, and 2.2.3.

2.2.1. *Tracing ICME Field Connections*

Particles with energies higher than those that constitute the core of solar wind distributions act as field line tracers. Like core particles, they are confined to gyrating motions about field lines; but their considerably higher velocity components result not only in larger gyroradii but in high field-aligned speeds that create particle beams that give nearly instant information about solar connections. For example, solar energetic particle (SEP) events observed inside magnetic clouds give incontrovertible evidence of field lines connected to the Sun at least on one end, as opposed to field lines detached at both ends or closing upon themselves in plasmoids (e.g., Richardson, 1997; Malandraki *et al.*, 2003; and references therein). Further discussion of ICME tracing with particles in the SEP energy range can be found in Section 4.6 of Wimmer-Schweingruber *et al.* (2006, this volume). This section focuses primarily on the lower-energy suprathermal electrons ($E \gtrsim 80$ eV) as ICME field-line tracers.

Because fluxes are higher at lower energies, suprathermal electrons constitute a continuous source of field-aligned particles from the Sun. They focus into beams as their pitch angles decrease owing to decreasing magnetic field strength with distance from the Sun. While scattering processes, shocks, and other inhomogeneities in the heliospheric magnetic field alter these beams as they propagate outward (Wimmer-Schweingruber *et al.*, 2006, this volume), informed use of suprathermal electron data have yielded a large body of information about ICME connections. Counterstreaming beams, used as one of the first widely-accepted signatures of ICMEs (Gosling *et al.*, 1987), are interpreted as a signature of closed field lines, connected to the Sun at both ends. Unidirectional beams signal open field lines, connected at only one end. The lack of beams, called a "heat flux dropout" (HFD) because suprathermal electrons carry heat flux away from the Sun, is a necessary but unfortunately not sufficient signature of field lines disconnected from the Sun at both ends (Crooker *et al.*, 2002; Crooker *et al.*, 2003; Pagel *et al.*, 2005; and references therein). Studies of counterstreaming suprathermal electrons as well as higher-energy particles conclude that ICMEs contain a mixture of open, closed, and, on rare occasions, disconnected field lines (Bothmer *et al.*, 1996; Larson *et al.*, 1997, 2000; Malandraki *et al.*, 2003; Crooker *et al.*, 2004). For example, in a study of 48 magnetic clouds at 1 AU, Shodhan *et al.* (2000) found counterstreaming only 59% of the time, on average, leaving the clouds 41% open.

2.2.2. *Conceptual Modeling of ICME Connections*

An explanation for how a coherent flux rope in the solar wind can contain a mix of open and closed field lines, as pictured in Figure 3a, has been provided by Gosling *et al.* (1995). The conceptual model is based upon an MHD simulation of flux rope release in Earth's magnetosphere (Hesse and Birn, 1991) in which reconnection between differently-connected field lines occurs seemingly randomly yet progressively disconnects closed field lines. The steps leading to disconnection are illustrated in Figure 3b: (1) closed loops with sheared footpoints reconnect to form a flux rope that is still connected to the Sun at both ends (i.e., closed); (2) an open field line reconnects with a field line in one leg of the flux rope to form an open coil; (3) an open field line reconnects with a field line in the other leg of the flux rope to form a disconnected coil; (4) two open field lines reconnect to form a U-shaped disconnected field line encasing the disconnected coil. Since

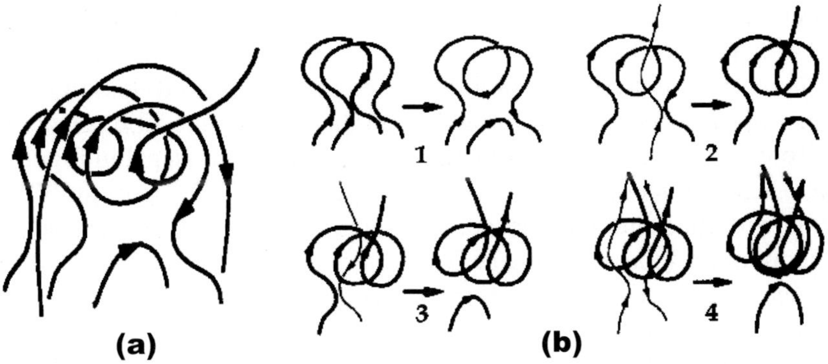

Figure 3. Schematic drawings of magnetic field lines in CME flux rope (Gosling *et al.*, 1995). (a) Coherent flux rope with open coil nested in a closed coil. (b) Four steps to disconnection: 1. partial disconnection, two closed loops reconnect to form coil; 2. interchange reconnection, open field line reconnects with closed coil to form open coil; 3. open field line reconnects with open coil to disconnect coil; 4. two open field lines reconnect to form U-shaped disconnected field line.

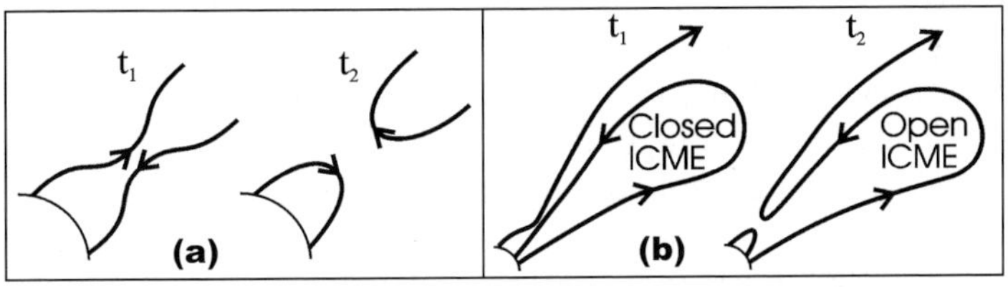

Figure 4. Before (t1) and after (t2) solutions to the problem of magnetic flux build-up from CMEs: (a) disconnection and (b) interchange reconnection (Crooker *et al.*, 2002).

observations show that disconnected field lines in ICMEs are rare, steps 3 and 4 are not important for CMEs. Steps 1 and 2, respectively called "partial disconnection" and "interchange reconnection," result in the configuration in Figure 3a and play an important role in the heliospheric magnetic flux budget (Crooker *et al.*, 2002), discussed in the following section.

2.2.3. *Heliospheric Magnetic Flux Budget*

Without some mitigating process, the closed flux that CMEs introduce to the heliosphere would result in a continuous build-up of magnetic flux, which is not observed. McComas (1995) argues that the only means of preventing flux build-up from CMEs is to disconnect fields elsewhere through reconnection of open field lines back at the Sun. Figure 4a illustrates the resulting U-shaped field with no connection to the Sun (cf. step 4 in Figure 3b). The problem with this solution is that true signatures of disconnection are rare, as mentioned in Section 2.2.1, not only within ICMEs but throughout the solar wind. About 90% of HFDs at time scales > 1 hr show electrons with reduced intensities and/or at higher energies still streaming from the Sun along what must be connected field lines (Lin and Kahler, 1992; Pagel *et al.*, 2005).

An alternative solution to the problem of magnetic flux build-up is that the closed field lines within ICMEs open through interchange reconnection (Gosling *et al.*, 1995; Crooker *et al.*, 2002). As illustrated in Figure 4b (cf. step 2 in Figure 3b), an open field line can reconnect with a closed field line in one leg of an ICME back at the Sun with the result that the closed loop in the heliosphere is exchanged for a closed loop in the solar atmosphere. This alternative solution is attractive because interchange reconnection generates no disconnected field lines, in agreement with the observation that they are rare, and it can continue to open CMEs well after they have left the Sun, until they are completely open and add no flux to the heliosphere.

If interchange reconnection is the means by which the flux budget is balanced, one might expect that ICMEs observed by Ulysses beyond 1 AU would be more open than those at 1 AU, but this seems not to be the case. Using counterstreaming electrons as a signature of closed fields, Riley *et al.* (2004) could detect no radial trend in the degree of openness in ICMEs encountered on the way to Jupiter, and (Crooker *et al.*, 2004) found that magnetic clouds near 5 AU were not significantly more open on average than those at 1 AU. Both papers conclude that the rate at which a CME opens by interchange reconnection must slow significantly as its leading edge moves out into the heliosphere and that it may take months to years rather than days to open completely, leading to a temporary flux build-up that is consistent with the factor of two solar cycle variation in heliospheric magnetic flux (e.g., Wang *et al.*, 2000). On the other hand, as discussed in detail by Crooker (2005), after months to years, closed loops moving out into the heliosphere will likely lose their counterstreaming signature and be indistinguishable from open field lines in spacecraft measurements. The interchange reconnection that eventually opens them will then give the signature of open field lines reconnecting, or disconnection, which reopens the problem of finding sufficient disconnection signatures. A different problem arises if one argues that ICMEs should be completely open by the time they reach 5 AU based upon estimates of the rate of interchange reconnection at the Sun (Reinard and Fisk, 2004). Although this eliminates the need for disconnection signatures, it casts doubt upon the relatively robust and widely-used interpretation of counterstreaming suprathermal electrons as signatures of closed fields. Clearly current understanding of these issues leads to dilemmas that remain to be resolved.

2.3. Imprint of Plasma Origins

Progress in understanding plasma characteristics of ICMEs in terms of what we know about CMEs has been limited owing to a number of constraints on observations. Two topics of interest concern the interpretation of elemental and ionic composition data from ICMEs and ICME manifestations of the three-part structure of CMEs observed in coronagraphs. The first is treated by Wimmer-Schweingruber *et al.* (2006, this volume), von Steiger and Richardson (2006, this volume), and Gazis *et al.* (2006, this volume). Here, relevant to the discussion in section 2.2.2, we note that the high charge state of heavy ions characteristic of ICMEs and indicative of high-temperature origins may well be a signature of magnetic fields reconnecting during CME liftoff, as argued by Lepri and Zurbuchen (2004).

The second topic, ICME manifestations of CME three-part structure, still raises more questions than it answers. The classic three parts are the bright outer rim, the dark cavity, and the bright core (see, e.g., Schwenn *et al.*, 2006, this volume). These have been loosely associated with the pile-up of plasma or streamer material at the leading edge, the flux rope, and the filament, respectively, but these associations raise unsettled issues, particularly about flux rope formation and filament structure. What is assumed to be evidence of cool filament material from low in the solar atmosphere, for example, the presence of He+, is only rarely found in the solar wind (Zurbuchen and Richardson, 2006, this volume; Wimmer-Schweingruber *et al.*, 2006, this volume), yet sometimes the bright core is a substantial fraction of the volume of an ICME. Suleiman *et al.* (2005) illustrate such a case and argue that although the bright core may be filament material, it may no longer reside on filament field lines. Through partial disconnection the filament material may gain access to the much larger flux rope formed by that process and thus lose both its magnetic coherence and the imprint of its cold origins (Crooker, 2005).

3. External Forces and Structures

The interaction of ICMEs with the ambient solar wind through which they propagate can significantly alter their properties as well as change the solar wind plasma itself. These interactions need to be understood in order to relate ICME properties to properties at their solar origins and thereby learn about what causes their generation and ejection. These interactions also tend to make ICMEs harder to identify and study. Significant additional effects of solar wind/ICME interactions include the energisation of particles by shocks (e.g., Reames, 1999), increased geoeffectiveness (e.g., Webb *et al.*, 2000; Siscoe and Schwenn, 2006, this volume), and the enhanced blocking of energetic particle propagation (e.g., Ifedili, 2004).

The study of ICMEs over the last few decades has led to an increasing appreciation of the complexity that can arise from the dynamics of ICME interactions. These interactions result in extremely structured objects which are highly undersampled with in situ spacecraft data, and it is therefore challenging to deduce their 3D structure. Nevertheless, considerable progress has been made. Increasingly sophisticated simulations of ICME dynamics have shown what behaviours are possible and help interpret in situ data (see Forsyth *et al.*, 2006, this volume). Advances have also been made in analytical models of magnetic flux ropes to take into account the effects of dynamical deformation.

We consider some of the most important consequences of dynamics in this paper. A number of related issues such as ICME deceleration and multi-spacecraft observations are discussed by Forsyth *et al.* (2006, this volume).

3.1. Kinematic Evolution

Kinematic aspects of the propagation of an ICME into interplanetary space result in changes to its shape, independent of any interaction with the ambient plasma. ICMEs are typically extended objects and cover a finite solid angle near the Sun. The propagation of the ICME plasma radially away from the Sun results in a preservation of this solid angle and a consequent increase in the extent of the ejecta perpendicular to the radial direction. Therefore, if the ICME retains its radial extent, it will expand into a "pancake" shape far from the Sun. This kinematic effect is shown schematically in Figure 5(a). Riley and Crooker (2004) show that this effect is significant by 1 AU for typical ICMEs. Radial expansion and the interaction with the ambient solar wind will obviously also alter the ICME shape, but this simple

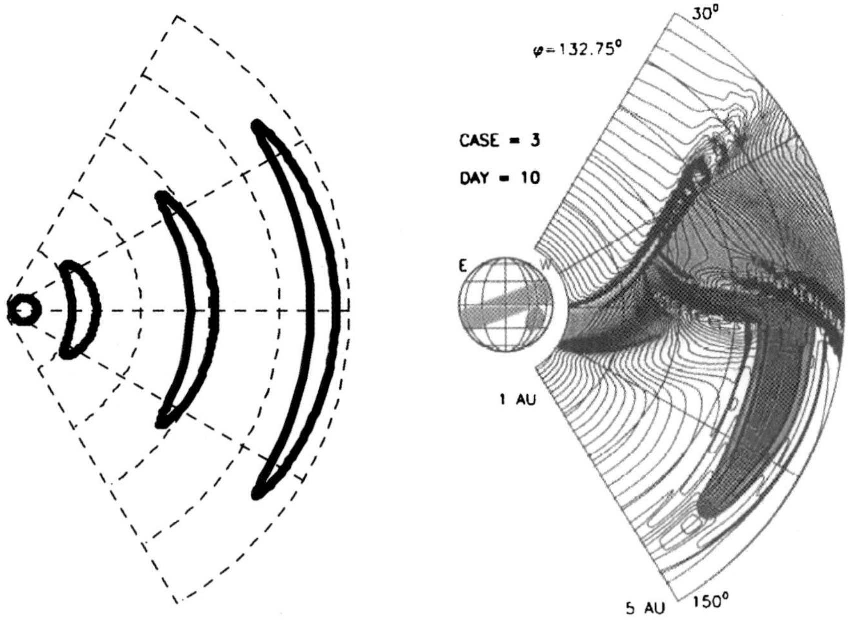

Figure 5. (a) Schematic of the kinematic effects of the radial expansion of ICMEs, leading to a "pancake" shape. (b) Results of a 3D simulation of an ICME propagating through a structured solar wind: the ICME is greatly distorted by its interaction with slow solar wind at low latitudes (after Odstrčil and Pizzo, 1999b).

geometrical effect implies that it is never possible to assume that ICMEs propagate unchanged into interplanetary space.

3.2. Dynamic Evolution

3.2.1. *Overexpanding ICMEs*

The simplest interplanetary signatures of ICMEs were in fact the last to be identified. Ulysses observations within steady, high-speed solar wind at high latitudes at several AU revealed (e.g., Gosling *et al.*, 1998) a class of transients lasting a few days, bounded by a forward and reverse shock, the latter being uncommon for low-latitude ICMEs. Their internal structure was remarkably uniform, and all the events were similar in their gross form. As with low-latitude ICMEs, around 1/3 contained magnetic flux ropes. Perhaps most surprisingly, these events tended to have a lower pressure inside than the ambient wind, although they were bounded by compressions and shocks. Gosling *et al.* (1998) showed that these signatures were consistent with ejecta with an initial overpressure relative to the ambient solar wind: this pressure drives the expansion of the ICME, producing a lower density cavity. In addition, simulations (e.g., Schmidt and Cargill, 2001) show that at least parts of ICMEs can propagate in latitude from the streamer belt into polar solar wind (see Section 3.2.3), so the observation of overexpanded ICMEs in high-speed wind does not imply that they originate in coronal holes. The magnetic field of flux rope ICMEs can act to prevent disruption of the large scale ICME structure (Cargill *et al.*, 2000).

The remarkable similarity of the observed events implies that, in the presence of uniform solar wind conditions, many or all ICMEs will exhibit this profile. Some events exhibit less symmetric time profiles than others: Gosling *et al.* (1998) showed that this was due to differences in the relative speeds of the solar wind and ejecta.

3.2.2. *Interaction with the Ambient Solar Wind*

While overexpanded ICMEs represent a particularly simple and regular class of ejecta signatures, most observed events are more complex. This is largely due to the complicated interactions between the ejecta and the ambient solar wind plasma. Since many ICMEs do not travel at the same speed as the solar wind in which they are embedded, compressions and rarefactions develop at the edges of the events. Even simple 1D simulations (e.g., Gosling and Riley, 1996) of solar wind dynamics show some of the possible consequences of these interactions, such as shocks and the acceleration or deceleration of ICMEs. The ICME shape can also be greatly distorted. Some of the consequences of these interactions are discussed in the remainder of this paper.

3.2.3. *Low- and High-Latitude Manifestations of the Same ICME*

The observation of relatively simple overexpanded ICMEs in high-latitude fast wind and much more complex structures at low latitudes raises the question as to whether these are two different classes of events or simply different manifestations of the same phenomenon. Observations of the same ICME at high and low latitudes (Hammond *et al.*, 1995) show that these can be the same phenomenon, highlighting the importance of the ambient solar wind in determining the in situ signature of an ICME. As mentioned in Section 3.2.1, simulations (Riley *et al.*, 1997; Schmidt and Cargill, 2001) show that ICMEs launched from within the streamer belt can partially penetrate the stream interface and enter high-speed polar wind, resulting in an ICME with different signatures in fast and slow wind, as observed (see Section 4.3 of Forsyth *et al.*, 2006, this volume). When an ICME propagates within streams of different speeds, shear of the structure results from the effect of drag to bring speeds closer to that of the ambient solar wind.

The complexity that can arise from ICME-solar wind interactions, and the different character of a single ICME at different locations, is shown in the 3D simulation result in Figure 5(b), taken from Odstrčil and Pizzo (1999b). At high latitudes, the ICME resembles the kinematic ICME in Figure 5(a), although with a larger extent due to expansion caused by internal overpressure. At lower latitudes, the ICME is heavily distorted by solar wind interactions. Such simulations highlight the difficulties in interpreting in situ ICME data.

3.2.4. *Folded Flux Ropes*

If the footpoints of an ICME flux rope are rooted in the Sun, as sketched in Figure 2 of Zurbuchen and Richardson (2006, this volume), then solar rotation would be expected to cause distortion in the structure, just as the large scale magnetic field tends to form Archimedean (Parker) spirals. Such effects are seen in 3D simulations (Vandas *et al.*, 2002). Consistent with this view, Owens *et al.* (2004) suggested that west flank passages through ICMEs were around twice as common as east flank. In principle, it could be possible for a single spacecraft to pass through both legs of the same magnetic cloud, as suggested by Crooker *et al.* (1998) on the basis of mirror symmetric patterns in magnetic field elevation angle coincident with counterstreaming electrons trailing magnetic clouds. However, since several ICMEs often exist close to each other, it is difficult unambiguously to distinguish two encounters with one cloud from two separate events. A necessary but not sufficient test is for both events to exhibit the same handedness. Rees and Forsyth (2004) describe two such examples in Ulysses data, while Kahler *et al.* (1999) found only one in 8 possible cases in ISEE 3 data.

3.2.5. *Modelling Dynamic Effects: Non-Circular Flux Rope Models*

Analysis of ICMEs has often concentrated on magnetic flux ropes, despite their occurrence in only around 1/3 to 1/2 of apparent events, for a number of reasons: the relative simplicity of identifying flux ropes; their presumed relation to magnetic structures at the Sun; and because by fitting analytical models to their profiles, it is possible to estimate parameters such as the location and orientation of the rope's axis. The earliest models of flux ropes (e.g., Burlag, 1988) assumed circular cross sections: these often result in good agreement with observations, but deformation from this shape will occur as a result of both kinematics and dynamics. There is evidence that this deformation can lead to systematic errors in estimates of flux rope parameters derived from circular cross section models. As a result, considerable efforts have been made to extend models to include elliptical cross sections (e.g., Mulligan *et al.*, 2001; Hidalgo *et al.*, 2002). A more generalised fitting method (Hu and Sonnerup, 2002), assuming $2\frac{1}{2}$D variations, has recently been developed and shows considerable promise. These models are discussed further by Forbes *et al.* (2006, this volume).

3.3. Sheaths and Shocks

Both ICME propagation at a speed different from the ambient solar wind and elevated internal pressure result in compressions and rarefactions. Passage of compressed solar wind plasma and magnetic field in sheath regions upstream of ICMEs at 1 AU can last for many hours. If this compression is strong, the magnetic field can be much larger than typical and, hence, geoeffective (e.g., Tsurutani *et al.*, 1999; Siscoe and Schwenn, 2006, this volume). The orientation of the plane of compression in which the magnetic field in the sheath is forced to lie can be determined by minimum variance analysis and used to estimate the local orientation of the leading edge of an ICME (Jones *et al.*, 2002; Section 4.3 of Wimmer-Schweingruber *et al.*, 2006, this volume).

The shocks driven by speed and pressure differences between the ICME and the surrounding solar wind can propagate significant distances away from the ejecta itself, both radially and perpendicular to the flow. Simulations (e.g., Odstrčil and Pizzo, 1999a) show that the shock and resulting compression can result in profiles in the solar wind which might be mistaken for passage through the ejecta itself. This may explain events such as that reported by Richardson *et al.* (1994) when two spacecraft encountered a shock but only one entered ejecta material. In principle, composition signatures can help to distinguish these cases, since the sheath, being compressed solar wind, should retain solar wind composition. For example, Borrini *et al.* (1982) used enhancements of He/H to identifiy ejecta following shocks and explained the large number of shocks without this marker (48 out of 91) in terms of the much larger extent of shocks compared to ejecta. It is highly likely, however, that some ejecta went undetected owing to the variability of composition patterns in ICMEs (Wimmer-Schweingruber *et al.*, 2006, this volume; Crooker, 2005).

3.4. Reconnection

Both simulations and some limited observations suggest that reconnection occurs around and within ICMEs. The large compression ahead of some ICMEs would be expected to trigger reconnection between ICME and sheath magnetic field if their orientations were favourable. McComas *et al.* (1994) presented suprathermal electron data which could be interpreted as signatures of reconnection ahead of an ICME. Simulations (Cargill and Schmidt, 2002) show that reconnection can occur at the flanks of ICMEs, particularly if they are traveling through the streamer belt. Simulations also imply that reconnection can occur within ICMEs owing to shear by background solar wind inhomogeneity (Schmidt and Cargill, 2001). (See Sections 4.2 and 4.3 of Forsyth *et al.* (2006, this volume) for examples of simulation results.) Farrugia *et al.* (2001) have discussed one possible signature of such an event, and more direct evidence has been reported recently by Gosling *et al.* (2005). Behind ICMEs, simulations by Riley *et al.* (2002) indicate that the in situ signatures of partial reconnection back at the Sun (section 2.2.2) would be a slight velocity and density increase trailing an ICME as a result of an outward reconnection jet. Such signatures have been seen in spacecraft data, but only rarely (Riley *et al.*, 2002).

3.5. Interactions of Multiple ICMEs

The ejection of multiple CMEs from the vicinity of individual active regions over several days, combined with their variable velocities and large angular extent, makes it inevitable that ICMEs will sometimes interact. Indeed, as ICMEs propagate into the outer heliosphere, they merge and interact with CIRs and other ICMEs to form global merged interaction regions (GMIRs) – these effects are discussed by Gazis *et al.* (2006, this volume). Like ICME/solar wind interactions, ICME/ICME interactions can also result in complicated structures and spacecraft signatures. For example, Kahler *et al.* (1999) used bidirectional electron fluxes to argue that some magnetic clouds are in fact multiple events. Hu *et al.* (2003) used the reconstruction technique of Hu and Sonnerup (2002) to infer a double rope structure of a magnetic cloud at 1 AU.

Burlaga *et al.* (2002) discussed three sets of multiple halo CMEs and their associated ejecta at 1 AU. They showed that the ejecta were "complex," being fast ($>$ 600 km/s) events that were not magnetic clouds. These events typically showed substructure in parameters such as composition and density, suggesting that they were formed from several structures. They emphasised the challenges in quantitatively describing such events.

Simulations, again, reveal some of the possible consequences of multiple ICME interactions, such as shocks propagating through ejecta (Odstrčil *et al.*, 2003) – and, if two flux ropes are of the same chirality and polarity, the merging and reconnection of ICMEs (Schmidt and Cargill, 2004).

3.5.1. *Interacting ICMEs as Particle Accelerators*

Gopalswamy *et al.* (2002a) showed that radio emission occurred at around 10 solar radii when two CMEs came into contact and argued that this was due to either reconnection or the formation of a shock at this location. Gopalswamy *et al.* (2002b) argued that when one CME overtakes a second, slower event, solar energetic particle acceleration is significantly increased. However, this conclusion was recently disputed by Richardson *et al.* (2003) and remains controversial.

4. Conclusion

There is little question that ICMEs are the interplanetary manifestations of CMEs, but both simulations of their propagation and observations of their complicated signatures indicate that they evolve substantially as they move out into the heliosphere. Magnetic field lines change their connections, the imprint of the magnetic field at their source weakens, shapes and structures distort, and particles accelerate. It appears that many aspects of that evolution can be understood in terms of phenomenological models – a first step toward the long-term goal of understanding in terms of fundamental physical processes – but a number of basic questions remain. Some of the more important of these questions concern how long field lines remain connected to the Sun at both ends, the fate of filament plasma, and the degree to which simulations represent the actual distortion of ICMEs.

Acknowledgements

The authors thank D. Odstrčil for providing Figure 5(b) and the International Space Science Institute, Bern, for their support of this work. T. Horbury is supported by a PPARC (UK) Fellowship and N. Crooker by the (US) National Science Foundation grant ATM-0119700.

References

Borrini, G., Gosling, J. T., Bame, S. J., and Feldman, W. C.: 1982, *J. Geophys. Res.* **87**(16), 4365.

Bothmer, V., Desai, M. I., Marsden, R. G., Sanderson, T. R., Trattner, K. J., Wenzel, K.-P., *et al.*: 1996, *Astron. Astrophys.* **316**, 493.

Bothmer, V. and Rust, D. M.: 1997, In: Crooker, N. U., Joselyn, J. A., and Feynman, J. (eds.): *Coronal Mass Ejections, Geophys. Monogr. Ser.*, Vol. 99. Washington, D. C.: AGU, pp. 137.

Bothmer, V. and Schwenn, R.: 1998, *Ann. Geophys.* **16**, 1.

Burlaga, L. F.: 1988, *J. Geophys. Res.* **93**(12), 7217.

Burlaga, L. F., Plunkett, S. P., and St. Cyr, O. C.: 2002, *J. Geophys. Res.* **107**(A10), 10.1029/2001JA000255.

Cargill, P. J., Schmidt, J., Spicer, D. S., and Zalesak, S. T.: 2000, *J. Geophys. Res.* **105**(A4), 7509.

Cargill, P. J. and Schmidt, J. M.: 2002, *Ann. Geophys.* **20**, 879.

Crooker, N. U.: 2000, *J. Atmos. Sol. Terr. Phys.* **62**, 1071.

Crooker, N. U.: 2005, In: *Solar Wind Eleven*. ESA Spac. Publ. 592, pp 289–295.

Crooker, N. U., Forsyth, R., Rees, A., Gosling, J. T., and Kahler, S. W.: 2004, *J. Geophys. Res.* **109**(A18), 10.1029/2004JA010426.

Crooker, N. U., Gosling, J. T., and Kahler, S. W.: 1998, *J. Geophys. Res.* **103**(12), 301.

Crooker, N. U., Gosling, J. T., and Kahler, S. W.: 2002, *J. Geophys. Res.* **107**(A2), 10.1029/2001JA000236.

Crooker, N. U., Larson, D. E., Kahler, S. W., Lamassa, S. M., and Spence, H. E.: 2003, *Geophys. Res. Lett.* **30**, 10.1029/2003GL017036.

Farrugia, C. J., Vasquez, B., Richardson, I. G., Torbert, R. B., Burlaga, L. F., Biernat, H. K., *et al.*: 2001, *Adv. Space. Res.* **28**, 759.

Forbes, T. G., Linker, J. A., *et al.*: 2006, *Space Sci. Rev.,* this volume, 10.1007/s11214-006-9019-8.

Forsyth, R. J., Balogh, A., Smith, E. J., and Gosling, J. T.: 1997, *Geophys. Res. Lett.* **24**, 3101.

Forsyth, R. J., Bothmer, V., *et al.*: 2006, *Space Sci. Rev.,* this volume, 10.1007/s11214-006-9022-0.

Gazis, P. R., Balogh, A., *et al.*: 2006, *Space Sci. Rev.,* this volume, 10.1007/s11214-006-9023-z.

Gopalswamy, N., Lara, A., Yashiro, S., and Howard, R. A.: 2003, *Astrophys. J.* **598**, L63.

Gopalswamy, N., Yashiro, S., Kaiser, M. L., Howard, R. A., and Bougeret, J.-L.: 2002a, *Geophys. Res. Lett.* **29**, 10.1029/2001GL013606.

Gopalswamy, N., Yashiro, S., Michalek, G., Kaiser, M. L., Howard, R. A., Reames, D. V., *et al.*: 2002b, *Astrophys. J.* **572**, L103.

Gosling, J. T., Baker, D. N., Bame, S. J., Feldman, W. C., Zwickl, R. D., and Smith, E. J.: 1987, *J. Geophys. Res.* **92**(11), 8519.

Gosling, J. T., Birn, J., and Hesse, M.: 1995, *Geophys. Res. Lett.* **22**, 869.

Gosling, J. T. and Riley, P.: 1996, *Geophys. Res. Lett.* **23**, 2867.

Gosling, J. T., Riley, P., McComas, D. J., and Pizzo, V. J.: 1998, *J. Geophys. Res.* **103**(12), 1941

Gosling, J. T., Skoug, R. M., McComas, D. J., and Smith, C. W.: 2005, *J. Geophys. Res.* **110**(A9), 10.1029/2004JA010809.

Hammond, C. M., Crawford, G. K., Gosling, J. T., Kojima, H., Phillips, J. L., Matsumoto, H., *et al.*: 1995, *Geophys. Res. Lett.* **22**, 11690

Hesse, M. and Birn, J.: 1991, *J. Geophys. Res.* **96**(15), 5683.

Hidalgo, M. A., Nieves-Chinchilla, T., and Cid, C.: 2002, *Geophys. Res. Lett.* **29**, 10.1029/2001GL013875.

Hu, Q., Smith, C. W., Ness, N. F., and Skoug, R. M.: 2003, *Geophys. Res. Lett.* **30**, 10.1029/2002GL016653.

Hu, Q. and Sonnerup, B. U. Ö.: 2002, *J. Geophys. Res.* **107**(A7), 10.1029/2001JA000293.

Hundhausen, A. J.: 1993, *J. Geophys. Res.* **98**(17), 113,177–13,200.

Ifedili, S. O.: 2004, *J. Geophys. Res.* **109**(A18), 10.1029/2002JA009814.

Jones, G. H., Rees, A., Balogh, A., and Forsyth, R. J.: 2002, *Geophys. Res. Lett.* **29**, 10.1029/2001GL014110.

Kahler, S. W., Crooker, N. U., and Gosling, J. T.: 1999, *J. Geophys. Res.* **104**(13), 9919–9924.

Larson, D. E., Lin, R. P., McTiernan, J. M., McFadden, J. P., Ergun, R. E., McCarthy, M., *et al.*: 1997, *Geophys. Res. Lett.* **24**, 1911.

Larson, D. E., Lin, R. P., and Steinberg, J.: 2000, *Geophys. Res. Lett.* **27**, 157.

Leamon, R. J., Canfield, R. C., and Pevtsov, A. A.: 2002, *J. Geophys. Res.* **107**(A9), 10.1029/2001JA000313.

Lepri, S. T. and Zurbuchen, T. H.: 2004, *J. Geophys. Res.* **109**(A18), 10.1029/2003JA009954.

Lin, R. P. and Kahler, S. W.: 1992, *J. Geophys. Res.* **97**(16), 8203.

Malandraki, O. E., Sarris, E. T., and Tsiropoula, G.: 2003, *Ann. Geophys.* **21**, 1249.

Martin, S. F., Bilimoria, R., and Tracadas, P. W.: 1994, In: Rutten, R. J. and Schrijver, C. J. (eds.): *Solar Surface Magnetism*, Norwell, Mass.: Kluwer Academic, pp. 303.

Martin, S. F. and McAllister, A. H.: 1997, In: Crooker, N. U., Joselyn, J. A., and Feynman, J. (eds.): *Coronal Mass Ejections, Geophys. Monogr. Ser.*, Vol. 99. Washington, D. C.: AGU, pp. 127.

Marubashi, K.: 1997, In: Crooker, N. U., Joselyn, J. A., and Feynman, J. (eds.): *Coronal Mass Ejections, Geophys. Monogr. Ser.*, Vol. 99. Washington, D. C.: AGU, pp. 147.

McAllister, A. H., Mackay, D. H., and Martin, S. F.: 2002, *Sol. Phys.* **211**, 155.

McComas, D. J.: 1995, *Rev. Geophys. Suppl.* **33**, 603.
McComas, D. J., Gosling, J. T., Hammond, C. M., Moldwin, M. B., Phillips, J. L., and Forsyth, R. J.: 1994, *Geophys. Res. Lett.* **21**, 1751.
Mikic, Z. and Lee, M. A.: 2006, *Space Sci. Rev.,* this volume, 10.1007/s11214-006-9012-2.
Mulligan, T., Russell, C. T., Anderson, B. J., and Acuna, M. H.: 2001, *Geophys. Res. Lett.* **28**, 4417.
Mulligan, T., Russell, C. T., and Luhmann, J. G.: 1998, *Geophys. Res. Lett.* **25**, 2959.
Odstrčil, D. and Pizzo, V. J.: 1999a, *J. Geophys. Res.* **104**(A12), 28225.
Odstrčil, D. and Pizzo, V. J.: 1999b, *J. Geophys. Res.* **104**(A1), 493.
Odstrčil, D., Vandas, M., Pizzo, V. J., and MacNeice, P.: 2003, In: *AIP Conf. Proc. 679: Solar Wind Ten*. pp. 699.
Owens, M. J., Rees, A., and Cargill, P. J.: 2004, submitted to *Ann. Geophys.* unpublished manuscript.
Pagel, C., Crooker, N. U., and Larson, D. E.: 2005, *Geophys. Res. Lett.* **32**, 10.1029/2005GL023043.
Pilipp, W. G., Muehlhaeuser, K.-H., Miggenrieder, H., Rosenbauer, H., and Schwenn, R.: 1987, *J. Geophys. Res.* **92**(11), 1103.
Schwenn, R., Raymond, J. C., *et al.*: 2006, *Space Sci. Rev.,* this volume, 10.1007/s11214-006-9016-y.
Reames, D. V.: 1999, *Space Sci. Rev.* **90**, 413.
Rees, A. and Forsyth, R. J.: 2004, *Geophys. Res. Lett.* **31**, 10.1029/2003GL018330.
Reinard, A. A. and Fisk, L. A.: 2004, *Astrophys. J.* **608**, 533.
Richardson, I. G.: 1997, In: Crooker, N. U., Joselyn, J. A., and Feynman, J. (eds.): *Coronal Mass Ejections, Geophys. Monogr. Ser.*, Vol. 99. Washington, D. C.: AGU, pp. 189.
Richardson, I. G., Farrugia, C. J., and Winterhalter, D.: 1994, *J. Geophys. Res.* **99**(18), 2513.
Richardson, I. G., Lawrence, G. R., Haggerty, D. K., Kucera, T. A., and Szabo, A.: 2003, *Geophys. Res. Lett.* **30**, 10.1029/2002GL016424.
Riley, P. and Crooker, N. U.: 2004, *Astrophys. J.* **600**, 1035.
Riley, P., Gosling, J. T., and Crooker, N. U.: 2004, *Astrophys. J.* **608**, 1100.
Riley, P., Gosling, J. T., and Pizzo, V. J.: 1997, *J. Geophys. Res.* **102**(11), 14677.
Riley, P., Linker, J. A., Mikić, Z., Odstrčil, D., Pizzo, V. J., and Webb, D. F.: 2002, *Astrophys. J.* **578**, 972.
Rust, D. M.: 1994, *Geophys. Res. Lett.* **21**, 241.
Schmidt, J. and Cargill, P. J.: 2004, *Ann. Geophys.* **22**, 2245.
Schmidt, J. M. and Cargill, P. J.: 2001, *J. Geophys. Res.* **106**(15), 8283.
Shodhan, S., Crooker, N. U., Kahler, S. W., Fitzenreiter, R. J., Larson, D. E., Lepping, R. P., *et al.*: 2000, *J. Geophys. Res.* **105**(14), 27261.
Siscoe, G. and Schwenn, R.: 2006, *Space Sci. Rev.,* this volume.
Suleiman, R. M., Crooker, N. U., Raymond, J. C., and Ballegooijen, A. V.: 2005, In: *IAU Symposium*. pp. 71.
Tsurutani, B. T., Kamide, Y., Arballo, J. K., Gonzalez, W. D., and Lepping, R. P.: 1999, *Phys. Chem. Earth (C)* **24**(1–3), 101.
Vandas, M., Odstrčil, D., and Watari, S.: 2002, *J. Geophys. Res.* **107**(A9), 10.1029/2001JA005068.
von Steiger, R. and Richardson, J. D.: 2006, *Space Sci. Rev.,* this volume, 10.1007/s11214-006-9015-z.
Wang, Y.-M., Lean, J., and Sheeley, N. R.: 2000, *Geophys. Res. Lett.* **27**, 505.
Webb, D. F., Cliver, E. W., Crooker, N. U., Cry, O. C. S., and Thompson, B. J.: 2000, *J. Geophys. Res.* **105**(14), 7491.
Wimmer-Schweingruber, R. F., Crookers, N. U., *et al.*: 2006, *Space Sci. Rev.,* this volume, 10.1007/s11214-006-9017-x.
Zhao, X. P. and Hoeksema, J. T.: 1997, *Geophys. Res. Lett.* **24**, 2965.
Zurbuchen, T. H. and Richardson, I. G.: 2006, *Space Sci. Rev.,* this volume, 10.1007/s11214-006-9010-4.

ICMES IN THE OUTER HELIOSPHERE AND AT HIGH LATITUDES: AN INTRODUCTION

R. VON STEIGER[1,*] and J. D. RICHARDSON[2]
[1]*International Space Science Institute, Hallerstrasse 6, CH-3012 Bern, Switzerland*
[2]*Center for Space Research, Massachusetts Institute of Technology, Cambridge, MA 02139, USA*
*(*Author for correspondence: E-mail: vsteiger@issibern.ch)*

(Received 24 August 2004; Accepted in final form 9 May 2006)

Abstract. Interplanetary coronal mass ejections (ICMEs) are observed at all latitudes and distances from which data are available. We discuss the radial evolution of ICMEs out to large distances and ICME properties at high latitudes. The internal pressure of ICMEs initially exceeds the ambient solar wind pressure and causes the ICMEs to expand in radial width to about 15 AU. Large ICMEs and series of ICMEs compress the leading plasma and form merged interaction regions (MIRs) which dominate the structure of the outer heliosphere at solar maximum. The distribution of high-latitude ICMEs is solar cycle dependent. A few overexpanding ICMEs are observed at high-latitude near solar minimum. Near solar maximum ICMEs are observed at all latitudes, but those above 40° do not have high charge states.

Keywords: interplanetary coronal mass ejections, solar wind, heliosphere

1. Introduction

Coronal mass ejections (CMEs) propel large quantities of solar material outward; the ejected magnetized plasma regions are called interplanetary CMEs (ICMEs). ICMEs are identified by a variety of signatures described elsewhere (Gosling, 1990; Gosling, 2000; Zurbuchen and Richardson, 2006, this volume). They are generally described as flux ropes, which are magnetically connected to the Sun while they are carried outward by the solar wind (e.g., Burlaga, 1988; Bothmer and Schwenn, 1998). Most ICME studies have been conducted near Earth, at 1 AU and within about 7° of the solar equatorial plane, because that is where most spacecraft are located. CMEs are observed at all solar latitudes, especially near solar maximum, so ICMEs should be present at all latitudes as well. ICMEs persist well beyond 1 AU, although as they interact with the ambient solar wind and perhaps lose their magnetic connection to the Sun they become harder to identify.

Merged interaction regions (MIRs) are regions where two or more interaction regions coalesce (Burlaga, 1995). They are generally high magnetic field strength and high-density regions and dominate the plasma structure in the outer heliosphere near solar maximum (Richardson *et al.*, 2003). These MIRs act as barriers for inward transport of energetic particles (Burlaga *et al.*, 1993) and form large pressure pulses which can produce motions of the termination shock (Wang and Belcher, 1999; Zank

Space Science Reviews (2006) 123: 111–126
DOI: 10.1007/s11214-006-9015-z

and Müller, 2003). MIRs form when fast ICMEs, or series of ICMEs, run into the solar wind and ICMEs ahead of them and compress the plasma and magnetic field.

This chapter provides a tutorial on ICME observations beyond 1 AU and at high latitudes. The first half discusses the radial evolution of ICMEs as they travel to the outer heliosphere and is based on the Voyager observations. The second half highlights the Ulysses observations of ICMEs at high latitudes. Since the detailed signatures of ICMEs are discussed elsewhere in this book (Crooker and Horbury, 2006, this volume; Forsyth *et al.*, 2006, this volume; Zurbuchen and Richardson, 2006, this volume), we will focus mostly on the signatures and properties that can be studied in the outer heliosphere, including magnetic field signatures, helium abundance enhancements and kinetic plasma properties (speed, shocks, etc.).

2. Radial Evolution of ICMEs

After CMEs lift off from the solar surface, they propagate outward through the heliosphere. They interact with the solar wind in front of, to the sides of, and behind them. Shocks often form on the ICME boundaries and propagate through the solar wind surrounding the ICME. Faster ICMEs can run into preceding slower ICMEs, merging and/or forming complex ejecta. The ICMEs in the inner heliosphere are generally not in equilibrium with the ambient solar wind. They often have a larger internal (plasma plus magnetic) pressure and a leading edge which is ejected faster than the trailing edge. Both these features lead to expansion of ICMEs with distance. ICMEs are not always simple to identify near 1 AU; the evolution of ICMEs due to both internal and external factors presents challenges for identifying these features as they move outward. This section discusses methods which have been used to identify ICMEs and their effects at places as far distant as the heliopause.

2.1. Identification of ICMEs

A variety of signatures have been used to identify ICMEs; low ion temperatures, alpha (He^{++}) enhancements, bidirectional electron streaming, abundance and charge state anomalies of heavy ion species, leading shocks, and smooth magnetic field rotations (Neugebauer and Goldstein, 1997, and references therein). Out to 5 AU (the aphelion of Ulysses), the spacecraft instrumentation allows all these methods to be used. The spacecraft that have gone beyond 5 AU (the Voyagers and Pioneers) cannot measure counterstreaming electrons or element abundance and charge-state anomalies.

2.2. Case Studies

2.2.1. *Magnetic Clouds*

The first studies of the radial evolution of ICMEs focused on magnetic clouds. Characterized by smooth magnetic field rotation, magnetic clouds, which represent a subset of ICMEs, are relatively easy to identify. The frequency of magnetic clouds observed at large radii decreases with distance from the Sun, suggesting that the magnetic cloud structure decays further out in the heliosphere.

Burlaga *et al.* (1981) compared data from Helios 1 and 2, IMP 8, and Voyagers 1 and 2, in order to examine the radial evolution of a magnetic cloud in early 1978. At the time the Helios and IMP spacecraft were all near 1 AU, while the Voyagers were near 2 AU; all five spacecraft were within a 30° sector in heliolongitude. Despite their varied distances, all of these spacecraft detected a shock, followed by a turbulent sheath region, followed by a magnetic cloud. The pressure inside the cloud was dominated by the magnetic field and was larger than that in the ambient solar wind, so the cloud expanded between 1 and 2 AU. Burlaga and Behannon (1982) identified four magnetic clouds between 2 and 3.5 AU which again had higher than ambient magnetic field strengths and total pressure and lower than ambient density, temperature, and momentum flux. These ICMEs were about twice as large in radial width as those observed at 1 AU, consistent with expansion at roughly half the Alfvén speed (Klein and Burlaga, 1982). Burlaga *et al.* (1985) identified a magnetic cloud at 11 AU with radial width of about 1 AU, consistent with continued expansion, and showed that MIRs formed by these ICMEs modulate the cosmic ray intensities.

The Bastille day event (July 14, 2001) at 1 AU comprised several shocks and two magnetic clouds. Earth and Voyager 2 were separated by 2.6° heliolongitude and 27° heliolatitude. Voyager 2 observed a shock on January 12, 2002; the timing and strength of this shock were consistent with model predictions based on the 1 AU data (Wang *et al.*, 2001). Burlaga *et al.* (2001) showed that data behind this shock have the characteristics of a magnetic cloud, which would make it the most distant magnetic cloud observed. The radial width of this cloud was about 1.8 AU and its duration about 6 days, suggesting substantial expansion of the magnetic cloud outside 1 AU. However, the cloud at Voyager 2 (at 62 AU) was right-handed whereas those at Earth were left-handed, so these were not the same magnetic clouds, but could have resulted from the same set of solar events.

2.2.2. *Helium Enhancements*

One characteristic used to identify ICMEs is the helium abundance; almost all events with $N(\mathrm{He})/N(\mathrm{H}) > 8\%$ are ICMEs (Neugebauer and Goldstein, 1997).

Voyager 2 is able to measure the helium abundance when its value is above the detection level of the instrument. Despite the radial fall of in the density of the solar wind ions, the relative abundance of helium should remain constant (to first order) at large distances from the Sun. As a result, the helium abundance is probably the best signature for tracking ICMEs through the heliosphere. The weakness of this method is that not all ICMEs have enhanced helium abundances, but this weakness can also be an advantage. Since helium enhancements are relatively rare, they can be used to trace ICMEs outward. Paularena *et al.* (2001) first used the technique of comparing helium abundances to trace an ICME from Ulysses to Voyager 2. Starting from the Helium abundance enhancement (HAE) list from Voyager of Wang and Richardson (2001), they looked for counterpart HAEs in the Ulysses data. They identified an event when Ulysses saw an HAE at 5.3 AU and Voyager 2 saw a similar event about seven months later. At Ulysses, the ICME had a leading shock, caused a decrease in cosmic ray intensity, and had increases in the average O and Fe charge states as well as the He^{++} enhancement. At 58 AU, the only measurement suggesting this event was an ICME was the helium abundance; other signatures had been lost as the solar wind evolved. Ulysses and Voyager 2 were at nearly the same heliolatitude but were separated by 130° in longitude, so this ICME had a large longitudinal extent. A 1-D MHD model was used to propagate the observed solar wind from Ulysses to Voyager 2; the timing of the arrival of the ICME at Voyager 2 verified this was the same ICME.

Richardson *et al.* (2002) reported that an ICME in September 1998, which passed Earth two days later, could be identified in the Ulysses data at 5.3 AU and in the Voyager 2 data at 58 AU. Figure 1 shows the helium abundance data from all three spacecraft for this event, where 10 days of data are shown for each spacecraft. The ejecta are identified on the basis of the enhanced He^{++} abundance, although at WIND and Ulysses other ICME signatures were also observed. Comparison with an MHD model shows that these events are likely the same ICME; latitudinal and longitudinal differences in spacecraft location account for the different helium abundances and profiles in the ejecta. The ICME took about 1.5 days to pass WIND at 1 AU and had a width (the duration times the average speed) of about 0.6 AU. The internal pressure, the thermal plus magnetic pressures, of the ICME was much larger than that of the ambient solar wind at 1 AU. As in most ICMEs, the internal pressure was dominated by the magnetic field. This overpressure caused the ICME to expand to the observed duration of 5.2 days and a radial width of 1.3 AU at Ulysses at 5.3 AU. The internal pressure of the ICME had nearly equilibrated with the background solar wind at Ulysses; thus the ICME stopped expanding and at Voyager 2 at 58 AU had a duration of 5.5 days and a radial width of 1.5 AU. The ICME also expands in the perpendicular directions; this expansion has not been quantified as it requires multiple spacecraft.

Burlaga (1995, and references therein) showed an example where five separate solar wind streams observed by Helios 2 at 0.85 AU merged into two MIRs at Voyager 1 at 6.2 AU and then into a single MIR at Pioneer 11 at 9.2 AU. Richardson *et al.* (2003) presented a case study of two ICMEs observed at Ulysses which bracketed a merged interaction region (MIR) at the distance of Voyager 2. Figure 2 shows the evolution of the solar wind structure with distance from 5 to 58 AU as predicted by a 1-D MHD model, including the effects of pickup ions, which slow

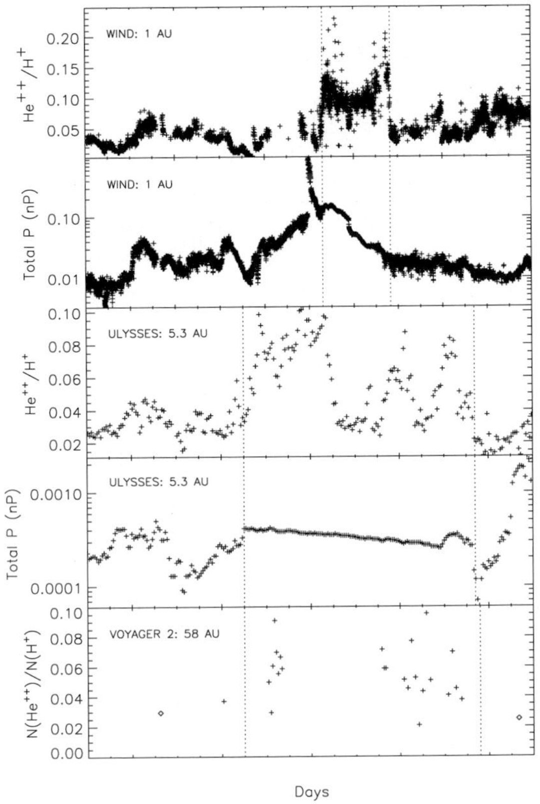

Figure 1. An ICME observed at 1, 5.3, and 58 AU. The times are shifted to align the ICMEs, where the ICME boundaries are determined from the regions of enhanced helium abundance.

the plasma (Wang *et al.*, 2000; Wang and Richardson, 2001). The locations of the two ICMEs are shown by the vertical dashed lines. The top trace shows Ulysses data; the ICME locations are determined by the enhanced helium abundance. These data are the input for the MHD model. The remaining traces show the solar wind densities predicted by the model every 10 AU and at the distance of Voyager 2; features in the density traces are used to track the ICME positions. The bottom panel shows the Voyager 2 density profile and the positions of the observed helium enhancements.

The first helium enhancement (A) is observed a few days after the model prediction; the second event occurs almost exactly where predicted. Both of these predictions are remarkably good given the 7 month solar wind propagation time from Ulysses to Voyager 2. The second ICME (B) starts out about 60 days behind the first ICME (A), but moves faster than the first ICME and at 58 AU is only 30 days behind. The converging ICMEs compress the plasma between them. The density increases by about a factor of two within the MIR and the magnetic field

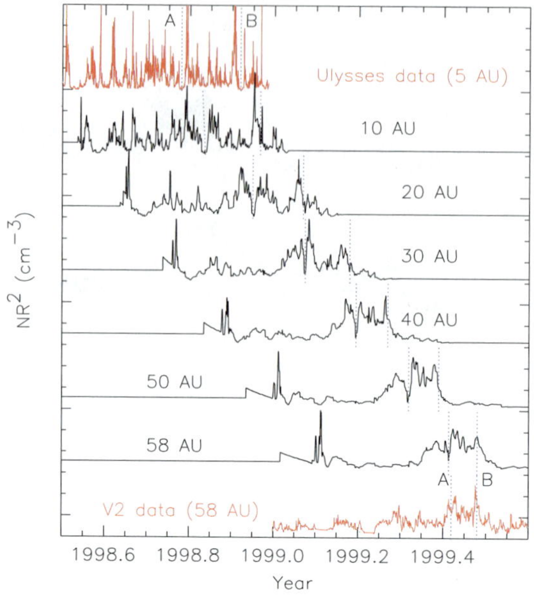

Figure 2. Using an MHD model to track ICMEs through the heliosphere. The top red trace shows Ulysses data, which were the model input, the black traces show the density profiles predicted by the model at various radial distances, and the bottom red trace shows the Voyager 2 data at 58 AU. The vertical blue lines show the locations of the two ICMEs A and B which converge to form a MIR.

strength also increases, the classic signatures of an MIR. This MIR also produced a decrease in the energetic particle fluxes at Voyager 2.

2.2.3. *Shocks*

Another way to trace the effects of ICMEs outward is to follow the fast-mode shocks that often precede them. These shocks propagate through the solar wind at the fast-mode speed, so in the outer heliosphere the shocks are well ahead of the ICME. The shocks often form the leading edges of MIRs and accelerate energetic particles. Several studies have traced shocks outward; the Bastille day ICME, discussed above in the context of magnetic clouds, produced a very strong shock at Earth and occurred when Earth and Voyager 2 were at nearly identical heliolongitudes (Wang *et al.*, 2001). Figure 3 shows the shock at 1 AU and how it evolves (based on the MHD model) until it reaches Voyager 2. The model and data again agree very well; the shock weakens with distance but was still strong enough to produce an enhancement in the >5 MeV/nuc particles.

The Bastille day event was an example of a single large ICME propagating to the outer heliosphere. Probably more common, and more able to produce large effects in the outer heliosphere, are cases where series of ICMEs merge. The top right panel of Figure 3 shows a series of ICMEs observed at Earth in April and May, 2001. The model predictions show that these features merge and by 60 AU

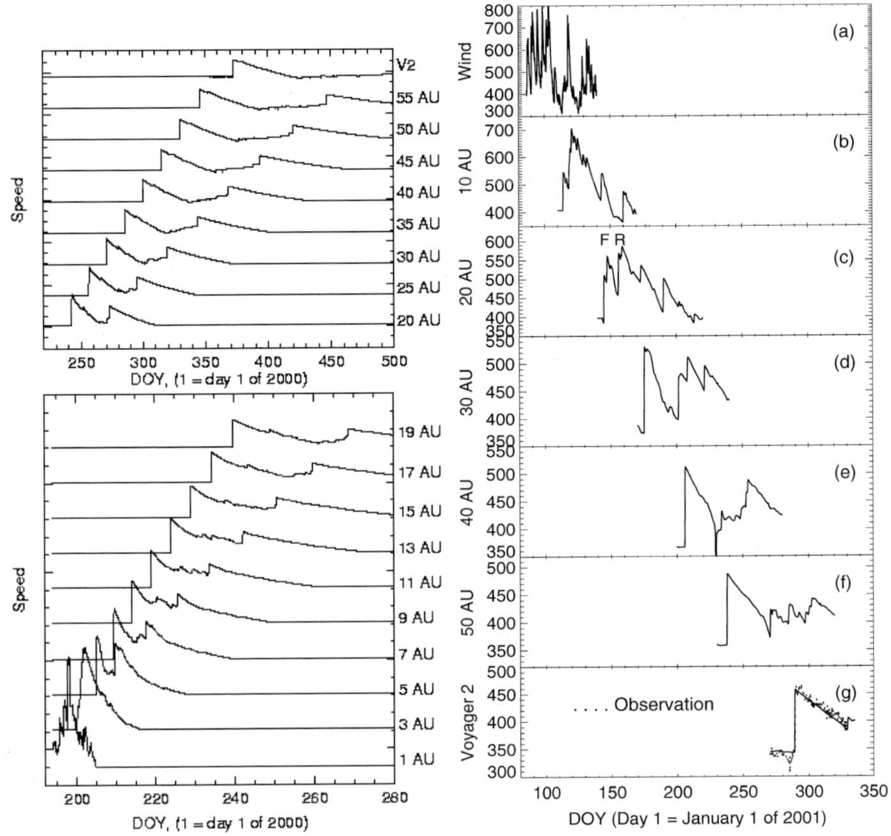

Figure 3. Using a 1-D MHD model to propagate shocks through the heliosphere. The left panel shows the Bastille day ICME, where the 1 AU data (bottom trace) is propagated to the distance of Voyager 2. The right panel shows the results of propagating the evolution of a series of ICMEs from April/May 2001 outward to Voyager 2.

form one large shock; a shock very similar to the model prediction was observed by Voyager 2 in October 2001 and is shown in the bottom panel. Although the individual shocks at 1 AU were weaker than the Bastille day shock, the resulting merged shock was much stronger in the outer heliosphere than the Bastille day shock and a better accelerator of energetic particles. Large ICME-driven shocks may trigger the 2–3 kHz radio emission observed by Voyager 2 in the descending phases of the past three solar cycles. Gurnett *et al.* (2003) suggest that the October 2001 shock triggered the first heliospheric radio emission of this solar cycle, in late 2002, when it reached the heliopause.

2.3. Statistical Studies

These case studies, combined with model results, give an intuitive feel for how ICMEs evolve with distance and how they affect the outer heliosphere. Statistical studies have also been performed looking at ICMEs from 0.3 to 58 AU. Wang and Richardson (2001), following earlier work (Borrini *et al.*, 1982), tabulated Voyager 2 observations where the He/H density ratio was over 10%. They found 56 events where the helium remained enhanced for over 12 hours. Their main results were 1) the solar cycle dependence of helium abundance enhancements (HAEs) persists in the outer heliosphere, 2) HAEs are clustered in time, 3) HAEs had higher speeds than the ambient solar wind, 4) temperatures in HAEs are generally lower than those in the solar wind, and 5) the magnetic field in HAEs is generally higher than that in the ambient solar wind. The difference between the speed, temperature, and magnetic field magnitude in the HAEs and in the ambient solar wind decreased with distance.

Liu *et al.* (2005) identified ICMEs in the Helios 1 and 2, WIND, ACE, and Ulysses data. They required their ICMEs to both meet the low-temperature criterion of Richardson and Cane, Cane and Richardson (1993, 2003) and to have helium abundances over 8%. Wang and Richardson (2004) identified ICMEs in the Voyager 2 data from 1-30 AU; they used the low-temperature criterion as their primary ICME-identifier but corroborated their picks using the other plasma and magnetic field data.

We combine the results from these two lists to investigate radial evolution of ICMEs over the radial range 0.3–30 AU. Figure 4 shows the radial width of 352 ICMEs (the duration of the ICME times the average speed of the ICME) as a function of distance. The average widths over 3 AU bins and the standard deviations in each bin are also shown so that the radial trend is easier to see. ICMEs expand from 0.3 AU to about 15 AU, after which the ICME width is relatively constant. The expansion stops when the ICMEs are in equilibrium with the background solar

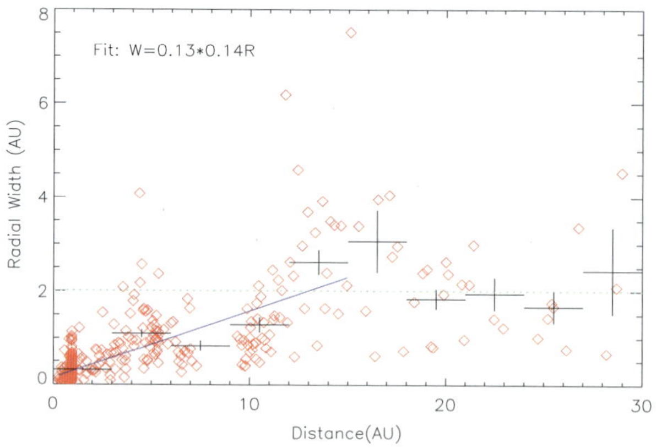

Figure 4. The radial width of observed ICMEs versus distance from the Sun. The diamonds show the widths of individual ICMEs, the black crosses show 3 AU averages of the width with the horizontal bar showing the size of the bin and the vertical bar the errors of the mean, and the blue line shows a linear fit to the data inside 15 AU.

wind; the peak widths near 15 AU may result from over-expansion of the ICMEs. The average radial width of ICMEs increases from about 0.3 AU at 1 AU to 2.5 AU at 15 AU, then averages about 2 AU from 15 to 30 AU.

An associated result is that the speed difference across the ICMEs decreases with distance, from about 75 km/s at 1 AU to near zero outside 20 AU. The expansion speed inside 15 AU is roughly 0.15 times the solar wind speed, or roughly the Alfvén speed (Wang and Richardson, 2004), consistent with the results of Klein and Burlaga (1982). As a result of this expansion, the ICMEs comprise a much larger percentage of the solar wind in the outer heliosphere than at 1 AU. Wang and Richardson (2004) showed that in the descending phase of the solar cycle, when Voyager 2 was at 15–20 AU, almost 40% of the solar wind was ICME material. Gosling *et al.* (1992) showed that at 1 AU near solar maximum, ICME plasma comprised about 15% of the solar wind. An expansion of ICMEs by a factor of 5–6 reconciles these two results.

3. ICMEs at High Heliographic Latitudes

The rate of occurrence of CMEs as a function of position angle (PA) at the Sun is highly dependent on the phase of the solar cycle. Near solar minimum, CMEs are concentrated around PA 90° and 270°, consistent with the location of the solar streamer belt at low latitudes. Conversely, near solar maximum CMEs are observed nearly uniformly at all position angles (Gopalswamy *et al.*, 2006, this volume). Consequently, a similar pattern is expected for ICMEs: They should be confined to low latitudes at solar minimum and occur at all latitudes near solar maximum. In this section we discuss observations of ICMEs at high latitudes in the light of that expectation.

3.1. Solar Minimum Conditions

When the Ulysses spacecraft, after flying by Jupiter in February 1992, traveled to high heliographic latitudes for the first time near solar minimum, it encountered a highly ordered heliosphere. High-speed streams emanating from the polar coronal holes filled the complete solid angle poleward of 30°, while slow, variable solar wind prevailed equatorward of 20° (McComas *et al.*, 1998). It was therefore not a small surprise when Gosling *et al.* (1994) discovered a new class of ICMEs that were fully embedded within the polar fast streams, termed overexpanding ICMEs. They are characterized by a forward-reverse shock pair driven into the ambient fast wind by virtue of their high internal pressure. Six such events were observed in more than two years of polar stream immersion, two of which are reproduced in Figure 5. One event was even observed both at low and at high latitudes (Gosling *et al.*, 1995). Neukomm (1998) investigated the ICME events of cycle 22 observed at Ulysses for the presence of compositional signatures (Zurbuchen and Richardson,

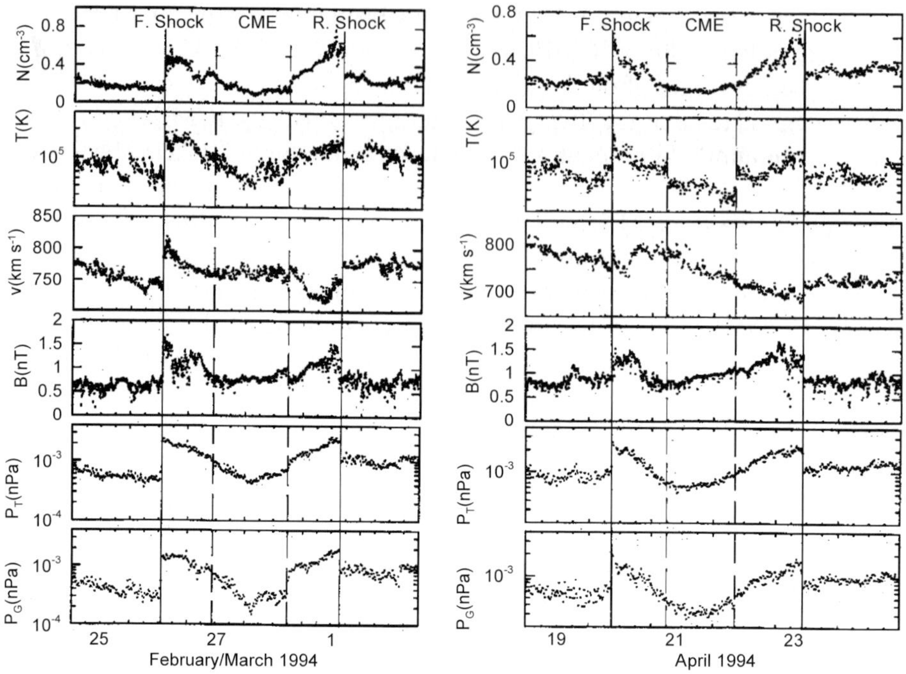

Figure 5. Two examples of overexpanding ICMEs at high latitudes (left: 54° south, right: 61° south), identified in the plasma data and by the presence of counterstreaming electrons. The main feature is the presence of a forward-reverse shock pair that is driven into the ambient (fast) solar wind due to the high internal pressure (figure from Gosling *et al.*, 1994).

2006, this volume), finding that all 6 high-latitude events were indistinguishable from the surrounding fast solar wind in these signatures. From this we may infer that these events represent the wake of an ICME traveling at lower latitudes but not containing genuine, hot CME material that would reveal itself by high charge state temperatures.

3.2. The Rise of Cycle 23

After solar minimum sometime in 1996 the activity cycle #23 started to rise, as illustrated in Figure 6 by McComas *et al.* (2001). The top row shows a series of LASCO C2 images that document the transition of the solar corona from a simpler configuration at solar minimum to a more complex solar maximum configuration with streamers no longer confined to the equator. The middle panel gives the solar wind speed at Ulysses as measured with the SWOOPS instrument. First, the spacecraft was still immersed in the north polar coronal hole, followed by a period of alternating slow and fast wind due to the tilt of the streamer belt combined with the solar rotation. What follows is almost a full year of exceptionally steady slow

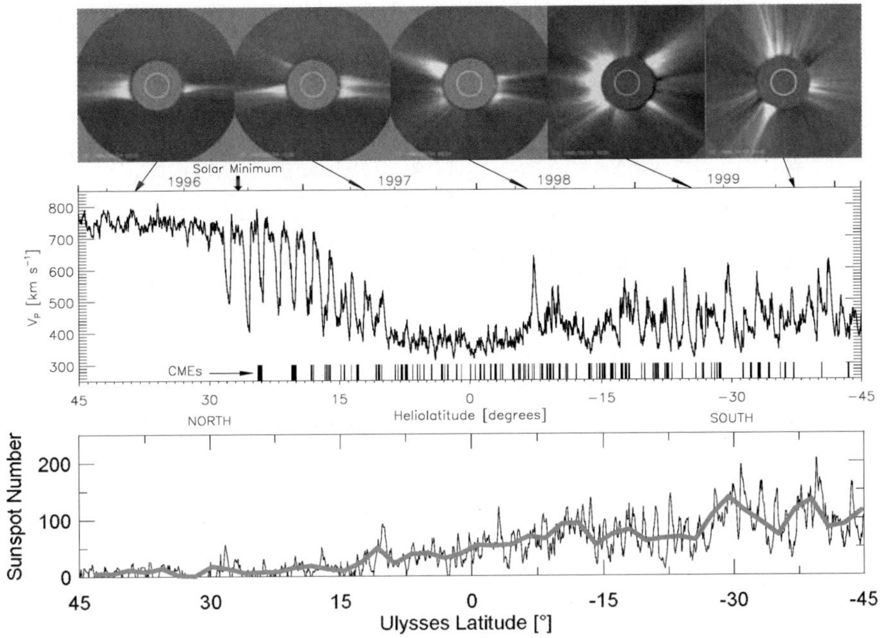

Figure 6. The rise of solar activity cycle #23 as seen from the SOHO-LASCO C2 coronagraph (top row) and Ulysses-SWOOPS (middle panel). The arrival times of ICMEs at Ulysses are marked near the bottom of the middle panel, and the sunspot number is given in the bottom panel. Note the apparent decrease in the ICME rate at Ulysses when it climbs to high southern latitudes even though solar activity continues to rise (figure adapted from McComas *et al.*, 2001).

solar wind until the first fast ICME encountered in May 1998. After that, the solar wind gradually changes from a bimodal, minimum configuration to a continuum of dynamic states (Zurbuchen *et al.*, 2002) typical for solar maximum.

The arrival of ICMEs at Ulysses is marked with vertical bars near the bottom of the middle panel. The ICME rate first increases with time, as expected from the increasing solar activity shown in the bottom panel (with time converted to Ulysses latitude in order to make the panels readily comparable). The surprising feature in this figure is the apparent drop in ICME rate in 1999 despite the fact that solar activity has risen to a broad maximum and remains high throughout that time. Do ICMEs occur less frequently at high latitudes at solar maximum even though CMEs occur uniformly around the solar disk?

3.3. Solar Maximum Conditions

Lepri and Zurbuchen (2004a,b) have investigated the rate of occurrence of a high average iron charge state as an ICME indicator (Lepri *et al.*, 2001; Zurbuchen and Richardson, 2006, this volume) both at Ulysses during the better part of its solar

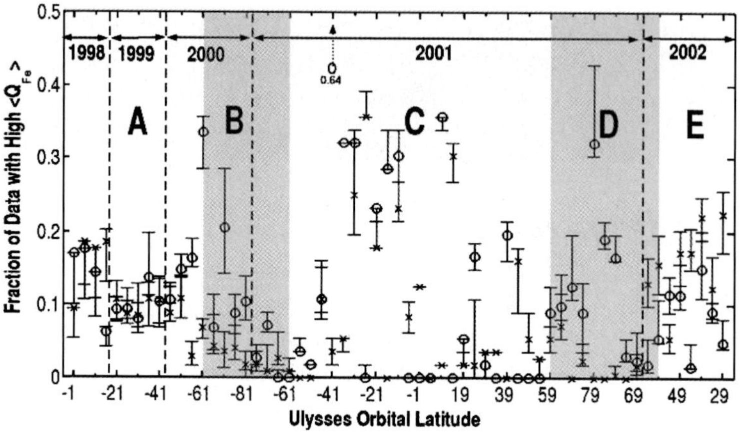

Figure 7. Fraction of the solar wind occupied by ICMEs with a high average iron charge state, $\langle Q_{Fe} \rangle \geq 12$, both at ACE (circles) and at Ulysses (crosses). The shaded periods B and D mark the times when Ulysses was at $>60°$ in latitude. During these periods it encountered a significantly lower number of high charge state ICMEs than ACE did near the ecliptic plane (figure from Lepri and Zurbuchen, 2004a).

maximum polar orbit and at ACE, which stayed at L1 during that time. Their result is summarized in Figure 7. During the shaded periods B and D, when Ulysses was at $>60°$ latitude, the ACE rate (circles) consistently exceeds the Ulysses rate (crosses) by a significant factor. The ACE and Ulysses rates are similar, however, when Ulysses was at low to mid latitudes (periods A, C, and E; the large scatter in period C is due to low statistics, as this period corresponds to the fast latitude scan and thus a $5°$ bin is traversed much more quickly than during the other periods). Of course the rate of high Fe charge states does not translate directly to the ICME rate, as a significant fraction of ICMEs (close to 50%) have charge states resembling the slow solar wind (Neukomm, 1998). The difference is attributed to the magnetic connectivity between the flaring site associated with the CME (if any) and the point of observation: Since active regions (and thus flares) rarely occur poleward of $45°$ even at solar maximum, ICMEs with hot charge state temperatures are rare at high latitudes. Of the 19 ICMEs observed during Ulysses' second fast latitude scan, all 8 with a high charge state temperature (save one marginal case) occur below $40°$ (Forsyth *et al.*, 2003). But overall, the 19 events are distributed more or less uniformly along the Ulysses trajectory between $80°$ north and $80°$ south, so the concentration at low latitudes only applies to ICMEs with a high charge state signature. This result is reminiscent of the apparent concentration of periods with a high first ionization potential (FIP) bias at low latitudes (von Steiger and Zurbuchen, 2002), i.e., strong signatures in both element abundances and charge state ratios are preferentially observed at low to mid latitudes.

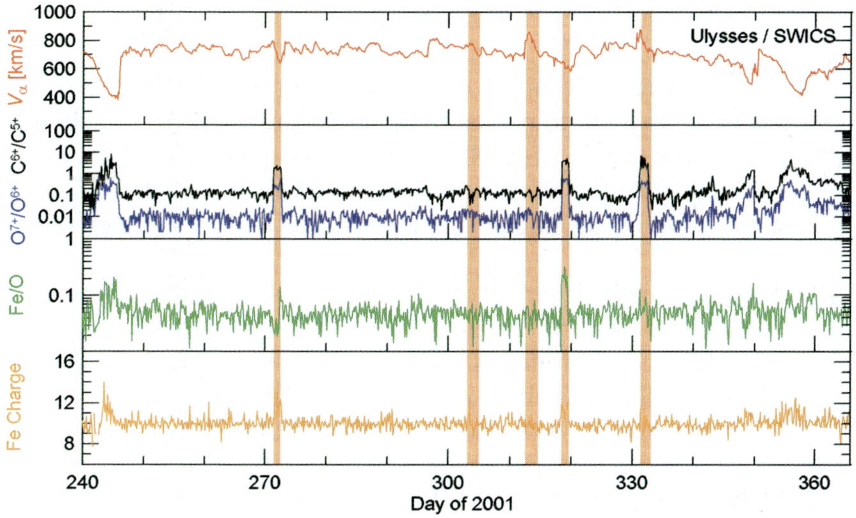

Figure 8. High-latitude ICME periods form the Ulysses-SWOOPS list (shaded bands) plotted over some of the Ulysses-SWICS archive parameters, when Ulysses was poleward of 65° north and immersed in the fast stream of the newly formed polar coronal hole. Composition signatures are easily visible in some of the events but completely absent in others.

Near the end of the fast latitude scan at solar maximum, when Ulysses was at high northern latitudes, it encountered steady, fast solar wind from the newly formed northern polar coronal hole from days 246 to 355, 2001 (McComas *et al.*, 2002). This stream was very similar to the large polar streams encountered on the previous orbit near solar minimum: fast, steady, and with a low charge state temperature, but with the opposite magnetic polarity than the north polar stream of cycle 22, thus clearly belonging to the new cycle. The reason this fast flow persists only a few months was probably that Ulysses traveled to lower latitudes more quickly than the newly formed polar coronal hole expanded. A difference from the solar minimum fast streams was that this new stream was interrupted by five ICMEs in just three months, whereas only six overexpanding ICMEs were observed in over two years in the solar minimum coronal hole flow (Reisenfeld *et al.*, 2003). This difference is illustrated in Figure 8, where the 5 ICMEs on the Ulysses-SWOOPS list at `http://swoops.lanl.gov/cme_list.html` are plotted as shaded bands over the Ulysses-SWICS archive parameters from `http://helio.estec.esa.nl/ulysses/archive/swics.html`. The diversity of compositional signatures seen in just these five ICMEs is striking. One has an extreme Fe/O signature, three have a high charge state temperature, but two do not show any composition signature, just like the overexpanding ICMEs at solar minimum. Note that the high charge state events are easily identified here, in particular in the O^{7+}/O^{6+} ratio, although they do not quite reach the threshold value (von Steiger and Zurbuchen, 2003) because of the low temperature of the surrounding fast stream.

4. Summary

ICMEs have been observed from 0.3 AU out to the distance of Voyager 2 at 70 AU and their effects are thought to persist to the heliopause and beyond. In the inner heliosphere the prime feature of ICME evolution is expansion. ICMEs increase in radial width by, on average, a factor of 5–6 from 1 to 15 AU until their internal pressures match those of the ambient solar wind. Beyond 15 AU they remain at a constant width.

ICMEs have also been observed at all heliographic latitudes, but their latitudinal distribution is solar cycle dependent. At solar minimum only very few and rather special ICMEs can be found at high latitudes within the steady fast streams, which are overexpanding and show no compositional signatures. At solar maximum, ICMEs occur more or less uniformly at all latitudes, but events with high charge states, i.e., from a hot source region, appear to be limited to $<40^\circ$.

The ICMEs and the shocks and MIRs spawned by them have great impact on the outer heliosphere. The solar wind structure at 60–70 AU near solar maximum is dominated by ICME driven MIRs with their associated pressure pulses. The MIRs modulate the cosmic ray intensities (Burlaga *et al.*, 1993) and drive motions of the termination shock which persist up to a year and pressure waves in the heliosheath which can persist for solar cycles (Zank and Müller, 2003). The large shocks preceding global, ICME-driven MIRs may trigger the heliospheric radio emissions when these shocks reach the heliopause (McNutt, 1988; Gurnett *et al.*, 1993), thus extending their influence to well over 100 AU. High-latitude ICMEs are particularly important to make their combined effect a truly global one, as opposed to corotating interaction regions (CIRs) which are limited to low- and mid-latitudes and thus cannot cause a global response of the heliosphere.

Acknowledgements

The work at MIT was supported under NASA contract 959203 from JPL to MIT.

References

Borrini, G., Gosling, J. T., Bame, S. J., and Feldman, W. C.: 1982, *J. Geophys. Res.* **87**, 7370–7378.
Bothmer, V., and Schwenn, R.: 1998, *Ann. Geophys.* **16**, 1–24.
Burlaga, L., and Behannon, K. W.: 1982, *Sol. Phys.* **81**, 181.
Burlaga, L., Sittler, E., Mariani, F., and Schwenn, R.: 1981, *J. Geophys. Res.* **86**, 6673–6684.
Burlaga, L. F.: 1988, *J. Geophys. Res.* **93**, 7217–7224.
Burlaga, L. F.: 1995, *Interplanetary Magnetohydrodynamics*. Oxford University Press, New York.
Burlaga, L. F., Goldstein, M. L., McDonald, F. B., and Lazarus, A. J.: 1985, *J. Geophys. Res.* **90**, 12,027–+.
Burlaga, L. F., McDonald, F. B., and Ness, N. F.: 1993, *J. Geophys. Res.* **98**, 1–11.

Burlaga, L. F., Ness, N. F., Richardson, J. D., and Lepping, R. P.: 2001, *Sol. Phys.* **204**, 399–411.
Cane, H. V., and Richardson, I. G.: 2003, *J. Geophys. Res. (Space Phys.)* **108**(A4), doi:10.1029/2002JA009817.
Crooker, N. U., and Horbury, T. S.: 2006, *Space Sci. Rev.*, this volume, doi: 10.1007/s11214-006-9014-0.
Forsyth, R. J., Rees, A., Reisenfeld, D. B., Lepri, S. T., and Zurbuchen, T. H.: 2003, Velli, M., Bruno, R., and Malara, F. (eds.), *AIP Conf. Proc. 679: Solar Wind Ten*, pp. 715–720.
Forsyth, R. J., Bothmer, V., *et al.*: 2006, *Space Science Reviews*, this volume, doi: 10.1007/s11214-006-9022-0.
Gopalswamy, N., Mikic, Z., *et al.*: 2006, *Space Sci. Rev.*, this volume, doi: 10.1007/s11214-006-9020-2.
Gosling, J. T.: 1990, *Washington DC Am. Geophys. Union Geophys. Monogr. Series* **58**, 343–364.
Gosling, J. T.: 2000, Dingus, B. L., Kieda, D. B., and Salamon, M. H. (eds.), *AIP Conf. Proc. 516: 26th International Cosmic Ray Conference, ICRC XXVI*, pp. 59–+.
Gosling, J. T., McComas, D. J., Phillips, J. L., and Bame, S. J.: 1992, *J. Geophys. Res.* **97**, 6531–6535.
Gosling, J. T., McComas, D. J., Phillips, J. L., Pizzo, V. J., Goldstein, B. E., Forsyth, R. J., *et al.*: 1995, *Geophys. Res. Lett.* **22**, 1753–1756.
Gosling, J. T., McComas, D. J., Phillips, J. L., Weiss, L. A., Pizzo, V. J., Goldstein, B. E., *et al.*: 1994, *Geophys. Res. Lett.* **21**, 2271–2274.
Gurnett, D. A., Kurth, W. S., Allendorf, S. C., and Poynter, R. L.: 1993, *Sci.* **262**, 199–203.
Gurnett, D. A., Kurth, W. S., and Stone, E. C.: 2003, *Geophys. Res. Lett.* **30**, doi:10.1029/2003GL018514.
Klein, L. W., and Burlaga, L. F.: 1982, *J. Geophys. Res.* **87**, 613–624.
Lepri, S. T., and Zurbuchen, T. H.: 2004a, *J. Geophys. Res. (Space Phys.)* **109**(A18), A06101, doi:10.1029/2004JA010455.
Lepri, S. T., and Zurbuchen, T. H.: 2004b, *J. Geophys. Res. (Space Phys.)* **109**(A18), A01112, doi:10.1029/2003/JA009954.
Lepri, S. T., Zurbuchen, T. H., Fisk, L. A., Richardson, I. G., Cane, H. V., and Gloeckler, G.: 2001, *J. Geophys. Res.* **106**, 29,231–29,238.
Liu, Y., Richardson, J. D., and Belcher, J. W.: 2005, *Planet and Space Sci.* **53**, 3–17.
McComas, D. J., Bame, S. J., Barraclough, B. L., Feldman, W. C., Funsten, H. O., Gosling, J. T., *et al.*: 1998, *Geophys. Res. Lett.* **25**, 1–4.
McComas, D. J., Elliott, H. A., Gosling, J. T., Reisenfeld, D. B., Skoug, R. M., Goldstein, B. E., *et al.*: 2002, *Geophys. Res. Lett.* **29**(9), 1290. doi:10.1029/2001GL014164.
McComas, D. J., Goldstein, R., Gosling, J. T., and Skoug, R. M.: 2001, *Space Sci. Rev.* **97**, 99–103.
McNutt, R. L.: 1988, *Geophys. Res. Lett.* **15**, 1307–1310.
Neugebauer, M., and Goldstein, R.: 1997, Crooker, N., Joselyn, J. A., and Feynman, J. (eds.), *Coronal Mass Ejections*, vol. 99 of *Geophysical Monograph*, pp. 245–251.
Neukomm, R. O.: 1998, Ph. D. thesis, Universität Bern.
Paularena, K. I., Wang, C., von Steiger, R., and Heber, B.: 2001, *Geophys. Res. Lett.* **26**, 2755–2758.
Reisenfeld, D. B., Gosling, J. T., Steinberg, J. T., Riley, P., Forsyth, R. J., and Cyr, O. C. S.: 2003, *AIP Conf. Proc. 679: Solar Wind Ten.*, pp. 210–213.
Richardson, I. G., and Cane, H. V.: 1993, *J. Geophys. Res.* **98**, 15295–+.
Richardson, J. D., Paularena, K. I., Wang, C., and Burlaga, L. F.: 2002, *J. Geophys. Res. (Space Phys.)* **107**(A4), doi:10.1029/2001JA000175.
Richardson, J. D., Wang, C., and Burlaga, L. F.: 2003, *Geophys. Res. Lett.* **30**, 2207, doi:10.1029/2003GL018253.
von Steiger, R., and Zurbuchen, T. H.: 2002, *AGU Fall Meeting Abstracts* p. A515.
von Steiger, R., and Zurbuchen, T. H.: 2003, *Solar Variability as an Input to the Earth's Environment*, vol. 535 of *ESA SP*, pp. 835–840.

Wang, C., and Belcher, J. W.: 1999, *J. Geophys. Res.* **104**, 549–556.
Wang, C., and Richardson, J. D.: 2001, *J. Geophys. Res.* **106**, 5683–5692.
Wang, C., and Richardson, J. D.: 2004, *J. Geophys. Res. (Space Phys.)* **109**(A18), A06104, doi:10.1029/2004JA010379.
Wang, C., Richardson, J. D., and Burlaga, L.: 2001, *Sol. Phys.* **204**, 413–423.
Wang, C., Richardson, J. D., and Gosling, J. T.: 2000, *Geophys. Res. Lett.* **27**, 2429–2432.
Zank, G. P., and Müller, H.-R.: 2003, *J. Geophys. Res. (Space Phys.)* **108**(A6), doi:10.1029/2002JA009689.
Zurbuchen, T. H., Fisk, L. A., Gloeckler, G., and von Steiger, R.: 2002, *Geophys. Res. Lett.* **29**, 66-1, doi:10.1029/2001GL013946.
Zurbuchen, T. H., and Richardson, I. G.: 2006, *Space Sci. Rev.*, this volume, doi: 10.1007/s11214-006-9010-4.

CORONAL OBSERVATIONS OF CMEs

Report of Working Group A

R. SCHWENN[1,*], J. C. RAYMOND[2], D. ALEXANDER[3], A. CIARAVELLA[1,4], N. GOPALSWAMY[5], R. HOWARD[6], H. HUDSON[7], P. KAUFMANN[8,9], A. KLASSEN[10], D. MAIA[11], G. MUNOZ-MARTINEZ[12], M. PICK[13], M. REINER[5], N. SRIVASTAVA[14], D. TRIPATHI[1], A. VOURLIDAS[6], Y.-M. WANG[6] and J. ZHANG[5]

[1]*Max-Planck-Institut für Sonnensystemforschung Katlenburg-Lindau, Germany*
[2]*Center for Astrophysics, Cambridge, MA, USA*
[3]*Dept. of Physics and Astronomy, Rice University, Houston, TX, USA*
[4]*INAF Osservatorio Astronomico di Palermo, Palermo, Italy*
[5]*School of Computational Sciences, George Mason University, Fairfax, VA, USA; NASA GSFC, Lab. for Extraterrestrial Physics, Greenbelt, MD, USA*
[6]*US Naval Research Laboratory, Washington, DC, USA*
[7]*Space Sciences Laboratory, University of California, Berkeley, CA, USA*
[8]*Universidade Presbiteriana Mackenzie, CRAAM, Sao Paulo, SP, Brazil*
[9]*Universidade Estadual de Campinas, CCS, Campinas, SP, Brazil*
[10]*Astrophysikalisches Institut Potsdam, Potsdam, Germany*
[11]*CICGE, Observatório Astronómico Professor Manuel de Barros, Faculdade de Ciências da Universidade do Porto, Vila nova de Gaia, Portugal*
[12]*Instituto de Geofísica, UNAM, Mexico*
[13]*LESIA, UMR 8109 CNRS, Observatoire de Paris, Meudon, France*
[14]*Udaipur Solar Observatory, Physical Research Laboratory, Udaipur, India*
*(*Author for correspondence, E-mail: schwenn@linmpi.mpg.de)*

(Received 9 January 2006; Accepted in final form 15 March 2006)

Abstract. CMEs have been observed for over 30 years with a wide variety of instruments. It is now possible to derive detailed and quantitative information on CME morphology, velocity, acceleration and mass. Flares associated with CMEs are observed in X-rays, and several different radio signatures are also seen. Optical and UV spectra of CMEs both on the disk and at the limb provide velocities along the line of sight and diagnostics for temperature, density and composition. From the vast quantity of data we attempt to synthesize the current state of knowledge of the properties of CMEs, along with some specific observed characteristics that illuminate the physical processes occurring during CME eruption. These include the common three-part structures of CMEs, which is generally attributed to compressed material at the leading edge, a low-density magnetic bubble and dense prominence gas. Signatures of shock waves are seen, but the location of these shocks relative to the other structures and the occurrence rate at the heights where Solar Energetic Particles are produced remains controversial. The relationships among CMEs, Moreton waves, EIT waves, and EUV dimming are also cloudy. The close connection between CMEs and flares suggests that magnetic reconnection plays an important role in CME eruption and evolution. We discuss the evidence for reconnection in current sheets from white-light, X-ray, radio and UV observations. Finally, we summarize the requirements for future instrumentation that might answer the outstanding questions and the opportunities that new space-based and ground-based observatories will provide in the future.

Keywords: solar corona, eruptive prominences, coronal mass ejections (CMEs), flares, solar wind, solar magnetic field, magnetic reconnection, interplanetary shock waves, ICMEs, space weather, solar energetic particles (SEPs), radio bursts

Space Science Reviews (2006) 123: 127–176
DOI: 10.1007/s11214-006-9016-y

1. Introduction

CMEs have been observed for over 30 years with a wide variety of instruments. White light coronagraphs in space provide the bulk of the observations. The examples shown in Figure 1 demonstrate how the capabilities of coronagraphs have developed over the years and the state of maturity they have reached. It is now possible to derive detailed and quantitative information on CME morphology, velocity, acceleration and mass.

Flares associated with CMEs are observed in white light, the Hα emission line, extreme ultraviolet light (EUV) and X-rays, and several different radio signatures are also seen. Optical and UV spectra of CMEs both on the disk and at the limb provide velocities along the line of sight and diagnostics for temperature, density and composition.

From the vast quantity of data available we attempt to synthesize the current state of knowledge of the properties of CMEs, along with some specific observed characteristics that illuminate the physical processes occurring during CME eruption. These include the common three-part structure of many CMEs, which is generally attributed to compressed material at the leading edge, a low density magnetic bubble and dense prominence gas. Signatures of shock waves are seen, but the locations of these shocks relative to the other structures and the occurrence rates at the heights where Solar Energetic Particles (SEPs) are produced remain controversial. The relationships among CMEs, Moreton waves, EIT waves (as discovered by the Extreme Ultraviolet Imaging Telescope on the Solar and Heliospheric Observatory, SOHO) and EUV dimming are also cloudy.

The close connection between CMEs and flares suggests that magnetic reconnection may play an important role in CME eruption and evolution. We discuss the evidence for reconnection in current sheets from white light, X-ray, radio and UV observations. Finally, we summarize the requirements for future instrumentation that might answer the outstanding questions and the opportunities that new space-based and ground-based observatories will provide in the near future.

2. Available Observations

2.1. Space-Based Coronagraphs

CMEs were discovered and have been mainly studied by space-based coronagraphs. Many observations described in this book were obtained by the Large Angle and Spectrometric Coronagraphs (LASCO; Brueckner *et al.*, 1995) operating aboard SOHO (Domingo *et al.*, 1995). LASCO is a wide-field white light and spectrometric coronagraph consisting of three optical systems having nested fields of view that together observe the solar corona from just above the limb at 1.1 $R_\odot$ out to very great

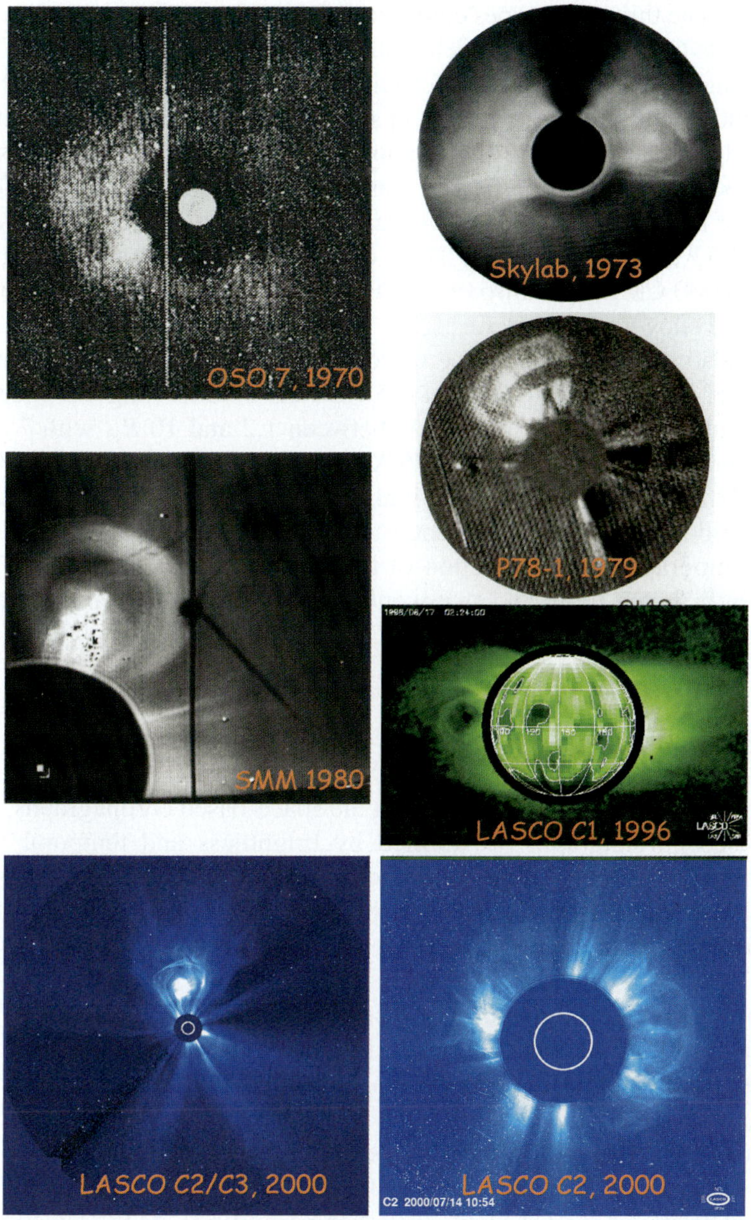

Figure 1. Examples of CME observations from space, from the first one (seen from OSO-7 on Dec. 14, 1971 by Brueckner *et al.* (1972, 1973)) to the still-ongoing LASCO. The lower right panel shows a "halo" CME pointed along the Sun-Earth line.

elongations. The three telescopes comprising LASCO are designated C1 (1.1 to 3.0 $R_\odot$), C2 (2.2 to 6.0 $R_\odot$) and C3, which spans the outer corona (4 to 32 $R_\odot$). C1 is fitted with an imaging Fabry-Perot interferometer, making possible spatially resolved high-resolution coronal spectroscopy in selected spectral lines. Unfortunately, C1 was severely damaged during the temporary loss of SOHO and no observations have been possible since June 1998. C2 and C3 mostly operate in synoptic mode at a cadence of about 24 min (C2) and 45 min (C3). The resolution is 11.9 arcsec/pix for C2 and 53 arcsec/pix for C3. Their observations have resulted in a database of about 10,000 CMEs (Yashiro *et al.*, 2004) as of mid 2005. Some characteristic examples are shown in Figure 1.

SOHO also carries another type of coronagraph, the Ultraviolet Coronagraph Spectrometer (UVCS; Kohl *et al.*, 1995). UVCS obtains long slit spectroscopic measurements of the corona at heights between 1.2 and 10 $R_\odot$ with 7 arcsec/pix resolution in a variety of lines (HI 1216 Å, OVI 1032, 1037 Å, Mg X 610/625 Å, Si XII 499/521 Å, Fe XII 1242 Å), and it also has a visible light channel. UVCS can rotate its field of view about Sun center to build maps of the full corona. Figure 2 shows a tomographic reconstruction of the coronal emission in the O VI λ1032 line as seen from the Earth and as it would be seen if the Sun's axis were tilted by increments of 40° (Panasyuk, 1999).

2.2. Ground-Based Coronagraphs

Ground-based coronagraphs complement the space-based coronagraphs. Although ground-based coronagraphs are limited by brightness and temporal variability of the sky, they allow a higher temporal resolution. This facilitates the understanding of the trigger mechanism of fast transient phenomena like CMEs and coronal waves in the lower corona as well as the density and temperature structure.

Ground-based white-light coronal observations can be obtained only for a small range of heliocentric heights and only with polarization measurements to remove most of the unpolarized sky background. Routine observations have been carried out at the Mauna Loa Solar Observatory (MLSO) for many years. The current coronagraph, Mark IV, has been operational since 1998. It records polarized brightness images of the corona from 1.08–2.85 $R_\odot$ every 3 minutes. The observations run from 17–22 UT daily (weather permitting). Mark IV complements nicely the LASCO observations by providing observations of the inner corona (Figure 3).

Finally, there are several ground-based spectroscopic coronagraph instruments operating within a single bandpass such as the Hα coronagraphs at Pic du Midi and MLSO, the [Fe XIV] and [Fe X] instruments in Sac Peak and Norikura solar observatories, the HeI coronagraph at MLSO, the MICA instrument in Argentina (Stenborg *et al.*, 2000) and some others.

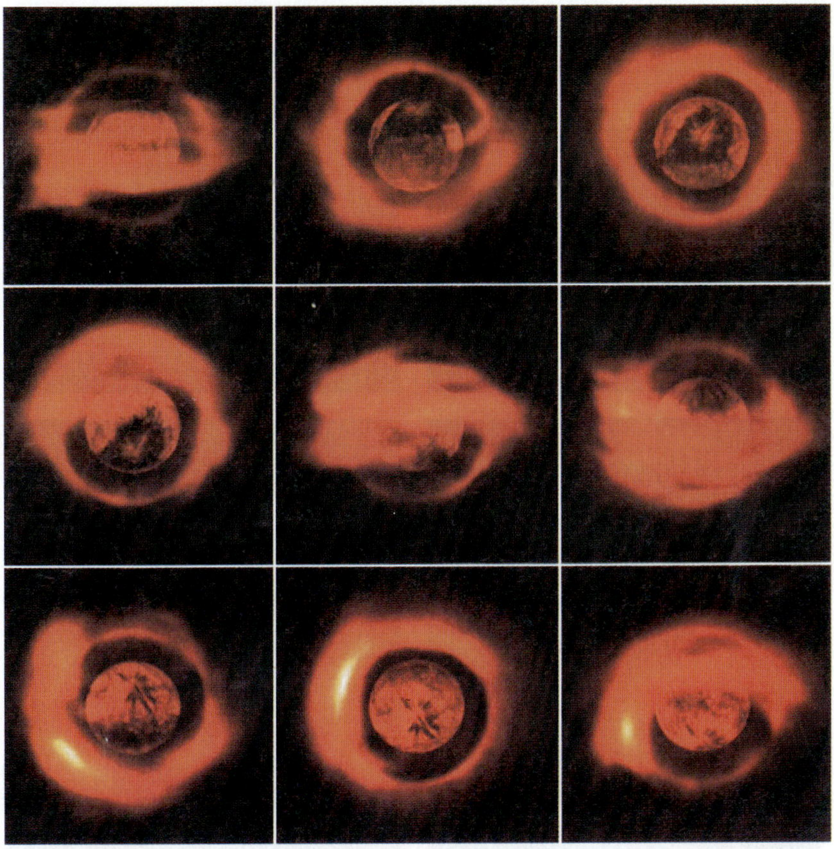

Figure 2. Tomographic reconstruction of the quiet Sun O VI λ1032 emissivity derived from UVCS observations during Whole Sun Month 1996 (Panasyuk, 1999). Successive panels left to right across each row show the corona as seen if the Sun's axis were tilted in the plane containing the Earth and the solar axis by increments of 40°. EIT brightness for this Carrington rotation is shown in the center.

2.3. EIT/TRACE, EUV SPECTROGRAPHS

Since the launch of SOHO, EIT (Delaboudinière *et al.*, 1995) has provided almost continuous full Sun images with 2.6″ pixels in 4 narrow EUV bands. The four bands reveal structures primarily in Fe IX, Fe XII, Fe XV and He II, covering a temperature range 10^5 to 2×10^6 K. In some circumstances other ions may dominate, as when Fe XXIV appears in flares or O V becomes strong in CMEs. TRACE (Handy *et al.*, 1999) has higher spatial resolution, with 0.5″ pixels, providing images such as that in Figure 9, and a much higher time cadence. It has 3 EUV bands similar to those of EIT, along with UV bands. The SPIRIT experiment on CORONAS-F produces images in EUV bands similar to those of EIT, along with images in the Mg XII Lyα X-ray line (Zhitnik *et al.*, 2003a).

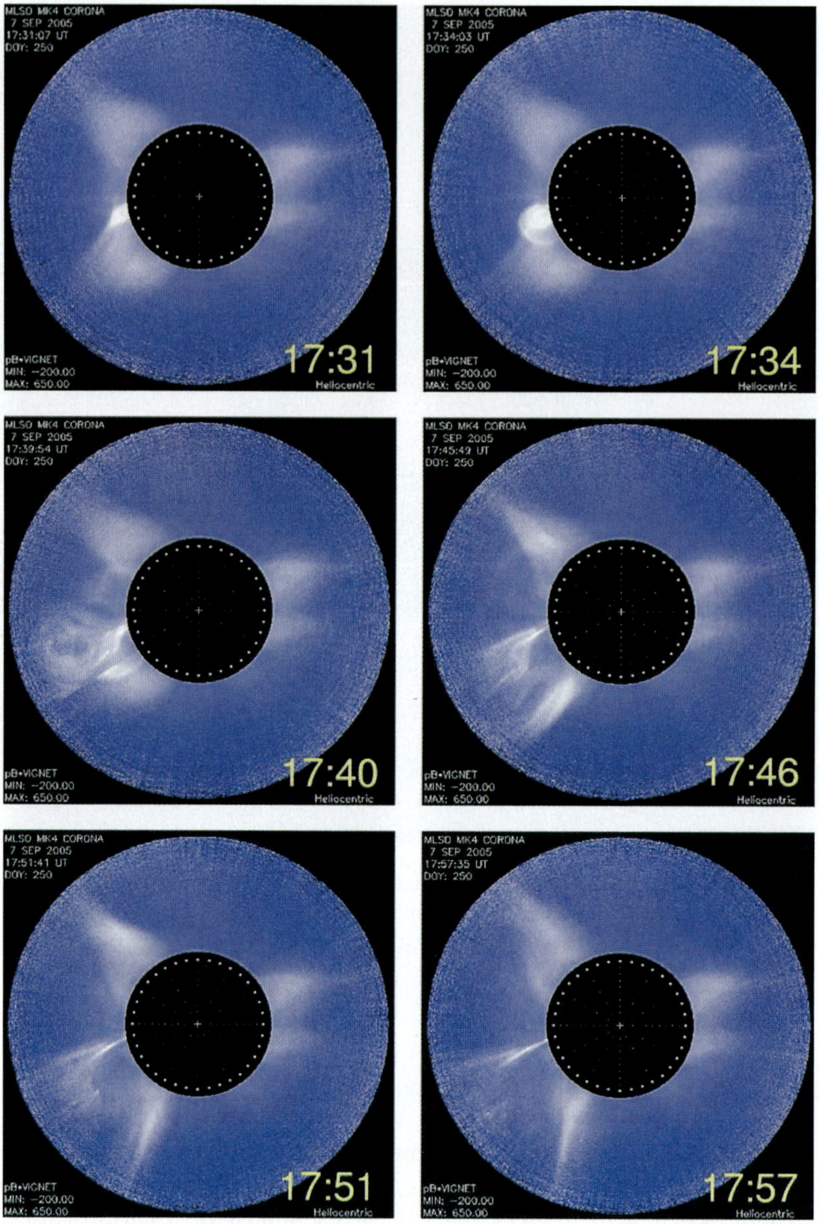

Figure 3. The big CME on September 7, 2005 as seen by the Mark IV coronagraph on Mauna Loa in Hawaii. The CME was accompanied by an X17 flare at S06E89 as reported by NOAA with a peak flux at 17:40. EIT and LASCO on SOHO were switched off because of an attitude maneuver. Despite the CME's origin at the east limb, the Earth was hit by a shock wave and a succeeding magnetic cloud that caused a Kp9 geomagnetic storm on September 11.

Three UV/EUV spectrographs currently operating are SUMER (Wilhelm *et al.*, 1995) and CDS (Harrison *et al.*, 1995) on the SOHO spacecraft and the SPIRIT experiment on CORONAS-F (Zhitnik *et al.*, 2003b). These instruments observe spectral ranges from 151 to 1610 Å, providing many lines useful as temperature, density and abundance diagnostics.

2.4. MDI

The Michelson-Doppler Imager (MDI) is part of the SOHO Solar Oscillations Investigation (SOI; Scherrer *et al.*, 1995). MDI provides full-Sun line-of-sight measurements of the photospheric magnetic field at 4″ resolution with a typical cadence of 96 minutes. A high resolution partial-Sun field of view has a 1.25″ resolution across a field-of view 11′ wide. Such measurements at lower cadence and with less spatial resolution are also provided by ground-based observatories such as Kitt Peak National Observatory (near Tucson, USA) and Wilcox Solar Observatory (near Los Angeles, USA).

MDI has contributed successfully to the understanding of CME initiation and propagation by providing the necessary global magnetic field context within which the CMEs occur. Knowledge of the time variability and distribution of the magnetic field on the solar surface is required as boundary conditions for models of the global solar corona through which the CMEs propagate (e.g., Linker *et al.*, 1999; Luhmann *et al.*, 2003) and for input to MHD models of the eruption process.

2.5. Soft X-Ray Imaging

X-ray light curves from GOES are the standard means for comparing CME and flare characteristics, and imaging data provide still stronger constraints. Soft X-ray imaging of the solar corona began in 1960 via a primitive pinhole camera on a sounding rocket; by the 1970s routine rocket-borne focusing-optics telescopes led to the revolutionary imaging from Skylab (e.g. Vaiana *et al.*, 1973). From 1991 the soft X-ray telescope SXT on *Yohkoh* began systematic CCD-based observations with excellent resolution, image cadence, and image dynamic range. Following *Yohkoh's* demise in 2001, the first NOAA soft X-ray imager (SXI on GOES-12; Hill *et al.*, 2005) began observations, and a sequence of follow-on SXI instruments will continue on future GOES spacecraft. Solar-B (2006 launch) will have an advanced soft X-ray telescope.

The soft X-ray images, as opposed to EUV images, give us a more broadband view of the optically thin corona. Phenomena that are well-observed by these instruments include coronal holes, filament channels, active regions, microflares, X-ray bright points, X-ray dimmings, sigmoids, trans-equatorial loops, hot ejecta, X-ray jets, flare loops, loop-top brightenings, *above*-the-loop-top sources, supra-arcade downflows, flare footpoints, Moreton-type shock fronts, and of course lovely flare arcades and spectacular cusp-shaped structures above them.

2.6. Hard X-Ray Imaging

At energies above a few keV, focusing optics have thus far been difficult to implement. Only these hard X-rays, though, can properly reveal particle acceleration and energy release in the low corona. Accordingly HXIS (on SMM), SXT on (*Hinotori*), HXT (on *Yohkoh*) and RHESSI (Lin *et al.*, 2002) have resorted to shadow-mask imaging, to provide arc-second resolution for hard X-rays. In the case of RHESSI even some γ-ray imaging has been possible.

The new data reveal a broad range of coronal sources as well as the footpoint sources of the flare impulsive phase. All of these sources require particle acceleration to high energies, and the particle acceleration produces a hard X-ray signature characteristic of CME sources (Kiplinger, 1995). The non-thermal energy release turns out to dominate even the weakest microflares as well. The signatures of solar particle acceleration are sometimes accompanied by SEP events, giving us the possibility of thus directly identifying the solar origin of the interplanetary field lines containing the SEP particles. Finally, and surprisingly, the RHESSI data show that ions and electrons occupy *different* flaring loops.

2.7. Ground-Based Emission Line Observations

From the ground, various features of mass ejections can be observed at visible and IR wavelengths. Such data help us to study in detail the pre-eruptive scenario to identify the chain of events and the conditions leading up to the eruption or the CME. A few important spectral lines include:

1. Hα: The Hα line has been used to record solar eruptions for a very long time. The lower chromosphere is the coolest layer in the Sun's atmosphere, and prominent features associated with solar activity such as flares and filaments are best observed in this line. In Hα images, the counterpart of the bright knot of the three part structure is observed, which corresponds to the features cooler than the frontal edge of the CMEs. Hα filaments on the disk are seen in absorption, but when they reach the solar limb and extend beyond it, they are seen in emission and are called prominences. When a filament in the course of a CME begins to rise, the Hα emission is Doppler shifted and may no longer be visible in narrow band images. These are the "Disappearing Filaments" (DFs; see Figure 4).
2. Ca II K: Because the Calcium K Line (393.3 nm) is sensitive to magnetic fields, magnetically active structures show up in high contrast against the surrounding chromosphere. Regions of moderate magnetic field appear bright, whereas high magnetic field regions are dark. In CaK images, one is able to see the brightness

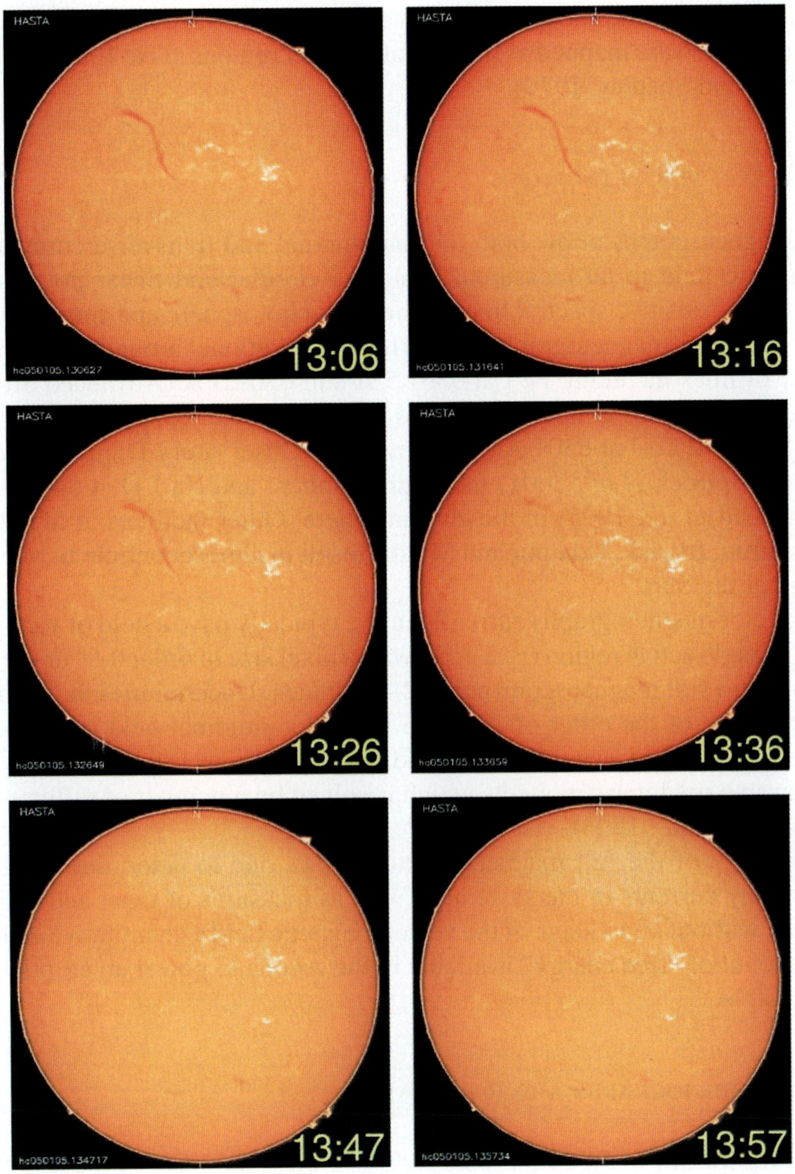

Figure 4. A disappearing filament in the northeast observed by the HASTA-telescope on January 5, 2005. It was associated with a minor X-ray brightening (B1.5) and a full halo CME. No associated shock was noted at the Earth, but a geomagnetic storm (Kp7) occurred late on January 7.

along the edges of large convection cells called supergranules and in areas called plages. Dark sunspots and filaments are also visible in this wavelength.

3. He 10830: The He 10830 line is very useful for indirect observations of the corona. The conditions in the overlying corona are mainly responsible for the

excitation of the chromospheric helium. Coronal holes are marginally detectable in this line and are mapped as somewhat brighter regions in He 1083.0 nm images (Harvey and Sheeley, 1979).

2.8. Vector Magnetographs

Vector magnetographs allow both the longitudinal and transverse components of the magnetic field to be measured, using the circular and linear polarization of magnetically-sensitive spectral lines in the solar photosphere and chromosphere. A number of vector magnetographs are currently operational around the world using a variety of lines including Fe I at 6301.5 Å and 6302.5 Å (Advanced Stokes Polarimeter, Skumanich *et al.*, 1997; Mees Imaging Vector Magnetograph, Mickey *et al.*, 1996) and Ca I at 6301 Å (Big Bear Solar Observatory Digital Vector Magnetograph, Spirock *et al.*, 2001) in the photosphere and Na I D at 5896 Å (Mees IVM, Metcalf *et al.*, 1995) in the chromosphere. Other facilities at MSFC, Potsdam, Huairou, Irkutsk, Udaipur and Mitaka result in almost continuous vector field coverage of the Sun.

The vector magnetographs currently in use typically have a field of view encompassing a single active region ($\sim 5' \times 5'$) with a pixel size of order $0.5''$ or greater and can generate a full magnetogram every 2–3 minutes. Uncertainties in the magnetic field tend to be of the order of 10–20 G in the longitudinal field and 25–50 G in the transverse field measurements. In addition, there is an 180° ambiguity issue in the transverse field measurements which is solved by a number of different means (e.g. Canfield *et al.*, 1993).

Knowledge of the full magnetic vector allows one to determine the size and nature of any currents in the system, as well as measures of twist and helicity. In particular, without knowledge of the vector magnetic field a quantitative assessment of the amount of "free energy" available in the corona to power an eruption would be impossible.

2.9. Radioheliographs and Radio Arrays

The radio imaging instruments map the corona over a range of altitudes depending on the observing frequencies. They provide observations of prominences, microwave activity during flares, coronal and CME-driven shocks.

There are only a very few dedicated solar imaging instruments currently under operation. They include the Nobeyama Radioheliograph (NoRH; Nakajima *et al.*, 1994; 17 and 34 GHz), the Owens Valley Radio Observatory (OVRO; Gary and Hurford, 1990; 1–18 GHz), the Nançay Radioheliograph (NRH; Kerdraon and Delouis, 1997; 450–150 MHz), the Gauribidanour Radioheliograph (Ramesh *et al.*, 1998; 40–150 MHz), the Siberian Solar Radio Telescope (SSRT, Zandanov *et al.*, 1999; 5.7 GHz), and the Ratan-600 radio telescope (Bogod *et al.*, 1998) (610 MHz

– 30 GHz). Large non-solar dedicated arrays, such as the Very Large Array (VLA) (operating at microwave frequencies and more recently at 75 MHz) in the United States (Erickson *et al.*, 2000) and the Giant Meterwave Radiotelescope (GMRT) operating at 327 MHz, 236 MHz, 600 MHz and 1420 MHz (Rao *et al.*, 1995; Swarup, 2000) in India provide occasional observations.

2.10. Submillimeter-Wave Solar Radio Astronomy

A new tool to observe the Sun in the submillimeter range of wavelengths has become available, operating on a daily basis since 2002 at El Leoncito observatory in the Argentina Andes (Kaufmann *et al.*, 2001). It has shown the association between the launch time of CMEs and the onset of the new kind of rapid subsecond pulsating bursts, discovered at 212 and 405 GHz (Kaufmann *et al.*, 2003).

3. CME Properties

White-light coronagraph images make it possible to derive the column density of CME plasma, and radio observations allow us to measure the density. Ultraviolet spectra from SUMER and UVCS provide the means to measure density, temperature, elemental composition and ionization state.

3.1. Statistical Properties

CMEs are characterized by speed, angular width, acceleration, and a central position angle in the sky plane. Measured speeds range from a few km/s to nearly 3000 km/s (e.g., Gopalswamy, 2004; see also previous studies by Howard *et al.*, 1985 and St. Cyr *et al.*, 2000), with an average value of $\sim$450 km/s (see Figure 5), which is slightly higher than the slow solar wind speed. The apparent angular width of CMEs ranges from a few degrees to more than 120 degrees, with an average value of $\sim 47^\circ$ (counting only CMEs with width less than 120°). The width and other parameters of a CME occurring close to the limb is likely to be the true width, whereas the width and source latitude of a CME occurring close to the disk center are severely affected by projection effects (Burkepile *et al.*, 2004). CME acceleration is discussed in Section 4.

3.1.1. *Mass and Energy*

The current LASCO database of CMEs (as of Fall 2005) contains more than 10000 events. CME observations are available over a significant part of a solar cycle, thereby allowing us to obtain a very reliable estimate of their mass and energy profiles. Such statistics for all events up to 2002 have been presented in Vourlidas *et al.* (2002). The measurement methods and assumptions used can also be found

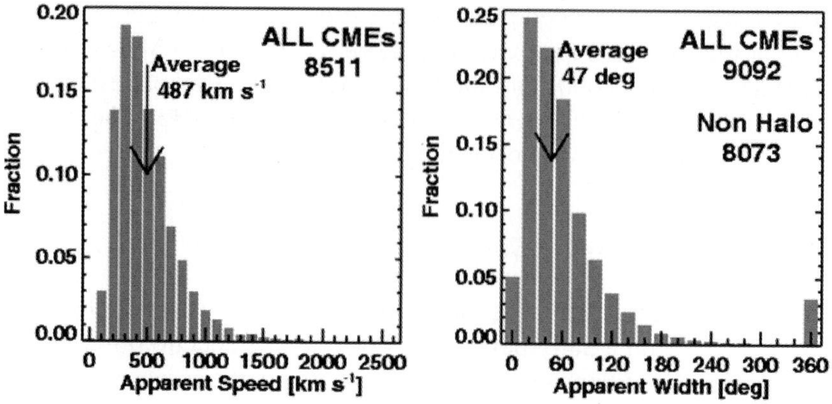

Figure 5. Speeds (left) and widths (right) of all CMEs observed by SOHO/LASCO from 1996 to the end of 2004 (Gopalswamy *et al.*, 2005b). The speed could not be measured for all the detected CMEs. The averages of the distributions are shown on the plots. The average width was computed from non-halo CMEs.

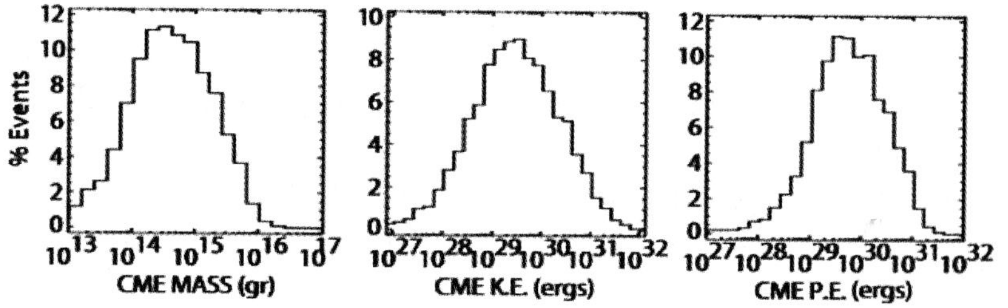

Figure 6. Statistics of mass and energy for 1996–2004 CMEs compiled by A. Vourlidas (6,335 events).

in the same paper. Here we update these statistics for all CMEs from 1996 to December 2004. Histograms of the mass, kinetic and potential energy of 6,335 CMEs are shown in Figure 6 and lead to several interesting observations: (i) There appears to be an upper bound of a few $\times 10^{16}$ g to the mass of a CME, (ii) about 3% of the CMEs are $< 5 \times 10^{13}$ g, (iii) the maximum kinetic energy of a CME is $10^{32} - 10^{33}$ erg, (iv) the "average CME" has mass of 1.4×10^{15} g and kinetic energy of 2.6×10^{30} erg. We compare the LASCO statistics to the data set from the Solwind instrument (Howard *et al.*, 1985) in Table I.

The difference between the two samples probably results from the larger number of small events in the LASCO database that results from the higher sensitivity, wider field of view and almost 100% coverage of LASCO compared with SOLWIND and SMM. A redefinition of the term "CME" is needed in order to distinguish "real" CMEs from small scale jets or fluctuations of the solar wind.

The total mass ejected in CMEs ranges from a few times 10^{13} g to more than 10^{16} g. Accordingly, the kinetic energy of CMEs with angular width $<120°$ ranges

TABLE I
Average CME Properties.

Parameter	LASCO	Solwind
Observing duty cycle	81.7%	66.5%
$\langle E_{\rm kin}\rangle$ (erg)	2.6×10^{30}	3.5×10^{30}
$\langle$Mass$\rangle$ (g)	1.4×10^{15}	4.1×10^{15}
Mass flux (g/day)	2.7×10^{15}	7.5×10^{15}

from $\sim 10^{27}$ erg to $\sim 10^{32}$ erg, with an average value of 5×10^{29} erg. Some very fast and wide CMEs can have kinetic energies exceeding 10^{33} erg, generally originating from large active regions and accompanied by powerful flares (Gopalswamy *et al.*, 2005a).

3.1.2. *CME Rate*

During solar minimum, the CME rate is typically 0.5/day. The rate during solar maxima is an order of magnitude higher. Figure 7 shows that the CME rate averaged over the Carrington rotation period (27.3 days) increased from less than 1 per day to more than 5 per day. The large spikes are due to active regions that are very active producers of CMEs. The correlation between the daily CME rate and sunspot number (SSN) is less than perfect, especially for large SSN (near solar maximum). CMEs occur around solar maximum with a relatively high rate from the polar crown filaments (PCFs), but have nothing to do with sunspots; hence, they need not be correlated with SSN. Both SSN and CME rates show a double maximum (late

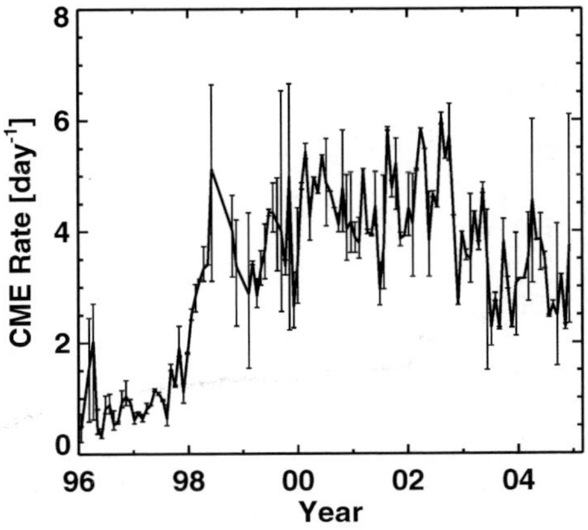

Figure 7. CME rate over the course of the SOHO mission from the LASCO CME catalog (Yashiro *et al.*, 2004).

2000 and early 2002). The low-latitude (LL) CME rate is generally higher than the high-latitude (HL) rate, but occasionally they can be very close. The cessation of HL CMEs coincided with the polarity reversal at the solar poles (Gopalswamy *et al.*, 2003b). The HL CMEs provide a natural explanation for the disappearance of PCFs, which need to be removed before the poles can acquire the open field structure of the opposite polarity.

3.1.3. *Variability of CME Speed*

The speed of a CME is usually measured by constructing a time-height diagram for the fastest moving feature of the CME front as it appears projected on the plane of the sky (POS). Inevitably, the POS values can deviate from the real radial speed of the CME front, depending on the actual direction of the motion. That may explain in part why different observers reached different conclusions. In the study by Howard *et al.* (1985) the mean POS speed of CMEs was found to increase with increasing solar activity. However, SMM data did not show a significant difference in the average speed of CMEs between solar activity minimum and maximum (Hundhausen, 1993). SOHO data confirmed an increase (Gopalswamy, 2004) as demonstrated in Figure 8. The annual mean speed increased from slightly below 300 km/s in 1996 to about 500 km/s in 2000. The speed showed a dip in 2001, as did the CME rate, and continued to increase to the second maximum in 2002. However, the speed did not decline after the 2002 maximum, but peaked in 2003. This is mainly because of the exceptional active regions (10484, 10486, and 10488) that produced fast and wide CMEs during October–November 2003. The speeds then started to decline with the CME rate.

These results must be taken with some caution. It is unclear at what distance the speed determinations were performed. Note that all histograms show CME speeds of less than 200 km/s, i.e. less than the minimum speed of even the slowest solar wind. That indicates that some of the CME speeds were determined close to the

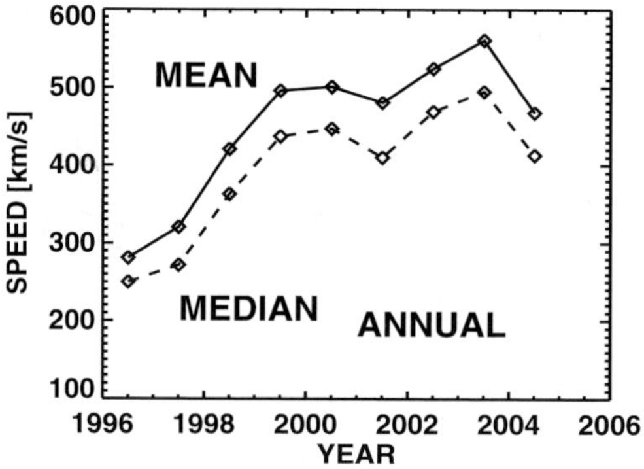

Figure 8. CME speeds over the course of the SOHO mission from the LASCO CME catalog (Yashiro *et al.*, 2004).

limb where CMEs are still being accelerated. The CME catalog maintained by Yashiro *et al.* (2004) shows how dramatically different the acceleration profiles for CMEs can be out to about 10 $R_\odot$. In order to quantify this huge diversity one has to make sure that the quantities under investigation (speed, width, latitude, mass) are all measured at the same "safe" distance from the Sun where they have reached constant speed, at about 15 $R_\odot$. Another reason for considering these results with some skepticism is the projection effect mentioned above. Burkepile *et al.* (2004) analyzed the role of these projection effects very carefully and based their study on 111 "limb" events observed by SMM. They found that these CMEs that are not affected by projection effects have a substantially higher average speed (519 ± 46 km/s) than that obtained from all SMM CMEs.

3.1.4. *CME Latitudes*

The latitude distribution of CME sources depends on how closed field regions are distributed on the solar surface. During the rising phase of cycle 23 (1997–1998), the CME latitudes were generally close to the equator and subsequently spread to all latitudes. During the maximum phase, there are many polar CMEs, and the number of such CMEs was larger in the southern hemisphere and occurred over a longer time period than in the north. This variation of CME latitude over the course of the solar activity cycle is consistent with previous measurements from Skylab/ATM (Munro *et al.*, 1979), P78-1/Solwind (Howard *et al.*, 1985) and SMM/CP (Hundhausen, 1993). However, Burkepile *et al.* (2004) noted the strong effect geometrical projection can have when relating CME apparent latitudes to source latitudes. They conclude that a much smaller percentage of limb CMEs were centered at high latitudes than was previously reported.

3.2. Morphological Properties

CMEs observed in white light are highly structured and are three-dimensional in nature. Many CMEs, especially the ones originating from filament regions show a three-part structure: a bright frontal structure, a dark void, and a bright core (Hundhausen, 1987). This is not seen in all CMEs. Even in prominence related CMEs, the three-part appearance depends on the location of the underlying prominence (Cremades and Bothmer, 2004). If the CME is super-Alfvénic, it can be expected to drive a shock ahead of it. Some CMEs have been interpreted as flux ropes (Chen *et al.*, 2000; Plunkett *et al.*, 2000). Some CMEs have voids with no prominence in them. Jets and narrow CMEs with no resemblance to the three-part structure have also been observed (Howard *et al.*, 1985; Gilbert *et al.*, 2001; Wang and Sheeley, 2002b; Yashiro *et al.*, 2003).

CMEs occurring close to the disk center often appear to surround the occulting disk of the coronagraph and are known as halo CMEs (Howard *et al.*, 1982). Only $\sim$3% of the SOHO CMEs are halos, but about 11% have a width exceeding 120°. CMEs with apparent widths between 120° and 360° are known as partial halos.

Halos can be "front-sided" or "back-sided," and for differentiation simultaneous disk observations are required. Some halos are asymmetric, heading predominantly above one limb with weak extensions on the opposite limb. These CMEs generally originate from locations closer to the limb than to disk center. The average speed of halo CMEs is 1000 km/s, more than 2 times the average speed of all CMEs (see Figure 5). This is clearly a result of bias. Halo CMEs are like ordinary CMEs that occur near disk center, but only the most powerful are detectable. (Yashiro *et al.*, 2004; Tripathi *et al.*, 2004). When front-sided, these "halo" CMEs can directly impact Earth causing geomagnetic storms, provided the magnetic field contained in the CMEs has a southward component (e.g. Gonzalez and Tsurutani, 1987).

The cone angles of CME expansion and, more generally, the shapes of the expanding CMEs are usually well maintained (Plunkett *et al.*, 1998). The CME shapes remain "self-similar." In other words, the ratio between lateral expansion and radial propagation appears to be constant for most CMEs. Low (1982) had already noticed the "often observed coherence of the large-scale structure of the moving transient," which could be explained by what he called "self-similar evolution." He studied the expulsion of a CME quantitatively on the basis of an ideal, polytropic, hydromagnetic description and found self-similar solutions that describe the main flows of white light CMEs as they are observed (see also Low, 1982, 1984, 2001; Gibson and Low, 1998).

The shapes of the vast majority of CMEs appear to be consistent with a nearly perfect circular cross section (Cremades and Bothmer, 2005). Indeed, halo CMEs moving exactly along the Sun-Earth line exhibit generally a circular and smooth shape. This observation is rather surprising in that CMEs are usually associated with the eruption of 2D elongated filament structures. Therefore, CMEs can be described in terms of "cone models" (for further discussion see the review by Schwenn (1986) and Cremades and Bothmer (2004). Thus the apparent lateral CME expansion speed can be considered independent of the viewing direction, and it is the only parameter that can be uniquely measured for any CME, be it on the limb or pointed along the Sun-Earth line, on the front or back side.

Schwenn *et al.* (2005) selected a representative subset of 57 limb CMEs and determined both the radial speed, $V_{\rm rad}$, of the fastest feature projected onto the plane of the sky, and the expansion speed, $V_{\rm exp}$, measured across the full CME in the direction perpendicular to $V_{\rm rad}$. They found a striking correlation between the two quantities, $V_{\rm rad} = 0.88\,V_{\rm exp}$, with a correlation coefficient of 0.86. This correlation holds for the slow CMEs as well as the fast ones, for the narrow ones as well as the wide ones. This means that the quantity $V_{\rm exp}$, which is always uniquely measurable for all types of CMEs, even halo and partial halo CMEs, can be used as a proxy for the radial propagation speed $V_{\rm rad}$ that is most often not accessible. Schwenn *et al.* (2005) established the lateral expansion speed $V_{\rm exp}$ as a pretty accurate though empirical tool for predicting the travel time of ICMEs to Earth.

There have been attempts to reveal the true 3D topology and internal structure of CMEs. Crifo *et al.* (1983) found the structure of a particular CME to be a 3D bubble rather than a 2D loop. Moran and Davila (2004) analyzed polarized images from LASCO. Their reconstruction of 2 halo CMEs suggests that these events were expanding loop arcades. Polarized measurements can indeed play a significant role for understanding CME structure and need to be pursued in the future, as Dere *et al.* (2005) stated. 3D structure can also be deduced by using Doppler shift to obtain line-of-sight velocities. This technique has been used to investigate the helical structure of a CME (Ciaravella *et al.*, 2000), the structure of the CME leading edge (Ciaravella *et al.*, 2003) and the relationship between a CME bubble and the hot plasma within it (Lee *et al.*, 2006).

3.3. Physical Properties

White-light coronagraphs detect electrons irrespective of the temperature, so spectral observations are needed to infer densities and temperatures, and radio observations provide density and magnetic field information.

3.3.1. *Density*

White-light images provide electron column densities integrated along the line of sight in different parts of the CME. For the scattering characteristics of the electrons it is generally assumed that they are in the plane of the sky, giving a lower limit to the mass (Vourlidas *et al.*, 2000). The white-light densities vary among events but they are generally of the order of 10^4–10^5 cm^{-3} for middle corona heights (4–7 $R_\odot$). They represent significant enhancements over the background corona ($\sim$10–100 times below 6 $R_\odot$ falling to $\sim$3–4 times at 30 $R_\odot$).

Radio observations of thermal emission provide density diagnostics of the prominence, the cavity and the leading edge. Multiwavelength observations of erupting prominences constrain their densities and temperatures (Irimajiri *et al.*, 1995). These estimates depend strongly on the surface filling factor assigned to the cool threads and the scale length of the filament. Akmal *et al.* (2001) found densities above 7×10^8 cm^{-3} at 1.3 $R_\odot$ for optically thick CME plasma at 10^4 K. At decimetric wavelengths, filaments are seen also as depressions on the disk (Marqué, 2004) corresponding to a low density structure surrounding the filament (a cavity), as also seen in white-light and X-ray (Hudson *et al.*, 1999) observations. The measurements taken with the Nançay radioheliograph are compatible with electron density depletions between 25 and 50% of the mean coronal density (75% for some filaments). CME leading edges observed at meter wavelengths require subtraction of quiet Sun emission (Bastian and Gary, 1997; Gopalswamy and Kundu, 1992; Kathiravan *et al.*, 2002). The densities inferred from these observations are of the order of those obtained from white-light coronagraphic observations.

Ultraviolet spectra provide two different types of density measurements. Classical density-sensitive line ratios of O IV lines (Wiik *et al.*, 1997) and OV lines (Akmal *et al.*, 2001) have been measured in a few events. At 3.5 $R_\odot$, Akmal *et al.* find densities from 1.4×10^6 cm^{-3} to more than 10^8 cm^{-3}. Both measurements pertain to fairly cool CME core material. The density can also be determined from the ratio of the collisionally and radiatively excited components of a UV spectral line. This method can be used at speeds below about 100 km/s, where a line is pumped by the emission in the same line from the solar disk, or at speeds near 369 and 172 km/s, where O VI λ1037 is pumped by C II $\lambda\lambda$1036.3, 1037.0 (e.g., Dobrzycka *et al.*, 2003). It can also be used at speeds near 1755 and 1925 km/s, where OVI λ1032 is pumped by Lyβ, and O VI λ1037 is pumped by O VI λ1032 (Raymond and Ciaravella, 2004). The latter study found densities ranging from 1.3×10^6 to 4×10^7 cm^{-3} at 3 $R_\odot$.

3.3.2. *Temperature*

YOHKOH images show gas at several million K electron temperatures. EIT and TRACE images are often interpreted under the assumption that the 195 Å image shows Fe XII emission formed at 1.5×10^6 K, though O V (2×10^5 K) or Fe XXIV (2×10^7 K) can contribute. It is often possible to discriminate between Fe XII and Fe XXIV on morphological grounds, however.

Spectra can constrain the CME temperature in several ways. The measured line widths give upper limits to the ion kinetic temperatures, and electron temperatures can sometimes be obtained from the ratio of two lines of a single ion if their excitation potentials differ (e.g. Ciaravella *et al.*, 2000). Lower limits on electron temperature can also be inferred from collisionally excited lines if the density and column density are known.

3.3.3. *Elemental Composition and Ionization State*

The composition of CME material is basically that of the pre-CME prominence or coronal plasma, though the charge state may be altered. Ciaravella *et al.* (1997) found a weak FIP (First Ionization Potential) enhancement, consistent with prominence abundances in cool CME core material. Structures identified as post-CME current sheets have been observed in a few cases, and the elemental composition matches that of the nearly active region (Ciaravella *et al.*, 2002).

Simply detecting a spectral line reveals the presence of that particular ionization state, and species from H I and C II to [Fe XVIII] and [Fe XXI] have been detected in various CMEs (e.g. Innes *et al.*, 2001; Raymond *et al.*, 2003). The faster, more powerful CMEs show little cool plasma, but the high temperature ions [Fe XVIII], [Fe XX] and [Fe XXI] are present (Raymond *et al.*, 2003; Innes *et al.*, 2001; 2003a,b).

The charge state of CME plasma is frozen in below 2 $R_\odot$ for typical CME speeds and densities. Cool prominence material often appears in absorption against background coronal emission in EIT images, indicating neutral gas. As it rises it often becomes bright in the 195 Å band, indicating emission in either O V or Fe XII (e.g. Ciaravella *et al.*, 2000; Filippov and Koutchmy, 2002).

3.3.4. *Magnetic Field*

At radio wavelengths, the analysis of gyroresonance emission, of polarization spectra of the thermal emission, of microwave QT-propagation (Quasi-Transverse), and of gyrosynchrotron emission, provides different techniques used to measure coronal magnetic fields in prominences, coronal holes, loops and CMEs (Gary and Keller, 2004). Radio observations indicate a magnetic field strength of 1 G in the corona at a heliocentric distance of 1.5 $R_{\odot}$ (see, e.g., Dulk and McLean, 1978). The field strength is 3–30 G in quiescent prominences and 20–70 G in active prominences (see e.g., Tandberg-Hanssen, 1995). The magnetic field in the cavity is virtually unknown, but a higher field strength may be required to compensate for the lower density.

3.4. Associated Phenomena

Eruptive prominences (EPs) observed in Hα or microwaves frequently accompany CMEs (Hundhausen, 1993; Hanaoka *et al.*, 1994; Gopalswamy *et al.*, 1996). They represent a good proxy for the configuration of the coronal magnetic field; they overlie polarity inversion lines, and the surrounding coronal magnetic fields are highly sheared. CMEs accompanied by an EP are particularly interesting to study in order to predict their magnetic topology (Rust and Kumar, 1994; Martin and McAllister, 1996; Bothmer and Schwenn, 1996; Rust, 2001).

Flares and CMEs may or may not be related to each other (Harrison, 1986, 2003), although when a CME does occur it usually has a close flare association. A strong X-ray flare can occur in the absence of CME and conversely a CME is not necessarily associated with a flare. However, when both flares and CMEs are produced conjointly, it seems that they share at their onset the same energetic processes for CME acceleration and flare energy release (Zhang *et al.*, 2004).

Long Duration Events (LDEs) observed at soft X-ray wavelengths are closely associated with CMEs (Sheeley *et al.*, 1983; Webb and Hundhausen, 1987). LDEs are observed as the appearance of large-scale loop systems, also called eruptive arcades (EAs) or post-eruptive arcades (PEAs), which often form and evolve in the aftermath of coronal eruptions. Their detection depends on their temperature, therefore on the wavelength in which observations are made. They are observed to occur in the lower corona close to the onset site of the eruption (Figure 9), which illustrates the bright loops of the Bastille Day flare). PEAs observed at 195 Å by EIT have almost one-to-one correlation with CMEs (Tripathi *et al.*, 2004).

Halo CMEs can be associated with many other manifestations seen on the solar disk. They are associated with EUV and soft X-ray dimmings, disappearing trans-equatorial loops, Hα Moreton waves and EIT waves (Pick *et al.*, 2006, this volume). It seems that for most of the large flare/CME events, all these manifestations are

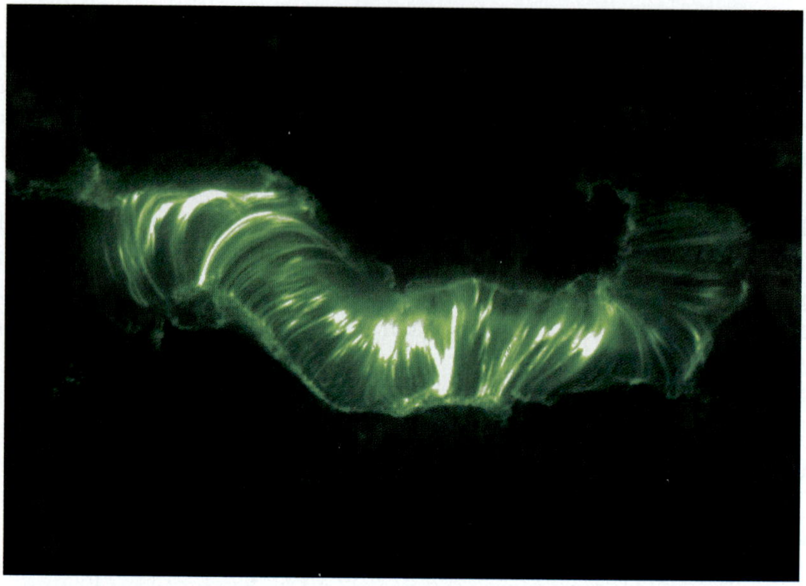

Figure 9. TRACE 195 Å image of the flare arcade during the Bastille Day event on July 14, 2000.

present. One interesting observational signature of CMEs on the solar disk is the evolution from a sheared structure to arcade structure (Sakurai *et al.*, 1992). The pre-eruptive scenario is often marked by the presence of an "S-shaped" sigmoid structure which denotes the dominant helicity in that hemisphere (Gopalswamy *et al.*, 2006, this volume).

Another important signature of an on-disk CME is the observation of transient dimming observed in X-rays and EUV (e.g., Hudson *et al.*, 1996; Gopalswamy *et al.*, 1997; Pick *et al.*, 2006, this volume). These dimmings have been interpreted as the opening of the initially closed field lines during the initial phase of a CME indicating mass loss of the order of 10^{14} g, an order of magnitude smaller than in a typical CME. Front-side halo events may start with the disappearance of trans-equatorial loops (TELs) observed in X-rays (Khan and Hudson, 2000; Pohjolainen *et al.*, 2001).

Radio bursts are produced by plasma emission processes when accelerated electrons travel along open field lines or loop systems. The U and J bursts exhibit a turn-over frequency at the top of the loop. Different components of type IV bursts are associated with different parts of the CMEs. Bastian *et al.* (2001) imaged CME loops, called radio CMEs, extending to about 3 $R_{\odot}$, behind the CME front (Figure 10). These radio emitting loops are the result of nonthermal synchrotron emission from 0.5–5 MeV electrons interacting with magnetic fields of about 0.1 to a few Gauss. Pick *et al.* (2006, this volume) discuss the relationships of coronal shocks to radio measurements in more detail.

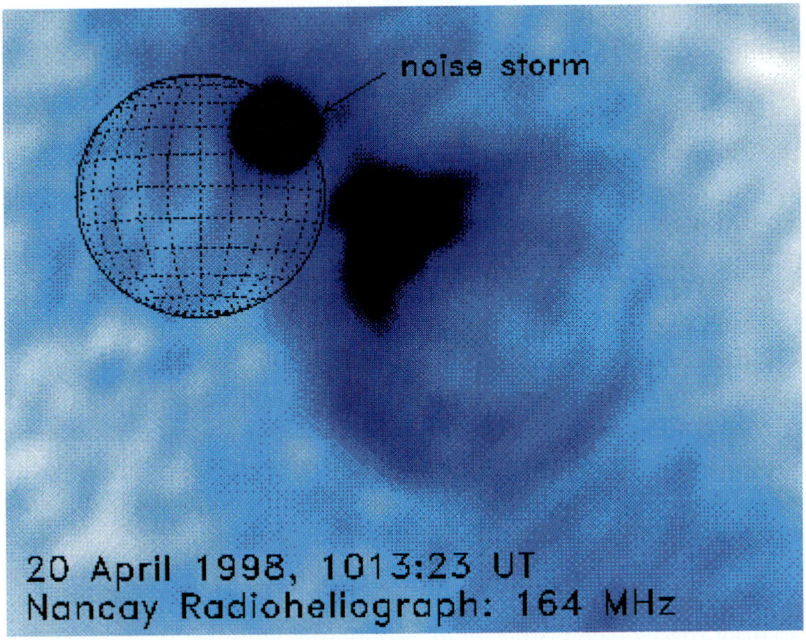

Figure 10. Snapshot of the radio CME observed on 20 April 1998 at 164 MHz at the time of maximum. The background emission from the Sun has been subtracted. Radio emission from a noise storm is present to the northwest. The brightness of the CME is saturated in the low corona because the map has been clipped at a level corresponding to a brightness temperature $T_b = 2.6 \times 10^5$ K. The radio CME is visible as a complex ensemble of loops extending to the southwest. (From Bastian *et al.*, (2001)).

4. CME Evolution and Dynamics

4.1. Kinematic Evolution of CMEs

A CME is basically a moving feature and, therefore, a key aspect of CME study is the kinematic evolution of a CME through the corona. One of the most interesting observational questions is when and where a CME gets accelerated in the corona. Both propelling and retarding forces acting on a CME need to be considered.

CMEs usually travel at a relatively constant speed with minor acceleration or deceleration. The statistical distribution of CME accelerations in the LASCO C2/C3 field of view shows a peak value close to zero and for a majority of events the acceleration within only ± 20 m/s^2 (Moon, 2002; Yashiro *et al.*, 2004). While the slowest CMEs tend to show positive acceleration, the fastest CMEs decelerate in the outer corona (Gopalswamy *et al.*, 2000a). This indicates that the major acceleration of a fast CME mainly occurs in the inner corona, i.e. below $2\,R_\odot$ and the subsequent evolution is primarily controlled by the interaction between the CME and the medium through which it propagates.

The complete kinematic evolution of a number of CMEs has been well observed by combining LASCO C1 with C2/C3 observations and the SOHO/EIT observations (Srivastava *et al.*, 1999; Gopalswamy and Thomson, 2000; Zhang *et al.*, 2001). One example is shown in Figure 11, where the kinematic evolution is represented by plots of height, velocity and acceleration against time. for the June 11, 1998 event. Based on a combined study of this and some other events, Zhang *et al.* (2001) described the kinematic evolution of a CME in terms of a three-phase scenario: (1) the initiation phase, (2) the impulsive (major) acceleration phase and (3) the propagation phase. The initiation phase is characterized by a slow ascension of certain coronal features (e.g, active region envelopes, filaments) for a period of up to tens of minutes, with a speed typically below tens of km/s. The major acceleration phase is characterized by fast acceleration (e.g, from a few hundred to a few thousand m/s^2). This phase lasts typically from a few minutes to tens of minutes (it can be even longer for certain events) in which the coronal features travel a distance ranging from a fraction of a solar radius to several solar radii. After the completion of the acceleration phase, a CME appears to be fully developed and travels at a more or less constant speed, a constant angular width and a constant position angle. This phase is therefore simply called the propagation phase, when the ICME is primarily subject to drag forces (see Forbes *et al.*, 2006, this volume).

Not all CMEs necessarily display a full three-phase evolution. Events which show persistently weak acceleration (i.e., <25 m/s^2) throughout the inner and the outer corona are known as gradual or balloon-type events (Srivastava *et al.*, 1999, 2000). These CMEs are usually slower than 100 km/s in the inner corona (i.e. below 2 $R_\odot$) and eventually reach terminal speeds no higher than 400–600 km/s in the LASCO /C3 field of view. Figure 12 clearly shows the difference in the kinematic evolution of the gradual events. The speed profiles of these "balloon-type" CMEs are very similar to the slow solar wind speed profiles that Sheeley *et al.* (1997) derived by tracing small density blobs floating along like "leaves in the wind." Similarly, the "balloon-type" CMEs appear to float along and follow the slow wind's acceleration. The CME acceleration values can differ by up to a factor of 1000, as Figures 11 and 12 demonstrate. Some authors take this as evidence for different types of CMEs driven by radically different release and acceleration mechanisms (e.g., Sheeley *et al.*, 1999).

4.2. White-Light and UV Signatures of Shock Waves

Many CMEs propagate with speeds >1000 km/s, much larger than the characteristic sound and Alfvén speed in the solar wind. Therefore, the ejected plasma clouds should drive shocks ahead of themselves. There is plenty of indirect evidence for waves associated with CMEs (Sheeley *et al.*, 2000), but there have been only two published detections of white light shocks (Sime and Hundhausen, 1987; Vourlidas *et al.*, 2003). This does not mean that CME shocks do not have a white-light

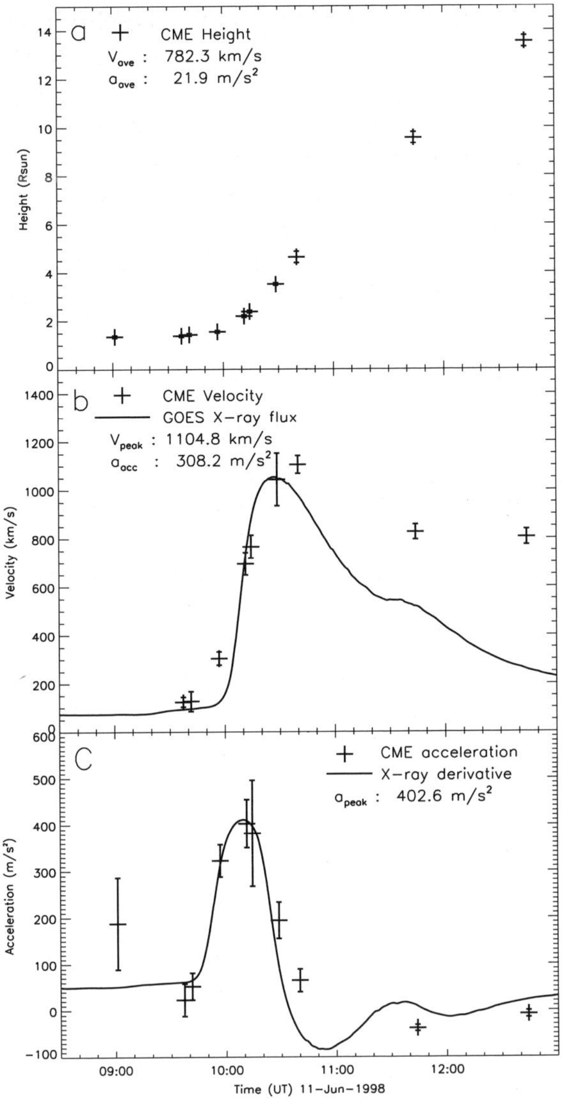

Figure 11. CME kinematic plots versus flare flux plots for the 1998 June 11 event. The three panels, from top to bottom, show CME height-time plot, velocity-time plot and acceleration-time plot (discrete plus signs with error bars), respectively. The solid curve in panel b shows the soft X-ray flux profile of the flare, which apparently correlates with the CME velocity profile during the rise phase. The solid curve in panel c shows the derivative profile of the soft X-ray flux (equivalent to the hard X-ray profile according to the Neupert effect), which apparently correlates with the CME acceleration profile in the impulsive phase. (Adapted from Zhang *et al.* (2004)).

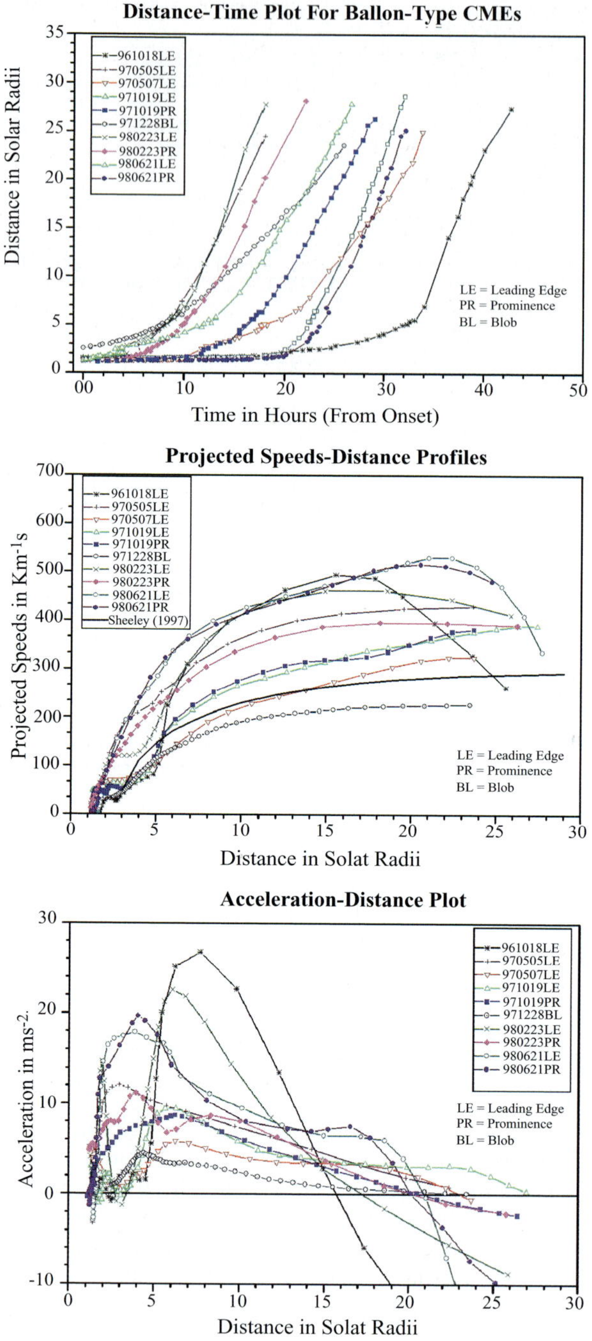

Figure 12. Kinematic plots for 7 balloon-type CMEs observed by LASCO. The three panels from top to bottom show CME height-time plots, velocity-time plots and acceleration-time plots, respectively. The solid curve in the middle panel shows the speed profile as derived from "leaves in the wind" by Sheeley *et al.* (1997), adapted from Srivastava *et al.* (1999).

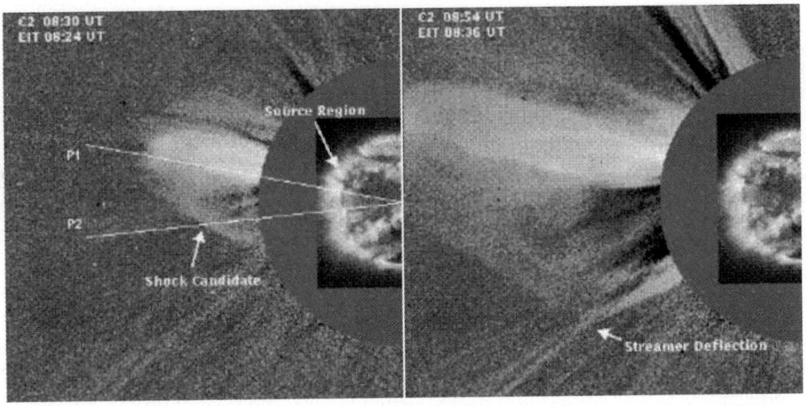

Figure 13. CME shock identification for the April 2, 1999 event. These are LASCO/C2 calibrated excess mass images. The EIT 195 image is also shown. Note the clear streamer deflection as the shock impinges on it. Adapted from Vourlidas *et al.* (2003).

signature. Rather, it is difficult to prove that a given sharp feature is indeed a shock front and not just a coronal loop or a wave front. Parker (1961, 1963) developed analytic models of interplanetary shock waves (see also Hundhausen, 1985) and interpreted one of the solutions as a compressed shell of fluid driven into the ambient medium, Simon and Axford (1966) showed that there is a second shock front that moves inwards with respect to the fluid. The complete solution consists of a shell bounded by an external shock that accelerates and compresses the ambient material and an internal shock that decelerates and compresses the fast solar wind.

Vourlidas *et al.* (2003) simulated a specific event with a detailed MHD model to show that the CME flank was actually a shock (Figure 13). Furthermore, a streamer was seen to bend as the flank impinged on it, thus providing compelling evidence that streamer deflections are indeed due to CME shocks. Although it is impractical to model every possible CME candidate to show that the front (or flank) is a shock, we know that many events exhibit similar overall morphology. Therefore, we can use the experience gained from modeling a single event and common sense to look for similar features in other events. As MHD models evolve and we understand better the white light CME morphology, we expect that the shock identification in coronagraph images may become more routine in the near future.

Shock waves are detectable in UV spectra of the corona through the high velocities, heating and compression that they produce. The shocks tend to be faint, however, and coronagraphic observations are required. UVCS has detected about 10 CME shocks so far (Raymond *et al.*, 2000; Mancuso *et al.*, 2002; Raouafi *et al.*, 2004; Ciaravella *et al.*, 2005, 2006). The most easily detectable signature is a sudden broadening of the O VI lines to a width comparable to the shock speed. Collisionally excited lines such as Si XII λ52.1 nm brighten because of the compression, while radiatively excited lines such as Lyα fade because of Doppler dimming at

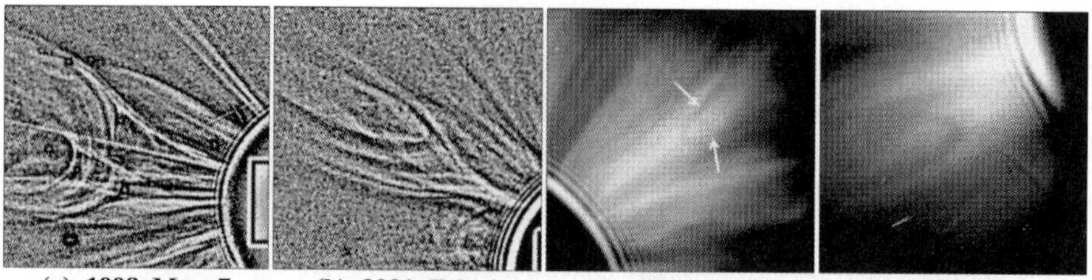

Figure 14. Examples of concave-outward structures observed with the LASCO C2 coronagraph during CME events. (a) CME with topology suggestive of a cylindrical or toroidal flux rope (edge-enhanced image). (b), (c), and (d) show dark U-shaped features. The U-loops in the December 28 and June 4 events are entrained in core material that is falling back toward the Sun (From Wang and Sheeley (2002a)).

high speeds. The UV spectra provide unique information about the heating of individual ion species and electrons in the collisionless coronal shocks. Spectra of the pre-CME corona provide the initial density, temperature and composition of the plasma, all necessary inputs for models of SEP production. More detail is given in the Working Group F chapter.

Another signature of coronal shocks is a hectometric-kilometric type II radio burst. This emission at the local plasma frequency and its harmonic is caused by energetic electrons accelerated at the shock front. The radio frequency is a direct measure of the density near the shock, and the drift rate provides a rough estimate of the shock speed. More detail is given in the paper by Pick *et al.* (2006, this volume).

4.3. Evidence for Magnetic Disconnection

During CME events, large-scale structures having a concave-outward topology are often observed in the white-light corona; examples are shown in Figure 14. Such structures are sometimes interpreted as evidence for magnetic disconnection (see, e.g., Illing and Hundhausen, 1983). In the majority of cases, however, we may be seeing three-dimensional flux ropes with their ends still anchored in the Sun rather than completely detached "U-loops" (Wang and Sheeley, 2002a). The *SOHO*/LASCO coronagraph has also recorded many events suggestive of ongoing disconnection, where an elongated streamer structure appears to split apart (Figure 14) or where a loop-shaped ejection with a hollow center separates into an outward- and an inward-moving component (Figure 15b). Again, the ejected component in such events is not necessarily completely detached but may represent a flux rope still connected at its far ends to the Sun. Even more frequently encountered are the multitude of small-scale inflows, in the form of dark collapsing loops and cusps, sinking columns of streamer material, and dark tadpole-shaped features

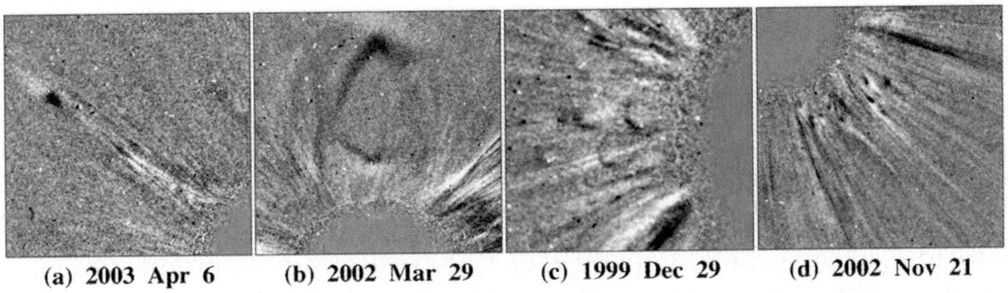

Figure 15. White-light coronagraph observations that suggest ongoing magnetic disconnection. Each panel represents the difference between two LASCO C2 images taken 20–30 min apart; white (black) means that the local coronal density has increased (decreased) during the elapsed interval. (a) Streamer detachment. (b) Looplike ejection with infalling counterpart (face-on view of a "streamer detachment"?) (c) Collapsing loops. (d) Collapsing cusps (sinking column inflows). (From Wang *et al.* (1999)).

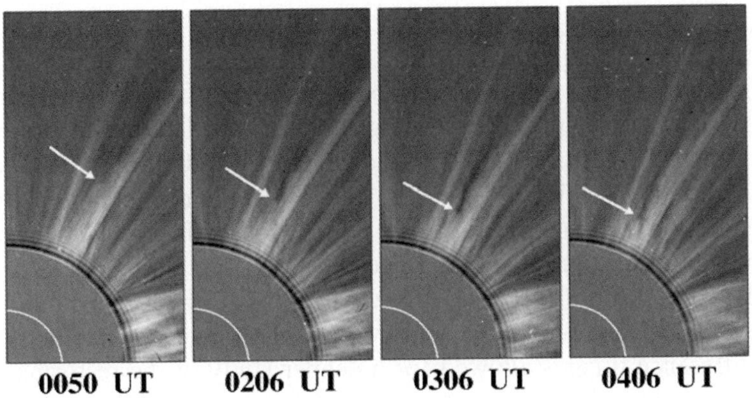

Figure 16. Inflow observed with the LASCO C2 coronagraph on October 25, 1999. The sinking column of streamer material leaves a dark depletion trail in its wake and takes on a cusp-like appearance below $r \sim 2.5R_{\odot}$ (From Sheeley and Wang (2002)).

Figures 15c, d, and 16. These events, which tend to occur well in the aftermath of CMEs (and are not always associated with them), are observed only at heliocentric distances below about $5R_{\odot}$. Although the inflows are highly suggestive of the pinching-off of open field lines at the heliospheric current sheet, no outgoing counterparts are detected, perhaps because they are below the sensitivity threshold of the LASCO coronagraph in the region beyond $5R_{\odot}$.

4.4. Post-Flare Inflows

Analogous inflow events have also been detected at much lower heights with *Yohkoh* soft X-ray observations, above post-flare loop systems that exhibited fan-like structures extending above the loop tops (McKenzie and Hudson, 1999). These inflows

(McKenzie, 2000) had the appearance of voids and, although moving faster than the LASCO inflows, still had speeds well below the inferred Alfvén speeds in the structures. *SUMER* confirmed these characteristics of the inflows spectroscopically (Innes *et al.*, 2003a), and the higher-resolution *TRACE* imaging showed sinuous motions as the voids appeared to slide down between the spike structures (Gallagher *et al.*, 2002). In contrast to the inflows observed by SXT, TRACE and SUMER and interpreted as plasma voids, Tripathi *et al.* (2006) discovered a bright coronal inflow observed by EIT above post-eruptive arcades after a CME eruption. This inflow provides another evidence of post-CME reconnection. Asai *et al.* (2004) found examples of such flows even in the impulsive phase of a major flare, increasing the likelihood that this phenomenon – ill-understood theoretically at present – has a direct relationship with magnetic reconnection and flare energy release.

4.5. Mass Replenishment, Post-CME Recovery

The observation of "transient coronal holes" using Skylab data (the term was introduced by Rust, 1983) provided the suggestion of an X-ray identification for the source regions of CME mass. This followed earlier white-light observations of "coronal depletions," which presumably showed the same phenomena in limb observations. In this picture, the post-CME recovery would appear in the low corona as the disappearance of the newly-formed holes.

Kahler and Hudson (2001) conducted a survey of transient coronal holes observed by SXT. Movie sequences with reasonable image cadence and the ability to make difference images allowed for a deeper study of the re-formation of the corona. In the conventional 2D picture of field-line opening followed by magnetic reconnection, the transient coronal holes represent the opened field; in 3D they would represent the footpoints of a large-scale flux rope. In either case the arcade resulting from reconnection would be expected simply to fill in the holes, starting at their inner edges (the outer edges of the flare ribbons) and proceeding outwards.

In the SXT data, this simple picture was not seen. The holes form in large-scale sigmoid bends of the filament channel, thus differing morphologically from the more cylindrical arcade structure. The arcade sources do advance into the hole regions, but the holes often also shrink inwards from the outside. There is no tendency for the holes simply to brighten in place, and the durations of the dimmings suggest that the open magnetic field lines may stay open for periods longer than the arcade growth. The limit to such observations is imposed by foreground/background confusion as the Sun rotates and the perspective changes.

It seems reasonable to believe that magnetic reconnection, either on a large scale or by reconnection interchange on smaller spatial scales, could contribute to the replenishment of the lost coronal mass by providing closed flux tubes capable of trapping mass. The geometry of these processes is not so clear, especially in cases without a clear conjugate pair of transient coronal holes.

4.6. Type IV Burst and Radio Ejecta: Observations of Trapped Electrons

Type IV bursts, first identified by Boischot (1957), are long lasting continuum radio emissions with large bandwidth associated with CMEs and flares (Stewart, 1985; Robinson, 1985; Kahler, 1992). Successive episodes in the development of type IV bursts are closely associated with different phases of CME development. Nonthermal electrons, which are accelerated at the vicinity of the flare site or by a shock wave, get trapped in moving or stationary structures and are thought to produce these various long-lasting continuum emissions.

The first stage, following the onset of the flare, corresponds to broad-band emission which extends from microwave to meter wavelengths (e. g. Kundu, 1965), and is associated with a gradual hard X-ray burst (Frost, 1974). This emission lasts between 10 min and about one hour, and can be accompanied by a moving type IV emission and rising structures with velocities up to 1000 km/sec or even more. The corresponding radio sources are complex, sometimes with two components located near the foot points and a third component near the top of the arch (e.g. Wild, 1969; Gopalswamy and Kundu, 1990). Limb observations have shown that these arches are behind the CME leading edge. Spectra of moving type IV bursts are often consistent with Razin-suppressed gyrosynchrotron emission from nonthermal electrons trapped in moving magnetic structures (e.g. Ramaty and Lingenfelter, 1967; Boischot and Clavelier, 1968; Gopalswamy and Kundu, 1989).

Bastian *et al.* (2001) reported radio thin loops behind the CME front (see Figure 10). Disk observations of halo CME events showed that radio observations image a set of ejected loops that lift as part of the CME and that EUV dimming regions are observed later in the same location as the radio emitting region (Pohjolainen *et al.*, 2001). The heated prominence material can also trap electrons and propagate as narrow ejecta (e. g. Sheridan, 1970; Riddle, 1970) or "isolated plasmoids" (e.g. Gopalswamy and Kundu, 1990).

Following this first stage, stationary type IV bursts (continuum storms) can last for several hours, shifting progressively toward lower frequencies and evolving gradually into noise storm activity associated with post-eruptive loops. Such long-lasting radio emission is the signature of suprathermal electrons in the corona, broadly associated with long-lasting soft X-ray emissions (Lantos *et al.*, 1981) and transient legs in white light (Gergely *et al.*, 1979; Kerdraon *et al.*, 1983). Kahler and Hundhausen (1992) identify type IV burst sources with newly formed streamers and suggest that the energetic electrons are produced during magnetic reconnection.

Stationary type IV bursts can occur in the absence of a flare or a gradual hard X-ray or microwave outburst. They are associated with magnetic field restructuring of the white light corona (Kerdraon *et al.*, 1983; Habbal *et al.*, 1996) and filament disappearance (e.g. Lantos *et al.*, 1981). The absence of flares in these events confirms the close association of type IV bursts with CMEs because many of these signatures imply a CME. The simplest interpretation is that moving type IV bursts correspond to magnetic structures associated with the CME, while the stationary type IV bursts correspond to quasi-stationary post-eruption arcades. In some events, moving and stationary sources are closely coupled: Figure 17 shows a broad band continuum (type IV burst) modulated by successive packets of fast sporadic bursts which coincidence closely with hard X ray peaks (Classen *et al.*, 2003). In the example shown in this Figure, the type IV burst is formed by two sources: a fast-moving one and a quasi-stationary one. The time coincidence found between the flux peaks of these two radio sources and the underlying hard X-ray source implies a causal link: they must be fed by electrons accelerated in the same region. The observations are consistent with an erupting twisted flux rope with the formation of a current sheet behind (Pick *et al.*, 2005).

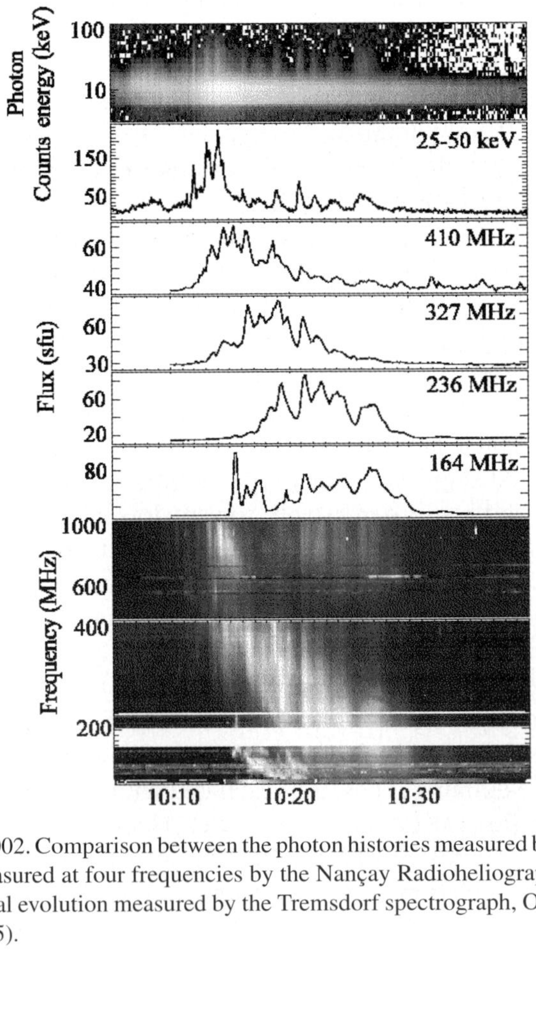

Figure 17. June 02 2002. Comparison between the photon histories measured by RHESSI (top panel), the flux evolution measured at four frequencies by the Nançay Radioheliograph, NRH, (middle four panels) and the spectral evolution measured by the Tremsdorf spectrograph, OSRA, (bottom panels). From Pick *et al.* (2005).

4.7. Sigmoids and Helicity

Active regions with pronounced soft X-ray sigmoids (S or reverse-S shaped structures; Figure 18) exhibit a greater tendency to erupt (Canfield *et al.*, 1999; Sterling *et al.*, 2000), and the eruptions produced tend to result in moderate geomagnetic storms (Leamon *et al.*, 2002). Relating the sigmoids at X-ray (and other) wavelengths to magnetic structures and current systems in the solar atmosphere is the key to understanding their relationship to CMEs. The correspondence between soft X-ray sigmoids and active region CMEs is not one-to-one. Sigmoids may also disappear and reappear without a detectable CME (see Gibson *et al.*, 2002) even while generating significant soft X-ray and EUV brightenings. Consequently, we need to understand more about the formation and evolution of sigmoid structures and explore the conditions that drive them to eruption. An important component of the physical nature of sigmoids and their relationship to CMEs is the role played by magnetic helicity.

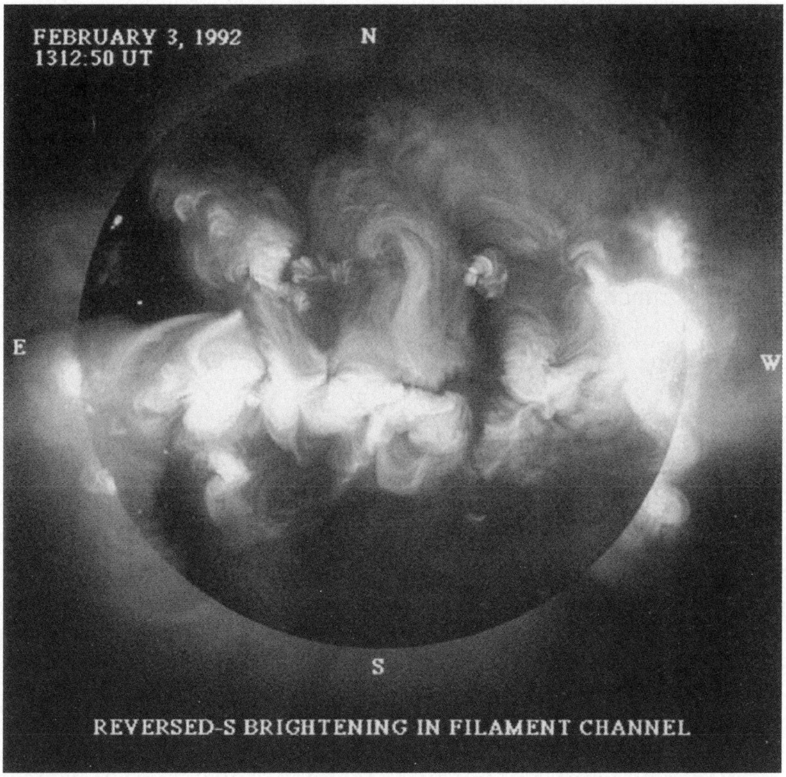

Figure 18. *Yohkoh* image of the X-ray corona on 1992 February 3 at 1312:50 UT. A faint, negative sigmoid brightening (center) extends across the equator.

The role of helicity injection, or "helicity-charging" (Rust and Kumar, 1994), has been a focal point in the discussion of eruptive events. In particular, it has been argued that CMEs are the means by which the solar corona expels magnetic helicity accumulated over hours and days by the combination of local shearing motions, differential rotation and the emergence of twisted flux systems (Low, 1996; DeVore, 2000; Démoulin *et al.*, 2002). The magnetic helicity is a useful quantity for studying the magnetic evolution of the CME-latent corona as it is an MHD quantity which is globally conserved on the time scales of interest (Berger, 1984).

In the solar corona, the magnetic helicity and free magnetic energy must be supplied by the photospheric or sub-photospheric activity, either in the form of the emergence of twisted structures or the in-situ shearing of previously emerged structures. Devore (2000) has argued that a significant quantity of magnetic helicity is injected by the action of differential rotation over the lifetime of an active region; enough to explain the total "ejected" helicity detected in interplanetary magnetic clouds, the interplanetary counterparts of CMEs (Gosling *et al.*, 1995). However, this assertion has been contested by Démoulin *et al.* (2002) and Green *et al.* (2002) who argue that the helicity injected by differential rotation is 5 to 50 times smaller than that inferred to be carried away in CMEs, requiring the emergence of sub-photospheric structures with significant twist as the dominant source of helicity in active regions.

In addition to the effect of differential rotation, strong local shearing may contribute significantly to the helicity injection into large but otherwise local structures associated with active regions. Recent studies by Kusano *et al.* (2002), Nindos and Zhang (2002) and Moon *et al.* (2002) have demonstrated that multiple flaring is often associated with regions of strongly sheared footpoint motions and local helicity injection. In the case discussed by Kusano *et al.* (2002) the shearing motions can contribute as much, if not more, helicity as the flux emergence.

4.8. EIT Waves

Observations with EIT revealed a new wave phenomenon, the so-called coronal flare or EIT-waves (Moses *et al.*, 1997; Thompson *et al.*, 1998). These waves appear as bright rims sometimes circularly expanding around the flaring active region and propagating over a hemisphere of the Sun at speeds of of 200 to 300 kilometers per second. The EIT-waves are reminiscent of Moreton waves (Moreton and Ramsey, 1960; Thompson *et al.*, 1998), but they appear in EUV spectral lines emitted by a hot corona, whereas the Moreton waves are seen in Hα at velocities in the range 440–1125 km/s with mean value 650 km/s (Smith and Harvey, 1971).

The relationships among EIT waves, Moreton waves, type II radio emission and CMEs, as well as theoretical models are discussed by Pick *et al.* (2006, this volume).

4.9. Interacting CMEs

The combination of CME data from SOHO/LASCO and radio data from the Wind/WAVES helped identify the energetic nature of colliding CMEs within the LASCO field of view. When a fast CME overtakes a slow CME from the same solar source or a neighboring solar source, often an enhanced radio emission is detected in the radio dynamic spectrum. If such an interaction results in a single resultant CME, the process is referred to as "CME cannibalism" (Gopalswamy *et al.*, 2001, see Figure 19). Given the high rate (6 per day) of CME occurrence during solar maximum and the observed range of speeds, one would expect frequent interaction between CMEs. In the best examples, the apparent interaction seen in LASCO images coincides with a radio enhancement at the frequency corresponding to the plasma frequency at the observed distance from the Sun (e.g. Burlaga *et al.*, 2002). The mechanism of radio enhancement can be interpreted as follows: when the shock driven by the second CME passes through the high density of the first CME, it encounters a region of low Alfven speed (inversely proportional to square-root of density) and hence its Mach number increases temporarily. A high Mach number shock accelerates more electrons resulting in the enhanced radio emission. There may be other mechanisms that boost the efficiency of shock acceleration depending on the magnetic topology of the preceding CME and the turbulence in its aftermath.

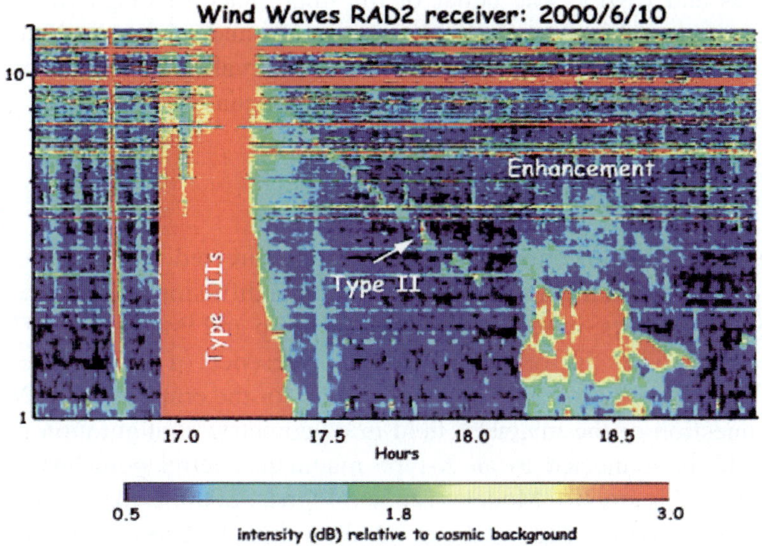

Figure 19. Wind/WAVES dynamic spectrum in the 1–14 MHz range obtained by the RAD2 receiver. The vertical features are type III-like bursts due to electron beams escaping from the shock along open field lines. The thin slanted feature is the type II burst. The enhancement in question is the bright emission between 18:12 and 18:48 UT. It corresponds exactly to the time of interaction between two ICMEs (Gopalswamy *et al.*, 2001).

The CME interaction is also relevant for SEPs because the same shock accelerates electrons and ions. If the shock propagates through a preceding CME, trapping of particles in the closed loops of preceding CMEs can repeatedly return the particles back to the shock, thus enhancing the efficiency of acceleration (Gopalswamy *et al.*, 2004; Kallenrode and Cliver, 2001). A systematic survey of the source regions of large SEP events of cycle 23 has revealed that the SEP intensity is high when a CME-driven shock propagates into a preceding CME or its aftermath originating from the same solar source (Gopalswamy *et al.*, 2004, 2005). The existence of preceding CMEs can greatly enhance the turbulence upstream of the shock, resulting in shorter acceleration time and higher intensities for SEPs (Li and Zank, 2005). In addition, the large scatter in the CME speed – SEP intensity plots (Kahler, 2001) is significantly reduced when the interacting and non-interacting cases are considered separately. The preceding CMEs could accelerate seed particles for the following stronger shock. Presence of seed particles seems to make a difference in the resulting intensity of the SEP events (Kahler, 2001).

5. Reconnection: Observable Signatures?

Magnetic reconnection is thought to play a fundamental role in driving or at least accompanying a number of transient phenomena in the solar atmosphere, including solar flares and CMEs. The belief of the importance of magnetic reconnection comes from a combination of theoretical models and a large number of indirect signatures. We present here a summary of the observable signatures pointing to the role of reconnection in CMEs and related phenomena.

5.1. Evidence for Current Sheets

In the classical reconnection model, oppositely directed magnetic field lines get stretched out to form a current sheet defined by a diffusion region where the magnetic field reconnects, releasing energy (e.g. Kopp and Pneumann, 1976). In most reconnection models the formation, or at least presence, of a current sheet is crucial for reconnection to occur. The standard picture for the eruption of a CME involves the reconnection of the magnetic field in a geometric configuration in which the rising CME is connected to an X-type magnetic neutral point via an extended current sheet (Figure 20). Models (Lin and Forbes, 2000; Linker *et al.*, 2001) predict that an extended, long-lived current sheet must be formed for any physically plausible reconnection rate. Formation of the current sheet in such a configuration drives conversion of the free energy in the magnetic field to thermal and kinetic energy, and can cause significant observational consequences, such as growing post-flare/CME loop system in the corona and fast ejections of the plasma and the magnetic flux.

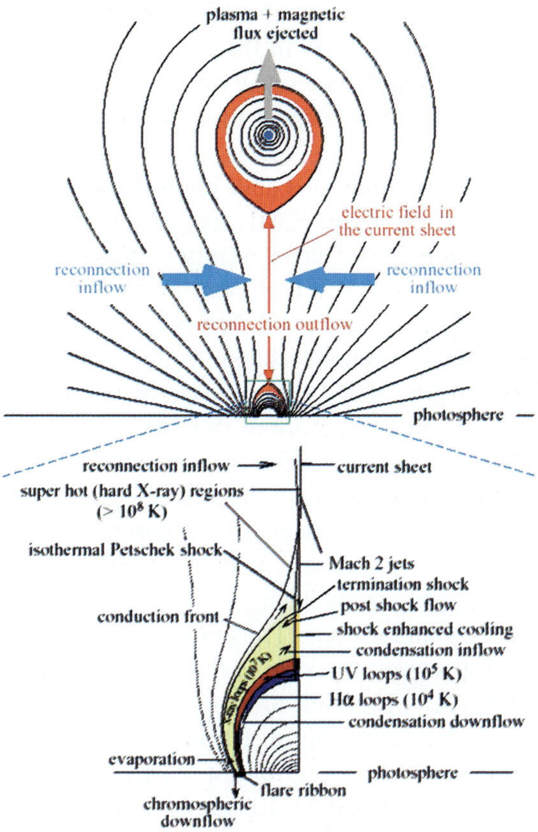

Figure 20. Schematic picture of flare loops, CME, and the current sheet between them (Lin *et al.*, 2004). *Upper part:* Sketch of the flux rope/CME model of Lin and Forbes (2000) showing the eruption of the flux rope, the current sheet formed behind it, and the postflare/CME loops below, as well as the inflows and outflows associated with the reconnection. *Lower part:* Enlarged view of the postflare/CME loops (adapted from Forbes and Acton, 1996). The upper tip of the cusp rises as reconnection happens continuously.

In order to confirm the role of reconnection in CME initiation it is therefore necessary (although not sufficient) to determine whether current sheets can be identified behind an erupting CME. Such a current sheet would be extremely thin due to the high electrical conductivity of the solar corona (see Ko *et al.*, 2003) making observation of this phenomenon difficult. On the other hand, because it is difficult to dissipate the current sheet once it forms it can exist as a well-defined structure for hours or even days (Lin, 2002); indeed, a coronal helmet streamer is normally stable. Despite many decades of indirect observations suggesting the presence of a current sheet, such as soft X-ray cusp structures (Tsuneta, 1996) and horizontal inflow (Yokoyama *et al.*, 2001) direct observations of the formation and evolution of a current sheet in the solar atmosphere have been lacking. However, the observations from white light data discussed above sometimes do suggest the presence of large-scale current sheets in the solar corona associated with CMEs. Joint EUV (EIT), hard X-ray and radio observations of a type IV bursts also suggest magnetic reconnection in a current sheet (see Section 4, subsection on type IV emission).

Recently, a number of investigations have confirmed the presence of large-scale narrow structures behind an erupting CME suggestive of the classic current sheet topology. Sui and Holman (2003) have reported the formation of a large-scale current sheet associated with an M1.2 flare observed by RHESSI on 15 April 2002. This conclusion is based on a number of different facets observed. In particular, the RHESSI images show a dramatic change in the apparent magnetic topology suggesting a transition from an X-type to a Y-type configuration as the current sheet formed. The temperature structure of this flare suggests that a current sheet formed between a source high in the solar corona and the top of the flaring loops.

CME eruptions observed in white light and UV coronagraph data have also pointed to the existence of current sheets in the corona. Webb *et al.* (2003) show that bright rays observed in conjunction with CMEs with SMM are consistent with the existence of current sheets lasting for several hours and extending more than 5 solar radii into the outer corona. The evidence for current sheets was further strengthened by UVCS observations which exhibited a very narrow, very hot feature in the Fe XVIII line between the arcade and the CME, consistent with the eruption-driven current sheet predicted by the initiation models (Ciaravella *et al.*, 2002). Finally, Ko *et al.* (2003) analyzed an eruption which occurred on January 8, 2002 and which was observed by a LASCO, EIT and CDS on SOHO and the Mark IV K-Coronameter on Mauna Loa. They found that the properties and behavior of the observed plasma outflow and the highly ionized states of the plasma inside these streamer-like structures expected from magnetic reconnection in a current sheet (see Pick *et al.*, 2006, this volume). Figure 21 shows a reconstructed view of the April 21, 2002 CME as it would be seen from the West limb in O VI and [Fe XVIII] (Lee *et al.*, 2006). This reconstruction, based on the intensities and Doppler shifts observed by UVCS, shows a bar-like structure in the high temperature [Fe XVIII] emission (dark red), which is most easily interpreted as the part of the current sheet just below the CME core, though other interpretations cannot be excluded.

The aftermath of reconnection often includes signatures of strong heating to many millions of degrees in the solar corona. Yohkoh/SXT demonstrated that in the late phase of CME-related solar flares, high temperature soft X-ray cusp-shaped features were common (Tsuneta *et al.*, 1992) and thus this morphology points to an energy source in the solar corona suggestive of large-scale reconnection expected from many flare models.

Potentially stronger soft X-ray evidence was presented by Sterling and Hudson (1997) who found a characteristic X-ray morphology to be associated with CME formation, as detected by the formation of "transient coronal holes," a form of dimming originally identified in Skylab soft X-ray images. The sigmoid, identified

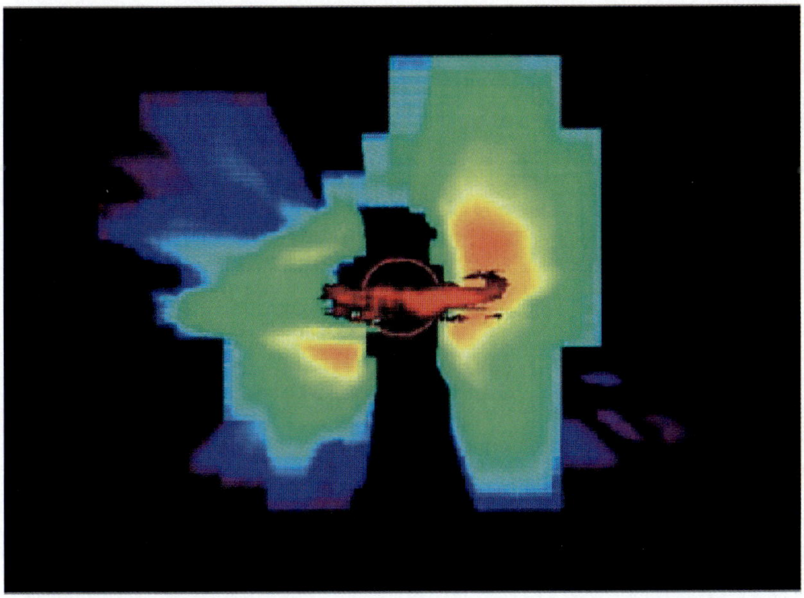

Figure 21. Image of the April 21, 2002 limb CME as it would be seen from the West limb based on a 3D reconstruction from UVCS data. The O VI emission (rainbow colors) shows the expected halo morphology, while the [Fe XVIII] emission (dark red) shows a bar-like structure which is just below the CME core when viewed from Earth or from solar north. The reconstruction uses Doppler velocities to determine the structure along the line of sight (Lee *et al.*, 2006).

with a hot component aligned with a filament channel, evolves during the event into an arcade perpendicular to it. This suggests "dipolarization" responsible for energy release following reconnection.

The cusp configuration in itself does not unequivocally imply that reconnection is ongoing. Magnetic reconnection in such a geometry also requires the presence of an inflow into the X-point (in a 2D representation). This has been surprisingly hard to detect, perhaps because the inflowing plasma is expected to be of extremely low-density being associated with open or opening field lines. Combining soft X-ray and EUV observations of a flare on the East limb of the Sun, Yokoyama *et al.* (2001) found a large-scale coronal cusp with strong evidence for an inflow of plasma into the cusp (X-point) region. Figure 22 shows a time-distance plot for the EUV emission integrated along a line passing through the observed location of the X-point. The figure clearly shows the 1 MK plasma moving inwards. Lin *et al.* (2005) show UV and visible-light evidence for both inflow and outflow in a post-CME current sheet.

An intriguing and perplexing dynamic phenomenon associated with erupting flares and CMEs has received attention in recent years. Modern soft X-ray and EUV observations frequently show a fan of spikes extending above the loop system of a flare arcade. McKenzie and Hudson (1999) found structures moving downward

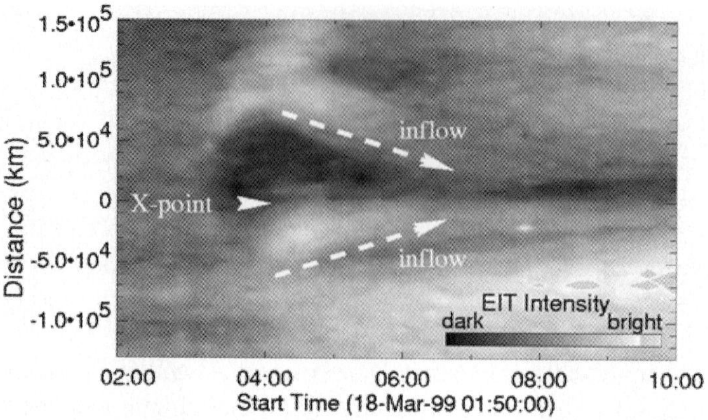

Figure 22. Time Evolution of the one-dimensional distribution of EUV intensity across the X-point, from Yokoyama *et al.* (2001).

through this fan, normally voids (Innes *et al.*, 2003a), during flare evolution. These downflows are most readily seen in recent TRACE observations of the 2002 Apr 21 event (Figure 23). It is not yet clear whether these downflows in the lower corona are the same phenomenon observed in white light at much greater distances from the Sun. However, given the apparent magnetic nature of the downflow structures observed by TRACE and SUMER, the association with post-reconnection dynamics is attractive. This ready identification with the reconnection outflow jet, expected from the models is problematic, however, because the speeds observed (up to a few

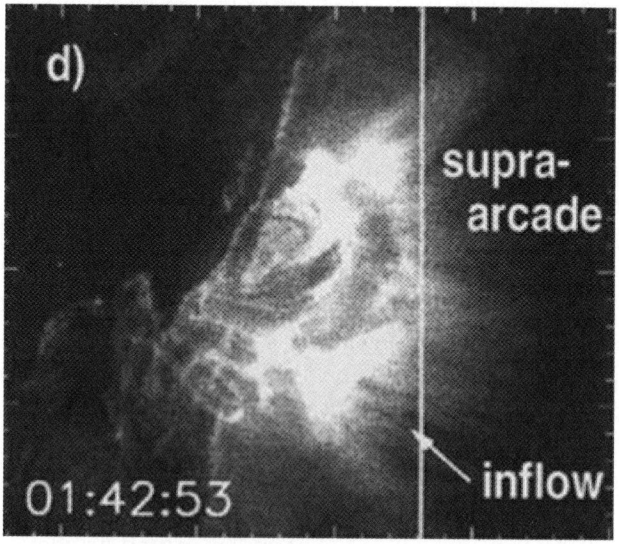

Figure 23. Dark lanes due to downflows as seen in a TRACE image (Innes *et al.*, 2003b).

hundred km/s) seem much slower than the super-Alfvénic flows expected from the models, as indicated in Figure 20.

One of the key "benefits" of reconnection in the corona is the conversion of magnetic energy into various other energetic manifestations. An important component of this energy release is the production of energetic particles which then interact with the ambient solar atmosphere to produce radiation signatures at radio, hard X-ray and gamma-ray energies. The relationship between this powerful particle acceleration and magnetic reconnection is not understood theoretically at present. In solar flares, the hard X-rays not only emanate from the chromosphere (the footpoint sources) but also, on occasion, from high up in the corona.

The hard X-ray footpoints exhibit strongly correlated fluctuations implying a common source of particles in an acceleration region high in the corona (e.g., Krucker and Lin, 2002). The geometry implied by the simultaneous and spectrally similar production of hard X-ray photons in the chromosphere points to an energy release at a favorable location in the corona where the free magnetic energy can be tapped. Further support for a relation between particle acceleration and reconnection comes from the discovery of an above-the-loop-top hard X-ray source (Masuda *et al.*, 1994) which suggests the reconnection geometry more directly.

In association with the hard and soft X-ray signatures discussed above, observations in the radio provide important evidence for reconnection in the impulsive phase of solar eruptive events. Radio emission at metric and decimetric frequencies provides direct observations of high energy electrons in the solar corona, with the temporal and frequency behavior of the radio data yielding diagnostic information about the origins of the particles. Recently, Karlický *et al.* (2004) have reported on a rare series of slowly drifting structures (drifting in the sense that the frequency decreases, which means an outward motion through a region of decreasing density) observed during two solar flares in the 0.8–4.5 GHz frequency range. These drifts are thought to map the flare magnetic field reconnection as magnetic plasmoids are formed in an extended current sheet due to tearing processes (see also Kliem *et al.*, 2000). Radio observations may also strengthen the case for reconnection in the erupting solar corona (Aurass *et al.*, 2002).

Finally, strong observational evidence for reconnection comes from the post-eruption emission and dynamics. In particular, the behavior of flare ribbons has provided strong support for the reconnection scenario. The behavior of the outer edges of the ribbons especially has pointed in this direction, since these outer edges exhibit irregular Hα line profiles. Recently SOHO/CDS observations have shown the presence of the required "chromospheric evaporation" directly via EUV spectroscopic imaging (Czaykowska *et al.*, 1999).

5.2. Reconnection Summary

The array of coronal and chromospheric manifestations associated with a CME throughout its build-up, initiation and evolution all clearly point to a major reconfiguration of the magnetic field in the solar corona. The energy required by these manifestations is unarguably provided by the magnetic field, yet we are to date still unable to determine exactly how this energy conversion occurs. Magnetic reconnection, as we presently understand it, seems to provide many of the characteristics required but we are limited both observationally by the extremely small scale on which the reconnection physics occurs and theoretically by the complexity of the reconnection processes in 3D. The many solar signatures presented above are consistent with the general idea of magnetic reconnection playing an important role in CMEs and solar flares. However, even if reconnection occurs as part of the CME process we are still not in a position to determine whether reconnection is required to initiate a CME or whether it results in response to a CME. Before we can state unequivocally that magnetic reconnection plays an important role in the dynamics of an event we need a better understanding of the physics of reconnection in 3D and how it occurs in the magnetic and plasma conditions of the solar corona.

6. Future Observations Needed

6.1. Coronagraphs

Looking into the future, we need to fill in the gaps in our knowledge of CMEs such as their initiation mechanism, their 3-dimensional structure and dynamics, their interaction with the solar wind, and their internal ionization state and elemental composition. Fortunately, efforts are already under way to attack the first two problems. STEREO is planned for launch in 2006. Solar Orbiter and Solar Polar Imager (SPI), two missions currently under study, will allow us to view the corona from viewpoints outside the ecliptic plane. This enables imaging of the entire streamer belt at once, which will immediately answer a number of questions related to interaction of CMEs with the solar wind.

What is further needed are high spatial resolution and high cadence observations at low coronal heights, as close as possible to the solar limb. Because the imaging quality deteriorates and the stray light increases rapidly for small distances from the occulter, future coronagraphs must locate the occulter on long booms (of the order of a few tens of meters) to achieve arcsec resolution for heights $<1.3\ R_{\odot}$. We know that the initial CME development occurs very rapidly in the low corona (<15 min) and involves plasma at many temperatures. Therefore, progress on the initiation and composition of the CMEs requires both white light and spectroscopic imaging observations at a cadence of about 1 minute and in several coronal and transition region lines. All these requirements point towards large aperture ($\sim$30 cm) instruments (for high throughput), in near Earth orbit (for high telemetry downlink rates) with a significant complement of mechanisms, filters and polarizers to acquire both white light and spectroscopic imaging observations of the inner corona. Only with such instruments will we be able to build a comprehensive picture of the very complex evolution of the CME in the lower corona.

6.2. EUV Telescopes and Spectrographs

Future EUV imaging experiments will require improved spatial resolution and time cadence in order to identify reconnection sites and relate them to CME initiation. To understand the sizes and dynamics, resolutions as high as 10–100 km will be needed. The combination of high spatial resolution and cadence implies a high enough sensitivity to detect a reasonable number of photons per pixel per exposure. Full Sun coverage is important because of the unpredictability of CMEs. Coverage of a broad temperature range is also needed in order to follow the rapidly changing temperatures as plasma is heated and as it cools adiabatically when it is ejected. This requires an adequate number of spectral bands and careful choice of their wavelengths. Temperatures as high as 2×10^7 K are accessible. Major progress on these issues can be expected from the Solar Dynamics Observatory (SDO) mission currently under development.

Future EUV spectral experiments also require high spatial resolution and cadence, though the cadence may not be as high as that achieved by the imaging experiments. Spectral resolution adequate to detect Doppler shifts of a few km/s will provide the third dimension for dynamical studies. The biggest challenge may be obtaining a broad enough spectral range to simultaneously determine densities and compositions over a broad temperature range. While there exist a number of spectral bands that include some density-sensitive line ratios and some lines formed over a broad temperature range, a complete characterization of the CME plasma requires quite a large number of spectral lines. As these lines vary greatly in strength and brightness, a large dynamic range is crucial. In addition, a coronagraphic capability is important for studying CME evolution as the eruption reaches its terminal mass and energy. The next EUV spectrograph planned is the EIS instrument on Solar-B, that is to be launched in late 2006.

6.3. Views from Multiple Perspectives

Our understanding of the CME phenomenon is still hindered by a fundamental observational limitation. Imaging observations are necessarily projections on the plane of the sky. We have no direct knowledge of the 3-dimensional distribution and evolution of the CME plasma along the line of sight. Even the radial propagation direction of the center of mass, a very important quantity for space weather studies cannot be reliably measured and therefore must be inferred indirectly (say from the location of the associated active region). The obvious approach on this problem is to obtain simultaneous observations from multiple perspectives and invert them using tomography techniques to obtain an estimate of the 3D morphology of an event.

This is precisely the objective of the STEREO mission to be launched in 2006. It consists of two identical spacecraft carrying a complement of EUV full disk imagers, white light coronagraphs and in-situ instruments. The two spacecraft will be placed at 1 AU orbits ahead and behind the Earth, respectively. The scale of solar structures accessible to reconstruction is proportional to the spacecraft separation angle (e.g., Wiegelmann, 2004). Therefore, the continuously changing perspective of the observations from the STEREO spacecraft will allow evaluation of the 3D structure of many solar features, from small scale loops to Earth-directed CMEs. STEREO also carries heliospheric imagers (HI) always observing space all along the Sun-Earth line which will enable us to follow CMEs all the way to the Earth.

6.4. Radio Arrays

Two ground based radio arrays, the Frequency Agile Solar Radio Telescope (FASR) and the Low Frequency Array (LOFAR), together with the space-based Solar Imaging Radio Array (SIRA) are presently being developed or planned to provide images in the frequency range 24 GHz to 30 kHz. FASR, LOFAR and SIRA operating jointly will be able to image electron beams, shocks and CMEs during their propagation through the corona and heliosphere up to 1 AU.

6.5. Solar Orbiter

The Solar Orbiter mission presently being studied by ESA (Marsch *et al.*, 2002) will provide the next major step forward in the exploration of the Sun and the heliosphere. It incorporates both a near-Sun and a high-latitude phase.

These are the scientific goals for the mission:

- Determine the properties, dynamics and interactions of plasma, fields and particles in the near-Sun heliosphere;
- Investigate the links between the solar surface, corona and inner heliosphere;
- Explore, at all latitudes, the energetics, dynamics and fine-scale structure of the Sun's magnetized atmosphere;
- Probe the solar dynamo by observing the Sun's high-latitude field, flows and seismic waves.

6.6. The Sentinel Mission

As part of NASA's Living With a Star (LWS) program the planned Sentinel mission will develop the physical understanding necessary to reliably model and predict the radiation environment for Lunar and Martian missions. Sentinels will accomplish this by discovering the physical conditions and mechanisms that govern heliospheric initiation, propagation and solar connection of those energetic phenomena that adversely affect space exploration and life and society here on Earth.

The Sentinels mission will comprise four spinning spacecraft that will approach the Sun to as close as 0.25 AU. They will be separated in solar longitude such as to allow measuring the spatial extent of shock fronts, CMEs and SEP fluxes. They will be assisted by a far-side Imaging Sentinel spacecraft orbiting at some 120° offset from the Earth. Its set of remote-sensing instruments will provide global solar photospheric magnetic field measurements and coronal plasma density and composition observations. A Near Earth Sentinel is the third component of the mission. Its white light and UV coronagraphic capabilities provide remote sensing of the corona.

6.7. Solar Polar Imager

The Solar Polar Imager (SPI) mission uses solar sail propulsion to place a spacecraft in a 0.48 AU circular orbit around the Sun with an inclination of 75°. The rapid 4 month polar orbit and the combined in situ and remote sensing instrument suite allow unprecedented studies of the link between the Sun and the solar wind and solar energetic particles. Moreover, SPI can serve as a pathfinder for a permanent solar polar sentinel for space weather prediction in support of the NASA Vision for Space Exploration.

The feasibility of this most desirable mission hinges upon the success of some new and highly ambitious technical developments that are presently underway.

6.8. The KuaFu Space Weather Explorer Mission

KuaFu is an initiative by China for a major scientific space mission. It will be an essential element of the International Living with a Star (ILWS) mission line-up. KuaFu is composed of three spacecraft: KuaFu-A will be located at the L1 libration point between Sun and Earth, and it will have instruments to continuously observe the solar disk in the chromospheric Lyman-alpha and coronal extreme ultraviolet emission, to register CMEs in white-light and Lyman-alpha radiation, and to measure in-situ radio waves, the local solar wind plasma and magnetic field, solar energetic particles, as well as the hard X-ray and Gamma-ray spectrum. KuaFu-B1 and KuaFu-B2 are a satellite pair in an Earth polar orbit chosen to allow continuous (24 hours a day) observation of the northern auroral oval and ring current regions.

KuaFu data will be used for the scientific study of space weather phenomena and to raise the standard of end-to-end observation of the Sun-Earth system.

6.9. Measurements of Magnetic Fields in the Chromosphere and the Corona

The dynamics of the solar atmosphere is governed by the magnetic field. In particular, it determines the coronal structure and dynamics from the upper chromosphere through the corona out into the heliospheric environment. The crucial role played by the magnetic field awards the development of measurement methods an extremely high priority in observational solar physics. However, the magnetic field in the corona is so weak that none of the well-known techniques (e.g., using the Zeeman effect) can easily be applied. Lin *et al.* (2004) report a measurement (using coronal Zeeman magnetometry) of the line-of-sight coronal magnetic field 100 arcsec above an active region implying a flux density of about 4 Gauss. They believe that such measurements will be some of the most exciting new opportunities that the Advanced Technology Solar Telesope (ATST) to be built in a few years time will realize. Using spectropolarimetry of the infrared He I line at 1083 nm, chromospheric vector fields of some tens to hundreds of Gauss (see, e.g., Solanki *et al.*, 2003) have been measured. Comparison with field extrapolations from a photospheric magnetogram showed that a non-linear force-free approximation reproduced the observations best (Wiegelmann *et al.*, 2005). Other techniques for inferring the coronal magnetic fields involve Faraday rotation of polarized solar radiation or radio gyrosynchrotron magnetometry, but these techniques are still developing. At this moment it is not known when such measurements will become available. However, it is clear that quantitative measurements of the magnetic field in the upper solar atmosphere are required to solve several of the burning issues on solar physics.

References

Akmal, A., *et al.*: 2001, *ApJ* **553**, 922.

Asai, A., Yokoyama, T., Shimojo, M., and Shibata, K.: 2004, *ApJ* **605**, L77.

Aurass, H., Vršnak, B., and Mann, G.: 2002a, *A&A* **384**, 273.

Bastian, T. S., and Gary, D. E.: 1997, *JGR* **102**, 14031.

Bastian, T. S., Pick, M., Kerdraon, A., Maia, D., and Vourlidas, A.: 2001, *ApJL* **558**, L65.

Bastian, T. S.: 2003, *Adv. Space Res.* **32**, 2705.

Berger, M.: 1984, *Geophys. Astrophys. Fl. Dyn.* **30**, 79.

Bogod, V. M., Grebinskij, A. S., Opeikina, L. V., and Gelfreikh, G. B.: 1998, *Three-Dimensional Structure of Solar Active Regions*, ASP Conf. Ser. 155, p. 279.

Boischot, A.: 1957, *Comptes Rendus Acad. Sci. Paris* **244**, 1326.

Boischot, A., and Clavelier, B.: 1968, *IAU Symposium 35: Structure and Development of Solar Active Regions*, p. 565

Bothmer, V., and Schwenn, R.: 1996, *Adv. Sp. Res.* **17**, 319.

Brueckner, G. E., Koomen, M. J., and Tousey, R.: 1972, *Sol. Phys.* **24**, 254.

Brueckner, G. E.: 1974, *IAU Symposium* **57**, 333.

Brueckner, G. E., *et al.*: 1995, *Sol. Phys.* **162**, 357.

Burkepile, J. K., Hundhausen, A. J., Stanger, A. L., St. Cyr, O. C., and Seiden, J. A.: 2004, *JGR* **109**, A03101.
Burlaga, L. F., Behannon, K. W., and Klein, L. W.: 1987, *J. Geophys. Res.* **92**, 5725.
Burlaga, L. F., Plunkett, S. P., and St. Cyr., O. C.: 2002, *JGR* **107**, SSH1B.
Canfield, R. C., *et al.*: 1993, *ApJ* **411**, 362.
Canfield, R. C., Hudson, H. S., and McKenzie, D. E.: 1999, *GRL* **26**, 627.
Chen, J., *et al.*: 2000, *Astrophys. J.* **533**, 481.
Ciaravella, A., *et al.*: 1997, *ApJ Lett.* **491**, 59.
Ciaravella, A., *et al.*: 2000, *ApJ* **529**, 575.
Ciaravella, A., Raymond, J. C., Li, J., Reiser, P., Gardner, L. D., Ko, Y.-K., *et al.*: 2002, *ApJ* **575**, 1116.
Ciaravella, A., Raymond, J. C., van Ballegooijen, A., Strachan, L., Vour-lidas, A., Li, J., *et al.*: 2003, *ApJ* **597**, 1118.
Ciaravella, A., Raymond, J. C., Kahler, S., Vourlidas, A., and Li, J.: 2005,*ApJ* **621**, 1121.
Ciaravella, A., Raymond, J. C., and Kahler, S. K.: 2006, *ApJ*, submitted.
Claßen, H. T., Mann, G., Klassen, A., and Aurass, H.: 2003, *A&A* **409**, 309.
Cremades, H., and Bothmer, V.: 2004, *A&A* **422**, 307.
Cremades, H., and Bothmer, V.: 2005, *IAU Symposium* **226**, 48.
Crifo, F., Picat, J. P., and Cailloux, M.: 1983, *Solar Phys.* **83**, 143.
Czaykowska, A., de Pontieu, B., Alexander, D., and Rank, G.: 1999, *ApJL* **521**, L75.
Delaboudiniére, J.-P., Artzner, G. E., Brunaud, J., *et al.*: 1995, *Solar Phys.* **162**, 291.
Démoulin, P., Mandrini, C. H., van Driel-Gesztelyi, L., Thompson, B. J., Plunkett, S., Kovári, Zs., *et al.*: 2002 *A&A* **382**, 650.
Dere, K. P., Wang, D., and Howard, R.: 2005, *ApJL* **620**, L119.
DeVore, C. R.: 2000, *ApJ* **539**, 954.
Dobrzycka, D., Raymond, J. C., Biesecker, D. A., Li, J., and Ciaravella, A.: 2003, *ApJ* **588**, 586.
Domingo, V., *et al.*: 1995, *Sol. Phys.* **162**, 1.
Dulk, G. A., and McLean, D. J.: 1978, *Solar Phys.* **57**, 279.
Erickson, W. C., Kassim, N. E., and Perley, R. A.: 2000, *Radio Astronomy at Long Wavelengths*, p. 303.
Filippov, B., and Koutchmy, S.: 2002, *Sol. Phys.* **208**, 283.
Forbes, T. G., and Acton, L. W.: 1996 *ApJ* **459**, 330.
Forbes, T. G., Linker, J. A., *et al.*: 2006, *Space Sci. Rev.* this volume, 10.1007/s11214-006-9019-8.
Frost, K. J.: 1974, *IAU Symp. 57: Coronal Disturbances*. p. 421.
Gallagher, P. T., Dennis, B. R., Krucker, S., Schwartz, R. A., and Tolbert, A. K.: 2002, *Solar Phys.* **210**, 341.
Gary, D. E., and Hurford, G. J.: 1990, *ApJ* **36**, 290.
Gary, D. E., and Keller, C. U.: 2004, *Solar and Space Weather Radiophysics – Current Status and Future Developments* (Dordrecht: Kluwer).
Gergely, T. E., Kundu, M. R., Munro, R. H., and Poland, A. I.: 1979, *ApJ* **230**, 575.
Gibson, S. E., *et al.*: 2002, *ApJ* **574**, 1021.
Gibson, S. E., and Low, B. C.: 1998, *ApJ* **493**, 460.
Gilbert, H. R., Serex, E. C., Holzer, T. E., MacQueen, R. M., and McIntosh, P. S.: 2001, *ApJ* **550**, 1093.
Gonzalez, W. D., and Tsurutani, B. T.: 1987, *PSS* **35**, 1101.
Gopalswamy, N., and Kundu, M. R.: 1989, *Solar Phys.* **122**, 145.
Gopalswamy, N., and Kundu, M. R.: 1990, *ApJL* **365**, L31.
Gopalswamy, N., and Kundu, M. R.: 1992, *ApJL* **390**, 37.
Gopalswamy, N., Kundu, M. R., Hanaoka, Y., Enome, S., Lemen, J. R., and Akioka, M.: 1996, *New Astron.* **1**, 207.

Gopalswamy, N., Hanaoka, Y., Kundu, M. R., Enome, S., Lemen, J. R., Akioka, M., and Lara, A.: 1997, *ApJ* **475**, 348.
Gopalswamy, N., Nitta, N., Manoharan, P. K., Raoult, A., and Pick, M. 1999, *A&A* **347**, 684.
Gopalswamy, N., Kaiser, M. L. Jones, D. L., and The Alfa Team: 1999, in: T. S. Bastian, N. Gopalswamy and K. Shibasaki, (eds.), *Proceedings of the Nobeyama Symposium*, held in Kiyosato, Japan, Oct. 27-30, 1998, NRO Report No. 479., p. 447.
Gopalswamy, N., and Thompson, B. J.: 2000, *J. Atm. Solar-Terres. Phys.* **62**, 16, 1457.
Gopalswamy, N., *et al.*: 2000a, *Geophys. Res. Lett.* **27**, 145.
Gopalswamy, N. Hanaoka, Y., and Hudson: 2000, *Adv. Space Res.* **25**(9), 1851.
Gopalswamy, N., Yashiro, S., Kaiser, M. L., Howard, R. A., Bougeret, J.-L.: 2001, *ApJ* **548**, L91–L94.
Gopalswamy, N., Yashiro, S., Michalek, G., Kaiser, M. L., Howard, R. A., Reames, D. V., *et al.*: 2002, *ApJ* **572**, L103–L107.
Gopalswamy, N., *et al.*: 2003, *GRL* **30**(12), SEP 3–1.
Gopalswamy, N., Lara, A., Yashiro, S., and Howard, R. A.: 2003b, *ApJ* **598**, L63.
Gopalswamy, N.: 2004a, in: Poletto, G., and Suess, S. (ed.), *The Sun and the Heliosphere as an Integrated System*, ASSL series, (Kluwer: Boston), p. 201.
Gopalswamy, N., Yashiro, S., Krucker, S., Stenborg, G., and Howard, R. A.: 2004, *J. Geophys. Res.* **109**, 12105.
Gopalswamy, N., Yashiro, S., Krucker, S., and Howard, R.A: 2005, in: Dere, K. P., Wang, J., and Yan, Y. (eds.), *Coronal and Stellar Mass Ejections*, International Astronomical Union, p. 367.
Gopalswamy, N., Yashiro, S., Liu, Y., Michalek, G., Vourlidas, A., Kaiser, M. L., *et al.*: 2005a, *J. Geophys. Res.* **110**, A09S15.
Gopalswamy, N., Fleck, B., and Gurman, J. B.: 2005b, in: Rao, M., and Murthy, R. L. N. (eds.),*IAA "Bringing Space Benefits to the Asia Region"* (Bangalore: Asronautical Society of India).
Gopalswamy, N., Mikić, Z., *et al.*: 2006, *Space Science Reviews*, this volume, 10.1007/s11214-006-9020-2.
Gosling, J. T., Birn, J., and Hesse, M.: 1995, *GRL* **22**, 869.
Green, L. M., López Fuentes, M. C., Mandrini, C. H., Démoulin, P., Van Driel-Gesztelyi, L., Culhane, J. L.: 2002, *Sol. Phys.* **208**, 43.
Habbal, S. R., Mossman, A., Gonzalez, R., and Esser, R.: 1996, *JGR* **101**, 19943.
Hanaoka, Y., *et al.*: 1994 *PASJ* **46**, 205.
Handy, B. N., *et al.*: 1999, *Sol. Phys.* **187**, 229.
Harrison, R. A.: 1986, *A&A* **162**, 283.
Harrison, R. A.: 2003, *Adv. Sp. Res.* **32**, 2425.
Harrison, R. A., *et al.*: 1995, *Sol. Phys.* **162**, 233.
Harvey, J. W., and Sheeley, N. R.: 1979, *SSRv* **23**, 139.
Hill, S. M., *et al.*: 2005, *Sol. Phys.* **226**, 255.
Howard, R. A., Michels, D. J., Sheeley, N. R., Jr., and Koomen, M. J.: 1982, *Astrophys. J.* **263**, L101.
Howard, R. A., *et al.*: 1985, *JGR* **90**, 8173.
Howard, R. A., *et al.*: 1997, in: Crooker, N., Joselyn, J., and Feynman, J. (eds.), *Coronal Mass Ejections*, (Washington DC: Amer. Geophys. Union), p. 17.
Howard, R. A., Moses, J. D., Socker, D. G., Dere, K. P., and Cook, J. W.: 2002, *Adv. Sp. Res.* **29**(12), 2017.
Hudson, H. S., Acton, L. W., and Friedland, S. L.: 1996, *ApJ* **470**, 629.
Hudson, H. S., Acton, L. W., Harvey, K. L., and McKenzie, D. E.: 1999, *ApJL* **513**, L83.
Hundhausen, A. J.: 1985, in: Stone, R. G., and Tsurutani, B. T. (eds.), *Collisionless Shocks in the Heliosphere, A Tutorial Review*, (Washington: AGU), p. 37.
Hundhausen, A. J.: 1987, *Sixth International Solar Wind Conference*, p. 181.
Hundhausen, A. J.: 1993, *JGR* **98**, 13177.
Illing, R. M. E., and Hundhausen, A. J.: 1983, *J. Geophys. Res.* **88**, 10,210.

Innes, D. E., Curdt, W., Schwenn, R., Solanki, S., Stenborg, G., and McKenzie, D. E.: 2001, *ApJL* **549**, L249.
Innes, D. E., McKenzie, D. E., and Wang, T.: 2003a, *Sol. Phys.* **217**, 267.
Innes, D. E., McKenzie, D. E., and Wang, T.: 2003b, *Sol. Phys.* **217**, 247.
Irimajiri, Y., Takano, T., Nakajima, H., Shibaska, K., Hanaoka, Y., and Ichimoto, K.: 1995, *Sol. Phys.* **156**, 363.
Kahler, S. W.: 1992, *Ann. Revs. Astron. Ap.* **30**, 113.
Kahler, S. W., and Hudson, H. S.: 2001, *JGR* **106**, 29,239.
Kahler, S. W., and Hundhausen, A. J.: 1992, *JGR* **97**, 1619.
Kahler, S. W.: 2001, *J. Geophys. Res.* **106**, 20947.
Kallenrode, M.-B., and Cliver, E. W.: 2001, *Proceedings of the ICRC 2001*, 3318.
Karlický, M., Fárnik, F., and Krucker, S.: 2004, *A&A* **419**, 365.
Kathiravan, C., Ramesh, R., and Subramanian, K. R.: 2002, *ApJL* **567**, L93.
Kaufmann, P., *et al.*: 2001, in: Pinho, J. T., Santos Cavalcante, G. P., and Oliveira, L. A. H. G. (eds.), *Proc. SBMO/IEEE MTT-S Int. Microwave and Optoelectronic Conf.*, (Piscataway, IEEE), p. 439.
Kaufmann, P., *et al.*: 2003, *J. Geophys. Res.* **108**, 1280.
Kerdraon, A., Pick, M., Trottet, G., Sawyer, C., Illing, R., Wagner, W., *et al.*: 1983, *ApJL* **265**, L19.
Kerdraon, A., and Delouis, J.: 1997, in: Trottet, G. (ed.), *Coronal Phsyics from Radio and Space Observations; Proceedings of the CESRA Workshop held in Nouan le Fuzelier*, (Berlin: Springer Verlag), p. 192.
Khan, J. I., and Hudson, H. S.: 2000, *GRL* **27**, 1083.
Kiplinger, A. L.: 1995, *ApJ* **453**, 973.
Kliem, B., Karlický, M., and Benz, A. O.: 2000, *A&A* **360**, 715.
Ko, Y.-K., Raymond, J. C., Lin, J., Lawrence, G., Li, J., and Fludra, A.: 2003, *ApJ* **594**, 1068.
Kohl, J., *et al.*: 1995, *Sol. Phys.* **162**, 313.
Kopp, R. A., and Pneuman, G. W.: 1976, *Sol. Phys.* **50**, 85.
Krucker, S., and Lin, R.: 2002, *Sol. Phys.* **210**, 229.
Krucker, S., Christe, S., Lin, R. P., Hurford, G. J., and Schwartz, R. A.: 2002, *Sol. Phys.* **210**, 445.
Kundu, M. R.: 1965, *Solar Radio Astronomy*, (New York: Interscience Publication).
Kusano, K., Maeshiro, T., Yokoyama, T., and Sakurai, T.: 2002, *ApJ* **577**, 501.
Lantos, P., Kerdraon, A., Rapley, C. G., and Bentley, R. D.: 1981, *A&A* **101**, 33.
Lara, A., Gopalswamy, N., Caballero-Lopez, R., Yashiro, S., and Valdes-Galicia, J.: 2005, *ApJ* **625**, 441.
Leamon, R. J., Canfield, R. C., and Pevtsov, A. A.: 2002, *JGR* **107**, SSH1.
Lee, J.-Y., Raymond, J. C., Ko, Y.-K., and Kim, K.-S.: 2006, *ApJ*, submitted.
Li, G., and Zank, G. P.: 2005, *GRL* **32**, L02101.
Lin, H., Kuhn, J. R., and Coulter, R.: 2004, *ApJ* **613**, L177.
Lin, J.: 2002, *ChJAA* **2**, 539.
Lin, J., and Forbes, T. G.: 2000, *JGR* **105**, 2375.
Lin, J., Raymond, J. C., and van Ballegooijen, A. A.: 2004, *ApJ* **602**, 422.
Lin, J., Ko, Y.-K., Sui, L., Raymond, J. C., Stenborg, G. A., Jiang, Y., *et al.*: 2005, *ApJ* **622**, 1251.
Lin, R. P., Dennis, B. R., Hurford, G. J., and 63 co-authors: 2002, *Solar Phys.* **210**, 3.
Linker, J., *et al.*: 1999, *JGR* **104**, 9809.
Linker, J., Lionello, R., Mikić and Amari, T.: 2001, *JGR* **106**, 25165.
Low, B.-C.: 1982, *ApJ* **254**, 796.
Low, B.-C.: 1984, *ApJ* **281**, 392.
Low, B.-C.: 1996, *Sol. Phys.* **167**, 217.
Low, B.-C.: 2001, *JGR* **106**, 25,141.
Luhmann, J. G., Li, Y., Zhao, X., and Yashiro, S.: 2003, *Solar Phys.* **213**, 367.
Mancuso, S., Raymond, J. C., Kohl, J., Ko, Y.-K., Uzzo, M., and Wu, R.: 2002, *A& A* **383**, 267.

Marqué, C.: 2004, *ApJ* **602**, 1037.

Martin, S. F., and McAllister, A. H.: 1996, *Magnetodynamic Phenomena in the Solar Atmosphere - Prototypes of Stellar Magnetic Activity*, IAU Colloq. 153, p. 497.

Marsch, E., Antonucci, E., Bochsler, P., Bougeret, J.-L., Fleck, B., Harrison, R., *et al.*: 2002, *Adv. Sp. Res.* **29**, 2027.

Masuda, S., Kosugi, T., Hara, H., Tsuneta, S., and Ogawara, Y.: 1994, *Nature* **371**, 495.

McKenzie, D. E., and Hudson, H. S.: 1999, *ApJ* **519**, L93.

McKenzie, D. E.: 2000, *Sol. Phys.* **195**, 381.

Metcalf, T. R., Jiao, L., McClymont, A. N., Canfield, R. C., and Uitenbroek, H.: 1995, *ApJ* **439**, 474.

Milkey, D. L., Canfield, R. C., Labonte, B. j., Leka, K. D., Waterson, M. F., and Weber, H. M.: 1996, *Sol. Phys.* **168**, 229.

Moon, Y.-J., Choe, G. S., Wang, H. M., Park, Y. D., Gopalswamy, N., Yang, G., *et al.*: 2002, *ApJ* **581**, 694.

Moran, T., and Davila, J.: 2004, *Science* **305**, 66.

Moreton, G. E., and Ramsey, H. E.: 1960, *PASP* **72**, 357.

Moses, D. Clette, F., Delabouniniere, J.-P., and 32 co-authors: 1997, *Sol. Phys.* **175**, 571.

Munro, R. H., *et al.*: 1979, *Solar Phys.* **61**, 201.

Nakajima *et al.*: 1994, *Proceedings of Kofu Symposium*, Kofu, Japan, Sept. 6–10, 1993, p. 185.

Nindos, A., and Zhang, H.: 2002, *ApJL* **573**, 133.

Panasyuk, A.: 1999, *JGR* **104**, 9721.

Parker, E. N.: 1961, in: Liller, W. (ed.), *Space Astrophysics*, (New York: McGraw-Hill), p. 157.

Parker, E. N.: 1963, in: Evans, J. W. (ed.), *Proceedings of IAU Symposium 16*, (New York: Academic Press), p. 11.

Payne-Scott, R., Yabsley D. E., and Bolton, J. G.: 1947, *Nature* **160**, 256.

Pick, M.: 1986, *Solar Phys.* **104**, 19.

Pick, M., Démoulin, P., Krucker, S., Malandraki, O., and Maia, D.: 2005, *ApJ* **625**, 1019.

Pick, M., Forbes, T. G., Mann, G., *et al.*: 2006, *Space Science Reviews*, this volume, 10.1007/s11214-006-9021-1.

Plunkett, S. P., *et al.*: 1998, *GRL* **25**, 2477.

Plunkett, S. P., *et al.*: 2000, *Solar Phys.* **61**, 201.

Pohjolainen, S., Maia, D., Pick, M., *et al.*: 2001, *Astrophys. J.* **556**, 421.

Ramaty, R., and Lingenfelter, R. E.: 1967, *JGR* **72**, 879.

Ramesh, R., Subramanian, K. R., Sundararajan, M. S., and Sastry, C. V.: 1998, *Sol. Phys.* **181**, 439.

Rao, A. P., Ananthakrishnan, S., Balasubramanian, V., and Coles, W. A.: 1995, *Solar Wind* **8**, 94.

Raouafi, N.-E., Mancuso, S., Solanki, S. K., Inhester, B., Mierla, M., Stenborg, G., *et al.*: 2004, *A&A* **424**, 1039.

Raymond, J. C., Thompson, B. J., St. Cyr, O. C., Gopalswamy, N., Kahler, S., Kaiser, M., *et al.*: 2000, *GRL* **27**, 1439.

Raymond, J. C., Ciaravella, A., Dobrzycka, D., Strachan, L., Ko, Y.-K., Uzzo, M., *et al.*: 2003, *ApJ* **597**, 1106.

Raymond, J. C., and Ciaravella, A.: 2004, *ApJL* **606**, L159.

Riddle, A. C.: 1970, *Sol. Phys.* **13**, 448.

Robinson, R. D.: 1985, *Solar Radiophysics: Studies of Emission from the Sun at Metre Wavelengths*, p. 385.

Rust, D. M.: 1983, *SSRv* **34**, 21.

Rust, D. M., and Kumar, A.: 1994, *Sol. Phys.* **155**, 69.

Rust, D. M.: 2001, *JGR* **106**, 25,075.

Sakurai, T., Shibata, K., Ichimoto, K., Tsuneta, S., and Acton, L. W.: 1992, *PASJ* **44**, 123.

Scherrer, P. H., *et al.*: 1995, *Sol. Phys.* **162**, 129.

Schwenn, R.: 1986, *Space Sci. Rev.* **44**, 139.

Schwenn, R., dal Lago, A., Huttunen, E., and Gonzales, W. D.: 2005, *Ann. Geophys.* **23**, 1033.
Sheeley, N. R., Howard, R. A., Koomen, M. J., and Michels, D. J.: 1983, *ApJ* **272**, 349.
Sheeley, N. R., Jr., Wang, Y.-M., Hawley, S. H., and other 15 authors: 1997, *ApJ* **484**, 472.
Sheeley, N. R., Jr., Walters, J. H., Wang, Y.-M, and Howard, R. A.: 1999, *JGR* **104**, 24,739.
Sheeley, N. R., Jr., Hakala, W. N., and Wang, Y.-M.: 2000, *JGR* **103**, 5081.
Sheeley, N. R., Jr., and Wang, Y.-M.: 2002, *ApJ* **579**, 874.
Sheridan, K. V.: 1970, *Proceedings of the Astronomical Society of Australia* **1**, 376.
Sime, D. G., and Hundhausen, A. J.: 1987, *JGR* **92**, 1049.
Simon, M., and Axford, W. I.: 1966, *PSS* **14**, 901.
Skumanich, A., Lites, B. W., Martinez, P. V., and Seagraves, P.: 1997, *ApJS* **110**, 357.
Smith, S. F., and Harvey, K. L.: 1971, in: Macris, C. J. (ed.) *Physics of the Solar Corona*, p. 156.
Solanki, S. K., Lagg, A., Woch, J., Krupp, N., and Collados, M.: 2003, *Nature* **425**, 692.
Spirock, T. J., Denker, C., Varsik, J., Shumko, S., Qiu, J., Gallagher, P., *et al.*: 2001, *AGUSM*, SP51B-06.
Srivastava, N., Schwenn, R., Inhester, B., Stenborg, G., and Podlipnik, B.: 1999, *Space. Sci. Rev.* **87**, 303.
Srivastava, N., Schwenn, R., Inhester, B., Martin, S. F., and Hanaoka, Y.: 2000, *ApJ* **534**, 468.
St. Cyr, O. C., *et al.*: 2000, *J. Geophys. Res.* **105**, 18,169.
Stenborg, G., Bagalá. L. G., Bauer, O. H., Borda, R. F., Francile, C., Haerendel, G., Rovira, M. G., *et al.*: 2000, *J. Atmos. Terrest. Phys.* **62**, 1553.
Sterling, A. C., and Hudson, H. S.: 1997, *ApJL* **491**, 55.
Sterling, A. C., Hudson, H. S., Thompson, B. J., and Zarro, D. M.: 2000, *ApJ* **532**, 628.
Stewart, R. T.: 1985, *Solar Radiophysics: Studies of Emission from the Sun at Metre Wavelengths*, p. 361.
Sui, L., and Holman, G. D.: 2003, *ApJL* **596**, L251.
Swarup, G.: 2000, *Radio Astronomy at Long Wavelengths*. p. 297.
Tandberg-Hanssen, E.: 1995, *The Nature of Solar Prominences* (Kluwer, Dordrecht).
Thompson, B. J., Plunkett, S. P., Gurman, J. B., Newmark, J. S., St. Cyr, O. C., and Michels, D. J.: 1998, *Geophys. Res. Lett.* **25**, 2465.
Thompson, B. J., *et al.*: 1999, *Astrophys. J.* **517**, L151.
Tripathi, D., Bothmer, V., and Cremades, H.: 2004, *ApJ* **422**, 337.
Tripathi, D., Solanki, S. K., Schwenn, R., *et al.*: 2006, *A&A*, in press.
Tsuneta, S.: 1996, *ApJ* **456**, 840.
Tsuneta, S., Hara, H., Shimizu, T., Acton, L. W., Strong, K. T., Hudson, H. S., and Ogawara, Y.: 1992, *PASJ* **44**, L63.
Vaiana, G. S., Davis, J. M., Giacconi, R., Krieger, A. S., Silk, J. K., Timothy, A. F., *et al.*: 1973, *ApJL* **185**, L47.
Vourlidas, A., *et al.*: 2002, in: Wilson, A. (ed.), *Solar Variability: From Core to Outer Frontiers*. 10th Eur. Sol. Phys. Mtg, Prague, Czech Republic. ESA SP-506, Vol. 1. Noordwijk: ESA Publications Division, p. 91.
Vourlidas, A., Subramanian, P., Dere, K. P., and Howard, R. A.: 2000, *ApJ* **534**, 456.
Vourlidas, A., Wu, S. T., Wang, A. H., Subramanian, P., and Howard, R. A.: 2003, *ApJ* **598**, 1392.
Wang, Y.-M., Sheeley, N. R., Jr., Howard, R. A., St. Cyr. O. C., and Simnett, G. M.: 1999, *Geophys. Res. Lett.* **26**, 1203.
Wang, Y.-M., and Sheeley, N. R. Jr.: 2002a, *ApJ* **567**, 1211.
Wang, Y.-M., and Sheeley, N. R. Jr.: 2002b, *ApJ* **575**, 542.
Webb, D. F., and Hundhausen, A. J.: 1987, *Sol. Phys.* **108**, 383.
Webb, D. F., Burkepile, J., Forbes, T. G., and Riley, P.: 2003, *JGR* **108**, SSH6-1.
Wiegelmann, T.: 2004, *Sol. Phys.* **219**, 87.

Wiegelmann, T., Lagg, A., Solanki, S. K., Inhester, B., and Woch, J.: 2005, *Astron. Astrophys.* **433**, 701.

White, S., Kassim, N. E., and Erickson, W. C.: 2003, in: Keil, S. L., and Avakyan, S. V. (eds.), *Innovative Telescopes and Instrumentation for Solar Astrophysics. Proceedings of the SPIE, Volume 4853*, p. 111.

Wiik, J. E., *et al.*: 1997, *Sol. Phys.* **175**, 411.

Wild, J. P.: 1969, *Sol. Phys.* **9**, 260.

Wilhelm, K., *et al.*: 1995, *Sol. Phys.* **162**, 189.

Yashiro, S., Gopalswamy, N., Michalek, G., and Howard, R. A.: 2003, *Adv. Space Res.* **32**, 2631.

Yashiro, S., Gopalswamy, N., Michalek, G., St. Cyr, O. C., Plunkett, S. P., Rich, N. B., *et al.*: 2004, *JGR* **109**, A07105.

Yokoyama, T., Akita, K., Morimoto, T., Inoue, K., and Newmark, J.: 2001, *ApJL* **546**, L69.

Zandanov, V. G., Altyntsev, A. T., and Lesovoi, S. V.: 1999, in: Bastian, T. S., Gopalswamy, N., and Shibasaki, K. (eds.), *Proceedings of the Nobeyama Symposium, Held in Kiyosato, Japan, Oct. 27-30, 1998*, NRO Report No. 479., p.425–428. p. 425.

Zhang, J., Dere, K. P., Howard, R. A., Kundu, M. R., and White, S. M.: 2001, *ApJ* **559**, 452.

Zhang, J., Dere, K. P., Howard, R. A., and Vourlidas, A.: 2004, *ApJ* **604**, 420.

Zhitnik, I., *et al.*: 2003a, *MNRAS* **338**, 67.

Zhitnik, I., *et al.*: 2003b, *Adv. Sp. Res.* **32**(12), 2573.

UNDERSTANDING INTERPLANETARY CORONAL MASS EJECTION SIGNATURES

Report of Working Group B

R. F. WIMMER-SCHWEINGRUBER[1,*], N. U. CROOKER[2], A. BALOGH[3], V. BOTHMER[4], R. J. FORSYTH[3], P. GAZIS[5], J. T. GOSLING[6], T. HORBURY[3], A. KILCHENMANN[7], I. G. RICHARDSON[8], J. D. RICHARDSON[9], P. RILEY[10], L. RODRIGUEZ[4], R. VON STEIGER[7], P. WURZ[11], and T. H. ZURBUCHEN[12]

[1]*Institut für Experimentelle und Angewandte Physik, Extraterrestrische Physik, Christian-Albrechts-Universityät zu Kiel, Germany*

[2]*Center for Space Physics, Boston University, Boston, MA, USA*

[3]*The Blackett Laboratory, Imperial College London, United Kingdom*

[4]*Max-Planck-Institut für Sonnensystemforschung, Lindau, Germany*

[5]*San Jose State University Foundation, NASA Ames Research Center, Moffett Field, CA, USA*

[6]*Laboratory for Atmospheric and Space Physicsm University of Colorado, Boulder, CO, USA*

[7]*International Space Science Institute, Bern, Switzerland*

[8]*NASA Goddard Space Flight Center, Greenbelt, MD, USA*

[9]*Massachusets Institute of Technology, Cambridge, MA, USA*

[10]*Science Applications International Corporation, San Diego, CA, USA*

[11]*Physikalisches Institut, Universität Bern, Switzerland*

[12]*Dept. of Atmosph. ,Oceanic, and Space Sciences, University of Michigan, Ann Arbor, MI, USA*

*(*Author for correspondence: E-mail: wimmer@physik.uni-kiel.de)*

(Received 30 November 2005; Accepted in final form 24 May 2006)

Abstract. While interplanetary coronal mass ejections (ICMEs) are understood to be the heliospheric counterparts of CMEs, with signatures undeniably linked to the CME process, the variability of these signatures and questions about mapping to observed CME features raise issues that remain on the cutting edge of ICME research. These issues are discussed in the context of traditional understanding, and recent results using innovative analysis techniques are reviewed.

Keywords: coronal mass ejections, interplanetary physics, solar wind

1. Introduction

Coronal mass ejections (CMEs) and their interplanetary manifestations, interplanetary coronal mass injections (ICMEs,) are still poorly understood entities. A decade ago, Schwenn (1996) gave an extensive list of unsolved problems and questions. Since then, the Solar and Heliospheric Observatory (SOHO) has much improved and to some extent revolutionized our understanding of CMEs and ICMEs. Our current knowledge of the relation of solar observables with ICMEs is discussed by Crooker and Horbury (2006) and Forsyth *et al.* (2006) in this volume. This paper focuses on recent progress in understanding in-situ signatures of ICMEs, building upon the introductory paper by Zurbuchen and Richardson (2006, this volume).

Space Science Reviews (2006) 123: 177–216
DOI: 10.1007/s11214-006-9017-x

ICME identification is not always straightforward, as will become apparent in this paper. Progress toward the unambiguous in-situ identification of ICMEs is of considerable scientific interest and is important to space weather applications (see Siscoe, 2006, this volume).

Ideally, the observation of CMEs and the developing ICMEs would take place in the innermost heliosphere, at a distance of a few solar radii using a fleet of several spacecraft. There, interactions with the ambient solar wind would not have had much chance of modifying the original plasma state of the CME, thus allowing a more direct linkage of remotely sensed structures with in-situ observations. In the real world, we have a plethora of observations, both in situ and remote, the latter mainly from the vicinity of the Earth, at 1 AU, with some observations at other heliocentric distances (e. g., Helios, Ulysses, Voyager, or Pioneer). This paper begins by discussing progress in understanding these various in situ signatures, mainly at 1 AU and briefly summarizes the ICME detection algorithm used on the Genesis spacecraft as an application. Section 3 discusses the detection of boundaries and multiple ICMEs, while Section 4 discusses inferences about the ICME's three-dimensional structure. Finally, Section 5 discusses other solar wind transients – not every unusual solar wind parcel is necessarily an ICME.

2. Signatures of ICMEs with In-Situ Data

This section addresses how different signatures can be used to identify ICMEs in situ and also mentions pitfalls that may arise when blindly applying these signatures. An ICME brings several structures past a spacecraft, all with their own signatures. Figure 2 of Zurbuchen and Richardson (2006, this volume) shows an idealized sketch of an ICME. Fast ICMEs will tend to drive a shock, which is not a signature of the ICME proper, but, nevertheless, a convenient, often-used and reasonably well-understood signature associated with many ICMEs. This shock can accelerate particles. The turbulence in the sheath following the shock modulates their propagation and is an important ingredient in the acceleration process. The boundary between the trailing edge of the sheath and the ICME can sometimes be difficult to identify. This may be due partly to the dynamic nature of ICME propagation, and possibly to evolution with time, e. g., by reconnection (e. g., Cargill and Schmidt, 2002; Gosling *et al.*, 2005). Moreover, different in-situ signatures do not necessarily give the same boundaries. The internal structure of ICMEs may be highly inhomogeneous, again resulting in difficulties identifying substructures and boundaries with different signatures.

2.1. Charge-State Composition

Solar wind in situ ionic observations provide a remote measure of the thermodynamic state of the solar wind source region. They therefore provide a unique, useful way to relate in-situ plasma and field observations to their respective source (e.g., Hundhausen, 1968; Geiss *et al.*, 1995). Early observations of heavy ions in the solar wind, such as singly ionized helium or oxygen charge states (e.g., Bame *et al.*, 1968), provided a glimpse of the richness of ionic compositional data (see also Zurbuchen and Richardson (2006), this volume, for more references).

The relationship between enhanced ion charge states and ICMEs was first established when high ionization states of oxygen and iron were detected following interplanetary shocks (Bame *et al.*, 1979; Fenimore, 1980; Ipavich *et al.*, 1986). Bame *et al.* (1979) attributed these enhanced charge states to flare-heated coronal gas, an explanation recently explored by Lepri and Zurbuchen (2004). The magnetic bottle ICME (see Alexander *et al.*, 2006, this volume) was used to explain these and other post-shock phenomena (see Figure 2 of Zurbuchen and Richardson, 2006, this volume). Fenimore (1980) showed that other types of solar wind flows were related to their ionization states, with the hottest periods being those showing ICME signatures. Nevertheless, only a few ICME-related periods showed high freezing-in temperatures. These temperatures are determined by a competition between ionization/recombination times and the expansion time for the solar wind (Hundhausen *et al.*, 1968). They are derived from ionic charge-state pairs (e.g., O^{7+}/O^{6+}) and usually based on the assumption of equilibrium ionization and recombination in a Maxwellian electron gas. The study of Fenimore (1980) was limited by the use of electrostatic analyzers. More recent, dedicated instruments (e.g., the Solar Wind Ion Composition Spectrometers (SWICS) on ACE (Advanced Composition Explorer) and Ulysses) can measure the solar wind ionic composition under all circumstances. Results from these instruments have revealed that, indeed, the majority of all ICMEs are associated with elevated ionic charge states (e.g., Galvin *et al.*, 1997; Wimmer-Schweingruber *et al.*, 1999; Henke *et al.*, 2001; Lepri *et al.*, 2001; Zurbuchen *et al.*, 2003, Richardson and Cane, 2004; Rodriguez *et al.*, 2004; Zurbuchen *et al.*, 2003; Lepri and Zurbuchen, 2004). Henke *et al.* (2001), Rodriguez *et al.* (2004), and Richardson and Cane (2004) showed that high oxygen charge states are preferentially associated with the subset of ICMEs showing a magnetic cloud (MC) structure (see Figure 2). The observed elevated charge states suggest a direct linkage between ICME plasma and flares accompanying the related CMEs at the Sun. Such associations were found in ICME-CME analyses by Lepri and Zurbuchen (2004) and Reinard (2005). On the other hand, ICMEs at high latitudes do not necessarily exhibit elevated charge states (Neukomm, 1998).

Although extremely rare, ICMEs with unusually low charge states have also been reported. These were mostly identified based on the occurrence of singly charged He. The events described by Schwenn *et al.* (1980) and Gosling *et al.* (1980) had a He^{+}/He^{2+} ratio of up to 0.3, that is, nearly two orders of magnitude larger than that found originally by Bame *et al.* (1968). The explanation provided for such low charge states, which imply low coronal temperatures, was the presence of prominence material of chromospheric origin. These and other periods of abnormally low charge states were also analyzed by Zwickl *et al.* (1982).

These authors found periods in which both high and low ionization states were present, indicating mixing of the plasma and spatial inhomogeneity within these events and raising questions about the prominence interpretation. The existence of mixed (hot and cold) charge states during a He^+ rich event was also described by Gloeckler *et al.* (1999) and Skoug *et al.* (1999). The heavy ion charge-state composition showed clear evidence of unusually elevated charge states, typical of those found in ICMEs, throughout the most extended and most intense interval of enhanced He^+/He^{2+} ratio ever measured. The interpretation of these mixed charge states is still pending. More recently, Zurbuchen *et al.* (2005) analyzed the January 9, 2005, ICME associated with a CME accompanied by a prominence eruption on January 5th. The relation between the ICME and eruptive filament is suggested by its unique ionic composition, indicating coronal source temperatures well below 10^5 K. The time period with unusually cold ionic composition exhibited unusual elemental composition, with enhancements of O relative to He and Fe and was associated with a flux-rope-like magnetic field. Observations of this type may prove to be critical in distinguishing between CME initiation models.

2.2. Elemental Abundances

The classic ICME composition signature is an enhanced He/H abundance ratio (He/H $\geq$ 6%–8%), which may reach $\sim$25% or more (Hirshberg *et al.*, 1970, 1972; Neugebauer, 1981; Borrini *et al.*, 1982). The He/H ratio in the normal solar wind ranges between 3% and about 5% (e.g., Neugebauer, 1981; Schwenn, 1990) and should be compared with the photospheric or solar value of around 10%. Obviously, some process is fractionating the solar wind He content and appears to act in the chromosphere and/or low corona (Laming and Feldman, 2001). One mechanism that has been proposed is inefficient Coulomb drag (Geiss *et al.*, 1970) in which helium experiences a smaller proton drag force than other heavy ions and is fractionated against the bulk solar wind protons. This in turn must lead to an enrichment of helium in the chromosphere and/or low corona since it is already ionized at this height and cannot be lost to the photosphere, where it would be incorporated into the near-infinite mass of the outer convective zone. The helium enhancement in ICMEs has been attributed to a "cleaning out" of helium left behind in the low corona by the solar wind. However, it is far from clear exactly how and where such a helium accumulation is produced and how it can persist for any length of time until it is ejected with a CME. Figure 2, discussed below, shows that about 30% of all ICMEs show this enhancement of He/H above 6% and that this fraction appears to be somewhat higher in magnetic clouds. Other compositional anomalies have also been reported (Bame *et al.*, 1979; Mitchell *et al.*, 1983; Ipavich *et al.*, 1986; Zurbuchen and Richardson, 2006).

Nevertheless, as can be seen in Figure 1, discussed below, many ICMEs, especially high-latitude ICMEs, do not show an elemental composition different from

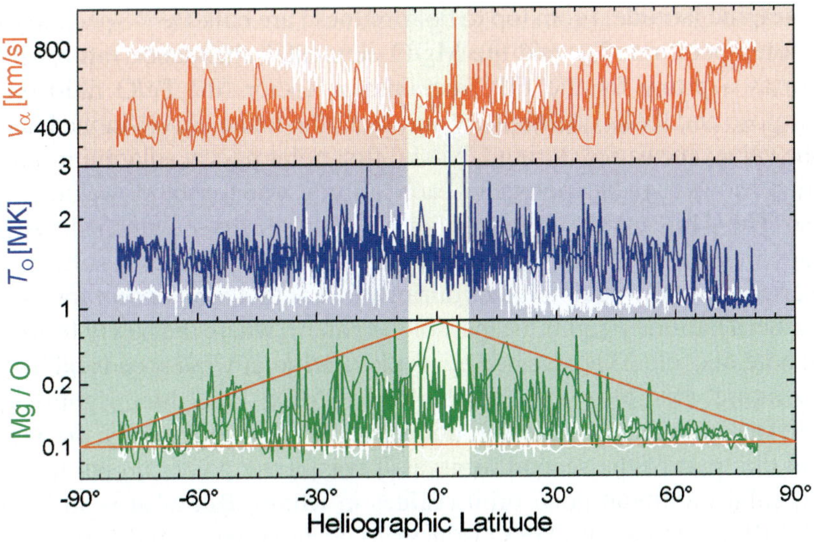

Figure 1. Ulysses-SWICS solar wind parameters from the second, solar maximum, polar orbit: alpha-particle speed (top), oxygen freeze-in temperature, T_O, (middle), and Mg/O abundance ratio (bottom). The distribution of fast and slow solar wind, which was well ordered by latitude for the first, solar minimum, pass (values shaded in white), looks drastically changed. The three "outliers" in T_O on the rightmost edge of the diagram correspond to three ICMEs detected by Ulysses at high latitudes. Remarkably, their unusually hot ionic composition is not reflected by the FIP enhancement in Mg/O that would be expected even in the normal slow solar wind.

the surrounding solar wind. Therefore, it seems fair to assume that the same basic first ionisation potential (FIP) fractionation mechanism applies to CME material as to the quasi-stationary solar wind (von Steiger, 1998). This process appears to separate neutral atoms from charged ions in the chromosphere and/or lower corona (Geiss, 1982). However, some ICMEs, at least some observed in the ecliptic plane, do show a much stronger fractionation (e.g., Wurz *et al.*, 1998; Gloeckler *et al.*, 1999; Wurz *et al.*, 2001), so there must be additional fractionation acting on the ejected material in addition to the normal solar wind fractionation. This may be either the same mechanism (the FIP effect) but operating more strongly or for a longer time (Widing and Feldman, 2001), or an additional, different mechanism that is operating on the pre-CME material. In some cases, especially for magnetic clouds, ICMEs show a mass-fractionated composition with an enhancement of heavy elements that cannot be understood in terms of FIP fractionation models (Wurz *et al.*, 2001). Wurz *et al.* (2000) have presented a model of how this mass fractionation might be explained. Nevertheless, such mass-fractionated ICMEs are rare events although they appear to be more frequent in magnetic clouds (Wurz *et al.*, 2001).

An interesting point in this context is illustrated in Figure 1, which shows three solar wind parameters obtained with the SWICS instrument on Ulysses as a function of heliographic latitude. From top to bottom:these are bulk He^{2+} speed, v_α, oxygen freeze-in temperature, T_O, and the Mg/O elemental abundance ratio, which may be taken as a proxy for the FIP fractionation factor (the Fe/O ratio serves the same purpose and looks qualitatively similar). The white curves in the background show the values measured during Ulysses' first polar pass at solar minimum. Close inspection shows two data points for each latitude, one for the slow and one for the fast pass. The white curves contrast with the colored curves, which show the same quantities but measured during the second polar orbit of Ulysses at solar maximum. Slow, variable solar wind dominates during most of that orbit except for a portion at high northern latitudes (rightmost part of the panel), where the newly forming polar coronal hole of cycle 23 is revealed by its elevated solar wind speed and low freeze-in temperature. One also sees 3 embedded ICMEs – the three rightmost spikes visible in freeze-in temperature (but not in Mg/O). Not surprisingly, the Mg/O ratio is low in that part of the second orbit, as it was in the polar coronal holes during the first, solar minimum polar orbit (values in white). But what is striking is that the high FIP fractionations (Mg/O values) all occur at low to mid latitudes. At high southern latitudes the FIP fractionation, be it in ICMEs or in just ordinary slow solar wind, is clearly less pronounced than it can be at low latitudes. This trend, illustrated by the red triangle surrounding the Mg/O values, is not pronounced, but the strongly FIP fractionated materials including ICMEs seem to avoid the polar regions. This is remarkable, as it appears to indicate a difference in the pre-CME state or onset mechanism of ICMEs.

Filaments are generally believed to have chromospheric composition (Schwenn *et al.*, 1980). The chromosphere is a thin layer of the Sun's atmosphere and is very likely intimately connected with the FIP effect (von Steiger, 1998; Laming and Feldman, 2001). It could be where the FIP effect acts on the solar wind flow, or it could lie beneath that location. Hence it is interesting to investigate filamentary material as a test of FIP-fractionation models. Neukomm (1998) has investigated ICMEs observed with Ulysses and found no evidence of photospheric composition in these ICMEs. This would argue that ICMEs (and the filaments within) are already (FIP) fractionated and certainly do not exhibit photospheric composition. This is borne out by the lowest panel of Figure 1 which shows the Mg/O elemental abundance ratio, a good proxy for the FIP effect. Except for few days at high southern latitudes during solar activity maximum, no solar wind was observed that had not been (FIP) fractionated. The analysis of the few days with unusually low Mg/O is still outstanding. As they do not coincide with periods of unusually low coronal temperature, a filamentary origin seems unlikely.

Some He isotopic anomalies in ICMEs have been reported. In the normal solar wind, the He isotope ratio $^3He^{2+}/^4He^{2+}$ has been determined as $(0.43 \pm 0.02) \cdot 10^{-3}$ (Geiss *et al.*, 1970b), while enhancements of $^3He^{2+}/^4He^{2+} \geq 10^{-3}$ always coincide with ICMEs (Ho *et al.*, 2000). Other isotopic anomalies of heavy elements have not yet been observed, mainly because of the low counting statistics of isochronous time-of-flight mass spectrometers (Wimmer-Schweingruber *et al.*, 1999b).

In summary, the range of variability of elemental abundances in ICMEs is large, both between different events as well as within one single event. It ranges from no compositional signature at all (with respect to the surrounding solar wind) to unusual events with compositions never seen in other contexts.

2.3. Relating Charge-State and Elemental Composition

In a study comprising about 200 ICMEs detected in the near-Earth environment, Richardson and Cane (2004) investigated various compositional signatures as a function of the location relative to the ICME boundaries. Results of this analysis are displayed in Figure 2. Periods with anomalous plasma composition are found to occur in non-cloud (left-hand panel) and in cloud (right-hand panel) ICMEs. Their criteria for "anomalous" composition are summarized in Table I. In general,

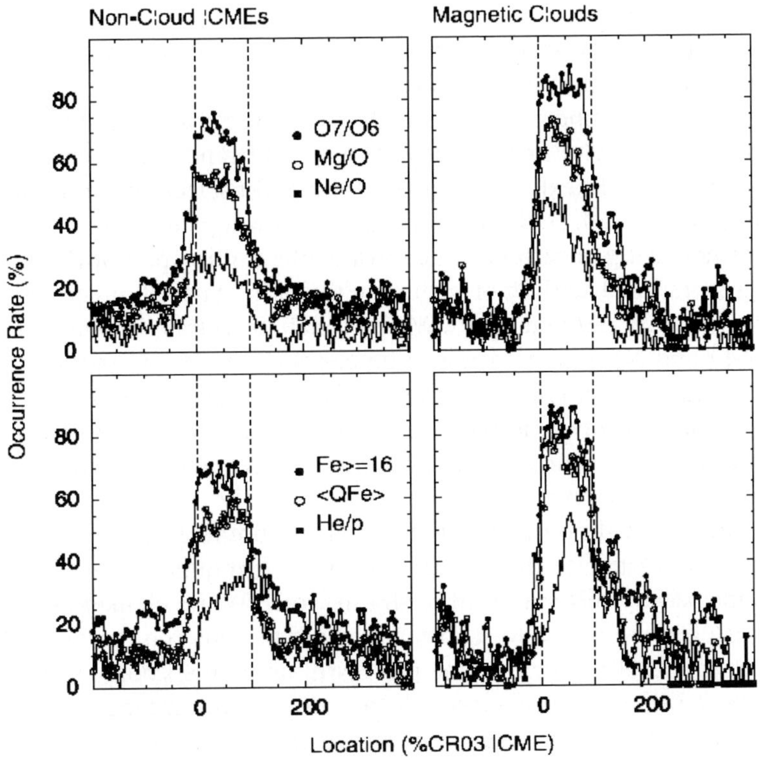

Figure 2. Occurrence rate of periods with anomalous composition for various signatures as a function of time/location with respect to the boundaries of cloud and non-cloud ICMEs. "Anomalous" is defined in Table I. An occurrence rate of 60% means that 60% of all ICMEs (non-cloud or cloud) exhibit that signature. 0% on the x-axis corresponds to the leading edge and 100% represents the trailing edge (From Richardson and Cane, 2004. © American Geophysical Union. Reproduced/modified by permission of American Geophysical Union).

TABLE I

Parameters characterizing "expected" and "anomalous" composition ratios in the solar wind; v is the solar wind speed.

Signature	Normal v-dependence	Anomalous	Reference
O^{7+}/O^{6+}	$3.004 \cdot \exp(-0.00587v)$	$2 \times$ normal	a
Mg/O	$0.491 \cdot \exp(-0.00367v)$	$2 \times$ normal	a
Ne/O	$0.295 \cdot \exp(-0.0017v)$	$2 \times$ normal	a
$Q_{Fe} \geq 16/\sum Q_{Fe,tot}$	$0.292 \cdot \exp(-0.00421v)$	$2 \times$ normal	a
$\langle Q_{Fe} \rangle$	$11.2 - 0.000857v$	$Q \geq \langle Q_{Fe} \rangle + 1$	a
He/p		≥ 0.06	a
$^3He/^4He$		$\geq 10^{-3}$	b

References: (a) Richardson and Cane (2004), (b) Ho *et al.* (2000).

elevated charge states provide higher occurrence rates when compared to elemental abundance ratios. The profiles presented in the top panels suggest a decrease in the occurrence rate of compositional anomalies from the leading to the trailing edge of ICMEs. Even though this may indicate a spatial trend, the possibility of a bias introduced by the ICME expansion and the criteria for selecting the anomalous periods cannot be discarded. Helium enhancements (lower panels) present the opposite behaviour (increasing occurrence rate towards the trailing part of ICMEs), consistent with gravitational settling of helium in the bulk material of coronal streamers (Geiss *et al.*, 1970), potential pre-CME coronal structures. However, the heavier elements Mg and Ne do not appear to be enriched relative to the lighter element oxygen towards the trailing edges of the ICMEs as would be expected in this picture. Their abundance, however, is consistent with fractionation by insufficient proton drag, as is seen in the vicinity of sector boundaries (Wimmer-Schweingruber, 1994) originating in coronal streamers (Borrini *et al.*, 1981).

The similarities in the overall spatial behavior of both elemental and ionic charge-state anomalies relative to the normal solar wind are quite remarkable when one considers the widely different time scales at work. While elemental fractionation processes need to be active for substantial time periods (on the order of days in the Wurz *et al.* (2000) model) in the pre-CME structures, charge states are determined along CME trajectories through the corona (on timescales of hours), and also very likely during the intense energy release associated with CME initiation (Reinard, 2005). Indeed, flare emissions clearly indicate that large amounts of thermal energy are released during the formation of a CME. Evidently, the low-density corona is also affected by this transfer from magnetic to thermal and kinetic energy, which is likely to be associated with magnetic reconnection and, hence, the CME initiation process (Forbes *et al.*, 2006, this volume). Frozen-in ionic charge states may therefore provide one of the most direct measures of the initiation process, while elemental composition probes pre-CME solar atmosphere conditions.

2.4. Bidirectional Electron Streaming

One of the first signatures used to identify ICMEs on a routine basis is counterstreaming of suprathermal (>80 eV) electron beams, or bidirectional electrons (BDEs) (e.g., Gosling *et al.*, 1987, 1990). Since the solar corona is a nearly continuous source of hot electrons, one normally observes a suprathermal electron beam called the "strahl" (e.g., Pilipp *et al.*, 1987) flowing away from the Sun along heliospheric magnetic field lines that are open (extending from the Sun to infinity). Counterstreaming beams form when field lines are connected to the Sun at both ends. These closed field lines can be either simple loops or coiled loops in flux ropes, the latter as sketched in Figure 2 of Zurbuchen and Richardson (2006, this volume). It is nearly always assumed that field lines closed in this manner arise from CMEs (see section 5 for exceptions).

On the other hand, not all field lines in ICMEs are necessarily closed. Some ICMEs, as identified by other signatures, are devoid of BDEs, and many contain intermittent intervals of BDEs, implying a mix of open and closed fields (e.g., Gosling *et al.*, 1995; Shodhan *et al.*, 2000; Crooker *et al.*, 2004a; Crooker and Horbury, 2006, this volume). Nevertheless, nearly all ICMEs observed within 5 AU of the Sun contain some closed fields and, on average, appear to contain more closed than open fields ($\sim$ 60% in magnetic clouds at 1 AU).

In using BDEs to identify ICMEs by their closed fields, one must avoid counterstreaming events generated on open field lines. Shocks are a well-known source because they accelerate electrons to suprathermal energies. As a result, magnetic connection to shocks beyond the observer can result in a beam of electrons flowing opposite to the electron outflow from the Sun. In the case of Earth's bow shock, this signature can usually be recognized and avoided by estimating whether the spacecraft is magnetically connected using simple geometrical models (e.g., Stansberry *et al.*, 1988). In addition, well away from Earth (say at L1), counterstreaming created by bow-shock connections usually can be distinguished from BDEs in ICMEs by the relatively short (<1 hr) connection duration owing to ever-present fluctuations in field direction. In the case of corotating shocks, electrons backstreaming toward the Sun are usually clear in pitch angle spectrograms because they characteristically stop or start with shock passage (e.g., Gosling *et al.*, 1993). In the case of shocks created by ICMEs, any electrons backstreaming toward the Sun would be found in the sheaths between the shocks and ICMEs (Figure 2 of Zurbuchen and Richardson, 2006, this volume), but few cases have been observed.

Counterstreaming on open field lines can also be generated by focusing and mirroring on field lines connected to regions of elevated magnetic field magnitude downstream of a spacecraft (Gosling *et al.*, 2001). These adiabatic actions create

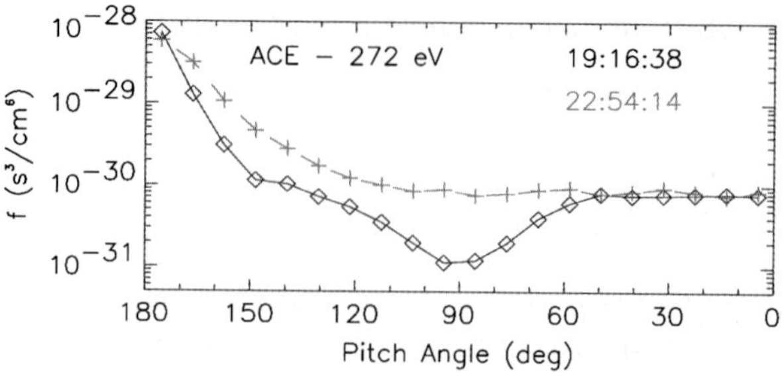

Figure 3. Two pitch angle distributions measured by ACE on 10 January 1999. The single beam at 180° indicates open field lines in both cases, while a depletion at 90° is apparent only in the distribution plotted with diamonds (from Gosling *et al.*, 2001. © American Geophysical Union. Reproduced/modified by permission of American Geophysical Union).

electron pitch angle distributions with a band of depletion centered on 90°. Gosling *et al.* (2001) estimate that depletions are present at least 10% of the time. Close examination of the shape of the pitch angle distribution can help distinguish depletions on open fields from BDEs on closed fields. Depletions produce a dip at 90° in a distribution that peaks either at 0° or 180° and is otherwise flat, as illustrated by the diamond symbols in Figure 3, whereas distributions on closed fields generally peak at both 0° and 180°. This distinction, however, is not always clear. Crooker *et al.* (2004a) report that their estimated average of 55% for closed fields in magnetic clouds at 5 AU is an upper limit and could be as low as 40% owing to mistaking depletions on open field lines for BDEs on closed field lines.

Sometimes counterstreaming suprathermal electrons can be confused with large temperature anisotropies in the core electrons, which commonly occur during periods of low plasma density (Phillips and Gosling, 1990), because at such times the core distribution extends up to energies much higher than 80 eV along the field. Commonly, the temperature parallel to the field is larger than the temperature perpendicular to the field. This sense of anisotropy results in a double field-aligned band of increased intensities in spectrograms at suprathermal energies below about 300 eV, mimicking the closed-field-line signature. These can be distinguished from each other by checking whether the double-banded pattern extends to energies well below 80 eV and if the apparent counterstreaming extends up to energies well above 300 eV.

Attempts to design quantitative routines for objectively selecting BDE events on closed field lines have met with little success. For example, Feuerstein *et al.* (2004) constructed scatter plots of suprathermal electron intensities parallel against antiparallel to the magnetic field, each intensity normalized by the intensity perpendicular to the field, and found no clearly separate BDE population in their plots. Instead, points in BDEs blend in with the unidirectional populations, owing at least in part to the fact that one counterstreaming beam is nearly always stronger than the other. This may reflect different source conditions at the base of each leg of the closed loops and/or a skewing of closed loops along the Parker spiral, thus biasing spacecraft interception toward one leg of the loop rather than its apex. The intensity of the beam from the base of the intercepted leg is expected to be stronger than the intensity of the beam from the more distant base of the far leg (Pilipp *et al.*, 1987). Other factors that undoubtedly contribute to the spread in BDE points are the effects of $90°$-depletions and temperature anisotropies in the core population, as discussed above.

In summary, while BDEs continue to be a widely accepted signature of closed magnetic fields in ICMEs, identifying them in pitch angle spectrograms continues to be a somewhat subjective process.

2.5. Magnetic Clouds

As discussed by Zurbuchen and Richardson (2006, this volume), magnetic clouds are a subset of ICMEs defined at 1 AU by the following criteria (Burlaga, 1991): (1) the magnetic field direction rotates smoothly through a large angle during an interval of the order of one day; (2) the magnetic field strength is higher than average; and (3) the temperature is lower than average. Other large-scale solar wind structures, such as interplanetary sector boundaries, co-rotating interaction regions (CIRs) or post-shock ICME flows, can exhibit any of the above features (e.g., Bothmer and Schwenn, 1992), but the combination of all three appears to be unique to magnetic clouds (e.g., Bothmer and Schwenn, 1998).

Goldstein (1983) introduced the now widely used force-free ($\nabla \times \mathbf{B} = \pm\mathbf{B}$) large-scale cylindrical magnetic flux rope model to explain the magnetic field variations in magnetic clouds. Later studies, however, show that not every cloud fits a force-free, cylindrically symmetric model (e.g., Mulligan and Russell, 2001; Hidalgo *et al.*, 2002; Forbes *et al.*, 2006, this volume).

Bothmer and Schwenn (1998) used Helios 1 and Helios 2 data in order to systematically study magnetic clouds between 0.3 and 1 AU. They found that, during the years 1974 to 1980, 74% of the clouds showed South to North rotations of the magnetic field vector. Using additional observational results for the average properties of prominences (disappearing filaments) and photospheric bipolar regions (sunspots), they explained the dominance of this type of rotation and introduced the solar cycle relationship of the field structure of magnetic clouds shown in Figure 4. The dominance of SN type magnetic clouds in odd cycles and of NS ones in even cycles was thus explained as a consequence of the predominant magnetic polarity of sunspots and bipolar regions in the two solar hemispheres in a given solar cycle, of the average orientation of neutral lines separating them, and of the orientation of filaments (see, also, Bothmer and Rust, 1997). This and related solar imprints on magnetic clouds are discussed further by Forsyth *et al.* (2006, this volume) and Crooker and Horbury (2006, this volume).

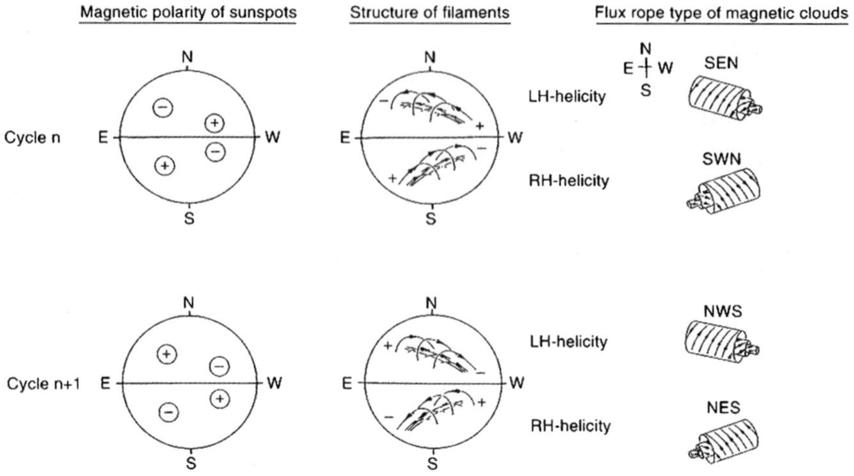

Figure 4. Solar cycle dependence of the magnetic field structure of filaments at the Sun and that of the corresponding magnetic clouds in the interplanetary medium. Adapted from Bothmer and Schwenn (1998). Note that for simplicity the magnetic clouds are oriented horizontally with respect to the ecliptic plane. Furthermore, the "cycle overlap" during the declining phase of the preceeding cycle is not considered here. In this complicated phase the old magnetic polarity is still prominant in near-equatorial regions, while the new polarity emerges at mid to high latitudes.

Recent cloud studies by Huttunen *et al.* (2005) and Wu *et al.* (2003) show that the overall frequency of magnetic clouds varies over the course of the solar cycle but was not in phase with either the sunspot cycle nor the total CME rate during 1996–2002. Richardson and Cane (2004b), however, noted that the total number of near-Earth ICMEs approximately follows the solar cycle and concluded that the fraction of ICMEs that are magnetic clouds varies from roughly 100% near times of solar minimum to about 15% around solar maximum.

2.6. A Case Study: The 2000 Bastille-Day Event

On July 14, 2000, the LASCO coronagraphs on board the Solar and Heliospheric Observatory (SOHO) observed a front-side full halo CME (top right-hand panel of Figure 5) with a speed of $\sim$ 1600 km/s in the plane of the sky at about 11 UT (Lepping *et al.*, 2001). SOHO's solar magnetic field (MDI, lower left-hand panel) and extreme ultraviolet (EIT, lower right) observations and Yohkoh's soft X-ray measurements (upper left) show that the CME originated from a bipolar magnetic region located in the northern solar hemisphere around central meridian. In soft X-rays, a sigmoid structure became prominent near the CME's onset time, indicative of coronal heating. The sigmoid was followed by a post-eruptive arcade (insert in EIT image), an unmistakable signature of CMEs (e.g., Sterling *et al.*, 2000; Tripathi *et al.*, 2004).

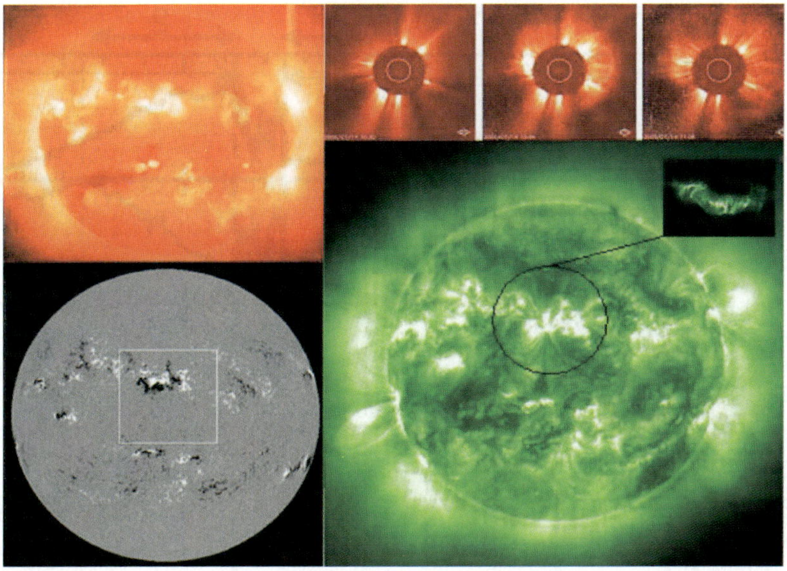

Figure 5. Combined SOHO and Yohkoh observations showing the July 14, 2000 halo CME and its corresponding source regions in the lower corona and photosphere. The EUV post-eruptive arcade demarks the CME source location on the visible solar disk. The square in the lower left-hand panel denotes the bipolar region of origin, the sigmoid structure is seen at the corresponding location in the panel above. The remains of the source region are seen as post-eruptive arcades in the EIT image (lower right). The resulting halo CME is clearly visible in the three LASCO images (upper right) (Adapted from Bothmer, 2003).

The ICME corresponding to the halo CME was detected by several spacecraft stationed near Earth. In-situ data from ACE of this and preceding ICMEs are shown in Figure 6. The top three panels show magnetic field data B, RTN θ_B, and ϕ_B (all from ACE level 2 data), the next three panels show proton data, v_p, T_p, and n_p (SWEPAM data, courtesy R. Skoug and D. J. McComas), and the three bottom panels show composition data (from SWICS). These are the ionic charge-state ratios C^{6+}/C^{5+} and O^{7+}/O^{6+}, average iron charge state, $\langle Q \rangle_{\mathrm{Fe}}$, and elemental abundance ratios Mg/O and Fe/O. Because of the high background due to penetrating particles, SWEPAM data has only $\sim$ 30 minute resolution from DOY 196/11:06–198/01:33, and He^{2+} data were unavailable. Halo electron measurements were also not available during 196/02:00–197/23:00 (Smith *et al.*, 2001). The time period shown here was preceded by an interval with two ICMEs. The end of the second of these is labeled a in the uppermost panel. The following ICME labeled b is indicated by a thin two-ended arrow bounded by two thick vertical dashed lines at 196/17:00 and 197/14:00. It is clearly identified by the low level of fluctuations in B and by the enhanced ionization temperatures in the second half. He/O shows a sharp increase from a low value to a very high value exceeding 200. ICME b contained a magnetic cloud indicated by the thick two-ended arrow labeled 1 above the panels and the

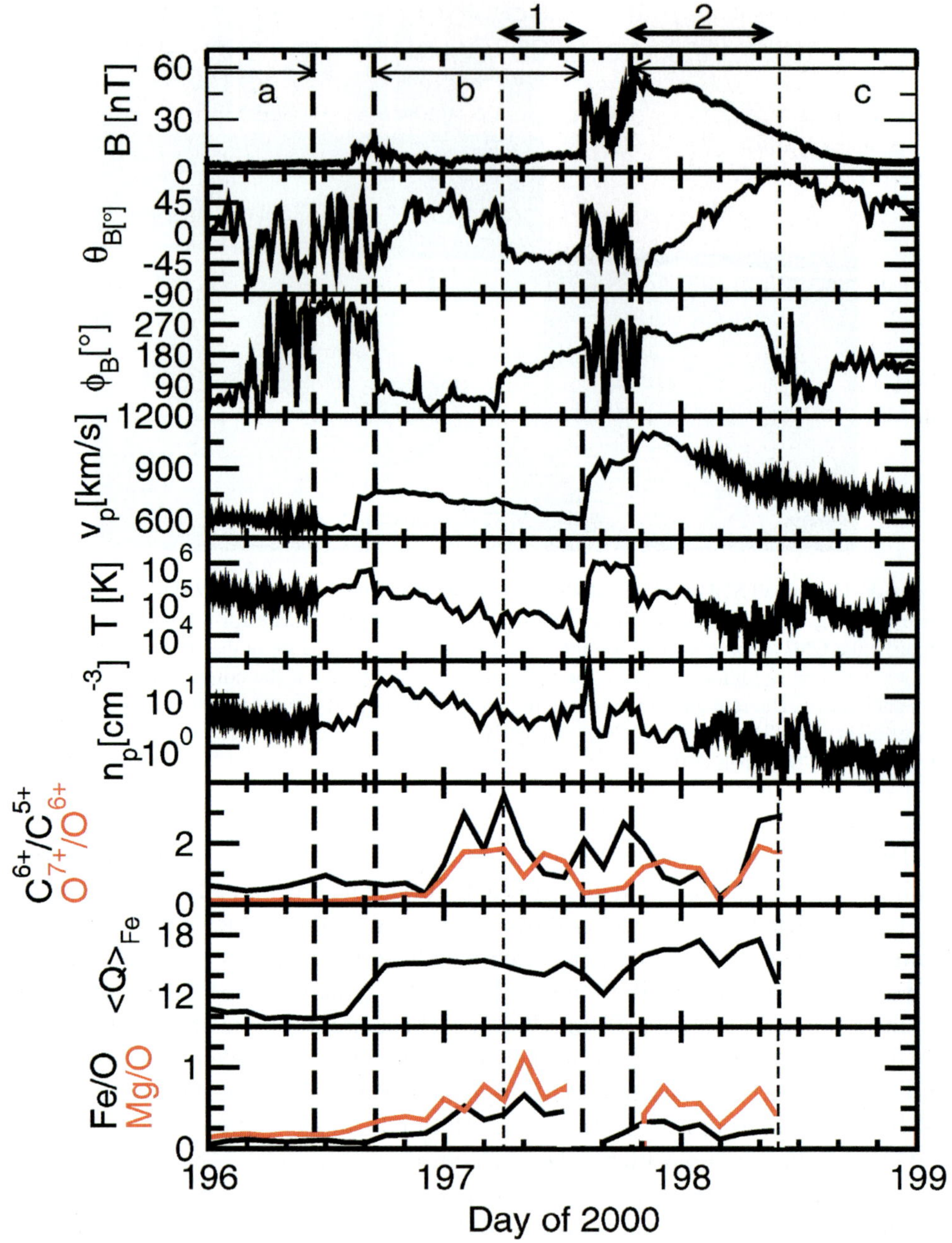

Figure 6. In-situ data from the ACE spacecraft for the ICMEs/magnetic clouds observed on July 14–16, 2000. From top to bottom: B, θ_B, and ϕ_B in RTN coordinates. Proton speed, temperature, and density are from SWEPAM, the lower three panels show charge-state ratios of carbon and oxygen, average iron charge state, and elemental abundances of Fe and Mg relative to O from SWICS. The thick double-ended arrows at the top of the panel denote the magnetic clouds, while the thin lines inside the topmost panel show the ICMEs. ICME c extended beyond the time period shown here.

thin vertical dashed line. It is clearly visible as a systematic rotation in ϕ_B and low proton temperature. Ionization states rose before the onset of the magnetic cloud. ICME *c* started with magnetic cloud 2 (again indicated by a thick double-ended arrow and a thin vertical dashed line marking its end). See Smith *et al.* (2001) for timing information.

ICME *c* corresponds to the halo CME in Figure 5, launched on July 14, 2000, described in the previous paragraph. From the MDI measurements in Figure 5 one may infer that ICME *c* originated from the region of opposite magnetic polarities inside the square, as is generally the case for the source regions of CMEs that can be tracked back to the solar surface (Cremades and Bothmer, 2004). White areas in the MDI image denote magnetic fields pointing away from the solar surface, and black areas denote fields pointing towards the surface. If, inside the square, one drew an arcade of field lines from one polarity to the other and let them expand radially outward and reconnect beneath, one would expect to observe a South to North turning of the magnetic field vector in space. This is the direction of turning predicted according to the classification scheme in Figure 4, and it matches the negative to positive rotation in θ_B in the second panel of Figure 6.

The high magnetic field strength and strong proton speed gradient in the first and fourth panels during cloud 2 in Figure 6 indicate that this cloud was still expanding as it reached ACE. The strong southward B component coincided with the high speed, exceeding 1100 km/s at the leading edge of the cloud, and triggered a large geomagnetic storm on July 15–16, 2000. The depressed proton temperature in the fifth panel together with the high field strength resulted in a low-β structure. The coincident low density seen in the sixth panel, possibly due to the expansion, led to a near sub-Alfvénic flow within magnetic cloud 2 (Smith *et al.*, 2001). In the third panel of Figure 6, magnetic field longitude, ϕ_B, shows large excursions towards the end of ICME *a*, between ICME *a* and *b*, and between magnetic clouds 1 and 2. These field polarity reversals are not unusual (Klein and Burlaga, 1982). ICMEs and magnetic clouds have been found to replace the heliospheric current sheet reversal in several instances (Crooker and Intriligator, 1996; Crooker *et al.*, 1998; Crooker and Horbury, 2006, this volume). In the seventh panel of Figure 6, the He/H ratio increases sharply at the onset of ICME *c*/cloud 2, (not shown, but see Smith *et al.*, 2001) as does O^{7+}/O^{6+}, but less so for C^{6+}/C^{5+}. The oxygen charge states, as well as the average iron charge state in the bottom panel, are highly elevated, corresponding to a source electron temperature of at least 2.5–3 MK. Together with the low in-situ proton temperature, this implies a strong heating at the source with a subsequent rapid cooling of the plasma as it expanded on its way out of the corona into interplanetary space. The similar elemental composition of ICMEs *b* and *c* indicates a common origin. The preceding ICME *a* had a different elemental composition, similar to the slow solar wind, hinting at a different origin. Alternatively, its ejection could have triggered a strong elemental fractionation in the source plasma of ICMEs *b* and *c*.

2.7. Application: The Genesis ICME Detection Algorithm

As an application of the use of in-situ signatures to identify ICMEs, we discuss the ICME detection algorithm used on the Genesis spacecraft. While it does not make use of all the signatures discussed so far, it appears to have worked rather well and is probably one of the best studied algorithms for this task. The primary purpose of the Genesis mission was to measure both the elemental and isotopic compositions of the outer layers of the Sun. To accomplish this, the spacecraft collected samples of the solar wind, which were returned to Earth in 2004 for analysis. To differentiate between different types of wind, collectors were exposed according to an on-board algorithm (Neugebauer *et al.*, 2003; Reisenfeld *et al.*, 2003). The algorithm attempted to distinguish between three fundamental solar wind flows: high-velocity streams emanating from coronal holes (CHs); slower, inter-stream flow (defined as flow observed between successive coronal hole streams); and ICMEs. Here we describe how the real-time, on-board data were used to identify the presence of ICMEs.

It is worth noting several limitations of the technique as implemented for the Genesis mission. First, the primary objective was to provide pristine samples of CH flow, at the expense of contaminating the interstream and CME samples. Second, the array-changing mechanisms had a design requirement of 400 regime changes over the entire mission. Third, only the ion and electron monitors supplied data to the algorithm. A magnetometer, ion mass spectrometer, and energetic particle detectors were not available on Genesis.

The algorithm took as input the following parameters: (1) Solar wind speed (v_p); (2) the ratio of expected proton temperature for normal solar wind to the measured proton temperature (T_{ex}/T_p); (3) the He^{2+}/p ratio ($n_{He^{2+}}/n_p$); and (4) a bidirectional streaming parameter (B_e). The expected temperature T_{ex} was computed as a linear function (Neugebauer *et al.*, 1997), a quadratic function (Burlaga and Ogilvie, 1973) or combination thereof (Lopez, 1987) of v_p, such that T_{ex}/T_p significantly exceeds 1 within CMEs, reflecting their unusually low temperature (Richardson and Cane, 1995). The bidirectional streaming parameter (B_e) was computed from the angular distribution function of suprathermal electrons. Operationally, the algorithm identified the peak count rate (C_{peak}), the minimum count rate ($\pm 90^\circ$ away from this peak, C_{min}), and the count rate 180° away from the peak (C_{180}). If the ratios C_{peak}/C_{min} and C_{180}/C_{min} exceeded some threshold, the algorithm assigned a value of 1 to B_e.

Running averages of the solar wind speed, temperature, alpha abundance, and bidirectional streaming parameter were computed over one hour windows. The algorithm began by assessing whether the spacecraft had encountered a forward shock within the last 12 hours. This does not preclude identifying slow CMEs (i.e., CMEs that do not drive a shock); however, the criteria for switching to the CME regime were made more difficult if no shock occurred. Using a simple fuzzy-logic scheme incorporating: (1) $n_\alpha/n_P > 0.06$; (2) $T_{ex}/T_p > 1.5$; (3) $B_e = 1$; (4) the identification of a shock within the last 24 hours; and (5) whether the CME criteria were met within the last 6 hours, a threshold was set, above which a CME regime is triggered (i.e., the CME array is exposed to the solar wind). Extra safety mechanisms were also built into the algorithm. For example, it was not possible to trigger a change into the CME regime based solely on B_e since connections to CIR-associated and planetary bow shocks can produce a bidirectional electron signature, as discussed in Section 2.4.

Given how difficult it can be for an "expert" to identify a CME (and particularly its boundaries) even when a full complement of data sets is available, one might question how successful such an automated procedure can be. After identifying and fixing a number of problems associated with the algorithm during the initial portion of the mission, the algorithm appears to have performed satisfactorily. The shock-identification portion of the algorithm worked well, and shocks observed at Genesis matched well with those identified by the SOHO spacecraft. Moreover, it is important to remember that the primary goal of the algorithm was not to identify CMEs but rather to exclude CME material from the coronal hole regime. As is discussed in Reisenberg *et al.* (2003), who compared the Genesis regimes with those identified with additional criteria such as solar wind composition from ACE/SWICS, the agreement between the two independent methods was remarkably good.

3. Boundaries and Multiple ICMEs

3.1. Relative Timings and Intercomparison of Signatures

As noted at the beginning of Section 2, the boundary between an ICME and the ambient solar wind might be expected to be a simple tangential discontinuity that encompasses a region with ICME-like signatures. In practice, ICME boundaries are often elusive, ambiguous, or complex. In particular, the various ICME signatures often do not indicate exactly the same boundaries (e.g., Zwickl *et al.*, 1983; Crooker *et al.*, 1990; Neugebauer and Goldstein, 1997; Richardson *et al.*, 2003), presumably since they arise from a variety of phenomena (Zurbuchen and Richardson, 2006, this volume). Plasma boundaries may also be identified within ICMEs.

One case study, by Crooker *et al.* (1990), is shown in Figure 7. They identified 11 magnetic field discontinuities in the vicinity of a magnetic cloud previously identified by Zhang and Burlaga (1988) and bounded by discontinuities 5 and 8. Discontinuities 2, 4, 5 and 7 were determined to be tangential discontinuities and aligned nearly parallel to each other, while others yielded ambiguous results. Discontinuity 1 is the ICME-driven shock. Near the leading edge of the following ICME, the decrease in proton temperature (at discontinuity 2) occurs earlier than the leading edge of the putative magnetic cloud. If the temperature decrease is taken as the true ICME boundary, then there is a distinct region of predominantly eastward magnetic field ($B_y > 0$) inside the ICME prior to the classic magnetic cloud signature (which has a westward field at the leading edge). Discontinuity 5 would

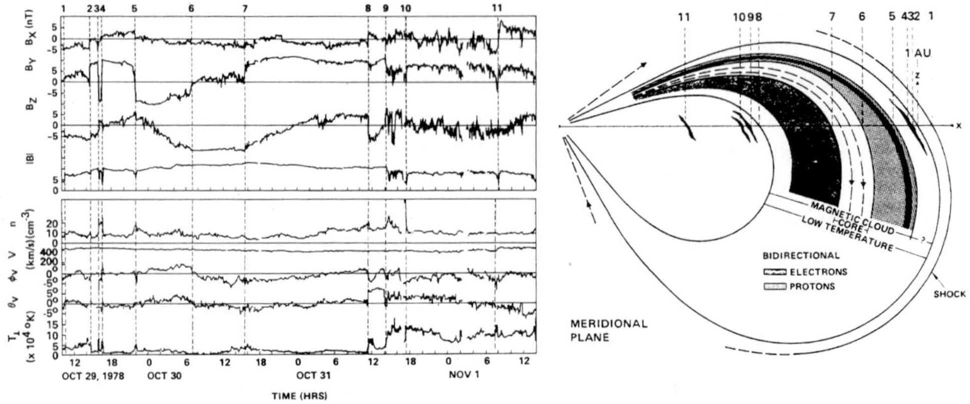

Figure 7. (Left) Magnetic field observations in the vicinity of a magnetic cloud in October, 1978. Discontinuities are numbered. (Right) Interpretation (in the meridional plane) of the structures shown in the left-hand panel and their relationship to BDE and bi-directional energetic ion intervals Crooker *et al.,* 1990. © American Geophysical Union. Reproduced/modified by permission of American Geophysical Union).

then indicate the boundary of flux-rope-like substructure within the ICME rather than the ICME leading edge. In addition, discontinuities 3 and 4 apparently bound a "magnetic hole" (another example may be present at discontinuity 5). Magnetic holes are frequently found near ICME leading edges (Burlaga, 1995). One example discussed by Farrugia *et al.* (2001) was a complex, pressure-balanced structure including a rotational discontinuity and slow shock, with evidence of reconnection between field lines within the ICME. Returning to Figure 7, discontinuities 6 and 7 define the core of the magnetic cloud in which there is little field-line twisting. At the trailing edge, although discontinuity 8 marks the end of the magnetic cloud signature, the proton temperature only recovers fully at discontinuity 9.

Figure 7 also shows a schematic of the ICME structure based on these observations (note that the cloud axis lies approximately perpendicular to the ecliptic, with the mid-plane below the ecliptic). Regions of BDEs and bidirectional energetic ions are also indicated. These predominantly occupy the leading or trailing regions of the magnetic cloud structure, respectively, and both are absent from the cloud core and the low temperature region ahead of the magnetic cloud. Hence, quite different "ICME" regions would be inferred depending on whether the magnetic cloud signature, low proton temperatures, BDEs, or bidirectional ions are considered.

Solar wind composition signatures are now routinely available from instruments such as ACE/SWICS and provide additional clues to the location of ICME boundaries. Although compositional boundaries are often reasonably consistent with those suggested by other signatures, again there may not be total agreement (e.g., cf. Figure 3 of Zurbuchen and Richardson, 2006; Richardson *et al.*, 2003). Boundaries within ICMEs may be associated with substructures of the ICME. For example, Osherovich *et al.* (1999) discuss a magnetic cloud (at Ulysses) which may be modeled as two intertwined helical flux tubes separated by a region of enhanced plasma pressure. Other case studies of the plasma and magnetic field structures within ICMEs show similar complications. These studies include analysis of the magnetic clouds of 18–20 October, 1995 (e.g., Lepping *et al.*, 1997; Janoo and Farrugia, 1998) and December 23–26, 1996 (e.g., Farrugia *et al.*, 2001; Vasquez *et al.*, 2001).

Multi-spacecraft observations of individual ICMEs also provide unique information on their structure and boundaries. Such observations are rather rare, but several were possible during the Helios 1 and 2 missions. One striking example is an ICME observed in January 1977 by Helios 1 and 2 associated with a filament eruption near E50° relative to the Earth (Cane *et al.*, 1986). Helios 1, located at 0.95 AU and near the filament longitude, observed a shock followed by an ICME

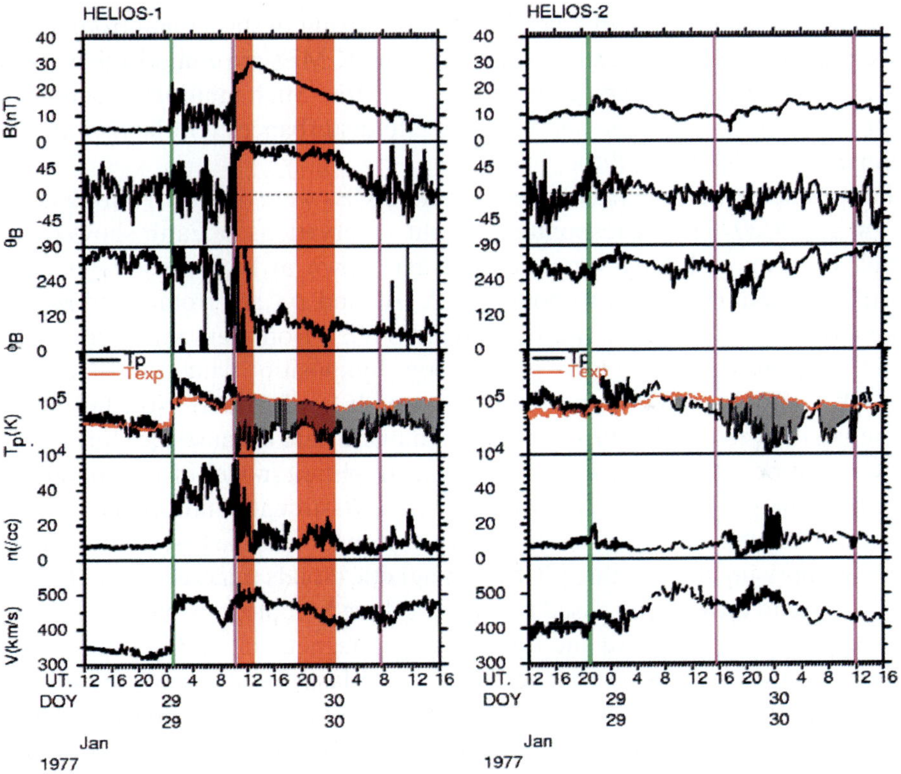

Figure 8. A shock (vertical green line) and ICME (bounded by thin vertical magenta lines) in January 1977 observed by Helios 1 (left) and Helios 2 (right) when the spacecraft were separated by 26° in longitude. The ICME is identified based on the abnormally low proton temperature ($T_p < T_{\text{exp}}$). Note the magnetic cloud-like structure within the ICME at Helios 1 was not encountered by Helios 2, suggesting that it was a substructure of the ICME (adapted from Cane *et al.*, 1997. © American Geophysical Union. Reproduced/modified by permission of American Geophysical Union). Regions in which enhanced levels of He^+ were observed by Schwenn *et al.* (1980) are shaded in red.

with a magnetic cloud signature (Figure 8).At Helios 2, 26° west of Helios 1 at 0.97 AU, the shock was also observed. However, the ICME, indicated by a region of abnormally low proton temperature (and also a cosmic ray depression Cane *et al.*, 1997), lacked a clear cloud-like field signature, suggesting that the magnetic cloud observed at Helios 1 was only a substructure of the ICME. Variations in the other plasma parameters, particularly the density, are also quite different at the two spacecraft. Note also the oppositely-directed magnetic field azimuths (ϕ_B; GSE-coordinates) and that the signature $T_p < T_{\rm exp}$ persists beyond the magnetic cloud boundary.

3.2. Signatures of Multiple ICMEs

SOHO observations show that CMEs can occur in close spatial and temporal proximity, for example, with several CMEs occurring in succession from the same active region. Thus it is not surprising that some ICMEs are comprised of multiple CMEs, and interactions can occur between ICMEs. Such multiple structures may result in strong heliospheric and geomagnetic disturbances (Burlaga *et al.*, 1987; Bothmer and Schwenn, 1995).

For example, multiple front-side fast halo CMEs were observed during November 24–26, 2000. The corresponding in-situ observations at Earth showed a very complex region of highly-structured plasma shown in Figure 9. Magnetic field and plasma data (from top to bottom $|B|$, θ_B, and ϕ_B are from Wind/MFI, and plasma data v_p, T_p, and n_p are from Wind/SWE). Both data sets were obtained from CDAWeb. The lowest three panels show composition data from ACE/SWICS (O^{7+}/O^{6+}, $\langle Q\rangle_{\rm Fe}$, and Fe/O and Mg/O from ACE level-2 data). No adjustment for a time lag between ACE and Wind has been made here because the shift would be too small to be visible in this plot. Using a simplified model of four undeformed cylindrical magnetic clouds of variable radius, B-strength and orientation, Wang *et al.* (2002) managed to give a satisfactory explanation of the first two thirds of this region comprising the first three ICMEs/magnetic clouds (marked 1–3 in Figure 9). They appear to have been caused by the first three homologous halo CMEs of that time period. In the last third and for the fourth magnetic cloud (4 in Figure 9), the explanation may be of lesser quality because the last cloud was not caused by one of the homologous CMEs, and the apparent fluctuation (5 in parentheses) cannot be modeled with their approach. On the other hand, the Cane and Richardson (2003) catalogue of ICMEs only contains one ICME in the time period considered here, although they do mention that it could be due to several CMEs. The time period identified by Cane and Richardson is indicated by CR03 above the top panel of Figure 9. It appears to agree better with the compositional data around 08:00 on day 332. Ionization states of oxygen and iron rise to an enhanced level at the onset of the CR03 period, indicating a hot coronal origin. Elemental abundance ratios Fe/O and Mg/O as well as He/H rise dramatically at the onset time, although Cane

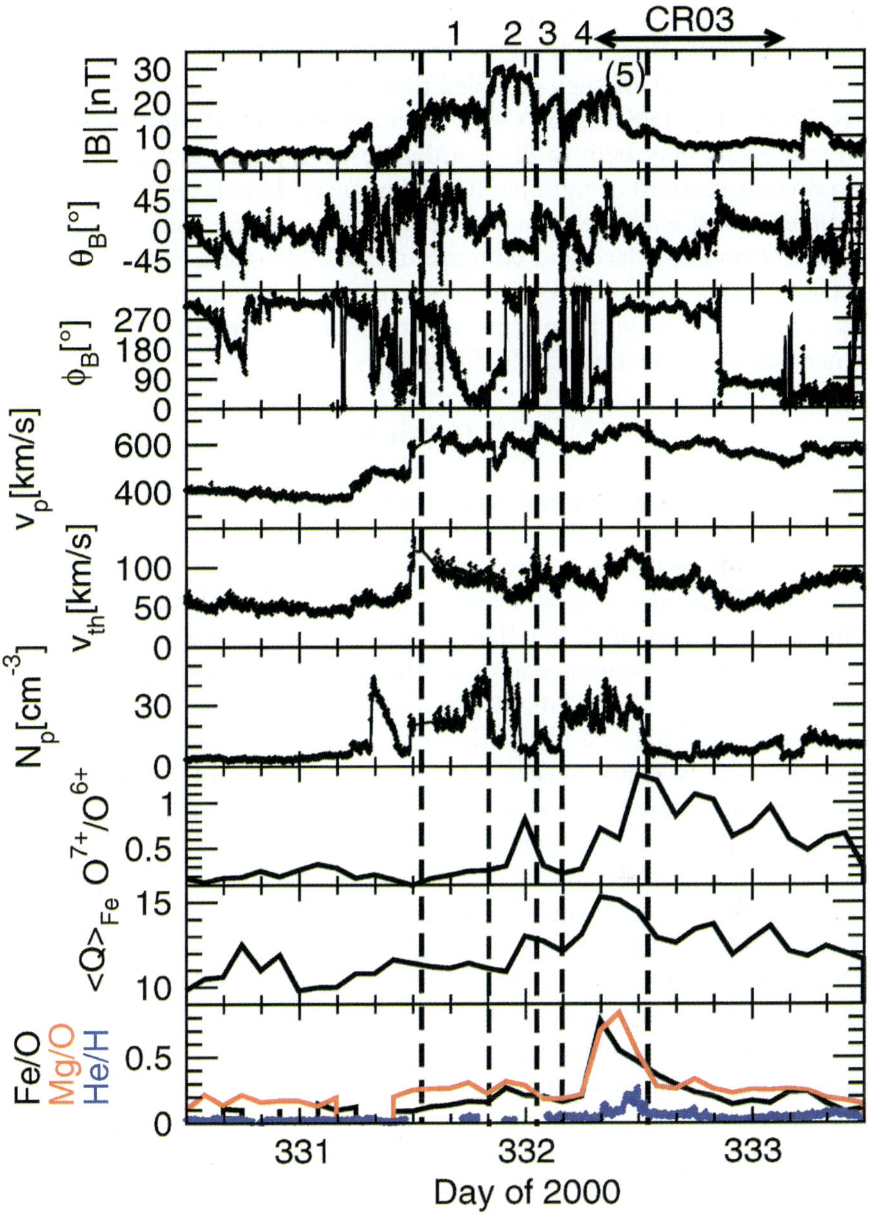

Figure 9. Observations from the November 25, 12:00 UT, to 28 November 2000, 12:00 UT, time period around Earth orbit. From top to bottom: magnetic field strength B, polar and azimuthal field angles θ_B and ϕ_B (in GSE coordinates), bulk solar wind speed v_p, proton thermal speed, v_{th}, and proton density N_p, all from Wind (data from CDAWeb). The lowest three panels show compositional data: oxygen charge-state ratio O^{7+}/O^{6+}, average iron charge state, and Fe and Mg to O abundance ratios (data from ACE level 2). Vertical dashed lines denote the borders of the four magnetic clouds (1 – 4) identified by Wang *et al.* (2002). The arrow marked CR03 shows the time period of an ICME identified by Cane and Richardson (2003).

and Richardson did not use these compositional signatures to derive their ICME list (Cane and Richardson, 2003).

Obviously, at times of such complex flows and multiple halo CMEs, it is frequently difficult to relate the flows with specific CMEs. Observations from widely-spaced spacecraft may help to discern the flow structure, such as in the Helios study (previous section) of multiple ICME/magnetic clouds and their interactions with ambient solar wind flows and with each other. Once understood in a more quantitative manner, solar wind composition will also be a very helpful tool in understanding such complex regions in space, much like in the complex corotating interaction regions with multiple stream-interface crossings at high latitudes (Wimmer-Schweingruber *et al.*, 1997,1999).

4. 3-D Structure

4.1. Correspondence to CME Structure

A particularly interesting question is whether one can identify the interplanetary counterparts of the three-part structure of CMEs observed by coronagraphs, i.e., a bright, dense leading edge, low-density (presumably magnetic field-dominated) void, and cool, dense prominence (e.g., Hudson *et al.*, 2006, this volume). The compressed material ahead of the ICME and the ICME proper, respectively, most likely correspond to the first two CME components. Prominence material, which may be indicated by dense plasma with unusually low ion charge states, is only occasionally encountered. However, as discussed in Sections 2.1 and 2.2, Schwenn *et al.* (1980) noted the intermittent presence of He^+ during the ICME observed at Helios 1 and shaded in red in Figure 8, and suggested that this was cold chromospheric material associated with a prominence. Interestingly, Schwenn *et al.* (1980) report He^+ predominantly near the leading edge and centre of the ICME. On the other hand, the prominent He^+ enhancement identified by Burlaga *et al.* (1998) occurred in exceptionally dense plasma ($\leq$185 protons cm^{-3}) at the trailing edge of a magnetic cloud in January 1997. Wurz *et al.* (1998) found dramatic mass fractionation in this filament material. Gloeckler *et al.* (1999) noted that both low and high ion charge states coexisted for extended intervals in the ICME on May 2, 1997. Hence, there appears to be considerable event-to-event variation in the distribution of apparent prominence material that may be due to its 3-D distribution inside the ICME or to true ICME variability.

4.2. Multi-Spacecraft Observations

A fundamental property of an ICME is its extent in helio-longitude and latitude. The longitudinal extents of ICMEs and the related shocks have been examined in several studies of individual ICMEs by multiple, widely-separated, spacecraft. For example, Burlaga *et al.* (1981) identified a magnetic cloud at four spacecraft

(Voyager 2, Helios 1 and 2 and IMP 8) spanning 40° in longitude. The cloud axis was estimated to curve on a scale of ~0.5 AU at 1 AU. Other ICMEs observed by Helios/IMP 8 were reported by Burlaga *et al.* (1987), Behannon *et al.* (1991), and Bothmer and Schwenn (1996), while Cane *et al.* (1997), noted that remarkably few ICMEs were observed by a pair of spacecraft, even when separated by only ~ 40° in longitude. They concluded that ICMEs typically extend in longitude only by ~ 50°, similar to the average *latitudinal* extent of CMEs observed in projection by coronagraphs (e.g., St. Cyr *et al.*, 2000). In such studies, care must be taken to ensure that unrelated shocks/ICMEs at different locations are not erroneously assumed to be part of the same event (e.g., Cane *et al.*, 1991).

Other studies have inferred the longitudinal extents of shocks/ICMEs by observing, at one or more locations, the succession of shocks and ICMEs originating in solar events associated with a major active region as it rotates in longitude due to solar rotation. For example, Figure 10 (Cane and Richardson, 1995) shows a sketch of the ICME and shock configuration inferred from cuts made by the IMP 8 and ISEE 3/ICE spacecraft (located 65°W of Earth) through a sequence of four shocks/ICMEs (numbered 1–4) in October, 1989 that originated in a single active region. The spacecraft trajectories (solid line = shock and ICME encountered; dashed line = ICME not encountered) are drawn relative to the solar event longitude. The ICME width of ±50° from the event location, and the shock width, extending over almost 180°, are consistent with the observations assuming symmetry about the event location. Thus, when the active region is beyond ~E50° relative to the observing spacecraft, only the western flank of the shock is detected. As the active region moves closer to, then crosses the longitude of the spacecraft, the ICME is detected following the shock. Finally, the active region moves beyond

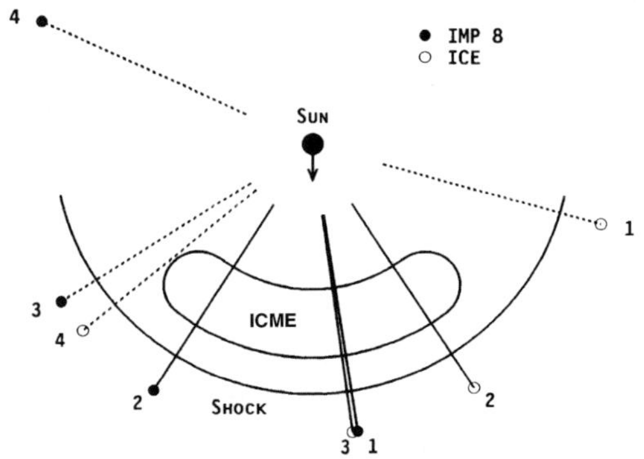

Figure 10. Summary of the trajectories of IMP 8 or ISEE 3/ICE (65°W of Earth) relative to four shocks/ICMEs in October, 1989 (after Cane and Richardson, 1995. © American Geophysical Union. Reproduced/modified by permission of American Geophysical Union), suggesting typical longitudinal widths of nearly 180° and ~ 100° for the shocks and ICMEs, respectively. (The ICME shape is arbitrary; solid lines indicate trajectories on which the ICME was detected).

$\sim 50^\circ$ west of the spacecraft and the eastern flank of the shock is detected, but not the ICME.

A related approach is a multi-event statistical study examining how the presence or absence of an ICME or shock depends on the longitude of a solar event relative to an observing spacecraft, typically near the Earth (e.g., Borrini *et al.*, 1982; Cane, 1988; Richardson and Cane, 1993; Cane and Richardson, 2003). Such studies again suggest that ICMEs may extend up to $\sim 50^\circ$ in longitude west and east of the solar event location (i.e., a total extent of $\sim 100^\circ$). Cane (1988) concluded from transit speeds to 1 AU that shocks are typically quasi-spherical over at least $\sim 100^\circ$ around the event longitude and extend well beyond the respective ICMEs, as also noted by Borrini *et al.* (1982). In particular, the flanks of shocks from near-limb events may occasionally be detected at Earth. Such studies tend to be dominated by energetic events for which unambiguous associations between interplanetary and solar phenomena can be made. If the related ICMEs are less extended for less energetic solar events, this probably accounts for the smaller size (total width $\sim 50^\circ$) suggested by studies of individual ICMEs.

Thus, ICMEs apparently have typical (full-width) longitudinal extents $\sim 50^\circ$ (which, however, may be larger ($\lesssim 100^\circ$) in particularly energetic events) that are similar to the latitudinal extents of CMEs against the plane of the sky. On the other hand, simple cartoons of flux-rope-like ICMEs (cf. Figure 2 of Zurbuchen and Richardson, 2006) might suggest an ICME structure that is far more extended in the plane of the flux-rope. We conclude that further multi-spacecraft studies (for example of the type performed by Riley *et al.*, 2003) are required to help elucidate the 3-dimensional structure of individual ICMEs and attempt to reconcile current concepts of ICME structure.

It has been estimated that between 20% (solar maximum) and 70% (solar minimum) of all ICMEs contain magnetic flux ropes (Richardson and Cane, 2004b). As such, force-free as well as more sophisticated techniques can provide important insight into their intrinsic properties (see Section 2), provided some basic limitations are taken into account (Riley and Crooker, 2004; Riley *et al.*, 2004). These models are discussed in more detail in Forbes *et al.* (2006, this volume).

4.3. Shocks

Shocks can be precursors of ICMEs throughout the heliosphere. ICMEs are often launched from the Sun at speeds greater than the solar wind speed. Since the solar wind is supersonic, a shock must form to allow the accommodation of these two flow regimes. More specifically, a shock will form when the speed difference exceeds the local fast magnetosonic speed. In the inner heliosphere, most shocks are ICME-related (Lindsay *et al.*, 1994), since most corotating interaction region shocks only form beyond 1 AU (Gosling *et al.*, 1976). Roughly 50% of ICMEs are associated with interplanetary shocks (Marsden *et al.*, 1987). These fast forward shocks propagate through the ambient solar wind ahead of the ICME. Spacecraft in the solar wind will first see the shock, then the region of shocked, compressed plasma and magnetic field called the sheath, and finally the ICME plasma itself (see Figure 2 of Zurbuchen and Richardson, 2006, this volume). Thus, ICME-driven shocks, while not a true signature of the ICME proper, are useful indicators of the possible presence of a following ICME and can be used as such successfully, e.g., in the Genesis algorithm (Section 2.7).

Some high-latitude ICMEs observed by Ulysses are bounded by forward-reverse shock pairs. Gosling *et al.* (1994) suggested that these ICMEs are "overexpanding," meaning that the internal ICME pressure is higher than that of the ambient solar wind. The higher pressure drives expansion into both the leading and trailing solar wind and thus produces shocks and sheaths on both the leading and trailing sides of the ICME. An alternate explanation (Manchester and Zurbuchen, 2006) is that the reverse shock in these events results from the bimodal speed structure of the solar wind near solar minimum, with slow flow at low-latitudes and fast flow at high latitudes. In front of the ICME the slow solar wind is deflected to higher latitude, and behind the ICME the fast flow is deflected to lower latitudes; the collision of these deflected flows results in the formation of a reverse shock.

When data quality is sufficient, the shock normal and the shock speed can be determined. This information helps reveal which part of the ICME is observed; shock normals are radial at the nose of the ICME and rotate away from the Sun-nose direction toward the flanks (Szabo *et al.*, 2001). Numerous methods of determining the shock normal are in the literature. Kasper (2002) used many different methods to determine the shock normal and speed at multiple spacecraft. He then tested the prediction from each model for the time of shock arrival at the downstream spacecraft with the actual shock-arrival time. He found that the Rankine-Hugoniot methods are most accurate.

Comparison of shock normals from multiple spacecraft shows that shocks are not always planar even on the scale of Earth's magnetosphere. Deformations in the shock surface can arise both from asymmetries in the internal structure of the ICME and in the ambient solar wind into which the ICME propagates (Russell *et al.*, 1983; Szabo *et al.*, 1999). Szabo *et al.* (2001) show that even shocks preceding magnetic clouds, which are large coherent structures, show this non-planarity. They find that larger, faster magnetic clouds have more planar shocks than smaller, slower magnetic clouds. The shocks flare away from the cloud's central axis, as expected.

The direction of the magnetic field in the ambient solar wind also influences the geo-effect of an ICME. The sheath regions, comprised of shocked solar wind, have high densities and magnetic fields which may be deflected out of the ecliptic by draping around the ICME (McComas *et al.*, 1989a). The compressed and possibly deflected magnetic field in the sheath make the sheath an important contributor to space weather (e.g., Huttunen *et al.*, 2005). If the shock is perpendicular, the compression of the magnetic field is especially strong. Hence nearly perpendicular shocks have a much larger probability of driving intense geomagnetic storms than more parallel shocks (Jurac *et al.*, 2002).

4.4. Sheath Planarity

It has long been known (Gosling and McComas, 1987) that compression of the solar wind plasma can lead to the magnetic field being draped around the ICME. This draping of magnetic field lines into the plane of compression within the sheath may lead to the formation of a "planar magnetic structure" (PMS). Planar magnetic structures are extended regions where magnetic field vectors, although of variable direction, lie within a common plane. They were first identified by Nakagawa *et al.* (1989), who suggested various mechanisms for their formation. Neugebauer *et al.* (1993), in a systematic study of PMS events, showed that many of them occurred in the sheath preceding an ICME, a result that is consistent with their generation by the compression of the upstream interplanetary magnetic field lines into a plane.

Clack *et al.* (2000) discussed the planarity of magnetic field lines due to compression within co-rotating interaction regions (CIRs). They pointed out that, since the plane in which the field lines lie is the same as the plane in which the stream interface lies, the orientation of the CIR can be estimated from this PMS plane. The minimum variance direction of the sheath magnetic field vectors should lie along the normal to the PMS plane, making its orientation straightforward to determine using this method. Jones *et al.* (2002) argued that the same effect should be present in the sheaths of ICMEs if they are travelling significantly faster than the preceding solar wind plasma, a scenario that is shown schematically in Figure 11. Therefore, they argued, it is possible to deduce the orientation of the local edge of an ICME from the orientation of the PMS.

Jones *et al.* (2002) showed that it is possible to estimate the position of a spacecraft (in this case Ulysses) relative to the centre of the ICME using PMS analysis.

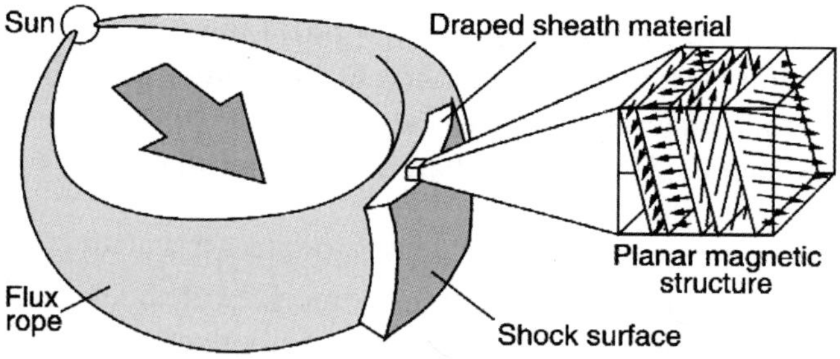

Figure 11. A fast-moving ICME compresses the solar wind ahead of it, leading to magnetic fields draped along this surface: a planar magnetic structure. The normal to this plane provides an estimate of the local orientation of the ICME (from Jones *et al.*, 2002. © American Geophysical Union. Reproduced/modified by permission of American Geophysical Union).

Their results agreed well with those resulting from flux rope fits to the magnetic field profile within ICMEs themselves. This suggests that this technique is a reliable estimator of the orientation of the local edge of the ICME and so could be used with data from several widely-separated spacecraft to estimate the global structure of an event or to determine the orientation of events without clear flux rope signatures.

4.5. Flow Deflections

Gosling *et al.* (1987b) reported flow deflections in the sheath of ICME-driven shocks and also in the ICME itself. In 17 of their 19 investigated shock events, an eastward deflection of the ICME was observed, with an average deflection angle of 3 degrees, corresponding to transverse velocities around 25 km/s. They usually observed an opposite flow deflection in the compressed or shocked ambient plasma ahead of the ICME. They interpreted this systematic pattern as a consequence of the magnetic pressure that builds up ahead of the ICME as the ambient magnetic field is draped around it on its western flank.

Owens and Cargill (2004) further established that, of intervals of non-radial solar wind flow (transverse velocity components >50 km/s) observed by ACE between 1998 and 2002, approximately one third occurred in regions upstream of fast ICMEs listed by Cane and Richardson (2003). The mean value of the maximum transverse flow component in the sheath regions of all fast ICMEs in the survey was of the order of 100 km/s.

In principle the magnitude and direction of these non-radial flows should be related to the shape and orientation of the ICME surface and its speed relative to the ambient solar wind flow. Specifially, for a spacecraft encounter passing through the axis of an ICME, where the axis runs through the length of the loop that is assumed to comprise the ICME, the flow deflection would be expected to be axis-aligned if it weren't for the built-up magnetic pressure of the draped IMF. Moreover, as the interception point of the spacecraft with the ICME moves away from the axis, the flow deflections should develop increasing velocity components perpendicular to the axis. Owens and Cargill (2004) report that typically the non-radial flows in ICME sheath regions are highly structured, containing both discontinuities and gradual rotations. However, for a subset of five events from the above survey where the upstream flow was relatively uniform, they were able to compare the flow deflections with the spacecraft position relative to the ICME axis estimated by variance analysis. They found a general consistency with the pattern of deflections relative to the axis suggested above. This suggests that there is merit in further exploring the possibility of using these deflected flows to make inferences about which part of an ICME a spacecraft is encountering and about the shape of the ICME leading edge and ellipticity of its cross section.

4.6. Energetic Particles and BDEs

The global magnetic structure of ICMEs has been a subject of much debate for a long time (e.g., Morrison, 1954; Cocconi *et al.*, 1958; Gold, 1959). Tongue, bottle, bubble, and connected or disconnected configurations have been proposed (Alexander *et al.*, 2006, this volume). Suprathermal particles serve as tracers of magnetic field lines, providing information on the global configuration of ICMEs. They serve as a tool for discerning between these different ICMEs topologies due to their small gyroradii, great speed and large particle scattering mean free paths in the smooth magnetic fields typical of ICMEs (e.g., Richardson, 1997; Crooker and Horbury, 2006, this volume; Klecker *et al.*, 2006, this volume). Even particles with very high energy, the galactic cosmic rays, are affected by ICMEs. Their intensity is observed to drop inside ICMEs, resulting in Forbush decreases (Cane and Lario, 2006, this volume; Klecker *et al.*, 2006, this volume).

For example, as discussed in section 2.4, bidirectional suprathermal electrons (BDEs, $\sim$100 eV) counterstreaming along magnetic field lines usually indicate that both ends of these field lines connect back to the corona. Earlier papers included the possibility that the field lines form a closed loop, entirely disconnected from the Sun (e.g., Montgomery *et al.*, 1974; Bame *et al.*, 1981; Gosling *et al.*, 1987a), but this is an unlikely extension of two-dimensional thinking. BDEs have also been used to explore magnetic cloud (MC) field polarities. If the dominant electron flow is away from the footpoint closer to the spacecraft, the relative directions of the field and electron flow indicate the field polarity, which should be constant during the magnetic cloud encounter assuming a single flux rope configuration. However, Kahler *et al.* (1999) found changes in polarity which cannot be explained by a single flux rope.

The streaming direction and the flux of electrons may vary extensively throughout a BDE event (e.g., Crooker *et al.*, 1990; Bothmer *et al.*, 1996; Shodhan *et al.*, 2000), indicating that connection of the magnetic field lines to the Sun is patchy. Shodhan *et al.* (2000) found a considerable variability in the duration of BDE events inside magnetic clouds and concluded that "magnetic clouds comprise a random mix of intertwined volumes of magnetically open and closed field lines".

Larson *et al.* (1997) used $\sim$0.1–100 keV electrons to deduce the magnetic topology, field line length and connectivity of a magnetic cloud observed by the WIND spacecraft. Figure 12a, panels A, B and C, show the magnetic field strength and polar and azimuthal angles. A clear rotation of the magnetic field vector is observed, characteristic for MCs. Solar wind speed and density are shown in panels D and E, respectively. Electrons streaming away from the Sun are displayed in panel F and in an energy vs. intensity (relative to quiet-time values) vs. time format in panel G. Several impulsive solar events can be seen above 20 keV. The faster electrons arrive earlier, as expected for injection at the Sun. The onset of each impulsive electron event coincides with a type III radio burst (panel H), which in turn is associated with a flare onset. Pitch angle distributions of 118 eV and 290 eV electrons are shown

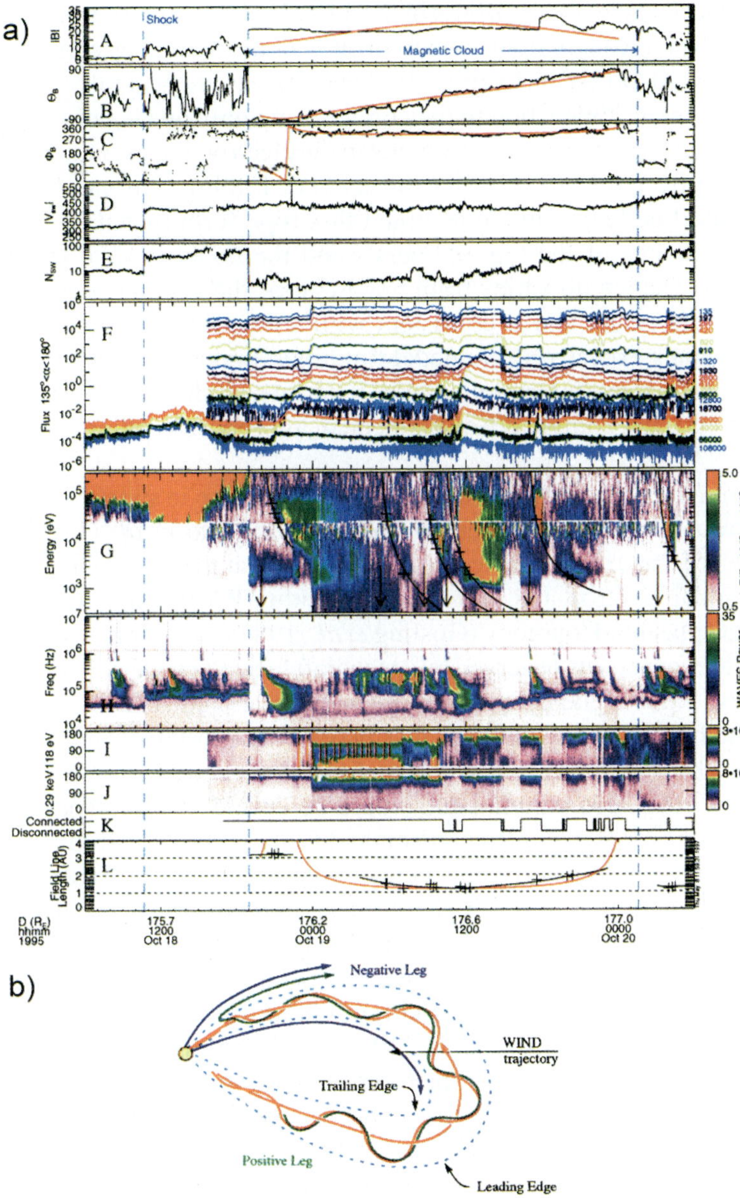

Figure 12. (a) Magnetic field, solar wind plasma, energetic electron, and radio observations from the Wind spacecraft during a magnetic cloud in October 1995; (b) schematic picture of possible topology (from Larson *et al.,* 1997. © American Geophysical Union. Reproduced/modified by permission of American Geophysical Union).

in panels I and J, respectively. BDEs in the 118 eV electrons (starting at ~0000 UT on October 19) indicate regions with both ends of the field lines connected back to the Sun. After ~0700 UT on October 19, the electrons are generally unidirectional, indicating magnetic connection to the corona along only one leg of the cloud. Also observable (panels I, J) are many dropouts in the electron fluxes, indicating possible disconnection from the corona (McComas *et al.*, 1989b). This can be also seen in panel K. Field line length in the assumed flux rope form (panel L) is determined from the onset time of the type III bursts and the travel time of these electrons from the Sun. The length varies from ~3 AU near the leading edge of the MC to $\gtrsim$1 AU near the center and is consistent with a flux rope configuration – the red line shows values expected for a model flux rope. Figure 12b illustrates the cloud topology. It shows intertwined magnetic field lines connected to the Sun at both ends, at one end, and completely disconnected, as proposed earlier by Gosling *et al.* (1995).

Bidirectional energetic particle flows similar to those in suprathermal electrons were first reported for solar energetic particles by Rao *et al.* (1967) and again suggest the presence of magnetic field loops rooted at the Sun. While early papers on bidirectional electron and proton events within ICMEs argued in favour of a disconnected plasmoid topology (Gosling *et al.*, 1987a; Marsden *et al.*, 1987), rapid solar particle event onsets observed by spacecraft located inside ICMEs (e.g., Kahler and Reames, 1991) argue in favour of the interpretation that ICME magnetic lines are rooted at the Sun. If disconnected plasmoids exist, most likely they form through reconnection between open coils and open field lines (step 3 in Figure 3b of Crooker and Horbury, 2006, this volume) and would be devoid of suprathermal electrons. Richardson *et al.* (1991) and Richardson and Cane (1996) noted that occasionally particles from normally poorly-connected eastern solar events arrived promptly in the vicinity of Earth along ICME field lines, also favouring the magnetic bottle configuration. In a study of 13 magnetic clouds detected by the Wind spacecraft, Mazur *et al.* (1998) detected impulsive flare particles in 4 of them. On the other hand, at greater distances from the Sun, Rodriguez *et al.* (2004) found no indication of impulsive electron onsets inside any of 40 magnetic clouds detected by the Ulysses spacecraft. They concluded that if the footpoints are still anchored back at the Sun, then the mechanisms accelerating the particles cannot be continuous from the time of the eruption to the time the MCs reach Ulysses.

Attempts have been made to obtain information on ICME topology using the velocity dispersion of energetic bidirectional ions, since stronger bidirectional fluxes might be expected with increasing particle energy (decreasing time to reach the loop feet) (e.g., Marsden *et al.*, 1987), but (Marsden *et al.*, 1985) and Rodriguez-Pacheco *et al.* (2003) found no such energy dependence. Moreover, in a case study of a cloud for which the flux-rope topology of Hidalgo *et al.* (2002) was assumed, Rodriguez-Pacheco *et al.* (2003) found no agreement with the expectation that, for a given particle energy, the bidirectional fluxes should be stronger at the cloud centre, where distance to the Sun is shortest. Almost the opposite pattern was found.

These authors concluded that mirroring at the cloud feet was not the cause of the bidirectional fluxes. Instead, they suggested that the particles exhibiting bidirectional flows in this event were accelerated at the shock ahead of the cloud and were injected into the ICME as a bidirectional flow. Furthermore, Popecki *et al.* (2001) analysed events in which the presence of particles showing impulsive characteristics was related to shock acceleration in interplanetary space, and not to a possible connection to a flaring region. Thus, the origin of bidirectional ion fluxes in ICMEs is still an open question.

In summary, there is considerable evidence from energetic particle observations that ICMEs consist primarily of a mix of magnetic loops or coils with one or both feet connected back to the Sun. On the other hand, the presence of disconnected plasmoids cannot be ruled out, and open questions remain regarding the interpretation of bidirectional ion signatures.

4.7. Trailing Velocity Increases

Of a more speculative nature, MHD models of flux rope initiation and evolution have been used to predict or verify observational signatures of the reconnection process occurring under the erupting flux rope (Riley *et al.*, 2002; Webb *et al.*, 2003). In particular, Riley *et al.* described how jetted outflow, driven by post-eruptive reconnection underneath the flux rope, would manifest itself as a speed enhancement trailing the ICME, and may remain intact out to 1 AU and beyond. They presented an example of a magnetic cloud with precisely these signatures and showed that the velocity perturbations are consistent with reconnection outflow. This may suggest that other velocity enhancements or unusual composition observed behind magnetic clouds are signatures of such reconnection and, in some cases, may not be associated with prominence material as has previously been suggested, although further analysis of in-situ observations is required to substantiate these tentative conclusions.

5. Other Solar Wind Transients

While ICMEs are the most conspicuous transient structures in the solar wind, they appear to be at one end of a continuum of transient structures that scale down to smaller size and/or to quieter, quasi-steady outflows. In the former category and closest in structure to ICMEs are the small flux rope structures identified by Moldwin *et al.* (1995, 2000). These occur on closed field lines, as signaled by BDEs, and contain the field signatures but not the low temperatures common to magnetic clouds. Moldwin *et al.* (2000) propose that the small flux ropes are created by reconnection in the solar wind rather than back at the Sun.

Other transient signatures, like ICMEs (see Forsyth *et al.*, 2006, this volume), tend to be associated with boundaries between sectors of opposite magnetic polarity, as suggested by Crooker *et al.* (1993), and with the highly variable slow wind in which the sector boundaries are imbedded. For example, large-scale magnetic field inversions immediately adjacent to sector boundaries, deduced from suprathermal electron measurements, have been interpreted in terms of quasi-steady outflows of quiet loops opened by interchange reconnection (Crooker *et al.*, 2004b). When a field line in the leg of a closed loop reconnects with an open field line, it creates a new open field line in which what was originally the leg of the loop becomes a segment that turns back toward the Sun, forming an inversion. The configuration is the same as that illustrated for opening closed fields in ICMEs in Figure 4b of Crooker and Horbury (2006, this volume) except that for inversions adjacent to sector boundaries the configuration must be represented in three dimensions. Crooker *et al.* (2004b) analyzed eight inversions with durations comparable to those of ICMEs. They found that some recurred from one rotation to the next and that a few displayed ICME signatures, most of a marginal nature. They suggest that the inversions may be the heliospheric counterparts of the quiet outflows of loops from active regions reported by Uchida *et al.* (1992).

Smaller-scale transient structures associated with sector boundaries are high-beta heliospheric plasma sheets. At first these were thought to be steady-state structures enveloping the heliospheric current sheet (Winterhalter *et al.*, 1994), which ideally constitutes the sector boundary. Crooker *et al.* (2004c), however, have found that plasma sheets are highly variable. Using suprathermal electron data to distinguish true sector boundaries from local current sheets created by field inversions, they found that plasma sheets are often missing at sector boundaries, sometimes owing to adjacent field inversions covering a wide range of scale sizes, and that plasma sheets often envelop local current sheets away from sector boundaries. Following (Wang *et al.*, 1998, 2000; Crooker *et al.*, 2004c) conclude that the heliospheric plasma sheet may consist entirely of transient plasma sheets and that these may be the heliospheric counterparts of the plasma blobs observed by coronagraphs to emanate from the tips of helmet streamers. If the blobs are released by interchange reconnection, as Wang *et al.* (1998, 2000) suggest, then interchange reconnection may be responsible for the inversions that create local current sheets in some plasma sheets.

Two additional observed patterns deserve mention as possible non-ICME transients: extremely low-density events and radial-field events. Like large-scale field inversions, low-density events can be recurrent, but they have the same kinds of pressure profiles as ICMEs, with magnetic field pressure dominating plasma pressure (Crooker *et al.*, 2000). These characteristics suggest a quiet but transient origin associated with sector boundaries. In radial field events (e.g., Jones *et al.*, 1998), the magnetic field deviates significantly from the Parker spiral toward a direction pointing radially toward or away from the Sun. Unlike the transients discussed above, which are attributed to outflows of spatial structures, radial events have been ascribed to temporal events, sudden changes in speed at the base of magnetic flux tubes (Gosling and Skoug, 2002; Neugebauer and Liewer, 2003; Wang *et al.*, 2003).

6. Conclusions and Discussion

As has become clear in this chapter, identifying ICMEs in situ is not straightforward. The key problem is that there is no single signature or a combination of signatures that is a foolproof ICME identifier. Different identification methods yield different results and are generally intermittent. The reason for this unsatisfactory state is unclear. Are ICMEs very inhomogeneous, are they individual entities from their onset, are they influenced strongly and differently by their evolution and propagation? These questions are especially important when trying to establish ICME boundaries. While CME boundaries appear reasonably sharp in white-light images, this is not at all the case for ICME boundaries. Why are they often elusive or ambiguous?

This report has clearly shown the potential of composition measurements in identifying ICMEs, but also in investigating their origin and possibly even their evolution and propagation through the solar corona and interplanetary space. We are beginning to understand the richness of elemental abundance and charge-state information, but several questions still remain.

Why is the range of compositional signatures so large? Why do they range from no signature in high-latitude ICMEs to otherwise unseen compositional oddities in a few selected ICMEs? Why are high-latitude ICMEs so different from low-latitude ones, at least compositionally? Is it due to a difference in the pre-CME state or in the onset/initiation mechanism? What is the relationship between ionic charge states and the CME initiation process, and how do we interpret mixed high and low charge states in the same bulk plasma? What leads to the He accumulation in the CMEs that we measure as He abundance enhancements in ICMEs?

The magnetic topology of ICMEs has long been a key in-situ signature for ICME identification. Both magnetic field and bidirectional electron measurements have allowed us to understand the global topology of ICMEs in some detail. Nevertheless, several puzzles remain. Remarkably, not all ICMEs are observed as flux ropes, as theory would predict. Is this simply due to an observational bias, or do ICMEs not necessarily need to be flux ropes? In either case, why does the fraction of magnetic clouds among all ICMEs vary with the solar cycle? While nearly all ICMEs within 5 AU contain some closed fields, somewhat less than half of the field lines within a typical ICME appear to be open, if the intermittency of bidirectional electrons is an indicator for a mix of open and closed fields. Where do these open field lines connect to? When and where do they disconnect? The still unclear magnetic connection of ICMEs with the Sun and the ambient solar wind must somehow be related to the origin of bidirectional ion fluxes observed in the interiors of ICMEs. We don't understand how.

Doubtlessly, some of the questions mentioned above will be addressed by the upcoming STEREO mission, which, in combination with other assets such as SOHO, Wind, and ACE will allow us to study the inhomogeneity and case-by-case variability of ICMEs in much more detail than previously possible. Nevertheless, the key to understanding the relation between ICMEs and CMEs, and thus ultimately understanding ICME signatures, lies in going closer to the Sun and studying them with a fleet of spacecraft with modern instrumentation such as envisioned with the Solar Orbiter and Sentinels missions.

Acknowledgements

This work was supported, in parts, by the Deutsche Forschungsgemeinschaft DFG. We thank the convenors for organizing a series of three stimulating workshops and the International Space Science Institute, ISSI, for hosting the final one.

References

Alexander, D., Richardson, I. G., and Zurbuchen, T. H.: 2005, *Space Sci. Rev.*, this volume, doi: 10.1007/s11214-006-9013-1.

Antiochos, S. K.: 1998, *Astrophys. J.* **502**, L181.

Bame, S. J., Hundhausen, A. J., Asbridge, J. R., and Strong, I. B.: 1968, *Astro J.* **73**, 55.

Bame, S. J., Asbridge, J. R., Feldman, W. C., Fenimore, E. E., and Gosling, J. T.: 1979, *Sol. Phys.* **62**, 179.

Bame, S. J., Asbridge, J. R., Feldman, W. C., Gosling, J. T., and Zwickl, R. D.: 1981, *Geophys. Res. Lett.* **8**, 173.

Behannon, K. W., Burlaga, L. F., and Hewish, A.: 1991, *J. Geophys. Res.* **96**, 21,213.

Bothmer, V.: 1999, *Solar Wind 9 Conference Proc.*, Published by American Institute of Physics (AIP), p. 119.

Bothmer, V.: 2003, in: A. Wilson (ed.), *Proc. ISCS 2003 Symp., Solar Variability as an Input to the Earth's Enviornment*, **ESA SP-535**, 419.

Bothmer, V., and Schwenn, R.: 1992, *Solar Wind Seven*, Pergamon, Oxford, p. 151.

Bothmer, V., and Schwenn, R.: 1995, *J. Geomagn. Geoelectr.* **47**, 1127.

Bothmer, V., and Schwenn, R.: 1996, *Adv. Space Res.* **17**, 319.

Bothmer, V., and Rust, D. M.: 1997, in: *Proc. of AGU Conf. on Coronal Mass Ejections: Causes and Consequences*, American Geophysical Union Geophys. Monogr. Series, p. 139.

Bothmer, V., and Schwenn, R.: 1998, *Ann. Geophys.* **16**, 1.

Bothmer, V., *et al.*: 1996, *Astron. Astrophys.* **316**, 493.

Borrini, G., Wilcox, J. M., Gosling, J. T., Bame, S. J., and Feldman, W. C.: 1981, *J. Geophys. Res.* **86**, 4565.

Borrini, G., Gosling, J. T., Bame, S. J., and Feldman, W. C.: 1982, *J. Geophys. Res.* **87**, 4365.

Burlaga, L.: 1988, *J. Geophys. Res.* **93**, 7217.

Burlaga, L. F.: 1991, in: R. Schwenn and E. Marsch (eds.), *Physics of the Inner Heliosphere* II, Springer-Verlag, Berlin, Heidelberg, p. 1.

Burlaga, L. F.: 1995, *Interplanetary Magnetohydrodynamics*, Oxford University Press, New York and Oxford.
Burlaga, L. F., and Ogilvie, K. W.: 1973, *J. Geophys. Res.* **78**, 2028.
Burlaga, L. F., and Behannon, K. W.: 1982, *Sol. Phys.* **81**, 181.
Burlaga, L. F., and Behannon, K. W.: 1987, *J. Geophys. Res.* **92**, 5725.
Burlaga, L. F., Sittler, E., Mariani, F., Lazarus, A., Schwenn, R.: 1981, *J. Geophys. Res.* **86**, 6673.
Burlaga, L. F., Behannon, K. W., and Klein, L. W.: 1987, *J. Geophys. Res.* **92**, 5725.
Burlaga, L., Fitzenreiter, R., Lepping, R., Ogilvie, K., Szabo, A., Lazarus, A. *et al.*, 1998, *J. Geophys. Res.* **103**, 277.
Cane, H. V.: 1988, *J. Geophys. Res.* **93**, 1.
Cane, H. V., and Richardson, I. G.: 1995, *J. Geophys. Res.* **100**, 1755.
Cane, H. V., and Richardson, I. G.: 2003, *J. Geophys. Res.* **108**(4), 1156, doi:10.1029/2002JA009817.
Cane, H. V., and Lario, D.: 2006, *Space Sci. Rev.* this volume, doi: 10.1007/s11214-006-9011-3.
Cane, H. V., Kahler, S. W., and Sheeley, N. R., Jr.: 1986, *J. Geophys. Res.* **91**, 13321.
Cane, H. V., Richardson, I. G., and Harvey, K.: 1991, *J. Geophys. Res.* **96**, 19525.
Cane, H. V., Wibberenz, G., and Richardson, I. G.: 1997, *J. Geophys. Res.* **102**, 7075.
Clack D., Forsyth, R. J., and Dunlop, M. W.: 2000, *Geophys. Res. Lett.***29**(5), 625.
Cocconi, G., Gold, T., Greisen, K., Hayakawa, S., and Morrison, P.: 1958, *Nuovo Cimento* **8**, 161.
Cremades, H., and Bothmer, V.: 2004, *Astron. Astrophys.* **422**, 307.
Crooker, N. U., Gosling, J. T., Smith, E. J., and Russell, C. T.: 1990, in: C. T. Russell, E. R. Priest and L. C. Lee (eds.), *Physics of Magnetic Flux Ropes*, Geophys. Monogr. Ser. Vol. 58, AGU., Washington D.C., p. 365.
Crooker, N. U., Siscoe, G. L., Shodhan, S., Webb, D. F., Gosling, J. T., and Smith, E. J.: 1993, *J. Geophys. Res.* **98**, 9371.
Crooker, N. U., and Intriligator, D. S.: 1996, *J. Geophys. Res.* **101**, 24343, doi:10.1029/96JA02129.
Crooker, N. U., Gosling, J. T., and Kahler, S. W.: 1998, *J. Geophys. Res.* **103**, 301.
Crooker, N. U., Shodhan, S., Gosling, J. T., Simmerer, J., Lepping, R. P., and Steinberg, J. T., 2000, *Geophys. Res. Lett.* **27**, 3769.
Crooker, N. U., Forsyth, R., Rees, A., Gosling, J. T., and Kahler, S. W.: 2004, *J. Geophys. Res.* **109**, A06110, doi:10.1029/2004JA010426.
Crooker, N. U., Kahler, S. W., Larson, D. E., and Lin, R. P.: 2004, *J. Geophys. Res.* **109**, A03108, doi:10.1029/2003JA010278.
Crooker, N. U., Huang, C.-L., Lamassa, S. M., Larson, D. E., Kahler, S. W., and Spence, H. E.: 2004, *J. Geophys. Res.* **109**, A03107, doi:10.1029/2003JA010170.
Crooker, N. U.: 2005, *Proc. Solar Wind 11, ESA-SP*, **592**, 289.
Crooker, N. U., and Horbury, T. S.: 2006, *Space Sci. Rev.*, this volume, doi: 10.1007/s11214-006-9014-0.
Farrugia, C. J., *et al.*: 2001, *Adv. Space Res.* **28**(5), 759.
Fenimore, E. E.: 1980, *Astrophys. J.* **235**, 245.
Feuerstein, W. M., Larson, D. E., Luhmann, J. G., Lin, R. P., Kahler, S. W., and Crooker, N. U.: 2004, *Geophys. Res. Lett.*, **31**, doi: 10.1029/2004GL020529.
Forbes, T. G., Linker, J. A., Chen, J., Cid, C., Kóta, J., and Lee, M. A., 2006, *Space Sci. Rev.*, this volume, doi: 10.1007/s11214-006-9019-8.
Forsyth, R. J., Rees, A., Reisenfeld, D. B., Lepri, S. T., and Zurbuchen, T. H.: 2003, in: M. Velli, R. Bruno and F. Malara (eds.), *Solar Wind Ten*, AIP Conf. Proc. **679**, Mellville, N.Y., p. 715.
Forsyth, R. J., Bothmer, V., Cid, C., Crooker, N. U., Horbury, T. S., Kecskemety, K., 2006, *Space Sci. Rev.*, this volume, doi: 10.1007/s11214-006-9022-0.
Galvin, A. B.: 1997, *Geophysical Monograph 99*. N. Crooker, J. A. Joselyn and J. Feymman (eds.), p. 253.
Geiss, J., Hirt, P., and Leutwyler, H.: 1970, *Sol. Phys.* **12**, 458.

Geiss, J., Bühler, F., Cerutti, H., Eberhard, P., and Filieux, C.: 1970b, *Sci. Rep. NASA SP* **315**, 14-1.
Geiss, J.: 1982, *Space Sci. Rev.* **33**, 201.
Geiss, J., Gloeckler, G., and von Steiger, R. : 1995, *Space Sci. Rev.* **72**, 49.
Gloeckler, G., Fisk, L. A., Hefti, S., Schwadron, N. A., Zurbuchen, T. H., and Ipavich, F. M. 1999, *Geophys. Res. Lett.* **26**(2), 157.
Gold, T.: 1959, *J. Geophys. Res.* **64**, 1665.
Goldstein, H.: 1983, in: M. Neugebauer (ed.), *Solar Wind Five*, NASA Conf. Publ. **22**, 731.
Gopalswamy, N., Lara, A., Yashiro, S., Nunes, S., and Howard, R. A.: 2003, in: A. Wilson (ed.), *Proc. ISCS 2003 Symp., Solar Variability as an Input to the Earth's Environment*, **ESA SP-535**, 419.
Gosling, J. T.: 1990, in: C. T. Russell, E. R. Priest and L. C. Lee (eds.), *Physics of Magnetic Flux Ropes*, American Geophysical Union, Geophysical Monograph **58**, 343.
Gosling J. T. and McComas, D. J.: 1987, *Geophys. Res. Lett.* **14**, 355.
Gosling, J. T., and Forsyth, R. J.: 2001, *Space Sci. Rev.* **97**, 87.
Gosling, J. T., and Skoug, R. M.: 2002, *J. Geophys. Res.* **107**(A10), 1327, doi::10.1029/2002JA009434.
Gosling, J. T., Hundhausen, A. J., and Bame, S. J.: 1976, *J. Geophys. Res.* **81**, 2111.
Gosling, J. T., Asbridge, J. R., Bame, S. J., Feldman, W. C., and Zwickl, R. D.: 1980, *J. Geophys. Res.* **85**, 3431.
Gosling, J. T., Baker, D. N., Bame, S. J., Feldman, W. C., Zwickl, R. D., and Smith, E. J.: 1987, *J. Geophys. Res.* **92**, 8519.
Gosling, J. T., Thomsen, M. F., Bame, S. J., and Zwickl, R. D.: 1987, *J. Geophys. Res.* **92**, 12399.
Gosling, J. T., Bame, S. J., McComas, D. J., and Phillips, J. L.: 1990, *Geophys. Res. Lett.* **17**, 901.
Gosling, J. T., Bame, S. J., Feldman, W. C., McComas, D. J., Phillips, J. L., and Goldstein, B. E.: 1993, *Geophys. Res. Lett.* **20**, 2335.
Gosling, J. T., Bame, S. J., McComas, D. J., Phillips, J. L., Scime, E. E., and Pizzo, V. J., 1994, *Geophys. Res. Lett.* **21**, 237.
Gosling, J. T., Birn, J., and Hesse, M.: 1995, *Geophys. Res. Lett.* **22**, 869.
Gosling, J. T., McComas, D. J., Phillips, J. L., Pizzo, V. J., and Goldstein, B. E., 1995, *Geophys. Res. Lett.* **22**, 1753.
Gosling, J. T., Skoug, R. M., and Feldman, W. C.: 2001, *Geophys. Res. Lett.* **28**, 4155.
Gosling, J. T., Skoug, R. M., McComas, D. J., and Smith, C. W.: 2005, *J. Geophys. Res.* **110**, A01107, doi:10.1029/2004JA010809.
Henke, T., Woch, J., Schwenn, R., Mall, U., Gloeckler, G., and von Steiger, *et al.*, 2001, *J. Geophys. Res.* **106**, 10597.
Hidalgo, M. A., Cid, C., Vinas, A. F., and Sequeiros, J.: 2002, *J. Geophys. Res.* **106**, DOI:10.1029/2001JA900100.
Hirshberg, J., Alksne, A., Colburn, D. S., Bame, S. J., and Hundhausen, A. J.: 1970, *J. Geophys. Res.* **75**, 1.
Hirshberg, J, Bame, S. J., and Robbins, D. E.: 1972, *Solar Phys.* **23**, 467.
Ho, G. C., Hamilton, D. C., Gloeckler, G., and Bochsler, P.: 2000, *Geophys. Res. Lett.* **27**, 309.
Hudson, H. S., Bougeret, J.-L., and Burkepile, J.: 2006, *Space Sci. Rev.*, this volume, doi: 10.1007/s11214-006-9009-x.
Hundhausen, A. J., Gilbert, H. E., and Bame, S. J.: 1968, *J. Geophys. Res.* **73**, 5485.
Huttunen, K. E. J., Schwenn, R., Bothmer, V., and Koskinen, H. E. J.: 2005, *Annales Geophysicae* **23**, 1.
Ipavich, F. M., Galvin, A. B., Gloeckler, G., Hovestadt, D., Bame, S. J. and Klecker, B., *et al.*, 1986, *J. Geophys. Res.* **91**, 4133.
Janoo, L. and Farrugia, C. J.: 1998, *J. Geophys. Res.* **103**, 17249.
Jian, L., Russell, C. T., Gosling, J. T., and Luhmann, J. G.: 2005, *American Geophysical Union*, Spring Meeting 2005, abstract #SH53A-11.
Jones, G. H., Balogh, A., and Forsyth, R. J.: 1998, *Geophys. Res. Lett.* **25**, 3109.

Jones G. H., Rees, A., Balogh, A., and Forsyth, R. J.: 2002, *Geophys. Res. Lett.* **29**(11), 10.1029/2001GL014110.
Jurac, S., and Richardson, J. D.: 2001, *J. Geophys. Res.* **106**, 29195.
Kahler, S.: 1977, *Astrophys. J.* **214**, 891.
Kahler, S. W. and Reames, D. V.: 1991, *J. Geophys. Res.* **96**, 9419.
Kahler, S. W., Crooker, N. U. and Gosling, J. T.: 1999, *J. Geophys. Res.* **104**, 9911.
Kasper, J. C.: 2003, 'Solar wind plasma : kinetic properties and micro-instabilities', Thesis (Ph. D.), Massachusetts Institute of Technology, Dept. of Physics, February 2003.
Klecker, B., Kunow, H., Cane, H. V., Dalla, S., Heber, B. and Kecskemety, K., *et al.*, 2006, *Space Sci. Rev.*, this volume, doi: 10.1007/s11214-006-9018-9.
Klein, L. W., and Burlaga, L. F.: 1982, *J. Geophys. Res.* **87**, 613.
Laming, J. M., and Feldman, U.: 2001, *Astrophys. J.* **546**, 552.
Lario, D., Hu, Q., Ho, G. C., Decker, R. B., Roelof, E. C., and Smith, C. W.: 2006, *Proc. Conf. Solar Wind 11, Whistler, Canada*, **ESA-SP-592**, 81.
Lario, D., Ho, G. C., Decker, R. B., Roelof, E. C., Desai, M. I., and Smith, C. W.: 2003, in: M. Velli, R. Bruno and F. Malara (eds.), *Proc. Conf. Solar Wind 10, Pisa, Italy*, **AIP-679**, American Institute of Physics, Melville, New York, p. 640.
Larson, D. E. *et al.*: 1997, *Geophys. Res. Lett.* **24**, 1911.
Lepping, R. P., Jones, J. A. and Burlaga, J. F.: 1990, *J. Geophys. Res.* **95**, 11957.
Lepping, R. P., *et al.*: 1997, *J. Geophys. Res.* **102**, 14049.
Lepping, R. P., Berdichevsky, D. B., Burlaga, L. F., Lazarus, A. J., Kasper, J., Desch, M. D., *et al.*: 2001, *Solar Physics* **204**(1–2), 285.
Lepri, S. T., Zurbuchen, T. H., Fisk, L. A., Richardson, I. G., Cane, H. V., Gloeckler, and G.: 2001, *J. Geophys. Res.* **106**(A12), 29231.
Lepri, S. T., and Zurbuchen, T. H.: 2004, *J. Geophys. Res.* **109**(A1), A01112.
Lindsay, G. M., Luhmann, J. G., Russell, C. T., and Gazis, P. R.: 1994, *J. Geophys. Res.* **99**, 11.
Lopez, R. E.: 1987, *J. Geophys. Res.* **92**, 11189
Low, B. C.: 1997, in: N. Crooker, J. A. Joselyn and J. Feynman (eds.), *Coronal Mass Ejections, Geophysical Monograph 99*, p. 39.
Low, B. C., and Zhang, M.: 2002, *Astrophys. J.* **564**, L53.
Manchester, W. B. and Zurbuchen, T. H.: 2006, *J. Geophys. Res.* **111**, A05101, doi:10.1029/2005JA011461.
Marsden, R. G., Sanderson, T. R., and Wenzel, K.-P.: 1985, *Proc. 19th Int. Cosmic Ray Conf.* **4**, 322.
Marsden, R. G., Sanderson, T. R., Tranquille, C., Wenzel, K.-P., and Smith, E. J.: 1987, *J. Geophys. Res.* **92**, 11009.
Martin, S. F., Bilimoria, R., and Tracadas, P. W.: 1994, *Solar Surface Magnetism*, Springer-Verlag, p. 303.
Marubashi, K.: 1986, *Adv. Space Res.* **6**, 335.
Marubashi, K.: 1997, in: N. Crooker, J. A. Joselyn and J. Feynman (eds.), *Coronal Mass Ejection, American Geophysical Union Geophys. Monogr.* Vol. 99, p. 147.
Mazur, J. E., Mason, G. M., Dwyer, J. R., and von Rosenvinge, T. T. 1998, *Geophys. Res. Lett.* **25**, 2521.
McComas, D. J., Gosling, J. T., Bame, S. J., Smith, E. J., and Cane, H. V.: 1989, *J. Geophys. Res.* **94**, 1465.
McComas, D. J., *et al.*: 1989, *J. Geophys. Res.* **94**, 6907.
Mitchell, D. G., Roelof, E. C., and Bame, S. J.: 1983, *J. Geophys. Res.* **88**, 9059.
Moldwin, M. B., Phillips, J. L., Gosling, J. T., Scime, E. E., McComas, D. J., and Bame, S. J., 1995, *J. Geophys. Res.* **100**, 19903.
Moldwin, M. B., Ford, S., Lepping, R., Slavin, J., and Szabo, A.: 2000, *Geophys. Res. Lett.* **27**, 57.
Montgomery, M. D., Asbridge, J. R., Bame, S. J., and Feldman, W. C.: 1974, *J. Geophys. Res.* **79**, 3103.
Morrison, P.: 1954, *Phys. Rev.* **95**, 646.

Mulligan, T., and Russell, C. T.: 2001, *J. Geophys. Res.* **106**, 10581.
Mulligan, T., Russell, C. T., and Luhmann, J. G.: 1998, *Geophys. Res. Lett.* **25**, 2959.
Nakagawa T., Nishida, A., and Saito, T.: 1989, J. Geophys. Res. **94**, 11761.
Neugebauer, M.: 1981, *Fundam. Cosmic Phys.* **7**, 131.
Neugebauer, M.: 1992, E., Marsch and R., Schwenn, (eds.), *Proceedings of Solar Wind 7*, p. 69.
Neugebauer, M., and Goldstein, R.: 1997, in N. Crooker, J. A. Joselyn and J. Feynman (eds.), *Coronal Mass Ejections*, AGU, Washington D.C., p. 245.
Neugebauer, M., and Liewer, P. C.: 2003, *J. Geophys. Res.* **108**(A1), 1013, doi:10.1029/2002JA009326.
Neugebauer M., Clay, D. R. and Gosling, J. T.: 1993, *J. Geophys. Res.* **98**(A6), 9383.
Neugebauer, M., Goldstein, R., and Goldstein, B. E.: 1997, *J. Geophys. Res.* **102**, 19743.
Neugebauer, M., *et al.*: 2003, *Space Science Reviews* **105**, 661.
Neukomm, R. O.: 1998, *Composition of Coronal Mass Ejections Derived with SWICS/Ulysses.* Ph. D. thesis, Universität Bern.
Osherovich, V. A., Fainberg, J., and Stone, R. G.: 1999, *Geophys. Res. Lett.* **26**, 401.
Owens, M., and Cargill, P., 2004, *Ann. Geophys.* **22**, 4397.
Phillips, J. L., and Gosling, J. T.: 1990, *J. Geophys. Res.* **95**, 4217.
Pilipp, W. G., Miggenrieder, H., Mülhäuser, K.-H., Rosenbauer, H., Schwenn, R., and Neubauer, F. M.: 1987, *J. Geophys. Res.* **92**, 1103.
Pizzo, V. J.: 1991, *J. Geophys. Res.* **96**, 5405.
Popecki, M., Desai, M., Skoug, R. M., Smith, C. W., Moebius, E., and Galvin, A. B., 2001, *Proc. 27th International Cosmic Ray Conference*, Hamburg, p. 3153.
Rao, U. R., McCracken, K. G., and Bukata, R. P.: 1967, *J. Geophys. Res.* **72**, 4325.
Reinard, A. A.: 2005, *Astrophys. J.* **620**, 501.
Reisenfeld, D. B., Steinberg, J. T., Barraclough, B. L., Dors, E. E., Wiens, R. C., and Neugebauer, M., 2003, *AIP Conf. Proc. 679: Solar Wind Ten* **679**, 632.
Richardson, I. G.: 1997, in: N. U. Crooker, J.-A. Joselyn and J. Feynman (eds.), *Coronal Mass Ejections*, Vol. 99, AGU., Washington D.C., p. 189
Richardson, I. G., and Cane, H. V.: 1993, *J. Geophys. Res.* **98**, 15295.
Richardson, I. G., and Cane, H. V.: 1995, *J. Geophys. Res.* **100**, 23397.
Richardson, I. G., and Cane, H. V.: 1996, *J. Geophys. Res.* **101**, 27521.
Richardson, I. G., and Cane, H. V.: 2004, *J. Geophys. Res.* **109**, A09104, doi:10.1029/2004JA010598.
Richardson, I. G., and Cane, H. V.: 2004, *Geophys. Res. Lett.* **31**, L18804, doi:10.1029/2004GL020958.
Richardson, I. G., Cane, H. V., and von Rosenvinge, T. T.: 1991, *J. Geophys. Res.* **96**, 7853.
Richardson, I. G., Cane, H. V., Lepri, S. T., Zurbuchen, T. H., and Gosling, J. T.: 2003, in: M. Velli, R. Bruno and F. Malara (eds.), *Solar Wind Ten*, AIP Conf. Proc. 679, Mellville, N.Y., p. 681.
Riley, P., and Crooker, N. U.: 2004, *Astrophys. J.*, in press.
Riley, P., *et al.*: 2002, *Astrophys. J.* **578**.
Riley, P., Linker, J. A., Mikic', Z., Odstrcil, D., Zurbuchen, T. H., Lario, D., Lepping, R. P.: 2003, *J. Geophys. Res.* **108**(A7), 1272, doi:10.1029/2002JA009760.
Riley, P., Linker, J. A., Lionello, R., Mikic, Z., Odstrcil, D., and Hidalgo, M. A., *et al.*, 2004, Accepted for publication in the *Journal of Atmospheric and Solar Terrestrial Physics*.
Rodríguez-Pacheco, J., Cid, C., Blanco, J. J., and Sequeiros, J.: 2003, *Sol. Phys.* **213**, 121.
Rodriguez, L., Woch, J., Krupp, N., Fänz, M., von Steiger, R., and Forsyth, R. J., *et al.*, 2004, *J. Geophys. Res.* **109**, A01108, doi:10.1029/2003JA010156.
Russell, C. T., Mellott, M. M., Smith, E. J., and King, J. H.: 1983, *J. Geophys. Res.* **88**, 4739.
Rust, D. M., and Kumar, A.: 1994, *Solar Phys.* **155**, 69.
Rust, D. M., Anderson, B. J., Andrews, M. D., Acuna, M. H., Russell, C. T., Schuck, P. W., *et al.*, 2005, *Astrophys. J.* **621**, 524.

Tripathi, D. K., Bothmer, V., and Cremades, H.: 2004, *Astron. & Astrophys.* **422**, 337.
Schwenn, R.: 1990, R., Schwenn, and E., Marsch, (eds.), *Physics of the Inner Heliosphere* I Springer-Verlag, Berlin, Heidelberg, p. 99.
Schwenn, R.: 1996, in: D. Winterhalter, J. T. Gosling, S. Habbal, W. S. Kurth and M. Neugebauer (eds.), *Solar Wind Eight*, Melville, New York, p. 426.
Schwenn, R., Rosenbauer, H., and Mühlhäuser, K.-H.: 1980, *Geophys. Res. Lett.* **7**, 201.
Sheeley, N. R. Jr., Howard, R. A., Koomen, M. J., Michles, D. J., Schwenn, R., and Mühlhäuser, K.-H.: 1985, *J. Geophys. Res.* **90**, 163.
Shodhan, S., *et al.*: 2000, *J. Geophys. Res.* **105**, 27261.
Siscoe, G.: 2006, *Space Sci. Rev.*, this volume.
Skoug, R. M., Bame, S. J., Feldman, W. C., Gosling, J. T., McComas, D. J., Steinberg, J. T., *et al.*: 1999, *Geophys. Res. Lett.* **26**(2), 161.
Smith, C. W., Ness, N. F., Burlaga, L. F., Skoug, R. M., McComas, D. J., Zurbuchen, T. H., *et al.*: 2001, *Sol. Phys.* **204**, 229.
Stansberry, J. A., Gosling, J. T., Thomsen, M. F., Bame, S. J., and Smith, E. J.: 1988, *J. Geophys. Res.* **93**, 1975.
St. Cyr, O. C., *et al.*: 2000, *J. Geophys. Res.* **105**, 18169.
Szabo, A., Smith, C. W., Tokar, R. L., and Skoug, R. M.: 1999, *Eos Transactions, American Geophysical Union* **80**(17) Supplement, S265.
Szabo, A., Lepping, R. P., Merka, J., Smith, C. W., and Skoug, R. M.: 2001, in: B. Battrick and H. Sawaya-Lacoste (eds.), *Solar Encounter, Proceedings of the First Solar Orbiter Workshop*, **ESA SP-493**, Noordwijk, ESA Publications Division, ISBN 92-9092-803-4, p. 383.
Vandas, M., and Romashets, E. P.: 2003, *Astron. Astrophys.* **398**, 801.
Vasquez, B. J., *et al.*: 2001, *J. Geophys. Res.* **106**, 29283.
von Steiger, R.: 1998, *Space Sci. Rev.* **85**, 407.
Wang, C., Richardson, J. D., Burlaga, L. F., and Ness, N. F.: 2003, *J. Geophys. Res.* **108**(A5), 1205, doi:10.1029/2002JA009809.
Wang, Y.-M., Sheeley, N. R., Jr., J. Walters, H., Brueckner, G. E., Howard, R. A., Michels, D. J., *et al.*: 1998, *Astrophys. J.* **498**, L165.
Wang, Y.-M., Sheeley, N. R., Jr., Socker, D. G., Howard, R. A., and Rich, N. B.: 2000, *J. Geophys. Res.* **105**, 25133.
Wang, Y. M., Wang, S., and Ye, P. Z.: 2002 *Sol. Phys.* **211**, 333.
Webb, D. F., Burkepile, J., Forbes, T. G., and Riley, P.: 2003, *J. Geophys. Res.* **108**(A12), 1440, 10.1029/2003JA009923,16 December 2003.
Webb, D. F., Cliver, E. W., Crooker, N. U., St. Cyr, O. C., and Thompson, B. J.: 2000, *J. Geophys. Res.* **105**, 7491.
Widing, K. G., and Feldman, U.: 2001, *Astrophys. J.* **555**, 426.
Wimmer-Schweingruber R. F.: 1994, 'Oxygen, Helium, and Hydrogen in the Solar Wind: SWICS/Ulysses Results', PhD thesis, University of Bern, Switzerland.
Wimmer-Schweingruber R. F., von Steiger, R., and Paerli, R.: 1997, *J. Geophys. Res.* **102**, 17407.
Wimmer-Schweingruber R. F., von Steiger, R., Geiss, J., Gloeckler, G., Ipavich, F. M., and Wilken, B.: 1998, *Space Sci. Rev.* **85**, 387.
Wimmer-Schweingruber R. F., Bochsler, P., Gloeckler, G., Ipavich, F. M., Geiss, J., Fisk, L. A., *et al.*: 1999a, *Geophys. Res. Lett.* **26**, 165.
Wimmer-Schweingruber R. F., Kern, O., and Hamilton, D. C.: 1999b, *Geophys. Res. Lett.* **26**, 3541.
Wimmer-Schweingruber R. F., von Steiger, R., and Paerli, R.: 1999c, *J. Geophys. Res.* **104**, 9933.
Winterhalter, D., Smith, E. J., Burton, M. E., Murphy, N., and McComas, D. J.: 1994, *J. Geophys. Res.* **99**, 6667.
Wu, C. C., Lepping, R. P., and Gopalswamy, N.: 2003, A., in: Wilson, (ed.), *Proc. ISCS 2003 Symp., Solar Variability as an Input to the Earth's Enviornment*, **ESA SP-535**, p. 429.

Wurz, P., Ipavich, F. M., Galvin, A. B., Bochsler, P., Aellig, M. R., and Kallenbach, R., *et al.*, 1998, *Geophys. Res. Lett.* **25**, 2557.

Wurz, P., Bochsler, P., and Lee, M. A.: 2000, *J. Geophys. Res.* **105**, 27239.

Wurz, P., Wimmer-Schweingruber, R. F., Issautier, K., Bochsler, P., Galvin, A. B., Paquette, J. A., and Ipavich, F. M.: 2001, *AIP conference proceedings* **598**, 145.

Zhang, G. and Burlaga, L. F.: 1988, *J. Geophys. Res.* **93**, 2511.

Zwickl, R. D., Asbridge, J. R., Bame, S. J., Feldman, W. C., and Gosling, J. T.: 1982, *J. Geophys. Res.* **87**, 7379.

Zwickl, R. D., Asbridge, J. R., Bame, S. J., Feldman, W. C., Gosling, J. T., and Smith, E. J.: 1983, in: M. Neugebauer (ed.), *Solar Wind Five; NASA Conference Proceedings 2280*, NASA, Washington, D.C., p. 711.

Zurbuchen, T., Fisk, L. A., Lepri, S. T., and von Steiger, R.: 2003, in: M. Velli, R. Bruno and F. Malara (eds.), *Solar Wind Ten*, AIP Conf. Proc. 679, Mellville, N.Y., p. 604.

Zurbuchen, T. H., Raines, J., Lynch, B., Lepri, S., Gloeckler, G., and Fisk, L.: 2005, 'In situ observation of filament plasma and their magnetic structure', *American Geophysical Union, Spring Meeting 2005*, abstract #SH54B-04.

Zurbuchen, T. H. and Richardson, I. G.: 2006, *Space Sci. Rev.*, this volume, doi: 10.1007/s11214-006-9010-4.

Zwickl, R. D., Doggett, K. A., Sahm, S., Barrett, W. P., Grubb, R. N., Detman, T. R., *et al.*: 1998, *Space Sci. Rev.* **86**, 633.

ENERGETIC PARTICLE OBSERVATIONS

Report of Working Group C

B. KLECKER[1,*], H. KUNOW[2], H. V. CANE[3], S. DALLA[4], B. HEBER[2], K. KECSKEMETY[5], K.-L. KLEIN[6], J. KOTA[7], H. KUCHAREK[8], D. LARIO[9], M. A. LEE[8], M. A. POPECKI[8], A. POSNER[10], J. RODRIGUEZ-PACHECO[11], T. SANDERSON[12], G. M. SIMNETT[13], and E. C. ROELOF[9]

[1] *Max-Planck-Institut für extraterrestrische Physik, 85740 Garching, Germany*
[2] *Institut für Experimentelle und Angewandte Physik, University of Kiel, 24118 Kiel, Germany*
[3] *School of Mathematics and Physics, University of Tasmania, Tasmania, Australia*
[4] *School of Physics and Astronomy, University of Manchester, Manchester M60 1QD, UK*
[5] *Hungarian Academy of Sciences, Central Research Institute for Physics, Budapest, Hungary*
[6] *Observatoire de Paris, LESIA-CNRS UMR 8109, Bat. 14, F-92195 Meudon, France*
[7] *Department of Physics, University of Arizona, Tucson, AZ 85721, USA*
[8] *University of New Hampshire, Durham, NH 03824, USA*
[9] *Applied Physics Laboratory, Johns Hopkins University, Laurel, MD, USA*
[10] *Southwest Research Institute, Space Science and Engineering Division, 6220 Culebra Rd., San Antonio, TX 78228, USA*
[11] *Space Research Group, Departamento de Fisica, Universidad de Alcalá, Ctra. Madrid-Barcelona, 28871 Alcalá de Henares, Espana*
[12] *Space Science Dept. of ESA, Postbus 299, 2200 AG Noordwijk, The Netherlands*
[13] *School of Physics and Space Research, University of Birmingham, B15 2TT, UK*
*(*Author for correspondence: E-mail: berndt.klecker@mpe.mpg.de)*

(Received 27 May 2005; Accepted in final form 2 June 2006)

Abstract. The characteristics of solar energetic particles (SEP) as observed in interplanetary space provide fundamental information about the origin of these particles, and the acceleration and propagation processes at the Sun and in interplanetary space. Furthermore, energetic particles provide information on the development and structure of coronal mass ejections as they propagate from the solar corona into the interplanetary medium. In this paper we review the measurements of energetic particles in interplanetary space and discuss their implication for our understanding of the sources, and of acceleration and propagation processes.

Keywords: solar energetic particles, energetic particle propagation, energetic particle acceleration, energetic particle composition, energetic particle ionic charge states, solar electrons, shock acceleration, interplanetary coronal mass ejections

1. Introduction

B. KLECKER, H.V. CANE, AND K.-L. KLEIN

Solar energetic particles (SEP), in their intensity-time profiles, energy spectra, elemental, isotopic, and ionic charge composition carry fundamental information on

Space Science Reviews (2006) 123: 217–250
DOI: 10.1007/s11214-006-9018-9

the source region and their acceleration and propagation processes. High-energy particles originating at the Sun were first reported by Forbush (1946). At that time there was little doubt that these high energy particles were closely related to contemporary solar flares. Later it became clear that acceleration at interplanetary shocks is also an efficient mechanism for particle acceleration (e.g. Bryant *et al.*, 1962). As anticipated (e.g. Gold, 1962), and confirmed by coronographic observations of CMEs, such shock waves are not blast waves from flares, but are driven by magnetic structures ejected from the Sun.

In the early seventies a new type of event was discovered (Cane and Lario, 2006, this volume) that showed enhanced ^{3}He abundances (Hsieh and Simpson, 1970) with ^{3}He/^{4}He-ratios >1 (Balasubrahmanyan and Serlemitsos, 1974), while the corresponding ratio in the solar corona and solar wind is 5×10^{-4}. Such events were later found to exhibit enhancements of heavy ions by about an order of magnitude (e.g. Hurford *et al.*, 1975; Mason *et al.*, 1986) relative to coronal abundances. These small events also showed high ionic charge states of Si ($\sim$14) and Fe ($\sim$20) that were interpreted as indicative of high coronal temperatures ($\sim 10^7$ K), compared to charge states compatible with $\sim 1.5 - 2 \times 10^6$ K in interplanetary (IP) shock related events. Based on these and other observations SEPs were classified as 'impulsive' and 'gradual', following a classification of flares based on the duration of soft X-rays (Pallavicini *et al.*, 1977). In this picture the 'impulsive' SEP events were related to flares and the 'gradual' SEP events were related to coronal mass ejection (CME) driven coronal and interplanetary shocks (see e.g. review by Reames, 1999).

However, new results from instruments with improved collecting power and resolution onboard several spacecraft (e.g. WIND, SAMPEX, SOHO, ACE) have shown that this two–class picture was oversimplified. The new composition and ionic charge measurements show that some solar particles have their origin in a dense plasma low in the corona, even in events classified as 'gradual', that enrichments in ^{3}He are also common in IP shock accelerated populations (Desai *et al.*, 2001), and that enrichments in heavy ions are often observed in large events at high energies. Whether these new findings are best explained by a suprathermal population from previous 'impulsive' events (Mason *et al.*, 1999), by the interplay of shock geometry and different seed populations (solar wind and flare suprathermals, Tylka *et al.*, 2005), or by direct injection from the flare acceleration process (e.g. Klein and Trottet, 2001, and references therein; Cane *et al.*, 2003), with or without further acceleration by a coronal shock, is now heavily debated.

In this chapter we will focus on SEP observations in gradual events. We provide in Section 2 an example of typical intensity versus time profiles, summarise the dependence of event characteristics on longitude and latitude, review SEP elemental, isotopic, and ionic charge composition, and summarise electron observations. In Section 3 we discuss acceleration and propagation processes and their relation to the observations. Section 4 relates energetic particles observed in interplanetary space to the electromagnetic signature of plasma and energetic particles in the solar atmosphere. In Section 5 we discuss energetic particle signatures associated with the passage of interplanetary CMEs (ICMEs) and the use of SEPs as a tool to infer the magnetic topology of these structures.

2. Solar Energetic Particle Observations

2.1. Spatial and Temporal Variations

S. Dalla, D. Lario, H. V. Cane, and T. R. Sanderson

The intensities of SEP events show a wide variety of spatial and temporal variations. They are the result of many factors, including the efficiency and time dependence of the acceleration, the particle transport conditions, the local plasma properties at the observing spacecraft (such as the presence of ICMEs, shocks and magnetic discontinuities), and the spacecraft location with respect to the associated flare and CME at the Sun. In this section we first present 'typical' observations of SEP intensity-time profiles and anisotropies at 1 AU from the Sun and in the ecliptic plane (Section 2.1.1). We then focus on a description of how event characteristics are influenced by the relative position of the observing spacecraft and the solar events source of the SEP (Section 2.1.2).

2.1.1. *Intensity-Time Profiles and Anisotropies at one Location*

ISEE-3, launched in 1978, was one of the first missions to study in detail the particle and field characteristics of ICMEs, as a consequence of its sophisticated set of instruments.

Figure 1 shows particle, magnetic field and solar wind data from ISEE-3 for an ICME event on 11 December, 1980 studied by Sanderson *et al.* (1983). The event of Figure 1 has typical intensity profiles and anisotropies at 1AU from the Sun and in the ecliptic plane, although it did not extend to high energies, i.e. above ~ 20 MeV.

There was a strong shock, a magnetic cloud with field rotation, and shock-accelerated energetic particles. It is important to realize that there is a great deal of variation in the characteristics of ICME events at 1 AU, and that not all events exhibit these characteristics. Other examples observed by ISEE-3 were studied by Klecker *et al.* (1981), van Nes *et al.* (1984) and Sanderson *et al.* (1985a).

The event shown in Figure 1 begins with a rapid onset in the intensity-time profiles, with the highest energy particles arriving first. These particles were accelerated close to the Sun and arrived within several hours of the start of the associated flare on December 9. At the onset there was a large first order field-aligned anisotropy. Similar events were discussed by Heras *et al.* (1994).

For around a day following onset, the intensities continued to rise at energies <1.6 MeV. The shape of this part of the intensity-time profile depends upon many parameters, such as the injection at the flare site, acceleration at the CME shock (Heras *et al.*, 1994), the position of the observer relative to the site (Sanahuja and

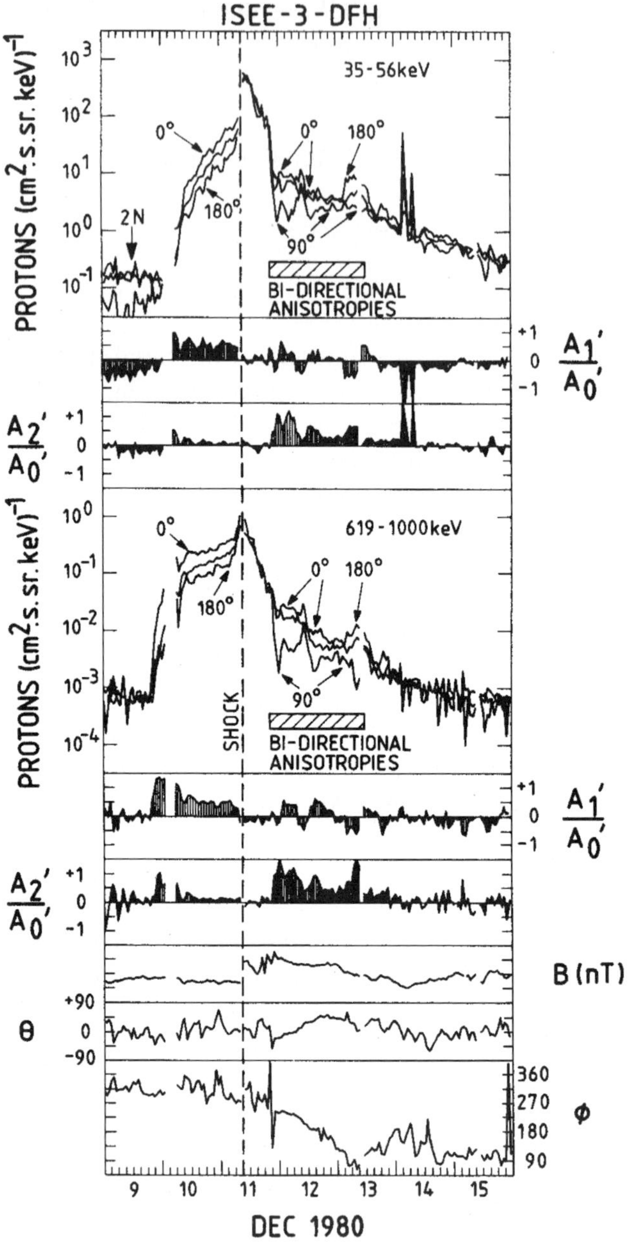

Figure 1. Intensity-time profiles together with anisotropy parameters, solar wind and magnetic field data for the 11 December, 1980 ICME event (from Sanderson *et al.*, 1983).

Domingo, 1987; Domingo *et al.*, 1989; Cane *et al.*, 1988) and the characteristics of the interplanetary medium (Klecker *et al.*, 1981; van Nes *et al.*, 1985; Beeck *et al.*, 1990).

The shock in front of this ICME arrived approximately two days after the solar event, together with a rapid increase in the lower energy particle intensity, the intensity rising around an order of magnitude in about an hour before the shock. In the model of Lee (1983), the later rise of the intensity is ascribed to the acceleration of particles by the interaction with upstream turbulence at quasi-parallel shocks.

At some shocks waves with frequencies <0.1 Hz were observed just upstream of the shock, i.e. in the frequency range where protons can be in cyclotron resonance with the waves (Kennel *et al.*, 1986; Sanderson *et al.*, 1985b), as predicted by Lee (1983). Needless to say, not all events follow this pattern. Some were preceded by complicated magnetic field structures such as planar magnetic structures (van Nes *et al.*, 1985), and others by magnetic loops (Balogh and Erdos, 1983), which may either assist or prevent the growth of these waves.

Traveling behind the 11 December, 1980 shock was a turbulent region lasting several hours, containing particles accelerated by the shock, with the maximum intensity occurring one or two hours after the shock and not at the shock. This is quite a common feature (e.g. Lario *et al.*, 2003a) that suggests that trapping of particles plays a role. This turbulent region is often responsible for a depression in particle intensity at very high energies, which is the first step of a two-step Forbush decrease (Sanderson *et al.*, 1992; Cane *et al.*, 1994).

In this event, the arrival of the ICME itself was heralded by a rapid drop in the low energy ion intensity coincident with the arrival of a discontinuity. A discontinuity, or multiple discontinuities (Sanderson *et al.*, 1990), are usually at the leading edge of ICMEs, which is also the start of the second step of the Forbush decrease.

Within the ICME, strong bidirectional ion anisotropy signatures were observed. In a similar event (Tranquille *et al.*, 1987) derived a large mean free path of $\sim$3 AU from the low magnetic field variance. Both low magnetic field variance and bidirectional ion anisotropies are typical signatures of this type of events (Marsden *et al.*, 1987).

2.1.2. *Dependence of Event Characteristics on Longitude and Latitude*

Several characteristics of an SEP event at 1 AU in the ecliptic plane are influenced by the relative separation in heliolongitude between the location of the associated flare and the magnetic footpoint of the observing spacecraft.

Looking at a large sample of near-Earth SEP data, it was recognised that peak intensities are reached earlier in SEP events following flares at western longitudes, than in those following eastern flares (Burlaga, 1967; van Hollebeke *et al.*, 1975). The onset time of SEP events was also shown to be ordered by heliolongitude (Barouch *et al.*, 1971).

The organisation of event profiles by longitudinal separation between flare location and spacecraft is most obvious by looking at the same event from several

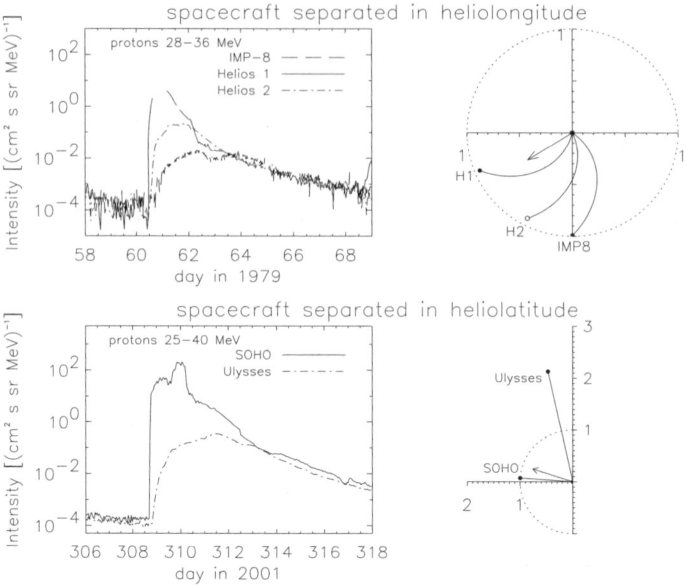

Figure 2. Upper panels: Left: SEP intensity-time profiles at spacecraft widely separated in heliolongitude: 1 March 1979 event as observed by Helios 1 and 2 and IMP–8. *Right:* A projection of the spacecraft locations onto the solar equatorial plane, with solid lines indicating Parker magnetic field lines connecting the spacecraft to the Sun. The arrow indicates the longitude of the flare associated with the event. Lengths are in units of AU. *Lower panels: Left:* SEP intensity-time profiles at spacecraft widely separated in heliolatitude: 4 November, 2001 event as seen by Ulysses and SOHO. *Right:* Projection of the positions of Ulysses and SOHO onto a meridional plane. The difference in longitude between the two spacecraft was 31°.

spacecraft far apart in longitude. The first multi-spacecraft studies were carried out during the Pioneer era (McCracken *et al.*, 1967; McKibben, 1972). Later, several studies combining Helios 1 and 2 and IMP-8 data (e.g. Cane *et al.*, 1988; Kallenrode, 1993) showed that, at energies >20 MeV, the spacecraft with apparent magnetic connection closest to the flare nearly always detects the highest peak intensity. In contrast, the peak intensities at the spacecraft farthest in longitude from the flare are typically orders of magnitude lower. These features are displayed in the left top panel of Figure 2, with the right top panel giving the spacecraft positions.

Data from the Ulysses passages over the solar poles during 2000–2002 allowed us for the first time to look at the dependence of SEP event characteristics on heliolatitude (see e.g. Simnett, 2001, for a review). Several high heliolatitude events were detected, at heliocentric radial distances between 1.6 and 3.2 AU. A comparison between particle arrival times at Ulysses and at near-Earth spacecraft showed very large delays in onsets at high heliolatitudes (~100–400 minutes) well ordered by the separation in latitude between flare and spacecraft (Dalla *et al.*, 2003a). Times of peak intensities were also found to be very delayed at high heliolatitude, and also well ordered by latitudinal separation (Dalla *et al.*, 2003b). On the other hand, during the 14 July, 2000 event (Zhang *et al.*, 2003), high energy protons arrived at Ulysses before relativistic electrons, and the arrival direction was not along the local magnetic field. These findings revived the discussion of cross-field diffusion in the interplanetary transport of solar energetic particles (e.g. Cane and Erickson, 2003). In contrast to measurements near the ecliptic plane, the high heliolatitude SEP data are characterised by very small anisotropies, which are field aligned during the onset phase (Sanderson *et al.*, 2003; Lario *et al.*, 2003b). The bottom panels of Figure 2 show an example of simultaneous observation of an SEP event at high and low latitudes.

A feature common to many SEP events is that intensities in the decay phase become very similar at widely separated spacecraft, as can be seen in the examples in Figure 2. This was first noticed in data from the Pioneer missions (McKibben, 1972). Roelof *et al.* (1992) reported periods of time when particle intensities at different radial distances from the Sun (1.0 and 2.5 AU) were essentially the same during the decay phase. They described the observation as the formation of particle reservoirs in the inner heliosphere. Reames *et al.* (1997) plotted SEP spectra at the two Helios spacecraft and IMP–8 during the decay phase of large events. The spectra were very similar at the three spacecraft, in the proton energy range from ~1–40 MeV, showing that similarities in the decay phase are present in a wide range of SEP proton energies. Because of these similarities, they introduced the term invariant spectra to describe the phenomenon.

Several explanations have been put forward to explain the multi-spacecraft decay phase observations. Magnetic trapping between the shock and the Sun, trapping by magnetic barriers produced by earlier solar events, as well as cross-field transport have been suggested (Roelof *et al.*, 1992; Reames *et al.*, 1997; Lario *et al.*, 2003b).

2.2. Element Abundances, Ionisation States, and Energy Spectra of SEPs

B. Klecker, H. V. Cane, M. A. Popecki

The variations of elemental and isotopic abundances by several orders of magnitude, or variations in the mean ionic charge, are the basis for dividing events into two classes (e.g. discussion in Reames, 1999, and Cane and Lario, 2006, this volume). (1) Gradual Events show large interplanetary ion intensities, small electron to proton ratios, on average coronal elemental abundances and ionic charge states consistent with source temperatures of $\sim 1.5 - 2 \times 10^6$ K, characteristic for the solar corona. These events are associated with interplanetary shocks. (2) Impulsive Events show high electron to proton intensity ratio, enhanced abundances of heavy elements, enhancements of ^{3}He relative to ^{4}He by up to a factor of 10^4, and high ionic charge states for Si (~14) and Fe (~20), which were interpreted as being due to temperatures of $\sim 10^7$ K. These events are usually associated with impulsive solar flares and the isotopic and elemental enhancements are interpreted as being due to resonant wave-particle interactions, for example selective heating of ^{3}He by electrostatic ion cyclotron waves (see Section 3.1.1). However, new results from the novel instrumentation with much improved sensitivity and resolution onboard several spacecraft (e.g. WIND, SAMPEX, SOHO and ACE) have shown that this picture was oversimplified. In this section we will concentrate on the new elemental, isotopic, and ionic charge measurements and their implications.

2.2.1. *Event Integrated Abundances*

Event integrated SEP abundances have been used extensively to infer solar or solar system abundances, that may not be accessible otherwise, or to relate SEP abundances to photospheric, coronal or solar wind abundances (e.g. Meyer, 1985). When comparing SEP abundances with coronal and photospheric abundances it was realized for many years that both, coronal and SEP abundances, show a dependence on the first ionisation potential (FIP) (Hovestadt, 1974; Meyer, 1985; Geiss, 1998), suggesting ion-neutral separation in the chromosphere as an important fractionation mechanism. For a review on FIP-fractionation mechanisms see e.g. Hénoux (1998). Furthermore, abundances in individual large SEP events generally show fractionation effects that monotonically depend on mass (M) or mass per charge (M/Q), usually approximated by a power law in M/Q (Breneman and Stone, 1985). This M/Q fractionation is not only observed for elemental abundances, but also for isotopic abundances (e.g. Leske *et al.*, 1999) and the correlation between isotopic and elemental abundances in individual events has been used to infer the abundances of the coronal source (Leske *et al.*, 2003).

Large enrichments of ^{3}He and heavy ions found in event-integrated abundances of impulsive events are their defining characteristic. However, Mason *et al.* (1999) reported small enrichments of ^{3}He in a survey of 12 large gradual events, with an average value of ^{3}He/^{4}He $= (1.9 \pm 0.2) \times 10^{-3}$. Even larger enrichments of ^{3}He with ^{3}He/^{4}He between 0.0014 and 0.24 have been found near the passage of 25 out of 56 IP shocks at 1 AU (Desai *et al.*, 2001). Furthermore, large abundances of Fe at energies of $\sim$1 MeV/amu (Mason *et al.*, 1999) and at energies $>$12 MeV/amu (Cohen *et al.*, 1999) and $>$25 MeV/amu (Cane *et al.*, 2003) have been observed in many large events. Thus, enrichments of ^{3}He and heavy ions are apparently not limited to small, impulsive, events as considered earlier. As a possible source of ^{3}He in large events Mason *et al.* (1999) proposed remnant suprathermal particles from previous impulsive events, serving as seed particles for the injection at the IP shock. This suggestion is qualitatively supported by the recent finding (Desai *et al.*, 2003) that the elemental abundances near 72 IP shocks observed during 1997–2002 were poorly correlated with (slow) solar wind abundances, but positively correlated with the elemental abundances upstream, suggesting the acceleration of suprathermals with a composition different from solar wind composition. In this scenario, high heavy ion abundances (e.g. Fe/O) could then be also interpreted as an admixture of suprathermal ions from previous impulsive events with heavy ion enrichment (Mason *et al.*, 1999). However, from a study relating suprathermal iron densities at $E > 0.01$ MeV/amu during quiet periods to the corresponding densities in Fe-rich events, Mewaldt *et al.* (2003a) concluded that there is not enough iron at quiet times to account for the overall enrichment of Fe. In fact, at high energies, where for most of the events the acceleration is close to the Sun, an alternative process may be more important: direct injection of particles from the CME-related flare (e.g. Klein and Trottet, 2001 and references therein; Cane *et al.*, 2003), with or without further acceleration by the coronal and/or IP shock.

2.2.2. *Compositional Variations During SEP Events*

The intensity–time profiles and elemental abundances during large SEP events show a complex variation with time. The intensity-time profiles strongly depend on the location of the observer relative to the passing IP shock and ICME (e.g. Figure 2 of Cane and Lario, 2006, this volume). A common feature is the large variation of elemental abundances of particles with the same velocity, but different rigidity (e.g., Fe/O) during the event onset, i.e., before the time of maximum (e.g., Klecker, 1982; Mason *et al.*, 1983; Tylka *et al.*, 1999; Reames *et al.*, 2000). The decrease of, for example, Fe/O during the rise of the event can be explained by the particles' mean free path λ increasing with rigidity in the background Kolmogoroff–type spectrum of the interplanetary magnetic field fluctuations. If λ is increasing with rigidity, then Fe/O at equal velocity will decrease during the rise of the event as a result of the larger M/Q of Fe. This has been reproduced by propagation models, assuming acceleration close to the Sun (e.g. Scholer *et al.*, 1978; Mason *et al.*, 1983). The very complex abundance variations later in the events, before and after the passage of the IP shock, can be understood in terms of models including proton-amplified waves, and scattering of all ions by these waves (Ng *et al.*, 1999; Tylka *et al.*, 1999; Ng *et al.*, 2003). Shock acceleration is heuristically (and not self-consistently) implemented in these models by continuous injection of particles with abundances derived from solar wind abundances and spectral slopes used as a model parameter. Nevertheless, a detailed comparison of the observed time variations with model calculations can be used as a tool to infer, as a function of distance of the IP shock from the Sun, important parameters such as injection efficiency, injection abundances, shock strength, and wave power near the shock.

2.2.3. *Energy Spectra*

The energy spectra as observed in interplanetary space near Earth are a result of the acceleration and propagation processes between the acceleration site and the observer. Typical energy spectra observed in large events can often be fitted by power laws with exponential roll-over at high energies. These spectral forms can be explained by shock acceleration: in the ideal case of an infinite and planar shock geometry and steady-state conditions, the energy spectra would be power laws and could be described as $dJ/dE \propto E^{-\gamma}$, where γ is related to the shock compression ratio (Axford *et al.*, 1978; Blandford and Ostriker, 1978). However, the IP shock driven by an ICME will not be planar, particularly close to the Sun. Whether steady-state conditions at a specific energy are reached will depend on the shock strength, the number of injected protons and the intensity of proton-amplified waves. Thus, steady state may be reached at low energies but not at high energies. Non steady-state conditions (Forman and Webb, 1985) and losses at the shock due to particles escaping upstream (e.g. Ellison and Ramaty, 1985) will result in a roll-over of the power-law spectra at high energies. This spectral form is frequently observed and can be fitted by $dJ/dE \propto E^{-\gamma}e^{-E/E_0}$, where E_0 depends on M/Q of the ions (Tylka *et al.*, 2000). The roll-over energy E_0 shows a large event-to-event variability ranging for protons from ~10 MeV (Tylka *et al.*, 2000) to ~800 MeV (Lovell *et al.*, 1998).

2.2.4. *Ionic Charge Composition in Interplanetary Shock-Related Events*

The ionic charge of solar energetic particles is an important parameter for the diagnostic of the plasma conditions at the source region in the solar corona. Furthermore, the acceleration and transport processes depend significantly on velocity and rigidity, i.e. on the mass and ionic charge of the ions. Several methods have been developed over the last ~30 years to determine the ionic charge of energetic ions (e.g. Popecki *et al.*, 2000a). At low energies of ~0.01–3 MeV/amu techniques involving electrostatic deflection provided the first direct ionic charge measurements (Gloeckler *et al.*, 1976; Hovestadt *et al.*, 1981) of SEPs. At higher energies, the SAMPEX instrumentation (Baker *et al.*, 1993) provided for the first time ionic charge measurements for many elements in the range C to Fe over the extended energy range of 0.3–70 MeV/amu, utilizing the Earth's magnetic field as a magnetic spectrometer.

Indirect methods use the rigidity dependence of diffusive interplanetary propagation to infer average ionic charge states of heavy ions from the time-to-maximum (O'Gallagher *et al.*, 1976; Dietrich and Tylka, 2003) or from the time profile in the decay phase (Sollitt *et al.*, 2003) of SEP events. Furthermore, the M/Q dependence of the roll-over energy E_0 (see above, Tylka *et al.*, 2000), and elemental fractionation depending on M/Q (Cohen *et al.*, 1999) have been successfully used to infer mean ionic charge states of heavy ions. Although these indirect methods are limited to the determination of average charge states and rely on the assumption of the mean ionic charge being independent of energy, they provide a valuable tool if direct measurements are not available.

The early measurements from about 25 years ago revealed for IP shock-related events an incomplete ionisation of heavy ions in the range C–Fe at energies of ~1 MeV/amu, with $Q(\text{Fe}) \sim 10 - 14$, indicative of source temperatures of about $1.5 - 2 \times 10^6$ K (Gloeckler *et al.*, 1976; Hovestadt *et al.*, 1981; Luhn *et al.*, 1984). It was also found that the mean ionic charge in ^{3}He- and heavy-ion-rich events was significantly higher (Klecker *et al.*, 1984; Luhn *et al.*, 1987), with $Q(\text{Fe}) \sim 20$ and $Q(\text{Si}) \sim 14$. This was interpreted as being indicative of a high temperature

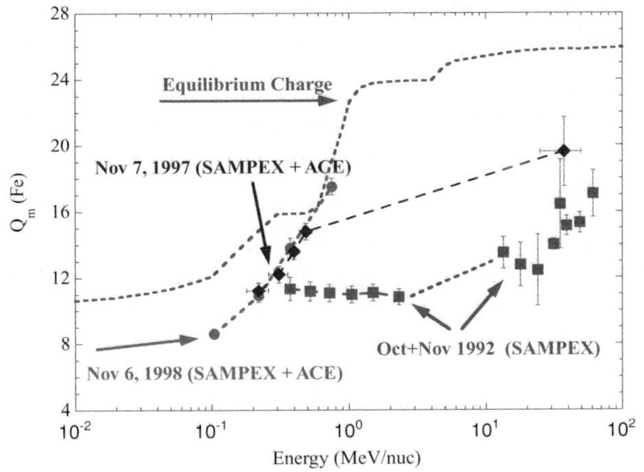

Figure 3. Mean ionic charge of Fe as a function of energy for 3 large SEP events (data from summary of Popecki *et al.* (2003) and energy dependence obtained for equilibrium conditions in a charge stripping model (Kocharov *et al.*, 2000).

of $\sim 10^7$ K in the source region. However, with SAMPEX, a significant increase of the mean ionic charge of heavy ions with energy was found in several large gradual events, with Q(Fe) increasing from $Q \sim 10$ at ~0.3 MeV/amu to ~18–20 at ~40 MeV/amu (Mason *et al.*, 1995; Leske *et al.*, 1995; Oetliker *et al.*, 1997; Leske *et al.*, 2001; Labrador *et al.*, 2003). Recent measurements with SOHO and ACE showed an increase of Q(Fe) even below 1 MeV/amu: The mean ionic charge of Fe in IP shock-related events at suprathermal energies (~0.01–0.1 MeV/amu) is consistent with solar wind charge states (Bogdanov *et al.*, 2000; Klecker *et al.*, 2000). At somewhat higher energies (0.2–0.6 MeV/amu) in many events the mean ionic charge increases with energy (Möbius *et al.*, 1999; Popecki *et al.*, 2003), with a large event-to-event variability (e.g. Möbius *et al.*, 2002 and Figure 3). It should be noted that in Fe-rich impulsive events a large increase of Q(Fe) with energy by ~4–6 charge units is systematically observed at $E < 0.6$ MeV/amu (Möbius *et al.*, 2003).

A small increase of the mean ionic charge with energy by 1–2 charge units in the energy range 0.2–1 MeV/amu, or a somewhat larger increase as observed above ~ 10 MeV/amu could be due to the acceleration process (Klecker *et al.*, 2000, 2003). However, a large increase of Q at energies <1.0 MeV/amu, as sometimes observed in gradual events, but systematically observed in impulsive events, requires a different mechanism and can be explained by ionisation in a dense environment. If the particles propagate in a sufficiently dense environment in the lower corona during or after the acceleration, a large increase of the mean ionic charge with energy is a natural consequence. This has been investigated quantitatively in recent years (Ostryakov and Stovpyuk, 1999; Barghouty and Mewaldt, 1999; Kocharov *et al.*,

2000; Stovpyuk and Ostryakov, 2001). The model calculations show that the mean ionic charge increases monotonically as a function of Nt, where N is the plasma density and t is the residence time. It approaches asymptotically an upper limit (the equilibrium mean charge) Q_{eq}, where $(Nt)_{eq}$ for 0.5 MeV/amu Fe ions is typically $\sim 10^{10}$ cm^{-3}s (Kocharov *et al.*, 2000). Figure 3 shows as an example the mean ionic charge of Fe as a function of energy, computed for the limiting case of equilibrium mean charge states caused by stripping in the corona. Ionization by electrons and protons, as well as radiative and dielectronic recombination, are taken into account in the calculation (Kocharov *et al.*, 2000). In these models, the large variability of the increase of Q with energy can be reproduced by variable acceleration rates and non-equilibrium conditions for charge stripping (e.g. Kovaltsov *et al.*, 2001). In the case of near-equilibrium conditions (as observed in the 6 November, 1997 event) and acceleration time scales of $\sim$1–100 seconds, this corresponds to coronal densities of 10^8–10^{10} cm^{-3}, i.e. altitudes of $<2R_\odot$ in the corona.

In summary, the interpretation that the presence of high ionic charge states of heavy ions in small impulsive events is an indication of high temperatures at the acceleration site must be seriously questioned. However, considering that the source plasma in these events may be far from thermodynamic equilibrium, it is not surprising that temperature may not be the organising parameter. The only explanation for the strong energy dependence of the ionic charge of Fe (as consistently observed in impulsive events) is acceleration in a dense environment in the low corona, i.e. high charge states (e.g. Fe^{20+}) can be used as a tracer for a source low in the corona. Whether these high charge states, when observed in large gradual events, are accelerated in the contemporary flare, or whether they are accelerated at the coronal or IP shock from a suprathermal distribution from previous flares, is presently under investigation.

2.3. Electrons in the Interplanetary Medium

G. M. Simnett

The electron intensity in the inner heliosphere as a result of solar activity is both very variable and hard to predict from observations of other transient activity. This is probably due to the variety of sources of electrons which range from those probably accelerated in the high corona (Potter *et al.*, 1980; Lin, 1985); those released in conjunction with CMEs (Haggerty and Roelof, 2002; Simnett *et al.*, 2002); to those accelerated in solar flares (Lin *et al.*, 1982; Evenson *et al.*, 1984; Moses *et al.*, 1989). Electrons are tied closely to magnetic field lines, so that it may be quite difficult at times for electrons to escape quickly from a flare site into space. Any observations at 1 AU, therefore, may be both difficult to interpret and arise from several of the above sources. Krucker *et al.* (1999) showed that some low energy events were released from the Sun at the time of a type III radio burst, which presumably indicates direct injection onto open magnetic field lines; other events may take up to half an hour to reach a spacecraft at 1 AU, presumably due to unfavourable propagation. Events with an inferred delayed onset relative to the type III radio bursts tend to be proton rich, which may well indicate that the protons escape more readily from closed magnetic field lines. For a discussion of electron onset times see also Section 6.4 of Pick *et al.* (2006, this volume).

Notwithstanding the electron origin, the transport from the site of acceleration (and release) is governed by the interplanetary magnetic field. It may happen that the inner heliosphere fills up with solar electrons which dissipate, either through escape or energy loss, only slowly. Therefore, at low energies below a few tens of keV, the observed intensity may be from a combination of *all* the sources outlined above. Therefore spectra such as presented by Lin (1985) (Figure 12 of that paper) for the 23 September, 1978 event are probably a superposition of these different sources. It is customary to characterise the differential electron energy spectrum as a power law, $dJ/dE \propto E^{-\gamma}$, although some events may have a variable γ. Potter *et al.* (1980) showed that some small, impulsive events have energy spectra that continue as unbroken power laws in kinetic energy down to 2 keV, with a power law index γ ranging from ~3.5 to 4.8; these tend to propagate from the Sun without scattering and therefore appear as a collimated beam. These events were also not flare-associated, which Potter *et al.* interpreted as indicating a high coronal origin. The observation of many impulsive solar electron bursts at energies below 1.4 keV with power law spectra extending down to ~0.14 keV (Gosling *et al.*, 2003) would also be compatible with a high coronal origin. However, because of their frequent association with type III radio bursts, Gosling *et al.* (2003) suggested as an alternative a low coronal origin, although they could not explain in this scenario that the power law spectra do not show any sign of energy loss effects at low energies.

Lin *et al.* (1982) carefully selected western hemisphere flare events to minimise the propagation effects, and showed that the events typically had differential energy spectra with $\gamma = 2.4-4.3$ above a few 100 keV, flattening to 0.6–2.0 at low energies. In addition, some events show a second spectral steepening above a few MeV.

At highly relativistic energies above a few MeV the situation becomes clearer. Evenson *et al.* (1984) investigated the association of intense events with solar gamma–ray production and showed that electron-rich events are likely to be associated with gamma-ray events, and have hard energy spectra. Such events are not associated with interplanetary shocks, which therefore are not a source of electron acceleration above a few MeV (Simnett, 2003). The electron/proton ratio (at the same energy) may vary by 5 orders of magnitude (Evenson *et al.*, 1984).

At large events, electrons with energies up to around 100 MeV are produced. Some typical spectra are shown in Figure 4. We have plotted the spectra as functions of kinetic energy. There are significant spectral differences between the various events, which may possibly be related to the different interplanetary and acceleration/release properties.

One of the first comprehensive studies of electron spectra was done by Moses *et al.* (1989) who showed that the spectra, when expressed in terms of rigidity, could

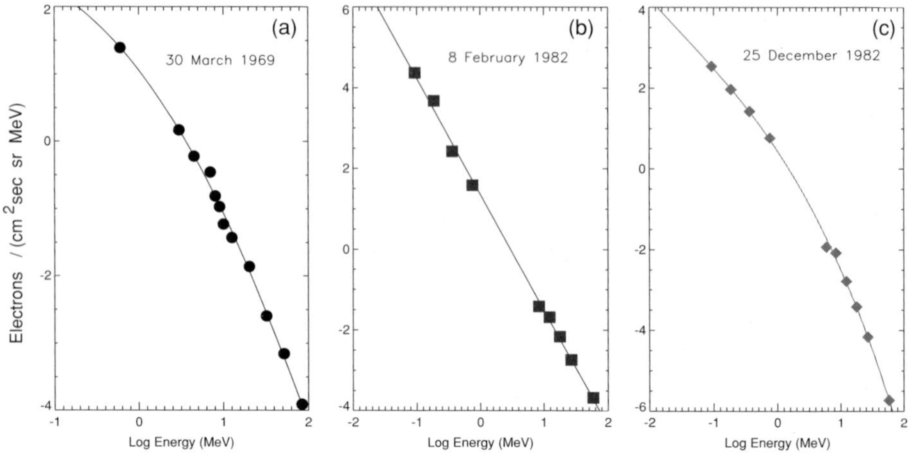

Figure 4. Interplanetary electron spectra for events extending to ~100 MeV. (a) A very long-lived event following a major flare on 30 March, 1969, where the inner heliosphere acted as a particle reservoir. (After Simnett, 1974). (b) An impulsive flare event from a W88° flare on 8 February, 1982. (c) A long duration event from an E45° flare associated with an interplanetary shock on 25 December, 1982. (After Moses *et al.*, 1989).

be ordered into two groups, those with single power laws in rigidity and those where the spectrum hardened with increasing rigidity. They found that those events with near power law spectra were well correlated with long duration soft X-ray events, whereas those with hardening spectra were correlated with short-duration events. We know that the long duration events are normally accompanied by coronal mass ejections, which in turn are accompanied by the injection into the interplanetary medium of near-relativistic electrons. Whether the latter are actually accelerated by the CME or merely gradually released is a matter for debate. Nevertheless, when analysing the spectrum generated by using the maximum flux at a given energy, this would tend to increase the lower energy part of the spectrum for the long duration events. We should point out that event (b) plotted as a function of rigidity would fall into the class of "hardening spectra" as defined by Moses *et al.* (1989).

Observations above 100 MeV are almost non-existent, so it is unclear if the solar electron spectrum continues to beyond 100 MeV. Longer duration events are illustrated by the spectrum from the 25 December, 1982 event, which exhibits the spectral steepening above a few MeV observed by Lin *et al.* (1982). This event is typical of those flare events which have a CME-driven shock. For the event shown in Figure 4 the inner heliosphere filled with electrons following a major flare on 30 March, 1969. The e-folding decay of the intensity at relativistic energies was 125 hours for over a week (Simnett, 1974), indicating that if conditions are favourable, even highly relativistic electrons have small losses. This flare event, which was from just behind the solar west limb, had a power law index γ of 3.04 above 5 MeV and would have certainly been accompanied by a major CME. Thus it would come into the category of events such as shown in Figure 4 c, except with impeded escape from the inner heliosphere.

The energy emitted in electrons is generally low in comparison to that in the protons at the same energy. At relativistic energies (above a few MeV) a typical electron/proton ratio is 10^{-3} (Evenson *et al.*, 1984), although as noted above there is a huge variation. Events associated with gamma–ray flares have a ratio closer to 10^{-2}. As the differential energy spectra have power law indices steeper than -2, the bulk of the energy resides at low energies. Here the propagation issues, plus the fact that low energy (<1 MeV) protons (a) have long travel times from the Sun and (b) are accelerated by strong interplanetary shocks means that an electron/proton ratio at low energies is hard to interpret in terms of production mechanisms. This merely emphasises that there are a variety of sources for both protons and electrons, having different efficiencies which are functions of time, energy and space.

3. Acceleration and Transport

3.1. Acceleration

H. Kucharek, M. A. Lee, B. Klecker, H. V. Cane, G. Simnett

There is a long history of studies of SEP events and their association with acceleration processes at the Sun and in interplanetary space. However, the relative importance of the various acceleration mechanisms is still controversial (see also discussion in Section 2). The questions are: How important are CMEs in producing the energetic particle population in different energy ranges measured in the heliosphere? Where does the acceleration occur and what are the sources of the energetic particles? How important are shocks associated with CMEs for accelerating ions? These are key science issues which are currently under intense investigation. In the following sections we will review the injection and acceleration models and discuss how well they are supported by observations.

3.1.1. *Acceleration Sites and Mechanisms*

For impulsive events several acceleration mechanisms have been proposed, including stochastic acceleration in the turbulence generated by the flare (Ryan and Lee, 1991; Holman, 1995), direct electric field acceleration at the site of magnetic reconnection (Holman, 1995; Litvinenko, 1996); resonant acceleration by electrostatic ion cyclotron (EIC) waves (Fisk, 1978) and by electromagnetic ion cyclotron (EMIC) and shear Alfvén waves (Miller and Vinas, 1993), and shock acceleration by the shock formed as the reconnection plasma jet impinges on the denser plasma below the flare arcades (Tsuneta and Naito, 1998; Aurass and Mann, 2004). A challenging test for any theory of impulsive events is to explain the observed large enrichments of ^{3}He (by several orders of magnitude), of heavy ions like Fe (by about a factor of ∼10), and of ultra-heavies in the mass range ∼80–200 (by a factor of ∼40–200, (Mason *et al.*, 2004)). Promising candidates are theories including resonant wave-particle interaction: for example, selective heating and acceleration of ^{3}He by EMIC waves and of heavy ions by cascading turbulence of shear Alfven waves (Miller and Vinas, 1993). For a recent review of particle acceleration in solar flares see Aschwanden (2002).

Particle acceleration also occurs at CME-driven shocks. In the idealised model of a planar, infinite shock under steady-state conditions, the particles' momentum spectra downstream of the shock are power laws, with a spectral exponent depending on the shock compression ratio (Axford *et al.*, 1978; Blandford and Ostriker, 1978). Particles are scattered on the proton-amplified waves, pick up energy by multiple encounters of the shock, and escape upstream at a distance depending on the scattering mean free path, λ. For the discussion of a quasi–parallel shock geometry see Lee (1983) and Gordon *et al.* (1999). However, due to the limited spatial extent of coronal and interplanetary shocks and the limted time available for acceleration, the spectra will roll over at some energy E_0 (see also discussion in Section 2.2.3). This leads us to the important questions of injection mechanisms and source populations injected into the acceleration process.

3.1.2. *Injection Mechanisms and Seed Particles*

Although the shock acceleration mechanism is well studied and widely accepted, the origin and injection of the seed particles into the acceleration process is much less clear. There are several seed particle populations: the solar wind, interstellar pickup ions, prominence material, ions that are pre-accelerated in interplanetary space (suprathermal tails, see below), and suprathermal particles from previous SEP events filling the inner heliosphere (in particular during solar maximum (Mason *et al.*, 1999)). A detailed description of the injection process is certainly a challenging task because wave dissipation, large amplitude waves, and the shock structure play an important role. In addition the shape of the particles' distribution function is a significant factor. While solar wind ions can be described by Maxwellian or kappa distributions in the solar wind frame of reference, interstellar pickup ions form a shell–like distribution. Thus, compared to solar wind ions, pickup ions are already suprathermal and therefore easier to inject into any acceleration process.

Numerical simulations such as hybrid simulations (Quest, 1988; Giacalone *et al.*, 1992; Liewer *et al.*, 1993; Kucharek and Scholer, 1995) or Monte Carlo codes (Ellison *et al.*, 1990) have been very useful over the last two decades in investigating shock acceleration for a range of Mach numbers and magnetic obliquities and in estimating injection rates. These codes allow us to study highly nonlinear processes which play an important role in the injection and acceleration of particles at shocks. Most of the analytical models are not able to include these nonlinear processes.

For injection at quasi-parallel shocks, Malkov (1998) introduced an analytical model in which particles are trapped in the large-amplitude magnetosonic wave of the shock and then may leak upstream. Hybrid simulations suggest a different mechanism. Solar wind ions perform a non-adiabatic motion when they reach the shock so that their velocity parallel to the magnetic field is small. Due to electric and magnetic forces in the shock ramp, particles may be trapped at the shock and gain energy in the electric field while in gyro-resonance with the upstream waves convected into the shock (e.g. Scholer *et al.*, 2000). Multiple encounters with the shock after scattering upstream and downstream then lead to large energy gains.

At quasi-perpendicular shocks a larger minimum energy is required for multiple shock encounters. At the first encounter with a quasi-perpendicular shock some particles are specularly reflected and gain energy while traveling along the motional electric field in the shock foot and ramp. During this process they gain up to twice the incoming velocity. However, the energy gain may not be enough for a particle to interact with the shock several times before it is convected downstream. Rapid scattering in the shock ramp may be one process to make this happen. Multi-dimensional hybrid simulations (Giacalone *et al.*, 1992; Scholer and Kucharek, 1999) suggest that specular reflection in combination with cross-field diffusion may be sufficient to inject and accelerate ions at quasi-perpendicular shocks. However, if the seed population is comprised of suprathermal particles they already start at higher energies and are more easily injected.

3.1.3. *Pickup Ions: A Tool for Studying Particle Injection and Acceleration*

Pickup ion distributions and suprathermal tails of the solar wind are frequently observed in Corotating Interaction Regions (CIRs) at distances of ~5.5 AU (Gloeckler *et al.*, 1994) and can be explained by statistical acceleration by fluctuations of the magnetic field magnitude within the CIR (Schwadron *et al.*, 1996). Significant increases of He^+ pickup ions at the injection speed of $2V_{SW}$ have also been found at ~1 AU in solar wind compression regions (Saul *et al.*, 2003). In fact, the abundance of energetic He^+ ions in the inner heliosphere can be rather large ($He^+/He^{2+} \sim 1$, Hovestadt *et al.*, 1984). In IP shock related events He^+ can constitute, after H^+ and He^{2+}, the third most abundant ion species in the energetic particle population in the inner heliosphere (e.g. Kucharek *et al.*, 2003, and references therein). The measurements of the He^+ pickup ion source and He^{2+} solar wind source populations and the distribution of accelerated particles can be used to study the influence of the distribution function on particle injection and acceleration.

3.1.4. *Comparison of Observations with Predictions from Theory*

The observations of energetic particles at IP shocks are in agreement with many of the predictions of the theory of shock acceleration (Forbes *et al.*, 2006; Mikić and Lee, 2006, this volume). For example, the exponential intensity increase upstream of quasi-parallel shocks, the power law spectra as observed at low energies and the increased power seen in the magnetic field fluctuations are predicted by the theory. However, quantitative comparisons show a highly variable degree of precision of the predictions (e.g. Kennel *et al.*, 1986; Bamert *et al.*, 2004). Furthermore, different processes of acceleration, injection, or both, might prevail in different energy ranges.

In this sub-section we will therefore limit the discussion to energies of less than a few tens of MeV and review more qualitative comparisons of shock acceleration theory with some of the observations.

Intensity-time profiles: The time profiles of SEP events depend on the magnetic connection of the acceleration site with the observer. Therefore, the large variability of the intensity-time profiles and the longitude distributions of SEP events can qualitatively be explained by the extended longitudinal range of CME-driven IP shocks and by the relative location of the observer to the presumed source location of the CME (Cane *et al.*, 1988, see also Figure 4 of Cane and Lario, 2006, this volume). Early in the event, much before the shock arrival, many large SEP events show a maximum-intensity plateau not exceeding several 100 protons per (cm^2 s sr MeV/amu) at $\sim$1 MeV. This plateau level can be explained by the scattering of escaping particles by the proton-amplified waves, limiting the intensity of escaping particles (the 'streaming limit') to a specific value (Reames and Ng, 1998, and references therein).

Composition: Shock acceleration theory usually assumes injection of ions at some threshold energy, E_{min}, with abundances given by the source population. Thus, in the idealised case of an infinite, planar shock and steady-state conditions, the composition of accelerated particles near the shock would reflect the composition of the seed particles, possibly modified by any mass or mass/charge dependent injection effects. However, the particle distributions usually measured at $\sim$1 AU are also influenced by the non-local effects of acceleration over the extended distance between the Sun and Earth (in a plasma environment that changes with space and time), and the effect of rigidity-dependent propagation. Therefore, the mass per charge dependent abundance variations found in large events are not surprising (see also Section 2.2.2). However, a prediction of observed abundance variations from the observed variations of shock parameters has so far not been successful.

Compositional variations during SEP events: The complex compositional variations of ions with different rigidity at the same velocity (e.g. Fe/C, He/p) during the onset phase of SEP events and before the shock passage can basically be understood in terms of a rigidity dependent scattering mean free path and by scattering of heavy ions by the proton-amplified waves near the shock (see discussion in Section 2.2.2).

Energy spectra: In general, IP shock related events do show power-law energy spectra, at least at low energies, and a roll-over at high energies (see also discussion in Section 2.2.3). However, in a statistical study using 72 events, Desai *et al.* (2004) compared the spectral slope predicted from the local shock compression ratio with the power law spectral index for C, O, and Fe and did not find a significant correlation.

3.1.5. *Theoretical Challenges*

The theoretical challenges for the ions are (1) a quantitative description of the acceleration in impulsive flares, including the energy dependent ionic charge states and ^3He- and heavy ion enrichment; (2) to unfold injection, acceleration and propagation processes for a better understanding of the fractionation effects observed in elemental and isotopic abundances; (3) to determine where and how energetic particles in large events are accelerated to high energies (>20 MeV/amu). This would shed light on the current debate on the origin of flare particles in large events: do these particles originate at the CME-related flare or from unrelated flares (suprathermal remnants), further accelerated by the CME-driven shock? Results from theory and numerical simulations, as well as from observational investigations, need to be combined to obtain a clearer view of the interplay of all these processes.

3.2. Energetic Particle Propagation

D. Lario, E. C. Roelof

Energetic particle time histories detected by instruments aboard spacecraft and on the ground are determined by the mechanisms that control the transport of energetic particles throughout the heliosphere and, in particular, by the effects that magnetic field structures have on this transport. Two transport mechanisms that are always operative are field-aligned transport and particle energy loss. A third mechanism is the transport across the moving magnetic field lines; i.e., in addition to the deterministic convection of the guiding center of the particles along with the field lines that are "frozen" into the solar wind.

Field-aligned transport results from weak pitch-angle scattering combined with the focusing effect of the outwardly decreasing interplanetary magnetic field (e.g. Roelof, 1969; Dröge, 2000). The amount of interplanetary scattering depends on both the level of turbulence existing in the interplanetary magnetic field (IMF) and the possible amplification of plasma waves produced by the energetic particles streaming along the IMF (Ng *et al.*, 2003). The energy loss is due to the gradient of the magnetic field in the diverging solar wind and the curvature drifts of the particles moving against the $\mathbf{V} \times \mathbf{B}$ electric field in the solar wind (Webb and Gleeson, 1979; Roelof, 2000). In the limit of very weak scattering, the energy loss process can be expressed in terms of a fractional momentum loss rate that depends only upon particle velocity and not on mass or charge (Roelof, 2000). The role of perpendicular diffusion across moving field lines is still under debate. One underlying physical process is the "random walk" or "braiding" of field lines arising from random diffusion of their photospheric foot points (Jokipii and Parker, 1968). An important mechanism in forming the global structure of the heliosphere stems from differential rotation of the solar magnetic field, the latitudinal flows of the photospheric magnetic field, combined with the tilt between the Sun's rotation axis and the magnetic dipole axis (Fisk, 1996). In strongly turbulent fields, it has been argued that guiding centers of the particles are displaced across the field in

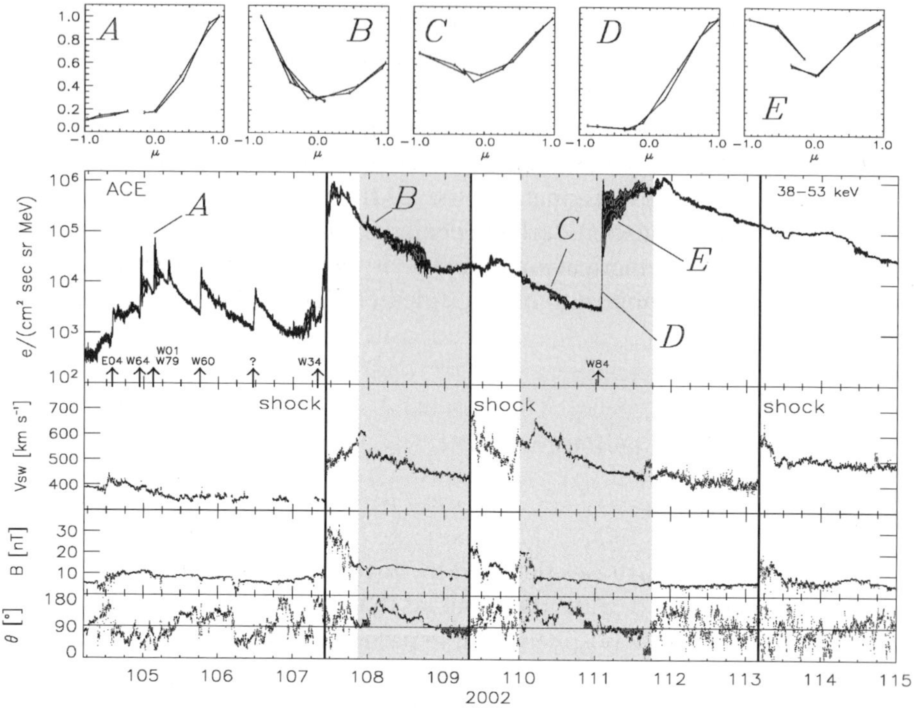

Figure 5. From top to bottom: 5 minute averages of normalized near-relativistic electron intensity distributions as a function of the pitch-angle cosine (μ) at five different time intervals (A–E). Time intensity profiles of the near-relativistic electrons as measured by ACE. Solar wind speed, magnetic field magnitude and elevation angle from 14 April, 2002 (day of year 104) to 25 April, 2002 (day of year 115). Solid vertical lines indicate the passage of shocks and gray vertical bars the passage of ICMEs. Vertical arrows show the occurrence of solar flares associated with the origin of the electron events.

single scattering processes by as much as their gyroradius (Jokipii, 1971). Persistent field-aligned anisotropies observed during solar energetic particle (SEP) events, however, suggest that cross-field diffusion is indeed small (Reinhard and Wibberenz, 1974; Sanderson *et al.*, 2003).

Increases (ΔB) in the IMF associated with ICMEs play a major role in controlling the intensity histories of SEP events (Barouch and Raguideau, 1970). Figure 5 shows 38–53 keV electron intensities observed by the ACE spacecraft during an intense period of solar activity in April 2002. At least eight electron events were observed throughout this 11-day period. Solar wind and magnetic field data show the passage of three interplanetary shocks (solid vertical lines) and two ICMEs (gray bars) as identified by Cane and Richardson (2003). On 15 April, 2002 (day of year 105) several small electron events were observed with rapid intensity increases and gradual slow decays characteristic of impulsive events. These impulsive SEP events usually exhibit pitch-angle distributions (PADs) that are strongly collimated outward along the magnetic field during the rise to maximum (panel A, top of Figure 5).

On day 107 an M2/2N flare occurred at 07:46 UT from the NOAA Active Region 9906 at S14W34 just before the arrival of an interplanetary shock at ACE at 10:21 UT on the same day. The arrival of the electrons injected at the time of the solar flare was modulated by the presence of the traveling CME-driven shock propagating between the Sun and the observer, reducing the pitch-angle anisotropy during the rising phase of the event. During the decay of the event bidirectional PADs were observed (panel B in top of Figure 5). These bidirectional flows have been alternatively interpreted as a result of either particle propagation in closed magnetic field configurations (e.g., Marsden *et al.*, 1987) or a reflection of particles by the magnetic field increase ΔB associated with the CME-driven shock and located beyond the spacecraft (e.g., Anderson *et al.*, 1992; Malandraki *et al.*, 2002, and references therein).

Bidirectional PADs were observed during the passage of the second ICME on day 110 (panel C in top of Figure 5). These lasted until the occurrence of an X1/1F flare from NOAA Active Region 9906 at S14W84 on day 111, producing the onset of a new SEP event with unidirectional outward PADs (panel D in top of Figure 5). These PADs gradually evolved into a distribution with a "seagull" profile (panel E). In less than 4 hours the intensities with pitch-angle cosine $\mu < 0$ gradually increased to reach values similar to those observed with $\mu > 0$. The flat intensity profile observed during the peak of this event, coupled with the fact that PADs do not show the U-shape typical of bidirectional flows but rather the characteristic "seagull" shape, suggests that nearly field-aligned particles ($\mu = \pm 1$) were indeed able to escape from this structure. These observations also suggest that particles were not completely confined in a closed magnetic field structure but they were also reflected by the magnetic field increase ΔB that crossed ACE on day 109 and that at the time of the X1 flare (W84) was located at $\sim$1.55 AU.

To summarize, the time interval shown in Figure 5 shows several examples of the effects that magnetic field (ΔB) increases produce on the properties of the SEP events. While an increase (ΔB) is en route to the observer, it can attenuate the intensity of the higher energy SEPs that are injected after the CME is launched, while possibly shock-accelerating the lower energy SEPs. When an increase (ΔB) is beyond the observer at the time of an SEP injection, it can enhance the SEP intensity by forming a reflecting barrier that impedes the escape of the SEPs into the outer heliosphere. When no significant increase (ΔB) is between the Sun and the spacecraft, we observe beam-like impulsive SEP events whose back-scatter depends, among other factors, upon whether or not there is an increase (ΔB) beyond 1 AU. Therefore, when using *in situ* SEP observations to interpret processes of particle acceleration and particle release into the interplanetary medium, it is important to analyze first the evolving global configuration of the IMF during the development of the SEP events and its effects on the particle transport.

4. Particle Release Time and the Electromagnetic Signature of SEP Events

A. Posner and K.-L. Klein

4.1. Determination of Solar Release Time

Because of interplanetary transport it is difficult to distinguish signatures of different particle sources and different accelerators in SEP time profiles at 1 AU. However, the timing of the event onset can be used to infer the solar release time of the *first* particles of an SEP event that reach a spacecraft, and to compare this time with electromagnetic signatures of energetic particles in the solar atmosphere. Particles of speed v that are released at the Sun at time t_{SRT} will arrive at a spacecraft at t_{arr} after travelling a path length S:

$$t_{arr} = t_{SRT} + \frac{S}{v}. \tag{1}$$

For parallel propagation along the nominal interplanetary magnetic field, S is the length of the archimedean spiral (at 1 AU: 1.04–1.3 AU). However, large excursions from this average geometry exist due to nonradial expansion of the solar wind, e.g. near the Sun, to transient disturbances such as CMEs, and to MHD wave activity that scatters the pitch angles of the particles. A linear relationship is indeed commonly observed between the first arrival of particles of a given speed range at a spacecraft and the inverse speed, with path lengths between 1 and 2 AU (e.g. Reames *et al.*, 1985; Krucker *et al.*, 1999; Krucker and Lin, 2000; Mewaldt *et al.*, 2003b; Tylka *et al.*, 2003). Examples are shown in Figure 6. The left figure shows an event where electrons and protons seem to be released simultaneously, but the protons have a longer path length (2 AU) than the electrons (1.3 AU). In the right figure both species have the same path length, but the protons seem to be released nearly 1 hour after the electrons. For a discussion of electron onset times see also Section 6.4 of Pick *et al.* (2006, this volume) in this volume.

Energy dispersion of the onset times of SEP events can hence be used to infer the initial solar release time of the particles, provided (i) all particles travel the same path, i.e. are released from a volume small compared with S, and (ii) all particles start essentially at the same time. A systematic error is induced in the release time estimation if the acceleration time is finite. Delayed release of high-energy (low-energy) particles would flatten (steepen) the onset time vs. $1/\beta$ function. Assuming a straight-line fit would imply an average release time for all particles and a path length which is different from the one actually travelled by the particles. Krucker and Lin (2000) explain the steeper slope of the proton graph in the 13 November, 1997 event (Figure 6) as a signature of the delayed release, at greater coronal heights, of low-energy protons in the course of acceleration at the bow shock of a CME.

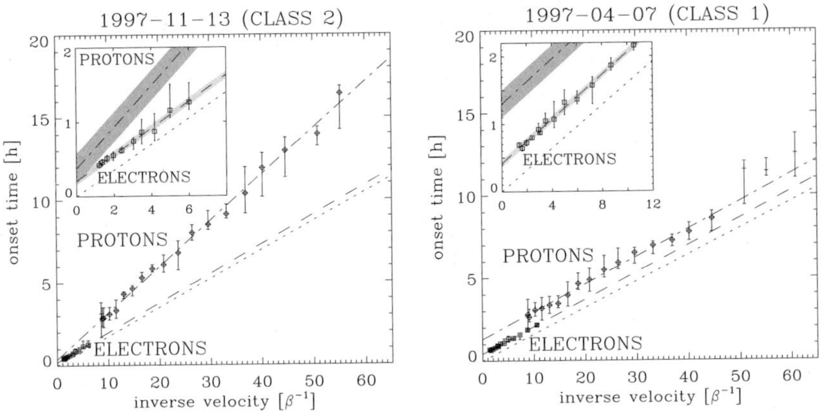

Figure 6. Plots of the onset time of protons between 0.1 and 6 MeV (diamonds) and electrons between 2 and 500 keV (squares; *Wind*/3DP) during two SEP events. The solar release time is the intercept of the straight lines fitted to the data with the vertical axis. The slope of the straight lines gives the path length travelled by the particles. (From Krucker and Lin, 2000).

Transport effects may alter onset times. Magnetic field irregularities lead to pitch angle scattering and drifts. Conservation of the magnetic moment in the expanding interplanetary magnetic field near the Sun leads to pitch angle focusing. The observed pitch angle distribution must therefore be field-aligned at the onset of an event. However, the scattering mean free path is uncertain, specifically close to the Sun (<0.3 AU), where no *in situ* observations exist. Large connection distances inferred from the onset time analysis can result from pitch angle scattering. This can be accounted for by assuming a non-zero average pitch angle for the first arriving particles.

Instrument-specific limitations are: (a) the width ΔE of the energy channels used, (b) the temporal resolution of the instrument and (c) the counting statistics in each channel. Larger instrumental geometric factors provide better statistics for a given event. On the other hand, the maximum size of events that can be analysed with large geometric factor instruments is limited. Electrons and ions behave differently in matter (straggling, bremsstrahlung) with the effect that ion channels are better defined in ΔE. Electrons with their largely higher speeds for any given kinetic energy have the advantage of smaller statistical uncertainty of inferred release times. Without corrections, these effects lead to delayed release times of ions over electrons.

In summary, with current instrumentation and single spacecraft observations, the release time determination cannot be better than a few minutes for measurements at 1 AU. This can be illustrated by the onset times and release times reported by different authors for a given event. Bieber *et al.* (2004) derived the injection function of relativistic protons during a ground level enhancement (GLE). The inferred release time is 3 to 4 min earlier than deduced from onset time analyses

(Kahler *et al.*, 2003; Tylka *et al.*, 2003). The solar release times reported for less energetic electrons and protons in the simple impulsive event of 1 May, 2000 show a scatter of 5–10 min (Tylka *et al.*, 2003; Mewaldt *et al.*, 2003b; Klein *et al.*, 2005; Klein and Posner, 2005). Similar uncertainties of release times were inferred from the analysis of simulated proton data (Lintunen and Vainio, 2004). Therefore only delays exceeding a minimum of, say, ~10 min between the release of particles of different species or between particles and photons are likely to be relevant indicators of the acceleration and transport of SEPs.

4.2. Comparison with Dynamic Processes in the Solar Corona

Knowing the release time of energetic particles allows us to study and identify several possible acceleration mechanisms when the times are compared with those derived from the solar electromagnetic signatures. Radio emissions from electrons accelerated in the corona were found to start up to several tens of minutes before the releases of protons and mildly relativistic electrons (Cliver *et al.*, 1982; Kallenrode, 1993; Krucker *et al.*, 1999; Haggerty and Roelof, 2002). Such delays are frequently ascribed to CME-shock acceleration of the escaping particles at heights of several solar radii (Lockwood *et al.*, 1990; Kahler, 2002; Simnett *et al.*, 2002). However, distinct radio signatures of late acceleration in the corona, well behind the CME and its presumed bow shock, were reported at the time of delayed releases of relativistic protons (Kocharov *et al.*, 1994; Akimov *et al.*, 1996; Klein *et al.*, 1999) and mildly relativistic electrons (Laitinen *et al.*, 2000; Maia and Pick, 2004; Klein *et al.*, 2005). This means that particles are accelerated in the corona for a time that may exceed the typical impulsive flare duration. If they are injected into different magnetic flux tubes extending into interplanetary space, a given spacecraft will detect only a portion of the escaping particles, and a one-to-one correspondence between particle acceleration in the corona and detection by the spacecraft becomes unlikely.

Altogether the onset time analyses reveal a complex timing of particles of different species or different energies. Time-extended particle acceleration in the aftermath of a CME or at its shock, particle transport in complex closed and open structures of the corona, and field line wandering or cross-field transport may all play a role in the timing.

5. ICMEs and Energetic Particles

J. Rodriguez-Pacheco, D. Lario, H. V. Cane, B. Heber and T. R. Sanderson

Energetic particle signatures associated with the passage of ICMEs over spacecraft located at 1 AU from the Sun and in the ecliptic plane have been described by several authors, e.g. Palmer *et al.* (1978), Bame *et al.* (1981), Kutchko *et al.* (1982), Sarris and Krimigis (1982), Sanderson *et al.* (1983), Tranquille *et al.* (1987), Richardson (1997), Torsti *et al.* (2004). These signatures include bidirectional ~1 MeV ion intensities, bidirectional cosmic ray intensities, energetic particle intensity depressions (i.e., Forbush decreases, see Section 2.1), and unusual flow directions of particles injected at the Sun at the time an ICME passed over an observer.

Energetic particle observations within and around the passage of an ICME can be used not only to detect the passage of the ICME but also to infer the magnetic topology of these structures (Richardson, 1997). For example, bidirectional particle flows, unusual solar particle event flows, and energetic particle intensity depressions are consistent with the presence of regions of looped magnetic field lines rooted at the Sun at both ends (Richardson, 1997). These signatures, however, may not be detected in all ICMEs, either because they are not present or as a result of either instrumental limitations or data gaps, or simply because no SEPs were injected at the Sun. Sometimes these signatures are only observed during a sub-interval within an ICME but not throughout its entire passage over an observer.

The origin of the energetic particles occasionally observed during the passage of ICMEs is uncertain. For events observed in the ecliptic plane near 1 AU, Richardson (1997) suggested three possible origins for the intra-ICME particles: (1) they are a fraction of particles accelerated by the CME-driven shock that are able to penetrate into the ICME, (2) particles accelerated at the time of the CME lift-off at the Sun (which would imply the existence of a particle acceleration mechanism different from the CME-driven shock), and/or (3) particles injected into the ICME by unrelated solar events, i.e. by flares not related to the CME lift-off. Energetic particle access into and out of ICMEs depends on both the magnetic topology of ICMEs and the way energetic particles propagate within and around ICMEs. Whereas open magnetic field topologies allow the relatively free access of shock-accelerated particles into ICMEs, the access into closed magnetic topologies is more difficult and requires particle diffusion perpendicular to field lines (Cane *et al.*, 1995).

In the case of intensity increases due to the injection of particles from unrelated impulsive solar flare events with their unique compositional and ionic charge signatures (i.e. case (3) above), these intensity increases can be used to probe the connection of the magnetic field lines to the Sun (Mazur *et al.*, 1998). Several energetic electron (Larson *et al.*, 1997) and ion observations (e.g. Popecki *et al.*, 2000b, and references therein) indicate that the footpoint of magnetic clouds may remain attached to the Sun long enough for a cloud to reach 1 AU.

To study both the origin of the intra-ICME energetic particles and the topology of ICMEs, it is also essential to analyse both the temporal and spatial variations of particles within and around ICMEs.

A subset of ICMEs have simple flux rope-like magnetic fields with enhanced magnetic fields that rotate over a large angle, known as magnetic clouds (MCs). Rodríguez-Pacheco *et al.* (2003) studied the energetic particles observed within the

MC illustrated in Figure 1 (Section 2.1). The magnetic field topology of this MC was inferred from fitting the magnetic field observations with the model developed by Hidalgo *et al.* (2002). Under the scenario of a flux rope anchored at the Sun and for a given particle energy, the bidirectional character of the particle distributions should be higher at the center of the cloud than at its borders, because there the ions encounter a much shorter way to the Sun (i.e., a shorter time in reaching the mirroring points closer to the footpoints of the magnetic loop). Nevertheless, the bidirectional flows observed within this specific MC did not fit with this scenario; moreover, they were independent of the ion energy. Therefore, Rodríguez-Pacheco *et al.* (2003) concluded that the energetic particle mirroring at the footpoints of the loop was not the cause of the bidirectional flows found in that event. Instead, they claimed that the intra-ICME particles came from the sheath region formed ahead of the MC and behind the CME-driven shock and that the intrinsic properties of the injection mechanism were responsible for the bidirectional properties of the event (see details in Rodríguez-Pacheco *et al.*, 2003).

Energetic particle measurements within ICMEs allow us to distinguish multiple components within these large-scale structures. Figure 7 shows energetic particle, plasma and magnetic field observations as measured by the ACE spacecraft during the passage of an ICME identified by Cane and Richardson (2003). The first three panels show from top to bottom the 1.9-4.8 MeV ion intensities, the first-order parallel (A_1) and second-order (A_2) anisotropy coefficients computed in the solar wind frame following the method described in Lario *et al.* (2004). Bidirectional 1.9–4.8 MeV ion flows were not observed throughout the passage of the ICME but only during the first half of day 303. Changes in the energetic particle distribution were observed in association with changes in both solar wind proton density and temperature as well as with a change in the magnetic field rotation, suggesting that the magnetic field configuration of this ICME was formed by a double flux rope. Similar examples of multiple flux-rope configurations have been also proposed by Vandas *et al.* (1999), Haggerty *et al.* (2000) and Hu *et al.* (2004). Simultaneous changes in both plasma, magnetic field and energetic particle distributions suggest spatial changes within the structure of the ICMEs and that those structures may be more complex than simple flux ropes anchored at the Sun.

6. Summary

B. Klecker, K.-L. Klein, and H. V. Cane

We have shown that there is now a wealth of observations of energetic electrons and ions in the atomic mass range from hydrogen to iron, covering a wide energy range from ~keV to ~100 MeV for electrons and 10s of keV/amu to 100s of MeV/amu for ions, respectively. New observations with instruments of much improved resolution and sensitivity onboard several spacecraft provide detailed information on the time-intensity profiles, energy spectra, elemental, isotopic, and ionic charge composition

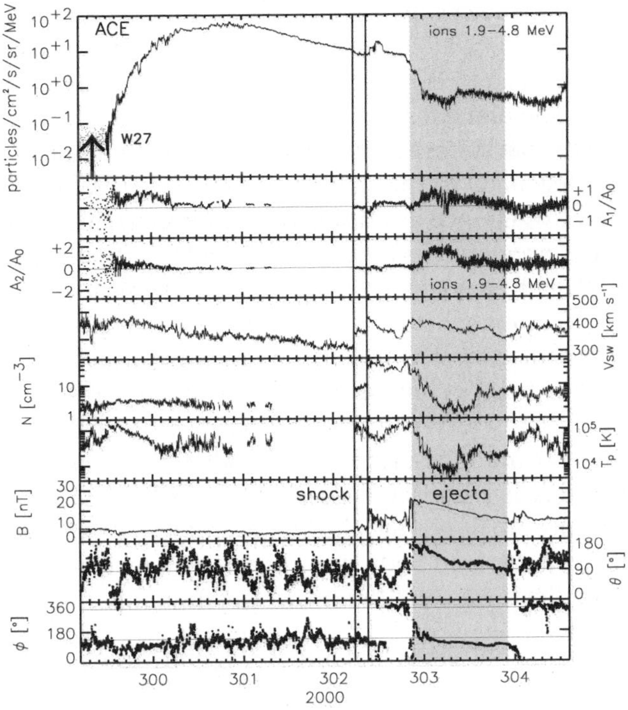

Figure 7. Observations of an event with experiments on the ACE spacecraft. From top to bottom: 1.9–4.8 MeV ion intensities measured by the EPAM instrument on board ACE. 1.9–4.8 MeV ion first-order parallel anisotropy coefficient in the solar wind frame. 1.9–4.8 MeV ion second-order anisotropy coefficient in the solar wind frame. Proton solar wind speed, density and temperature as measured by the SWEPAM instrument. Magnetic field magnitude and directions (in the GSE coordinate system) as measured by the MAG instrument. Solid vertical lines mark the arrival of the CME-driven shocks and gray bar indicates the passage of the ICME.

of energetic particles, accelerated either in solar flares and/or at CME-driven coronal and interplanetary shocks. These particle signatures carry fundamental information on the particle source, and on the injection, acceleration and propagation processes. However, with the particle observations being remote from the actual acceleration site, we always observe the combined effect of all these processes that are in general difficult to untangle.

Although there is considerable progress in our understanding of SEPs, there are a number of open questions that need to be addressed in the future, for example:

1. The systematic M/Q dependence of the elemental and isotopic composition in the energy range of a few MeV/amu is well documented. However, we are not able yet to relate this to the physical parameters of the plasma environment or the acceleration conditions.

2. The elemental abundances in the MeV/amu energy range at interplanetary shocks are strongly related to the suprathermal population upstream, suggesting acceleration at the shock. However, the spectral slopes derived from the local shock parameters are not consistent with the measured spectra.
3. Ionic charge states can strongly depend on energy. This implies that ionic charge states are not determined by the single parameter temperature, but rather by the combined effects of the plasma environment, impact ionisation during or after acceleration, the acceleration process, and interplanetary propagation. Thus, high charge states of Fe, e.g. Fe^{20+}, can be used as tracers for acceleration low in the corona. However, whether ions in high ionic charge states as observed in gradual events are accelerated from a suprathermal distribution from previous flare events or whether they are injected in the contemporary flare, needs further investigation.
4. We have time-resolved measurements of both SEPs (electrons and ions) and particles interacting in the solar atmosphere (hard X-rays, gamma-rays, radio), but due to transport to 1 AU the SEP time profiles are smeared out and only their start can be compared with electromagnetic signatures. This prevents us from identifying the traces of different acceleration mechanisms.

Further improvement in our understanding will require more modelling efforts, in particular:

1. Three dimensional simulation of CMEs and ICMEs, including the effect of particle acceleration in the dynamically evolving magnetic field configuration with parallel and perpendicular shock geometries;
2. to unfold injection, acceleration and propagation processes for a better understanding of the fractionation effects observed in elemental and isotopic abundances;
3. a quantitative description of the acceleration in impulsive flares, including impact ionization and propagation in interplanetary space for a better understanding of the abundance enhancements of ^{3}He and heavy ions, the correlation between heavy ion abundances and charge states, and the energy-dependent charge states.

Significant progress can also be expected from future missions, for example

1. the STEREO mission (Solar Terrestrial Relation Observatory), to be launched in 2006, with its two spacecraft separated in solar longitude will provide a stereoscopic view of CMEs and improve our understanding of their three-dimensional structure;
2. the Solar Orbiter mission (e.g. Marsden and Fleck, 2002) with its perihelion at $\sim$0.2 AU, in sync with the solar rotation for several days, will provide a close-up look at CMEs and allow much better correlation of the electromagnetic signatures and the characteristics of ions and electrons, because interplanetary propagation effects are minimal at this distance.

Acknowledgements

We thank the International Space Science Institute for their hospitality during the 2004 CME Workshop.

References

Akimov, V. V., Ambroz, P., Belov, A. V., Berlicki, A., Chertok, I. M., Karlicky, M., *et al.*: 1996, *Solar Phys.* **166**, 107–134.
Anderson, K. A., Chaizy, P., Lin, R. P., and Sommers, J.: 1992, *Geophys. Res. Lett.* **19**, 1283–1286.
Aschwanden, M. J.: 2002, *Space Science Reviews* **101**, 1–227.
Aurass, H., and Mann, G.: 2004, *Astrophys. J.* **615**, 526–530.
Axford, W. I., Leer, E., and Skadron, G.: 1978, *Proc. 15th Int. Cosmic Ray Conf., Plovdiv, Bulgaria*, Vol. 11, pp. 132–137.
Baker, D. N., Mason, G. M., Figueroa, O., Colon, G., Watzin, J. G., and Aleman, R. M.: 1993, *IEEE Transactions on Geoscience and Remote Sensing* **31**, 531–541.
Balasubrahmanyan, V. K., and Serlemitsos, A. T.: 1974, *Nature* **252**, 460–462.
Balogh, A., and Erdos, G.: 1983, *Planetary and Space Science* **31**, 1389–1395.
Bame, S. J., Asbridge, J., Feldman, W., Gosling, J., and Zwickl, R.: 1981, *Geophys. Res. Lett.* **8**, 173.
Bamert, K., Kallenbach, R., Ness, N. F., Smith, C. W., Terasawa, T., Hilchenbach, *et al.*: 2004, *Astrophys. J. Lett.* **601**, L99–L102.
Barghouty, A. F., and Mewaldt, R. A.: 1999, *Astrophys. J. Lett.* **520**, L127–L130.
Barouch, E., Gros, M., and Masse, P.: 1971, *Solar Phys.* **19**, 483–493.
Barouch, E., and Raguideau, J.: 1970, *Astron. Astrophys.* **9**, 146–150.
Beeck, J., Mason, G. M., Hamilton, D. C., Marsden, R. G., and Sanderson, T. R.: 1990, *J. Geophys. Res.* **95**(14), 10279–10290.
Bieber, J., Evenson, P., Droege, W., Pyle, R., Ruffolo, D., Rujiwarodom, M., Tooprakai, P., and Khumlumlert, T.: 2004, *Astrophys. J.* **601**, L103–L106.
Blandford, R. D., and Ostriker, J. P.: 1978, *Astrophys. J. Lett.* **221**, L29–L32.
Bogdanov, A. T., Klecker, B., Möbius, E., Hilchenbach, M., Kistler, L. M., Popecki, M. A., Hovestadt, D., and Weygand, J.: 2000, in R. A. M. *et al.* (ed.), *Acceleration and Transport of Energetic Particles Observed in the Heliosphere: ACE 2000 Symposium*, pp. 143–146.
Breneman, H. H., and Stone, E. C.: 1985, *Astrophys. J. Lett.* **299**, L57–L61.
Bryant, D. A., Cline, T. L., Desai, U. D., and McDonald, F. B.: 1962, *J. Geophys. Res.* **67**(16), 4983–5000.
Burlaga, L. F.: 1967, *J. Geophys. Res.* **72**(11), 4449–4466.
Cane, H. V., and Erickson, W. C.: 2003, *J. Geophys. Res.* **108**, 8–1, DOI:10.1029/2002JA009488.
Cane, H. V., and Lario, D.: 2006, *Space Sci. Rev.*, this volume, DOI:10.1007/s11214-006-9011-3.
Cane, H. V., Reames, D. V., and von Rosenvinge, T. T.: 1988, *J. Geophys. Res.* **93**(12), 9555–9567.
Cane, H. V., and Richardson, I. G.: 2003, *J. Geophys. Res.* **108** (A4), doi: 10.1029/2002JA009817.
Cane, H. V., Richardson, I. G., von Rosenvinge, T. T., and Wibberenz, G.: 1994, *J. Geophys. Res.* **99**(18), 21429–21441.
Cane, H. V., Richardson, I. G., and Wibberenz, G.: 1995, *Proc. 24th Int. Cosmic Ray Conf.*, Vol. 4, pp. 377–380.
Cane, H. V., von Rosenvinge, T. T., Cohen, C. M. S., and Mewaldt, R. A.: 2003, *Geophys. Res. Lett.* **30**, SEP 5–1, DOI 10.1029/2002GL016580.
Cliver, E. W., Kahler, S. W., Shea, M. A., and Smart, D. F.: 1982, *Astrophys. J.* **260**, 362–370.

Cohen, C. M. S., Cummings, A. C., Leske, R. A., Mewaldt, R. A., Stone, E. C., Dougherty, B. L., *et al.*: 1999, *Geophys. Res. Lett.* **26**, 149–152.

Dalla, S., Balogh, A., Krucker, S., Posner, A., Müller-Mellin, R., Anglin, J. D., *et al.*: 2003a, *Annales Geophys.* **21**, 1367–1375.

Dalla, S., Balogh, A., Krucker, S., Posner, A., Müller-Mellin, R., Anglin, J. D., *et al.*: 2003b, *Geophys. Res. Lett.* **30**, 9–1.

Desai, M. I., Mason, G. M., Dwyer, J. R., Mazur, J. E., Gold, R. E., Krimigis, S. M., *et al.*: 2003, *Astrophys. J.* **588**, 1149–1162.

Desai, M. I., Mason, G. M., Dwyer, J. R., Mazur, J. E., Smith, C. W., and Skoug, R. M.: 2001, *Astrophys. J. Lett.* **553**, L89–L92.

Desai, M. I., Mason, G. M., Wiedenbeck, M. E., Cohen, C. M. S., Mazur, J. E., Dwyer, J. R., *et al.*: 2004, *Astrophys. J.* **611**, 1156–1174.

Dietrich, W., and Tylka, A.: 2003, *Proc. 28th Int. Cosmic Ray Conf., Tsukuba, Japan*, Vol. 6, pp. 3291–3294.

Domingo, V., Sanahuja, B., and Heras, A. M.: 1989, *Adv. Space Res.* **9**, 191–195.

Dröge, W.: 2000, *Space Science Reviews* **93**, 121–151.

Ellison, D. C., Moebius, E., and Paschmann, G.: 1990, *Astrophys. J.* **352**, 376–394.

Ellison, D. C., and Ramaty, R.: 1985, *Astrophys. J.* **298**, 400–408.

Evenson, P., Meyer, P., Yanagita, S., and Forrest, D. J.: 1984, *Astrophys. J.* **283**, 439–449.

Fisk, L. A.: 1978, *Astrophys. J.* **224**, 1048–1055.

Fisk, L. A.: 1996, *J. Geophys. Res.* **101**, 15547–15554.

Forbes, T. G., Linker, J. A., *et al.*: 2006, *CME Theory and Modells, Space Sci. Rev.*, this volume.

Forbush, S. E.: 1946, *Physical Review* **70**, 771–772.

Forman, M. A., and Webb, G. M.: 1985, *Washington DC American Geophysical Union Geophysical Monograph Series* **35**, 91–114.

Geiss, J.: 1998, *Space Science Reviews* **85**, 241–252.

Giacalone, J., Burgess, D., Schwartz, S. J., and Ellison, D. C.: 1992, *Geophys. Res. Lett.* **19**, 433–436.

Gloeckler, G., Geiss, J., Roelof, E. C., Fisk, L. A., Ipavich, F. M., Ogilvie, K. W., *et al.*: 1994, *J. Geophys. Res.* **99**(18), 17637–17643.

Gloeckler, G., Sciambi, R. K., Fan, C. Y., and Hovestadt, D.: 1976, *Astrophys. J. Lett.* **209**, L93–L96.

Gold, T.: 1962, *Journal of the Physical Society of Japan, Vol. 17, Supplement A-II, Proceedings of the International Conference on Cosmic Rays and the Earth Storm, held 5–15 September, 1961 in Kyoto. Volume II: Joint Sessions. Published by the Physical Society of Japan,* **17**, 607.

Gordon, B. E., Lee, A. M., Möbius, E., and Trattner, K. J.: 1999, *J. Geophys. Res.* **104**, 28263–28277.

Gosling, J. T., Skoug, R. M., and McComas, D. J.: 2003, *Geophys. Res. Lett.* **30**, 30–1.

Hénoux, J.: 1998, *Space Science Reviews* **85**, 215–226.

Haggerty, D. K., and Roelof, E. C.: 2002, *Astrophys. J.* **579**, 841–853.

Haggerty, D. K., Roelof, E. C., Smith, C. W., Ness, N. F., Skoug, R. M., and Tokar, R. L.: 2000, in R. A. M. *et al.* (ed.), *Acceleration and Transport of Energetic Particles Observed in the Heliosphere: ACE 2000 Symposium*, American Institute of Physics, Vol. 528, pp. 266–269.

Heras, A. M., Sanahuja, B., Sanderson, T. R., Marsden, R. G., and Wenzel, K.-P.: 1994, *J. Geophys. Res.* **99** (18), 43–51.

Hidalgo, M. A., Cid, C., Viñas, A. F., and Sequeiros, J.: 2002, *J. Geophys. Res.* **106** (A1), 1002, DOI:10.1029/2001JA900100.

Holman, G. D.: 1995, *Astrophys. J.* **452**, 451–456.

Hovestadt, D.: 1974, *Solar Wind Conference*, pp. 2–92.

Hovestadt, D., Höfner, H., Klecker, B., Scholer, M., Gloeckler, G., Ipavich, F. M., *et al.*: 1981, *Adv. Space Res.* **1**, 61–64.

Hovestadt, D., Klecker, B., Scholer, M., Gloeckler, G., and Ipavich, F. M.: 1984, *Astrophys. J. Lett.* **282**, L39–L42.

Hsieh, K. C., and Simpson, J. A.: 1970, *Astrophys. J. Lett.* **162**, L191–L194.
Hu, Q., Smith, C. W., Ness, N. F., and Skoug, R. M.: 2004, *J. Geophys. Res.* **109**, A03102, DOI: 10.1029/2003JA010101.
Hurford, G. J., Mewaldt, R. A., Stone, E. C., and Vogt, R. E.: 1975, *Astrophys. J. Lett.* **201**, L95–L97.
Jokipii, J. R.: 1971, *Rev. Geophys. Space Phys.* **9**, 27–87.
Jokipii, J. R., and Parker, E. N.: 1968, *Phys. Rev. Lett.* **21**, 44–47.
Kahler, S.: 2002, *Astrophys. J.* **428**, 837–842.
Kahler, S. W., Simnett, G., and Reiner, M.: 2003, *Proc. 28th Intern. Cosmic Ray Conf.*, Vol. 6, pp. 3415–3418.
Kallenrode, M.: 1993, *J. Geophys. Res.* **98**(17), 5573–5591.
Kennel, C. F., Coroniti, F. V., Scarf, F. L., Livesey, W. A., Russell, C. T., and Smith, E. J.: 1986, *J. Geophys. Res.* **91**(11), 11917–11928.
Klecker, B.: 1982, *Adv. Space Res.* **2**, 285–292.
Klecker, B., Bogdanov, A. T., Popecki, M., Wimmer-Schweingruber, R. F., Möbius, E., Schaerer, R., *et al.*: 2000, in R. A. M. *et al.* (ed.), *Acceleration and Transport of Energetic Particles Observed in the Heliosphere: ACE 2000 Symposium*, AIP Vol. 528, pp. 135–138.
Klecker, B., Hovestadt, D., Scholer, M., Gloeckler, G., Ipavich, F. M., Fan, C. Y., *et al.*: 1984, *Astrophys. J.* **281**, 458–462.
Klecker, B., Popecki, M., Möbius, E., Desai, M., Mason, G., and Wimmer-Schweingruber, R.: 2003, *Proc. 28th Int. Cosmic Ray Conf., Tsukuba, Japan*, Vol. 6, pp. 3277–3280.
Klecker, B., Scholer, M., Hovestadt, D., Gloeckler, G., and Ipavich, F. M.: 1981, *Astrophys. J.* **251**, 393–401.
Klein, K., and Trottet, G.: 2001, *Space Science Reviews* **95**, 215–225.
Klein, K.-L., Chupp, E. L., Trottet, G., Magun, A., Dunphy, P. P., Rieger, E., *et al.*: 1999, *Astron. Astrophys.* **348**, 271–285.
Klein, K.-L., Krucker, S., Trottet, G., and Hoang, S.: 2005, *Astron. Astrophys.* **431**, 1047–1060.
Klein, K.-L., and Posner, A.: 2005, *Astron. Astrophys.* **438**, 1029–1042.
Kocharov, L., Kovaltsov, G. A., Torsti, J., and Ostryakov, V. M.: 2000, *Astron. Astrophys.* **357**, 716–724.
Kocharov, L. G., Kovaltsov, G. A., Kocharov, G. E., Chuikin, E. I., Usoskin, I. G., Shea, M. A., *et al.*: 1994, *Solar Phys.* **150**, 267–283.
Kovaltsov, G. A., Barghouty, A. F., Kocharov, L., Ostryakov, V. M., and Torsti, J.: 2001, *Astron. Astrophys.* **375**, 1075–1081.
Krucker, S., Larson, D. E., Lin, R. P., and Thompson, B. J.: 1999, *Astrophys. J.* **519**, 864–875.
Krucker, S., and Lin, R. P.: 2000, *Astrophys. J.* **542**, L61–L64.
Kucharek, H., Möbius, E., Li, W., Farrugia, C. J., Popecki, M. A., Galvin, A. B., *et al.*: 2003, *J. Geophys. Res.* **108** (A10), 8040, DOI: 10.1029/2003JA009938.
Kucharek, H., and Scholer, M.: 1995, *J. Geophys. Res.* **100**(9), 1745–1754.
Kutchko, F. J., Briggs, P., and Armstrong, T.: 1982, *J. Geophys. Res.* **87**, 1419.
Labrador, A., Leske, R., Mewaldt, R., Stone, E., and von Rosenvinge, T.: 2003, *Proc. 28th Intern. Cosmic Ray Conf., Tsukuba, Japan*, Vol. 6, pp. 3269–3272.
Laitinen, T., Klein, K.-L., Kocharov, L., Torsti, J., Trottet, G., Bothmer, V., *et al.*: 2000, *Astron. Astrophys.* **360**, 729–741.
Lario, D., Decker, R. B., Roelof, E. C., Reisenfeld, D. B., and Sanderson, T. R.: 2004, *J. Geophys. Res.* **109**, A01107, DOI:10.1029/2003JA010071.
Lario, D., Ho, G. C., Decker, R. B., Roelof, E. C., Desai, M. I., and Smith, C. W.: 2003a, *AIP Conf. Proc. 679: Solar Wind Ten*, pp. 640–643.
Lario, D., Roelof, E. C., Decker, R. B., and Reisenfeld, D. B.: 2003b, *Adv. Space Res.* **32**, 579–584.
Larson, D. E., Lin, R. P., McTiernan, J. M., McFadden, J. P., Ergun, R. E., McCarthy, M., *et al.*: 1997, *Geophys. Res. Lett.* **24**, 1911–1914.

Lee, M. A.: 1983, *J. Geophys. Res.* **88**(17), 6109–6119.
Leske, R. A., Cummings, J. R., Mewaldt, R. A., Stone, E. C., and von Rosenvinge, T. T.: 1995, *Astrophys. J. Lett.* **452**, L149–152.
Leske, R. A., Mewaldt, R. A., Cohen, C. M. S., Christian, E. R., Cummings, A. C., Slocum, P. L., *et al.*: 2003, *AIP Conf. Proc. 679: Solar Wind Ten*, pp. 616–619.
Leske, R. A., Mewaldt, R. A., Cohen, C. M. S., Cummings, A. C., Stone, E. C., Wiedenbeck, M. E., *et al.*: 1999, *Geophys. Res. Lett.* **26**, 2693–2696.
Leske, R. A., Mewaldt, R. A., Cummings, A. C., Stone, E. C., and von Rosenvinge, T. T.: 2001, in Wimmer-Schweingruber, R. F. (ed.), *AIP Conf. Proc. 598: Joint SOHO/ACE Workshop "Solar and Galactic Composition"*, pp. 171–176.
Liewer, P. C., Goldstein, B. E., and Omidi, N.: 1993, *J. Geophys. Res.* **98**(17), 15211–15220.
Lin, R. P.: 1985, *Solar Phys.* **100**, 537–561.
Lin, R. P., Mewaldt, R. A., and van Hollebeke, M. A. I.: 1982, *Astrophys. J.* **253**, 949–962.
Lintunen, J., and Vainio, R.: 2004, *Astron. Astrophys.* **420**, 343–350.
Litvinenko, Y. E.: 1996, *Astrophys. J.* **462**, 997–1004.
Lockwood, J. A., Debrunner, H., and Flueckiger, E. O.: 1990, *J. Geophys. Res.* **95**, 4187–4201.
Lovell, J. L., Duldig, M. L., and Humble, J. E.: 1998, *J. Geophys. Res.* **103**(12), 23733–23742.
Luhn, A., Klecker, B., Hovestadt, D., and Möbius, E.: 1987, *Astrophys. J.* **317**, 951–955.
Luhn, A., Klecker, B., Hovestadt, D., Scholer, M., Gloeckler, G., Ipavich, F. M., *et al.*: 1984, *Adv. Space Res.* **4**, 161–164.
Maia, D., and Pick, M.: 2004, *Astrophys. J.* **608**, 1082–1097.
Malandraki, O. E., Sarris, E. T., Lanzerotti, L. J., Trochoutsos, P., Tsiropoula, G., and Pick, M.: 2002, *J. Atmos. Solar-Terr. Phys.* **64**, 517–525.
Malkov, M. A.: 1998, *Phys. Rev. E* **58**, 4911–4928.
Marsden, R., and Fleck, B.: 2002, *ESA SP-477: Solspa 2001, Proceedings of the Second Solar Cycle and Space Weather Euroconference.*, pp. 359–360.
Marsden, R. G., Sanderson, T. R., Tranquille, C., Wenzel, K.-P., and Smith, E. J.: 1987, *J. Geophys. Res.* **92**(11), 11009–11019.
Mason, G. M., Gloeckler, G., and Hovestadt, D.: 1983, *Astrophys. J.* **267**, 844–862.
Mason, G. M., Mazur, J. E., and Dwyer, J. R.: 1999, *Astrophys. J. Lett.* **525**, L133–L136.
Mason, G. M., Mazur, J. E., Dwyer, J. R., Jokipii, J. R., Gold, R. E., and Krimigis, S. M.: 2004, *Astrophys. J.* **606**, 555–564.
Mason, G. M., Mazur, J. E., Looper, M. D., and Mewaldt, R. A.: 1995, *Astrophys. J.* **452**, 901–911.
Mason, G. M., Reames, D. V., von Rosenvinge, T. T., Klecker, B., and Hovestadt, D.: 1986, *Astrophys. J.* **303**, 849–860.
Mazur, J. E., Mason, G. M., Dwyer, J. R., and von Rosenvinge, T. T.: 1998, *Geophys. Res. Lett.* **25**, 2521–2524.
McCracken, K. G., Rao, U. R., and Bukata, R. P.: 1967, *J. Geophys. Res.* **72**(11), 4293–4324.
McKibben, R. B.: 1972, *J. Geophys. Res.* **77**, 3957–3984.
Mewaldt, R., Cohen, C., Mason, G., Desai, M., Leske, R., Mazur, J., *et al.*: 2003a, *Proc. 28th Intern. Cosmic Ray Conf., Tsukuba, Japan*, Vol. 6, pp. 3229–3232.
Mewaldt, R., Cohen, C. M. S., Haggerty, D. K., Gold, R., Krimigis, S., Leske, R., *et al.* : 2003b, *Proc. 28th Intern. Cosmic Ray Conf., Tsukuba, Japan*, Vol. 6, pp. 3313–3316.
Meyer, J.-P.: 1985, *Astrophys. J. Suppl.* **57**, 151–171.
Mikić, Z., and Lee, M. A.: 2006, *Space Sci. Rev.*, this volume, DOI:10.1007/s11214-006-9012-2.
Miller, J. A., and Vinas, A. F.: 1993, *Astrophys. J.* **412**, 386–400.
Möbius, E., Popecki, M., Klecker, B., Kistler, L. M., Bogdanov, A., Galvin, A. B., *et al.*: 1999, *Geophys. Res. Lett.* **26**, 145–148.
Möbius, E., Popecki, M., Klecker, B., Kistler, L. M., Bogdanov, A., Galvin, A. B., *et al.*: 2002, *Adv. Space Res.* **29**, 1501–1512.

Möbius, E., Cao, Y., Popecki, M., Kistler, L., Kucharek, H., Morris, D., *et al.*: 2003, *Proc. 28th Int. Cosmic Ray Conf., Tsukuba, Japan*, Vol. 6, pp. 3273–3276.

Moses, D., Droege, W., Meyer, P., and Evenson, P.: 1989, *Astrophys. J.* **346**, 523–530.

Ng, C. K., Reames, D. V., and Tylka, A. J.: 1999, *Geophys. Res. Lett.* **26**, 2145–2148.

Ng, C. K., Reames, D. V., and Tylka, A. J.: 2003, *Astrophys. J.* **591**, 461–485.

Oetliker, M., Klecker, B., Hovestadt, D., Mason, G. M., Mazur, J. E., Mewaldt, R. A., *et al.*: 1997, *Astrophys. J.* **477**, 495–501.

O'Gallagher, J. J., Hovestadt, D., Klecker, B., Gloeckler, G., and Fan, C. Y.: 1976, *Astrophys. J. Lett.* **209**, L97–L100.

Ostryakov, V. M., and Stovpyuk, M. F.: 1999, *Solar Phys.* **189**, 357–372.

Pallavicini, R., Serio, S., and Vaiana, G. S.: 1977, *Astrophys. J.* **216**, 108–122.

Palmer, I., Allum, F., and Singer, S.: 1978, *J. Geophys. Res.* **83**, 75–90.

Pick, M., Mann, G., Forbes, T. G., *et al.*: 2006, *Space Sci. Rev.*, this volume, DOI:10.1007/s11214-006-9021-1.

Popecki, M. A., Möbius, E., Klecker, B., Galvin, A. B., Kistler, L. M., and Bogdanov, A. T.: 2000a, in R. A. M. *et al.* (ed.), *Acceleration and Transport of Energetic Particles Observed in the Heliosphere: ACE 2000 Symposium*, AIP Vol. 528, pp. 63–70.

Popecki, M. A., Mazur, J. E., Möbius, E., Klecker, B., Bogdanov, A., Mason, G. M., *et al.*: 2003, *Proc. 28th Int. Cosmic Ray Conf., Tsukuba, Japan*, Vol. 28, pp. 3283–3286.

Popecki, M. A., Zurbuchen, T. H., Skoug, R. M., Smith, C. W., Galvin, A. B., Lee, M., *et al.*: 2000b, in R. A. M. *et al.* (ed.), *Acceleration and Transport of Energetic Particles Observed in the Heliosphere: ACE 2000 Symposium*, AIP Vol. 528, pp. 139–142.

Potter, D. W., Lin, R. P., and Anderson, K. A.: 1980, *Astrophys. J. Lett.* **236**, L97–L100.

Quest, K. B.: 1988, *J. Geophys. Res.* **93**(12), 9649–9680.

Reames, D. V.: 1999, *Space Science Reviews* **90**, 413–491.

Reames, D. V., Kahler, S. W., and Ng, C. K.: 1997, *Astrophys. J.* **491**, 414–420.

Reames, D. V., and Ng, C. K.: 1998, *Astrophys. J.* **504**, 1002–1005.

Reames, D. V., Ng, C. K., and Tylka, A. J.: 2000, *Astrophys. J. Lett.* **531**, L83–L86.

Reames, D. V., von Rosenvinge, T., and Lin, R.: 1985, *Astrophys. J.* **295**, 716–724.

Reinhard, R., and Wibberenz, G.: 1974, *Solar Phys.* **36**, 473–494.

Richardson, I. G.: 1997, in Crooker, N., Joselyn, J. A., and Feynman, J. (eds.), *Coronal Mass Ejections*, AGU Geophysical Monograph, Vol. 99, pp. 189–198.

Rodríguez-Pacheco, J., Cid, C., Blanco, J. J., and Sequeiros, J.: 2003, *Solar Phys.* **213**, 121–145.

Roelof, E. C.: 1969, in Ögelmann, H., and Wayland, J. (eds.), *Lectures in High Energy Astrophysics*, NASA, SP-199.

Roelof, E. C.: 2000, in R. A. M. *et al.* (ed.), *Acceleration and Transport of Energetic Particles Observed in the Heliosphere: ACE 2000 Symposium*, American Institute of Physics, Vol. 528, pp. 242–249.

Roelof, E. C., Gold, R. E., Simnett, G. M., Tappin, S. J., Armstrong, T. P., and Lanzerotti, L. J.: 1992, *Geophys. Res. Lett.* **19**, 1243–1246.

Ryan, J. M., and Lee, M. A.: 1991, *Astrophys. J.* **368**, 316–324.

Sanahuja, B., and Domingo, V.: 1987, *J. Geophys. Res.* **92**(11), 7280–7288.

Sanderson, T. R., Beeck, J., Marsden, G. R., Tranquille, C., Wenzel, K.-P., McKibben, B. R., *et al.*: 1990, *Proc. 21st Intern. Cosmic Ray Conf.*, Vol. 6, pp. 251–254.

Sanderson, T. R., Heras, A. M., Marsden, R. G., and Winterhalter, D.: 1992, *Solar Wind Seven*, pp. 663–666.

Sanderson, T. R., Marsden, R. G., Reinhard, R., Wenzel, K.-P., and Smith, E. J. : 1983, *Geophys. Res. Lett.* **10**, 916–919.

Sanderson, T. R., Marsden, R. G., Tranquille, C., Dalla, S., Forsyth, R. J., Gosling, J. T., *et al.*: 2003, *Geophys. Res. Lett.* **30**(19), 8036, DOI:10.1029/2003GL017306.

Sanderson, T. R., Reinhard, R., van Nes, P., and Wenzel, K.-P.: 1985a, *J. Geophys. Res.* **90**(9), 19–27.
Sanderson, T. R., Reinhard, R., van Nes, P., Wenzel, K.-P., Smith, E. J., and Tsurutani, B. T.: 1985b, *J. Geophys. Res.* **90**(9), 3973–3980.
Sarris, E., and Krimigis, S.: 1982, *Geophys. Res. Lett.* **9**, 167–170.
Saul, L., Möbius, E., Litvinenko, Y., Isenberg, P., Kucharek, H., Lee M., *et al.*: 2003, *AIP Conf. Proc. 679: Solar Wind Ten*, pp. 778–781.
Scholer, M., Hovestadt, D., Klecker, B., Gloeckler, G., and Fan, C. Y.: 1978, *J. Geophys. Res.* **83**(12), 3349–3354.
Scholer, M., and Kucharek, H.: 1999, *Geophys. Res. Lett.* **26**, 29–32.
Scholer, M., Kucharek, H., Krasnosselskikh, V. V., and Trattner, K.-H.: 2000, in R. A. M. *et al.* (ed.), *Acceleration and Transport of Energetic Particles Observed in the Heliosphere: ACE 2000 Symposium*, AIP Vol. 528, pp. 250–257.
Schwadron, N. A., Fisk, L. A., and Gloeckler, G.: 1996, *Geophys. Res. Lett.* **23**, 2871–2874.
Simnett, G. M.: 1974, *Space Science Reviews* **16**, 257–323.
Simnett, G. M.: 2001, *Space Science Reviews* **97**, 231–242.
Simnett, G. M.: 2003, *Solar Phys.* **213**, 387–412.
Simnett, G. M., Roelof, E. C., and Haggerty, D. K.: 2002, *Astrophys. J.* **579**, 854–862.
Sollitt, L., Stone, E., Mewaldt, R., Cohen, C., Leske, R., Wiedenbeck, M., *et al.*: 2003, *Proc. 28th Int. Cosmic Ray Conf., Tsukuba, Japan*, Vol. 6, pp. 3295–3298.
Stovpyuk, M. F., and Ostryakov, V. M.: 2001, *Solar Phys.* **198**, 163–167.
Torsti, J., Riihonen, E., and Kocharov, L.: 2004, *Astrophys. J. Lett.* **600**, L83–L86.
Tranquille, C., Sanderson, T. R., Marsden, R. G., Wenzel, K.-P., and Smith, E. J.: 1987, *J. Geophys. Res.* **92**(11), 6–14.
Tsuneta, S., and Naito, T.: 1998, *Astrophys. J. Lett.* **495**, L67–L70.
Tylka, A., Cohen, C. M. S., Dietrich, W. F., Krucker, S., McGuire, R. E., Mewaldt, R., *et al.*: 2003, *Proc. 28th Int. Cosmic Ray Conf.*, Vol. 6, pp. 3313–3316.
Tylka, A. J., Boberg, P. R., McGuire, R. E., Ng, C. K., and Reames, D. V.: 2000, in R. A. M. *et al.* (ed.), *Acceleration and Transport of Energetic Particles Observed in the Heliosphere: ACE 2000 Symposium*, AIP Vol. 528, pp. 147–152.
Tylka, A. J., Cohen, C. M. S., Dietrich, W. F., Lee, M. A., Maclennan, C. G., Mewaldt, R. A., *et al.*: 2005, *Astrophys. J.* **625**, 474–495.
Tylka, A. J., Reames, D. V., and Ng, C. K.: 1999, *Geophys. Res. Lett.* **26**, 2141–2144.
van Hollebeke, M. A. I., Ma Sung, L. S., and McDonald, F. B.: 1975, *Solar Phys.* **41**, 189–223.
van Nes, P., Reinhard, R., Sanderson, T. R., Wenzel, K.-P., and Roelof, E. C.: 1985, *J. Geophys. Res.* **90**(9), 3981–3994.
van Nes, P., Reinhard, R., Sanderson, T. R., Wenzel, K.-P., and Zwickl, R. D.: 1984, *J. Geophys. Res.* **89**(18), 2122–2132.
Vandas, M., Fischer, S., and Geranios, A.: 1999, in Habbal, S. R., *et al.* (eds.), *Solar Wind Nine*, American Institute of Physics, Vol. 471, pp. 127–130.
Webb, G. M., and Gleeson, L. J.: 1979, *Astrophys. Space Science* **60**, 335–351.
Zhang, M., McKibben, R., Lopate, C., Jokipii, J., Giacalone, J., Kallenrode, M.-B., *et al.*: 2003, *J. Geophys. Res.* **108**, 4.1–4.13 (DOI: 10.1029/2002JA009531).

CME THEORY AND MODELS

Report of Working Group D

T. G. FORBES[1,*], J. A. LINKER[2], J. CHEN[3], C. CID[4], J. KÓTA[5], M. A. LEE[2], G. MANN[6], Z. MIKIĆ[1], M. S. POTGIETER[7], J. M. SCHMIDT[8], G. L. SISCOE[9], R. VAINIO[10], S. K. ANTIOCHOS[3] and P. RILEY[2]

[1]*Institute for Earth, Oceans, and Space, University of New Hampshire, Durham, NH, USA*

[2]*Science Applications International Corp., San Diego, CA, USA*

[3]*Naval Research Laboratory, Washington, DC, USA*

[4]*Departamento de Física Universidad de Alcala, Alcala de Henares, Madrid, Spain*

[5]*Lunar and Planetary Laboratory, Department of Planetary Sciences, University of Arizona, Tucson, AZ, USA*

[6]*Astrophysikalisches Institut Potsdam, Potsdam, Germany*

[7]*Unit for Space Physics and School of Physics, North-West University, Potchefstroom, South Africa*

[8]*Imperial College, Space and Atmospheric Physics, The Blackett Lab., London, U.K.*

[9]*Boston University, Center for Space Physics, Boston, MA, USA*

[10]*Department of Physical Sciences, University of Helsinki, Helsinki, Finland*

*(*Author for correspondence: E-mail: terry.forbes@unh.edu)*

(Received 6 February 2006; Accepted in final form 6 April 2006)

Abstract. This chapter provides an overview of current efforts in the theory and modeling of CMEs. Five key areas are discussed: (1) CME initiation; (2) CME evolution and propagation; (3) the structure of interplanetary CMEs derived from flux rope modeling; (4) CME shock formation in the inner corona; and (5) particle acceleration and transport at CME driven shocks. In the section on CME initiation three contemporary models are highlighted. Two of these focus on how energy stored in the coronal magnetic field can be released violently to drive CMEs. The third model assumes that CMEs can be directly driven by currents from below the photosphere. CMEs evolve considerably as they expand from the magnetically dominated lower corona into the advectively dominated solar wind. The section on evolution and propagation presents two approaches to the problem. One is primarily analytical and focuses on the key physical processes involved. The other is primarily numerical and illustrates the complexity of possible interactions between the CME and the ambient medium. The section on flux rope fitting reviews the accuracy and reliability of various methods. The section on shock formation considers the effect of the rapid decrease in the magnetic field and plasma density with height. Finally, in the section on particle acceleration and transport, some recent developments in the theory of diffusive particle acceleration at CME shocks are discussed. These include efforts to combine self-consistently the process of particle acceleration in the vicinity of the shock with the subsequent escape and transport of particles to distant regions.

Keywords: Sun: Coronal Mass Ejections, CMEs, ICMEs, flares magnetic reconnection, shocks, solar energetic particles (SEPs)

Space Science Reviews (2006) 123: 251–302
DOI: 10.1007/s11214-006-9019-8

1. Introduction

T. G. FORBES, G. L. SISCOE

The life cycle of a CME encompasses a wide range of plasma processes in which the magnetic field plays a dominant role. Dynamo activity in the solar interior creates the magnetic field which builds up in the corona. Ultimately, this field erupts as a result of an instability or loss-of-equilibrium process which is yet to be identified. Once a CME is underway, a whole host of additional processes are triggered. These include magnetic reconnection, shock formation, and particle acceleration, among others. One of the main objectives of this chapter is to assess the current state of theoretical understanding of the various physical process involved in the life cycle of CMEs. Typically this understanding is brought about by using theoretical principles to construct mathematical models which describe both the form and evolution of CMEs as inferred from observations.

All models, whether numerical or analytical, require an initial state to be specified. For an MHD model, this means specifying eight variables (the three components of the magnetic field, the three components of velocity, density, and temperature) throughout the heliosphere prior to CME onset. Of these variables, the magnetic field components are the most critical. Because the magnetic field is inertially line-tied at the base of the corona (van der Linden *et al.*, 1994), only the magnetic field associated with coronal currents is available to drive CMEs. Unfortunately, this magnetic field is extremely difficult to measure. The best one can do at the present time is to estimate the field based on extrapolations of the vector fields at the photospheric and chromospheric levels. In practice, both measurement uncertainties and modeling limitations make it exceedingly difficult to deduce the pre-eruptive field solely from observations. Thus one is forced to make an initial guess for the pre-eruptive field and then evolve it in a manner consistent with the observed surface field to see if it leads to a CME-like eruption.

Because of the complexity of the equations which govern CME dynamics, much effort has been devoted in recent years to developing models using numerical methods (e.g. Mikić *et al.*, 1988; Biskamp and Welter, 1989; Forbes, 1990; Chen *et al.*, 2001; Gibson and Low, 1998; Antiochos *et al.*, 1999; Wu *et al.*, 2001; Amari *et al.*, 2000; Odstrcil *et al.*, 2002; Tokman and Bellan, 2002; Linker *et al.*, 2003; Roussev *et al.*, 2003, 2004; Kusano *et al.*, 2004). One of the hurdles that numerical models must cope with is the enormous range of spatial and temporal scales involved in the CME phenomenon, as shown in Figure 1. These scales range over 16 orders of magnitude from centimeters to hundreds of AU and from micro seconds to years. Considerable progress in covering this range has been made in the last few years by taking advantage of the fact that the flow of information in the solar wind, beyond the point where the fast mode Mach number exceeds one, is always outwards. This allows MHD codes to be chained together so that the output from one code is used as the input for another code. However, a large gap in coverage still exists between

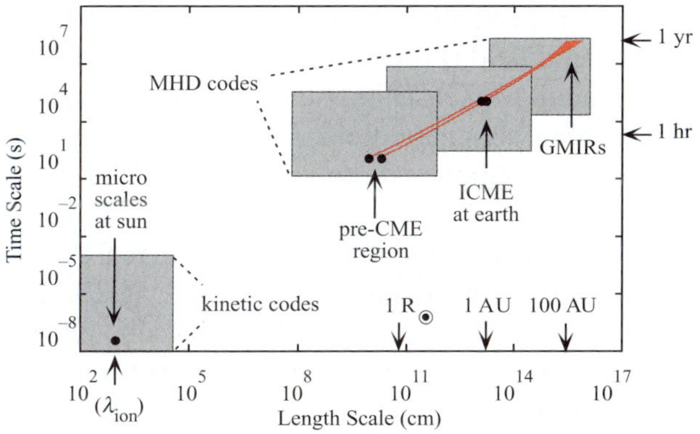

Figure 1. Time and length scales associated with CMEs as they propagate outwards through the solar system. The term "micro scales" refers to such parameters as the inertial length and gyro radius. The shaded rectangles indicate the ranges covered by existing numerical codes, and the curved lines indicate the leading and trailing edges of the CME as it progresses outwards through the solar system, first as a CME, then as an ICME (Interplanetary CME), and finally as part of a global merged interaction region (GMIR). Although the CME expands as it moves away from the Sun, it appears to contract in this log-log plot because it expands more slowly than the rate at which it moves away from the Sun.

the scale of the MHD codes and the kinetic codes required to model the physical processes occurring in current sheets (*i.e.*, reconnection) and shocks (*i.e.*, diffusive shock acceleration).

The heavy emphasis at the present time on the development of numerical models may lead to the impression that progress in modeling CMEs is just a matter of developing improved algorithms. While this may be true in some cases, it is not true for models of CME initiation, where the physical mechanism that triggers CMEs is unknown. Progress in this area depends on using both numerical and analytical models to develop a deeper understanding of the physical conditions that lead to an eruption. Although analytical models cannot cope with the same level of complexity as numerical models can, they are not restricted by resolution or scale limitations. Because analytical models provide a deeper level of insight into the underlying physics, they are often used in association with numerical models, either as a starting point or as an interpretive tool. An example of the former is the analytical model of Titov and Démoulin (1999), which has served as a starting point for the fully three-dimensional numerical simulations by Roussev *et al.* (2003, 2004) and Kliem *et al.* (2004).

Early models of CMEs developed in the 1970s were based on different principles than those used in present day models. Some of these early models, such as those of Steinolfson and Nakagawa (1977) and Dryer *et al.* (1979), assumed that CMEs are simply the result of a flare-generated blast wave. Today we know that, while some CMEs are flare associated, not all are. Furthermore, in many cases where there is a flare association, the CME can precede the flare (Wagner, 1981; Harrison, 1986).

Most importantly, the gas pressures generated during a flare are simply too small to blow open the magnetic field (Low, 1981; Emslie *et al.*, 2004).

While there is a general consensus that the energy that drives CMEs and flares originates from the coronal magnetic field, CME initiation has long been an area of substantial debate and continues to be so. Section 2 discusses three different models of CME initiation.

The dynamics of CMEs after initiation involves several factors. These include acceleration, expansion, drag, and distortion. Although acceleration and expansion are an integral part of the initiation process, they may also play a role in the long-term evolution of the CME either through a sustained operation of the forces which initiate the CME or through the interaction of the CME with the ambient solar wind. Drag and distortion result from the interaction of the CME with the ambient solar wind, corotating interaction regions (CIRs), and other CMEs. Section 3 explores our understanding of CME evolution and propagation through both analytical and numerical approaches.

A particularly important aspect of the evolution and propagation of CMEs are is the determination of the internal structure of ICMEs from in-situ measurements of the magnetic field. The strengths and limitations of these flux-rope fitting models are discussed in Section 4.

Large solar eruptions which generate high speed (1000–3000 km/sec) CMEs are the principal source of energetic particles produced in the solar system. A key aspect of the generation of solar energetic particle (SEP) events is the formation of shock waves, which is described in Section 5. SEP events can be subdivided into gradual and impulsive events. (The terms "gradual" and "impulsive" are used to refer to the evolution of the particle flux and do not refer to the X-ray profile of any associated flare.) Section 6 deals primarily with the theory of particle acceleration and transport of gradual events, as these are believed to be associated with CMEs.

2. CME Initiation

J. A. LINKER, T. G. FORBES, S. ANTIOCHOS, J. CHEN

Most CME initiation models today are based on the premise that CMEs and flares derive their energy from the coronal magnetic field. The currents that build up in the corona as a result of flux emergence and surface flows slowly evolve to a state where a stable equilibrium is no longer possible. Once this happens, the field erupts. If the eruption is sufficiently strong and the overlying fields not too constraining, plasma is ejected into interplanetary space. If strong magnetic fields exist in the erupted region, then bright, flare-like emissions occur. The latter is true, even if the field does not erupt (Švestka and Cliver, 1992).

At the present time, there is no consensus as to what the mechanism is that leads to the loss of a stable equilibrium. It may be purely ideal or involve non-ideal processes like reconnection (Forbes, 2000; Low, 2001). The possibility also remains that very slow (<150 km/s) CMEs which undergo weak acceleration over a period lasting as long as a day (Sirivastava *et al.*, 2000; Zhang *et al.*, 2004) may not involve a release of stored magnetic energy at all. For these CMEs the observed rates of flux emergence in the photosphere is of the same order as that required by the flux injection model (Krall *et al.*, 2000). Unlike most CME models, this model does not involve the storage of magnetic energy in the corona prior to onset. Instead, it injects magnetic flux and energy into the corona during the eruption.

One of the difficulties that all storage models face is explaining how it is possible to decrease the magnetic energy in the corona even though the ejection of the CME stretches the magnetic field as it moves outwards into interplanetary space. The stretching of the field creates an apparent paradox, since it implies that the magnetic energy of the system is increasing, whereas storage models require it to decrease. Aly (1984, 1991) and Sturrock (1991) have argued that for a simply-connected field, the fully-opened field-configuration always has a higher magnetic energy than the corresponding force-free field. This constraint seems to imply that storage models are energetically impossible (see Mikić and Lee, 2006, this volume). However, there are several possible ways around it. First, the magnetic field may not be simply connected and contain knotted field lines. Second, it may contain field lines that are completely disconnected from the surface. Third, an ideal-MHD eruption can still extend field lines as long as it does not open them all the way to infinity. Fourth, an ideal-MHD eruption may be possible if it only opens a portion of the closed field lines. Fifth, small deviations from a perfectly force-free initial state might make a difference. And finally, a non-ideal process, specifically magnetic reconnection, invalidates the constraint.

In the following subsections we describe three examples of CME initiation models. See Mikić and Lee (2006) in this volume for a discussion of the similarities and differences in the underlying physics of these models.

2.1. Flux Rope Models and Flux Cancellation

Coronal mass ejections are frequently associated with prominence eruptions as well as solar flares. Prominences (called filaments when observed on the solar disk) support cool, dense chromospheric material ($\sim 10^4$ K and $10^{10}-10^{11}$ cm^{-3}) against solar gravity in the surrounding hot, tenuous corona ($\sim 10^6$ K and 10^7-10^9 cm^{-3}). They are observed to lie above magnetic neutral lines in the photosphere and near the base of helmet streamers (regions of closed magnetic field that have confined the coronal plasma). The magnetic field in the prominence often exhibits "inverse polarity," meaning that when the coronal magnetic fields embedded in the prominence cross over the neutral line, they point in the direction opposite to that indicated by the photospheric magnetic field polarity (Leroy *et al.*, 1983, 1984). The prominence magnetic field is itself nearly aligned with the filament channel (Martin *et al.*, 1994;

Martin and Echols, 1994), indicating a highly sheared (and therefore magnetically energized) configuration.

The idea that a flux rope could explain the inverse polarity of a prominence dates back to the Kuperus and Raadu (1974) prominence model. In that model, a current filament (in two dimensions) produces closed magnetic loops that can support prominence material above the photosphere. Since that time there have been a number of authors who have focused on the support of prominence material by helical field lines and/or the disruption of these configurations as the possible cause of prominence eruptions and coronal mass ejections (van Ballegooijen and Martens, 1989; Forbes and Isenberg, 1991; Lin *et al.*, 1998; Titov and Démoulin, 1999; Amari *et al.*, 2000; Linker *et al.*, 2001; Low, 2001; Sturrock *et al.*, 2001; Low and Zhang, 2002; Amari *et al.*, 2003a,b; Birn *et al.*, 2003; Linker *et al.*, 2003; Roussev *et al.*, 2003, 2004; Kliem *et al.*, 2004).

Two possibilities exist for the formation of the flux rope. The flux rope could emerge intact from below the photosphere (Rust and Kumar, 1994; Lites *et al.*, 1995; Fan, 2005) or be formed as the result of motions at the photosphere or above. In this review, we focus on the second possibility, that flux ropes are first formed, and subsequently erupt, as a result of flux cancellation at the photosphere (defined below). Once a flux rope structure has formed in the corona, the susceptibility of the structure to eruption shouldn't depend on its origin.

2.1.1. *What is Magnetic Flux Cancellation?*

Martin *et al.* (1985) defined flux cancellation observationally as the mutual disappearance of magnetic fields of opposite polarity at the neutral line separating them. Martin *et al.* (1985) chose the term "flux cancellation" carefully, so as not to convey any theoretical interpretation of the process; they recognized that flux elements might be submerging, annihilating, or being expelled upward. Flux cancellation occurs everywhere on the Sun, (Livi *et al.*, 1985); observations have shown it to be active at filament sites (Litvinenko and Martin,, 1999; Wang, 2001) and in active regions as they disperse (Martin *et al.*, 1985). During this time, filaments are frequently observed to form along the neutral line. At times, these filaments disappear, presumably due to eruption, and may even reform in the same location later. This dispersal of magnetic flux is thought to occur on a small spatial scale by annihilation and submergence of magnetic dipole elements and has been modeled as a convective-diffusive process on a large scale (Wang *et al.*, 1989; Wang and Sheeley, 1990).

Associations of flux cancellation with solar flares have been noted previously (Livi *et al.*, 1989). More recently, flux cancellation has been associated with CMEs (Lin *et al.*, 2004). A particularly striking example is the "Bastille Day" event, an X5.7 flare and associated fast CME that occurred in an active region (NOAA AR 9077) on July 14, 2000. Flux cancellation was observed throughout this event

(Kosovichev and Zharkova, 2001) and has been interpreted as the cause of the massive eruption (Somov *et al.*, 2002).

2.1.2. *Flux Cancellation: Theoretical Interpretation*

The recognition that flux cancellation is active at filament sites and during the eruptive process led to the interpretation that the cancellation was in fact the annihilation of magnetic flux at the photosphere through reconnection. Using this interpretation, van Ballegooijen and Martens (1989) investigated the consequences of flux cancellation at the neutral line of a 2.5D sheared arcade configuration. They computed sequences of force-free equilibria to show that flux cancellation leads to the formation of a flux rope. The helical field lines of the model flux rope contained dips capable of supporting prominence material, and the rise in the equilibrium height of the flux rope with increased flux cancellation suggested possible eruptive behavior.

Calculations by Forbes and Isenberg (1991), Forbes *et al.* (1994) and Lin *et al.* (1998) investigated the stability of flux ropes embedded in a background field. They found that that once a flux rope is formed, continuation of the flux cancellation process can result in a loss of equilibrium. The new lower-energy equilibrium contains a current sheet and a higher height for the flux rope. While the energy release in this ideal process is relatively small, the new equilibrium height of the flux rope can be many solar radii from the Sun. The reason for this transition in equilibria can be understood as follows: The magnetic pressure forces in the flux rope want the rope to expand; these forces are restrained by tension in the surrounding fields. Flux cancellation converts the restraining field into magnetic flux in the rope, increasing the magnetic pressure. Eventually the system reaches a point that no nearby equilibrium is accessible (see Figure 2). In reality, the new equilibrium with a flux rope high above the photosphere is untenable; the flux rope would be pulled outward by the solar wind. Significant magnetic energy release could then occur through magnetic reconnection at the current sheet. Exploring this aspect of the problem is intractable analytically and must be studied with numerical simulations.

2.1.3. *Flux Cancellation: MHD Computations*

Amari *et al.* (2000) investigated the formation and eruption of a magnetic flux rope by flux cancellation in localized 3D Cartesian geometry. These calculations neglected the plasma pressure and were thus an appropriate approximation for an active region at heights low in the corona. They showed the formation and subsequent eruption of a flux rope when a sheared arcade was subjected to flux cancellation. Amari *et al.* (2003a,b) studied further examples of eruptions in this geometry and showed that the flux cancellation process causes little change to helicity of the sheared configuration.

Linker *et al.* (2003) performed numerical calculations that appear to have a close correspondence to the theoretical work of Lin *et al.* (1998). These calculations

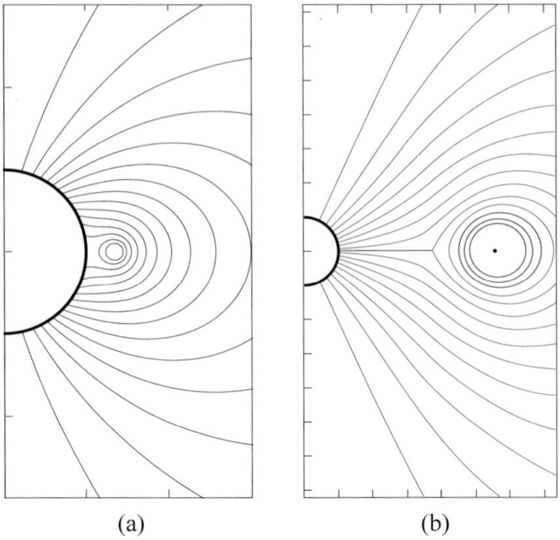

Figure 2. Azimuthally-symmetric flux rope model showing the ideal MHD transition from a flux rope in equilibrium close to the Sun (a) and the same flux rope at a large distance from the Sun after the transition (b). A current sheet forms as a result of this transition (after Lin *et al.*, 1998).

solved the full MHD equations and included the important effect of the solar wind. First, a helmet streamer configuration is developed by combining a spherically symmetric solar wind solution with a potential magnetic field and integrating the MHD equations in time until the solution settles down to an equilibrium. A helmet streamer with closed field lines forms, surrounded by open field lines along which the solar wind flows outward. To provide a source of free magnetic energy for the eruption, a shear flow is introduced near the neutral line of the streamer. This shear flow is not intended to model actual flows on the Sun. It is just a convenient mechanism for producing strongly sheared field lines that are nearly aligned with the neutral line, a frequently observed characteristic of filaments (Martin *et al.*, 1994). Figure 3 at $t = 1300\tau_A$ (τ_A, the Alfvén time, is 12 minutes), shows projected magnetic field lines and current density for the sheared helmet streamer. The investigation of the effect of flux cancellation begins at this point in the calculation. The change in flux is applied by specifying the tangential component of the electric field at the boundary, $\mathbf{E}_{t0}$. For example, when $\mathbf{E}_{t0} = 0$, B_{r0} (the radial magnetic field at the solar boundary) remains fixed in time. In order to specify a desired change in the magnetic flux, a nonzero $\mathbf{E}_{t0}$ is specified. This electric field drives converging flows and reduces the flux at the neutral line, as is believed to occur in the flux cancellation process (van Ballegooijen and Martens, 1989).

Flux cancellation first forms a stable flux rope configuration within the helmet streamer (Figure 3, $t = 1350\tau_A$). Without further flux cancellation or other imposed changes to the configuration, the flux rope will remain stable indefinitely. Therefore,

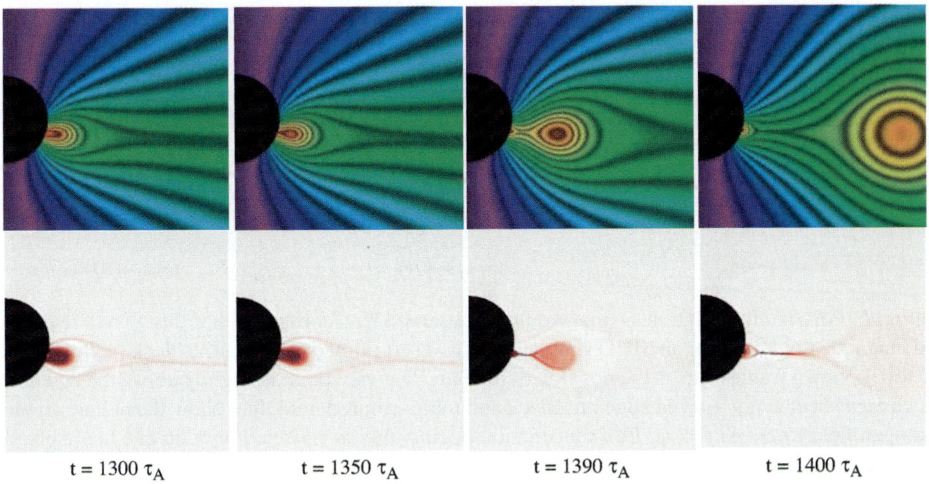

Figure 3. MHD Simulation of a helmet streamer eruption triggered by flux cancellation (after Linker *et al.*, 2003). The stripes in the top panels show projected field lines (there is also a B_ϕ component of the magnetic field out of the plane). The bottom panels shows the current density J_ϕ out of the plane. Time in Alfvén scale times ($\tau_A = 12$ minutes) is indicated. Note the similarity of the flux rope eruption and current sheet formation to that shown in Figure 2.

on the real Sun, prominences could form by this mechanism and remain stable for many days or weeks. Lionello *et al.* (2002) and van Ballegooijen (2004) have studied flux rope formation via this mechanism for more realistic field configurations. With continued flux cancellation, the helmet streamer is destabilized, subsequently erupting into the outer corona as shown in the last two frames of Figure 3. The eruption of the flux rope and the formation of the current sheet is reminiscent of the ideal calculation by Lin *et al.* (1998). However, for the resistive MHD calculation, energy is released rapidly through reconnection at this sheet. Linker *et al.* (2003) found that about 1.75×10^{32} ergs of magnetic energy was released, with about 1/2 of the energy going into kinetic energy.

Figure 4 shows the same eruptive process for a 3D simulation (flux cancellation begins at t_0 in the calculation). In this case the same helmet streamer configuration is used, but flux is reduced only along one side of the Sun. In the 3D eruption, the ends of the flux rope are attached to the photosphere. The polarization brightness is shown in the topmost panels. The images show, albeit in a very idealized way, the three part structure often seen in CMEs (Illing and Hundhausen, 1985). Roussev *et al.* (2004) have also demonstrated a flux-cancellation initiated eruption for a more realistic 3D configuration; they used low-resolution synoptic maps as the basis for their configuration.

2.1.4. *Discussion and Future Developments*

The flux cancellation mechanism is an attractive hypothesis for explaining both prominence formation and the initiation of CMEs with associated prominence

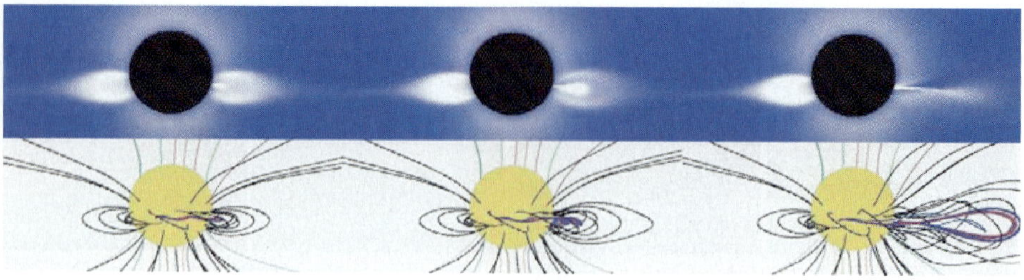

Figure 4. Polarization brightness (that would be observed by a coronagraph if this were a real CME) and magnetic field lines for an MHD simulation of a 3D eruption (see text). The black (yellow) disk in the top (bottom) frames shows the position of the Sun. The viewpoint is slightly above the equator, so the current sheet is not viewed edge on. Black and multi-colored field lines show the helmet streamer and open field lines. At $t = t_0$, flux cancellation begins; flux is canceled only on one hemisphere of the Sun. The blue and red field lines show the flux rope. The eruption in 3D is very similar to the 2D case, although the flux rope field lines are now line-tied to the Sun. (After Linker *et al.*, 2003.)

eruptions. The mechanism assumes that the frequent cancellation events that occur during the lifetime of an active region annihilate some of the surface magnetic flux and convert sheared fields in the active region into a flux rope. The flux rope, which supports the cool, dense material observed in prominences, can be stable for hours, days or weeks until cancellation increases the magnetic pressure in the flux rope to the point that it exceeds the surrounding tension, causing the violent eruption. The simulation results by Linker *et al.* (2003) indicate that the magnetic fields associated with the initial filament are a small fraction of the volume of the flux ropes observed in interplanetary space by in-situ spacecraft. This is because the resulting eruption ejects a portion of the overlying streamer belt. The flux cancellation mechanism avoids any problems with the Aly-Sturrock energy limit. The ideal process identified by Lin *et al.* (1998) does not violate the limit (the fields always remain closed); the reconnection that rapidly releases energy (Linker *et al.*, 2003) is a non-ideal process not accounted for by the theorem.

While the mechanism is certainly plausible, it is at this point far from being predictive. The amount of cancellation required to cross the threshold for eruption depends strongly on the details of the magnetic configuration, which are not easily deduced from presently available solar observations. To verify or rule out this mechanism, calculations need to be performed that start from a detailed model of an observed active region. Then one needs to study whether observed motions and flux changes can trigger an eruption. It is also important for analytic formulations of the mechanism to provide more insights into the loss of equilibrium process in 3D, particularly when line-tying is present. Modeling of the emergence of fields through the photosphere can help to clarify what subphotospheric processes occur that lead to the phenomena that we observe as flux cancellation.

When more detailed computations are available, new observations will undoubtedly be crucial for progress. Detailed sequences of vector magnetograms in filament channels and active regions, coupled with high-resolution X-ray and EUV imaging, can help to substantiate (or not) theoretical and computational models of the flux cancellation mechanism. Future missions such as *Solar-B*, STEREO, and SDO, as well as ground-based observations from SOLIS will provide these measurements. The confluence of more sophisticated models and improved observational capabilities over the next few years will likely provide the opportunity to verify or invalidate the role of flux cancellation as a trigger for CMEs.

2.2. The Breakout Model

2.2.1. *Physical Mechanism*

As discussed above, the most widely accepted models for CME/eruptive flares are those in which the energy for the eruption is stored in coronal magnetic fields, specifically, the strongly sheared/twisted field of a filament channel (see reviews by Forbes (2000), Klimchuk (2001), Low (2001), Wu *et al.* (2001) and Lin *et al.* (2003)). The basic picture is that a CME represents the catastrophic disruption of the force balance between the upward magnetic pressure of the highly-sheared filament-channel field and the downward tension of overlying quasi-potential field. Since the upward pressure force is constrained to increase only slowly, either by flux emergence or by photospheric motions, explosive eruption must be due to the fast decrease of the downward tension.

Three general types of reconnection models for CME initiation have been proposed, differing primarily in magnetic topology and in location of the reconnection. In this section we focus on the so-called breakout model, in which reconnection is postulated to occur external to the filament channel, between the quasi-potential overlying flux and neighboring flux systems (Antiochos *et al.*, 1999; Antiochos and DeVore, 1999). Consequently, an essential requirement for the breakout model is that the coronal magnetic topology is due to a multipolar flux distribution at the photosphere and that it contains at least one null point where reconnection can occur.

The basic mechanism is shown in Figure 5, which presents results from recent 2.5D (axisymmetric, with two spatial dimensions and three components of the vector fields) simulations by MacNeice *et al.* (2004). The initial field (first panel) is potential and contains four flux systems, with a coronal null "point" (latitudinal circle, in 2.5D). The photospheric flux distribution consists of four polarity regions separated by three neutral lines. For clarity, only field lines originating in the northern hemisphere are shown in the figure, but the system is symmetric about the equator. In order to produce an eruption, a filament channel must be added to this potential field. Observations suggest that the Sun creates filament channels through some not-yet understood process involving flux emergence, cancellation, and/or post-emergence subsurface motions (Martin *et al.*, 1984, 1994). In this simulation, the filament channel was created by simply imposing a slow, photospheric shear-flow localized at the equatorial neutral line. In fact, recent theoretical and observational (Welsch *et al.*, 2004) results appear to support such a flow. Furthermore,

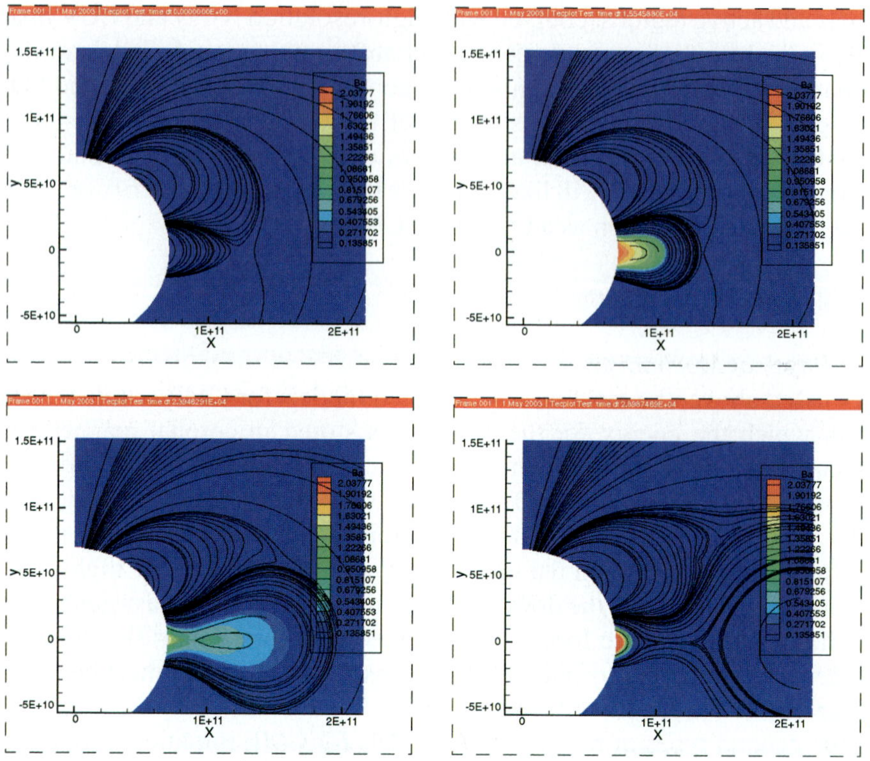

Figure 5. Selected field lines at four times during an axisymmetric simulation of the breakout model. The four panels correspond to times $t = 0$, 4.3, 6.6, and 8.1 hours. Also plotted are contours of B_ϕ, which show the location and magnitude of the applied shear.

the breakout model is expected to be insensitive to the details of the filament channel formation process. Unlike most other models, which require a particular form for the photospheric evolution and the filament field in order to obtain eruption, breakout should work for either flux emergence or cancellation and either a sheared arcade or a twisted flux rope.

The effect of the photospheric shear flow is to generate a large magnetic component parallel to the neutral line, (B_ϕ), which produces an upward magnetic pressure (second panel). This added pressure causes the overlying potential field lines to expand outward and increase their net downward tension, leading to the basic pre-eruption force balance described above. Another effect of the outward expansion is to stretch radially the field near the null so that the null region deforms into a current sheet structure (second and third panels of Figure 5). As long as the width of the current sheet is large compared to the grid scale of the simulation, the effective diffusion is negligible, and the system maintains a true stable equilibrium. It should be emphasized that such a stable energy buildup phase is necessary for all explosive eruption models.

As the shearing continues, however, the width of the current sheet structure at the null eventually decreases to the grid scale, and reconnection begins. Once reconnection appears, the outward expansion rate grows exponentially, even with no further shearing, because reconnection removes the overlying flux, thereby decreasing the downward tension. The tension decrease allows the sheared field to expand outward faster, which, in turn, drives faster reconnection. Energetically, the breakout mechanism can be considered to yield explosive eruption by minimizing the amount of flux that must open along with the sheared filament channel field.

For this simulation, the trigger for the eruption is the turn-on of numerical resistivity due to its grid dependence. It is likely that a scale-dependent resistivity due to collisionless effects or current-driven instabilities also operates in the Sun's corona. The important point for the model, however, is not that the breakout reconnection has a rapid turn-on, but that once it is on, the subsequent global evolution of the system causes the current sheet width to decrease and to drive the reconnection at an ever-faster rate.

One consequence of the eruption is that originally low-lying field lines become so expanded radially that they begin to approach the open state and, consequently, a vertical current sheet forms deep inside the sheared field region leading to reconnection there (last panel of Figure 2.1). It should be emphasized that this reconnection is completely distinct from the breakout reconnection ahead of the eruption. The reconnection that begins deep inside the erupting field corresponds to the usual flare reconnection and is common to nearly all CME models. In the breakout model, the flare reconnection does not initiate the eruption, but it may help accelerate it. Furthermore, the twisted flux rope that forms due to this flare reconnection (last panel) is a consequence of eruption, not its cause.

2.2.2. *Relation to Observations*

The major distinguishing predictions of the model are the breakout reconnection itself and the required multipolarity of the pre-eruption topology. A number of recent studies have identified observed CMEs/eruptive-filament events that appear to be explained by the breakout model (Aulanier *et al.*, 2000; Sterling and Moore, 2001, 2004; Sterling *et al.*, 2001; Wang *et al.*, 2003; Manoharan and Kundu, 2003; Subramanian *et al.*, 2003; Gary and Moore, 2004), but these studies are mainly qualitative in nature. They present evidence for particle acceleration, or for heating attributed to reconnection that is far removed from the flare ribbons, and/or for a multipolar magnetic topology in the corona. Perhaps the best example is that of Aulanier *et al.* (2000), who present compelling evidence for reconnection at a coronal null point prior to and during the July 14, 1998 flare. There has also been some quantitative analysis of the 2.5D simulation results for comparison with generic coronagraph and in-situ observations. Lynch *et al.* (2004) find that the breakout model reproduces the three-part structure commonly seen in coronagraph images, and the linear force-free flux rope magnetic structure that is often measured in ICMEs at 1 AU. At present, however, it is fair to state that there has not been a rigorous quantitative test of the model against a real event. The reasons are that all actual events are fully 3D and generally have a complex topology. Only now are 3D simulations of breakout being undertaken, and only for idealized topologies.

Although a definitive, quantitative comparison between the breakout model and data is not yet available, there are several aspects of the model that make it unique in being able to explain the observations. First, it has long been known that solar eruptions occur predominately in magnetically complex regions. In fact, the magnetic complexity of an active region is often used as a predictor of whether it will produce major flares. The breakout model naturally explains why complexity leads to eruption. Second, it is also well known that eruptions can occur under all kinds of photospheric conditions, from strong active regions to very quiet high-latitude filaments, and during flux emergence, or flux cancellation, or during periods of apparently fixed flux. Unlike other models which require a particular form for the photospheric evolution, the breakout mechanism requires only that the filament channel builds up free energy and, hence, the model is compatible with a wide range of observed photospheric conditions. Finally, the height at which the eruption begins exhibits a great deal of variability. In some events the eruption extends far down into the chromosphere (as determined by the locations of the flare ribbons), but in others only the corona well above a visible filament/prominence erupts. Again it is difficult to reconcile these observations with many of the models; but since breakout places no special requirement on the filament channel field, it is completely flexible as to how much of the field erupts in any one event.

On the other hand, it should be emphasized that the model has no flexibility with respect to magnetic complexity. Breakout cannot operate in a truly bipolar field. Furthermore, even if the field is complex, breakout requires the occurrence of external reconnection that transfers a substantial amount of flux from the overlying erupting filament channel. It is far from evident that such a flux transfer is present in all eruptions.

2.2.3. *Future Developments*

The necessary developments of the model are straightforward. First, the model must be extended to fully 3D topologies. Some success has already been demonstrated in 3D (Antiochos and DeVore, 1999), but this was for a topology containing a separator line in the corona and unlikely to be common on the real Sun. The extension of the 2.5D axisymmetric spherical models to a fully 3D geometry is clearly required. Furthermore, this type of work needs to be performed by more than one group, because the results of complex 3D simulations require verification by multiple codes.

Second, the sensitivity of the model to the assumed form for the resistivity needs to be determined. As discussed above, the resistivity must have a switch-on nature so that the breakout reconnection does not start too early and so that once the eruption is underway, the reconnection can keep pace. All the simulations to date have been performed with numerical resistivity, which has a strong scale dependence. Calculations need to be undertaken with different resistivity models and with different values for the resistivity in order to confirm that the model can yield fast eruptions even for large magnetic Reynolds number. Some of this work has already been performed in 2.5D (MacNeice *et al.*, 2004), but the 3D case is completely unexplored.

If these two developments are successful, the next step would be to test the model against real solar events by using observed photospheric magnetic and velocity fields as input and comparing the resulting eruption with coronal observations. Of course, these will be challenging simulations even with an adaptive mesh refinement code; but the codes and the hardware are now in hand. We expect that the next few years will prove to be decisive for determining the validity of not only the breakout model but all the present models for CME initiation.

2.3. Flux Injection Model

The theoretical model of Chen (1989, 1996) hypothesizes that the underlying magnetic field of a CME is that of a three-dimensional magnetic flux rope. The initial flux rope is assumed to be in MHD equilibrium, and the eruption occurs in response to the "injection" of poloidal magnetic flux Φ_p into the flux rope. The model then uses an approximate form of the MHD equations to calculate the evolution of the flux rope. No explicit prescription of the magnetic field is provided by this model, although it does include the effects of inertial line-tying at the solar surface. It also assumes that reconnection occurs so readily that any current sheets which form during the evolution have a negligible effect on the dynamics of the flux rope.

Since the launch of SOHO, the model has been extensively tested against EIT and LASCO data. It was shown by direct comparisons with data that the model solutions closely match the observed height-time profiles of CMEs within the LASCO field of view (Chen *et al.*, 1997, 2000; Wood *et al.*, 1999; Krall *et al.*, 2001). In particular, by analyzing the dynamics and morphology of 11 LASCO CMEs Krall *et al.* (2001) showed that flux-rope CMEs constitute a significant identifiable class. A prediction of the model is that the aspect ratio of a CME, defined as the ratio of the leading-edge height and the transverse width, is nearly constant except during the initial acceleration phase. This constancy of the height-to-width ratio at large distances from the sun is a general property of flux rope models, and it follows from the conservation of magnetic flux (Kumar and Rust, 1996). Significantly, this property, representing the dynamics in two orthogonal directions, was verified in these LASCO CMEs. In addition, synthetic coronagraph images (Chen *et al.*, 2000) show that the 2-D projections of 3-D flux ropes exhibit the generic morphological features of the prototypical 3-part CMEs (Illing and Hundhausen, 1986). It was also found that a model CME evolves into an interplanetary flux rope that resembles observed magnetic clouds at 1 AU and beyond (Chen, 1996).

The above studies show that the flux rope hypothesis is quantitatively consistent with observed CMEs. However, the coronal magnetic field is not directly measurable at this time. Thus, our understanding of magnetic structure, which is based on the morphology of projected density features, depends on various assumptions. The existing CME models assume as initial structures either magnetic arcades (*e.g.*, Forbes and Priest, 1995; Linker and Mikić, 1995; Antiochos *et al.*, 1999; Chen and Shibata 2000; Cheng *et al.*, 2003) or flux ropes (*e.g.*, Chen, 1989, 1996; Wu *et al.*, 1999; Amari *et al.*, 2000; Roussev *et al.*, 2003, 2004). Both model constructs lead to flux ropes after the eruption. In the former scenario, the initial arcade evolves into a flux rope via macroscopic reconnection. The key question is at what stage of the eruption the flux rope is formed. In this respect, Chen and Krall (2003) showed that if the initial structure is a flux rope or becomes a flux rope before the main acceleration phase, the acceleration peaks at a critical height Z_* given by $Z_* = S_f/2$, where Z refers to the height of the centroid of the apex from the solar surface and S_f is the distance between the two stationary footpoints at the base of the corona. This scaling law is universal in that it depends only on the 3-D toroidal flux rope geometry with constant S_f regardless of the speed of eruption. For typical CMEs, S_f is of the order of 1/2–1 R_s so that the acceleration maximum occurs below 2–3 R_s, while CMEs associated with polar crown prominence having maximum acceleration in the 3–4 R_s height range. This was shown to be true for a number of CMEs well-observed by LASCO, implying that the initial structures underlying these CMEs were flux ropes before the onset of the main acceleration phase. The theoretical result is also supported by an MHD simulation of a toroidal flux rope with fixed footpoints (Roussev *et al.*, 2003, 2004).

The basic physics embodied in the theory of Chen (1989) is that the flux rope motion is determined by the Lorentz force, pressure gradient, drag on the ambient coronal plasma, and gravity. The main driver is a specific form of Lorentz force, the so-called hoop force acting on curved current-carrying plasma (Shafranov, 1966; Anzer, 1978). This force is proportional to κ^2, where $\kappa = 1/R$ is the major radial curvature. The initial flux rope is set into motion by increasing the poloidal flux Φ_p of the system. In the model of Chen (1989, 1996), this is interpreted as injection of magnetic energy from subphotospheric sources ultimately determined by the solar dynamo at the base of the convection zone. However, the model by itself does not require a subphotospheric source. It is equally possible to interpret the mathematical function $\Phi_p(t)$ as derived from magnetic energy stored in the corona. For example, if the initial structure is an arcade that evolves into a flux rope via reconnection, a flux rope is created in this process so that its poloidal flux Φ_p increases from zero, or some initial value, to the final value when the flux rope is fully formed. In such a model, the increase in the function $\Phi_p(t)$ can be interpreted as energy injected into the flux rope via macroscopic reconnection. The ensuing flux motion is determined by how the flux rope is formed, *i.e.*, by the reconnection process.

3. CME Evolution and Propagation

J. SCHMIDT, G. L SISCOE

3.1. CME DYNAMICS

3.1.1. *Statement of Problem*

After an abrupt formation in the solar corona, a CME propagates into and through the interplanetary medium as an ICME, and then by assimilation into merged interaction regions in the outer heliosphere loses its identity. To describe this process quantitatively defines the CME-propagation problem, which is the subject of this section. Two approaches have been pursued, analytical formulation and MHD simulation. In the analytical approach the problem is to specify the equations that describe the motion through the spatially varying solar wind of a compressible (ICME) body subject to acceleration, deceleration, and deformation forces. Ordinary differential equations determine the position of the ICME and its geometry as a function of time. In MHD simulations, by contrast, the motion field and the force field are specified at every point of a simulation grid (and not at the center of mass of an ICME as in the analytic formulation), which requires partial differential equations (actually finite-difference representations of them). A CME and its interplanetary trajectory can be traced by following a closed contour of a scalar field such as normalized density.

This section focuses mainly on the analytical approach because, unlike simulations, it treats a CME as a distinct object and, thus, it explicitly describes the forces that propel, expand, and deform the object. We use a parameterized equation of motion that captures a CME's rapid acceleration from rest followed by slow deceleration. Comparing the results of different choices of parameters against observations favors some choices over others. The choices entail specifying the initial parameters of the CME, the nature of the force that propels a CME from the Sun, the nature of the drag force that couples it to the solar wind, and the aspect of "virtual mass" (explained below). We begin by reviewing selected observations that bear on discriminating between choices and end by comparing results of the analytical approach against those of an MHD simulation.

3.1.2. *Observations to Test Models*

The relevant observations are the initial acceleration of CMEs near the Sun, the subsequent acceleration or deceleration between the Sun and Earth, the size and rate of expansion of the ICME at 1 AU, the shape of the cross section of the ICME at 1 AU, and the typical values of the magnetic field strength and mass density within the ICME at 1 AU. Regarding CME acceleration near the Sun, in a statistical analysis of 24 CMEs, Zhang (2005) found typical values for acceleration and duration of acceleration to be respectively 200 m/s^2 and 40 min. These numbers give 0.82 Rs as a typical acceleration distance, whereupon the CME has reached a speed of 480 km/s. But the range in these numbers between events was considerable; for example, the acceleration ranged from a few m/s^2 to nearly 1000 m/s^2.

In a separate analysis of 28 CMEs, Gopalswamy *et al.* (2000) found speeds of CMEs near the Sun ranging from 124 km/s and 1,056 km/s. In other cases, speeds as high as 2500 km/s have been observed (Gopalswamy, private communication, 2004). These speeds were derived by measuring the plane-of-the-sky speed of the fastest CME feature as seen by the LASCO coronagraphs. Thus, these speed values may differ substantially from the radial CME speed, depending upon which way the CME was headed (see the discussion in Schwenn *et al.*, 2006, this volume). This uncertainty, however, does not detract from using the speeds to illustrate model forces. Between the Sun and Earth, fast CMEs decelerate and slow CMEs accelerate so that they arrive at 1 AU with speeds closer to that of the ambient solar wind. The same Gopalswamy *et al.* (2000) analysis correlated plane-of-the-sky CME speeds (u) as measured in coronagraph images with the speed of the resulting ICME at 1 AU to derive the following empirical formula for their in-transit acceleration a (deceleration being negative acceleration)

$$a[\mathrm{m/s^2}] = 1.41 - 0.0035\, u[\mathrm{km/s}] \tag{1}$$

We will refer to velocity-versus-distance profiles based on this formula as the Gopalswamy *et al.*, template against which we will compare predictions of the equation of motion to be discussed in the next section. (See Section 3.2 of Forsyth *et al.*, 2006, this volume, for more discussion of CME speeds.)

Support for the validity of the Gopalswamy *et al.*, template comes from Reiner *et al.* (2003), who used simultaneous radio and white light observations to document the deceleration of fast CMEs between the Sun and Earth. Their observations are consistent with a constant deceleration (such as is assumed in the construction of the Gopalswamy *et al.*, template) and inconsistent with a deceleration that decreases with distance from the Sun (a result that in the next section we shall use as a discriminator between model options).

The next observational discriminator listed above is the size and expansion speed of an ICME at 1 AU. The radial (from the Sun) dimension of an ICME at 1 AU is typically 0.2 to 0.25 AU (Klein and Burlaga, 1982; Hu and Sonnerup, 2002). Regarding the expansion speed, in an analysis of 37 ICMEs with clearly marked boundaries, Owens *et al.* (2005) find the following empirical relation between the rate at which the ICME radius is increasing (V_{EXP}) and the speed of the leading edge of the ICME (V_{LE}):

$$V_{\mathrm{EXP}}[\mathrm{km/s}] = 0.266\, V_{\mathrm{LE}}[\mathrm{km/s}] - 70.61 \tag{2}$$

For example, if $V_{\mathrm{LE}} = 500$ km/s, $V_{\mathrm{EXP}} = 62$ km/s, which means that at 1 AU the radial thickness of the ICME is growing at the rate of 124 km/s.

The next discriminator in the list is the cross-sectional shape of an ICME at 1 AU. The cross section of a CME as seen in a coronagraph near the Sun is roughly circular; thus, a zero-order expectation is that an ICME at 1 AU might be roughly circular, too. But the forces that act on an ICME in the direction of its motion are different than those acting perpendicular to its motion, so it is useful to consider an elliptical cross section as a first-order departure from circularity. Indirect methods suggest that the ratio of the major to minor axes of the cross section of an ICME at 1 AU might typically be less than 2 (based on a novel fitting procedure using the Grad-Shafranov equation, Hu and Sonnerup, 2002), whereas other analyses find a typical value closer to 4 [based on fitting multispacecraft observations (Mulligan and Russell, 2001) and on the ratio of the shock standoff distance to the ICME radial thickness (Russell and Mulligan, 2002)]. The ratio of major to minor ICME axes enters importantly in an analytical model in determining the values of the magnetic field strength and mass density within an ICME at 1 AU. Observationally, based on an average over 19 ICMEs, these numbers are $\sim$13 nT for field strength and $\sim$11 protons/cm^3 for number density (Lepping *et al.*, 2003), which completes the survey of relevant observations.

3.1.3. *An Analytical Model*

To repeat the main point, our goal is to give an equation of CME motion that captures the essential features of rapid acceleration from rest followed by slow deceleration, then to use it to show how different choices among its terms affect a CME's motion. The paradigm to follow for this Sun-to-Earth analytic approach is the treatments of van Tend and Kuperus (1978), van Ballegooijen and Martens (1989), Forbes and Isenberg (1991), and Chen (1996, 1997), which established the basic idea that a coronal magnetic flux rope anchored in the solar photosphere and in force-balance equilibrium can be destabilized by adding sufficient magnetic flux circulating around the central axis of the flux rope. Once destabilized, the flux tube expands in cross section and moves away from the Sun. The models mentioned above are more comprehensive than is needed here, since we are not concerned with the approach to destabilization (only what happens afterwards), and consequently we gain considerable simplicity by letting the flux tube encircle the Sun as a torus instead of anchoring of it in the photosphere. The approach is similar to those of Anzer (1978) and Kumar and Rust (1996). Under excess internal magnetic pressure the flux tube expands, and the force of expansion is balanced against the inertial reaction of the medium into which it is expanding. Under excess ambient pressure (particle plus magnetic) pushing up on its lower surface over that pushing down on its upper surface, the flux tube accelerates away from the Sun against the force of gravity and aerodynamic drag. This bottom-to-top ambient pressure excess minus the force of gravity can be thought of as a generalized buoyancy in the sense that it includes the magnetic pressure besides the usual hydrostatic pressure of the atmosphere. Thus, in words, the equations that describe the expansion and propagation of the flux tube are

Expansion :

$$\text{(Ambient Mass Density)} \times \text{(Rate of Expansion)}^2$$
$$= \text{Pressure Inside} - \text{Pressure Outside} \qquad (3)$$

Acceleration :

$$\text{(Mass of CME + Virtual Mass)} \times \text{Acceleration} = \text{Force of Gravity}$$
$$+ \text{Outside Magnetic \& Particle Pressure on Lower Surface Area}$$
$$- \text{Same on Upper Surface Area} - \text{Drag Term} \qquad (4)$$

The two equations are coupled since the "Pressure Outside" term in the first equation changes as the flux tube moves through the ambient medium as governed by the second equation, and the bottom-to-top pressure differences that propel the CME in the second equation are determined by the expansion of the flux tube as governed by the first equation. "Virtual Mass" is a concept from hydrodynamics that allows one to express, by an appropriate increase in the mass of the body, the force needed to move the ambient medium out of the way as the body accelerates. For a cylindrical body, such as our ICME torus approximates, the virtual mass is the volume of the cylinder times the mass density of the ambient medium (*e.g.*, Le Méhauté, 1976, p. 177).

Although the physics behind the coupled expansion and propagation equations is simple, replacing the word terms with mathematical terms entails choices among various options in formulation. Most of these choices are fairly pedestrian, but the drag term and the ratio of major-to-minor axes of the ICME cross-section merit description.

3.1.4. *The Drag Term*

As mentioned in Section 3.1.2, Reiner *et al.* (2003) found that radio observations of the propagation of fast CMEs away from the Sun are inconsistent with drag terms that cause deceleration to decrease with distance from the Sun. The pertinent distinction is between a velocity-versus-distance profile that is concave downward (the Reiner *et al.* template) and one that is concave upward (after the peak in velocity near the Sun). Reiner *et al.* then point out that the standard form of the drag term gives profiles that are incorrectly concave upwards, which leads to the important conclusion that the standard form of the drag term evidently does not apply to CME propagation. The standard drag term has the form $C_D A\rho\,|V_{\mathrm{CME}} - V_{\mathrm{SW}}|\,(V_{\mathrm{CME}} - V_{\mathrm{SW}})$, where C_D is the drag coefficient (typically around unity), A is the cross-sectional area of the body, and $V_{\mathrm{CME}} - V_{\mathrm{SW}}$ is the relative velocity between the body and the medium though which it moves, in this case the CME moving through the solar wind. Vršnak and Gopalswamy (2002) show that the disagreement just mentioned between the Gopalswamy *et al.*/Reiner *et al.* templates and the standard drag term occurs also if the drag term is linear in the velocity difference instead of quadratic. In either case, the primary reason that velocity-versus-distance profiles are concave upward after the velocity peak is that near the Sun the density, ρ, is large and the velocity difference, $V_{CME} - V_{SW}$, is also large, whereas these terms are small far from the Sun. That is, drag is strong near the Sun and weak farther out. The problem might go away, therefore, if for some reason the drag coefficient, C_D, should be compensatingly small near the Sun. Cargill *et al.* (1996) suggested a reason: C_D might be small near the Sun because the magnetic pressure greatly exceeds plasma pressure in this region. The idea is that a body moving fast through an ordinary fluid experiences drag as a result of the flow separating from the flanks of the body, leaving a low-pressure wake behind and so, consequently, a strong braking, fore-to-aft pressure difference. This is the situation to which the standard form of the drag term applies. By contrast when a strong magnetic field drapes fore-to-aft over the body as it moves, the field could force the plasma to flow all the way around the body without separating from it, thus leaving no low-pressure wake, which translates into a small drag coefficient.

Figure 6 illustrates the difference in velocity-versus-distance profiles that the two assumptions about C_D (variable or fixed) give. The three thin and (more or less) straight lines slanting downward to the right show examples of the Gopalswamy *et al.* template. They were obtained from equation (1) with initial velocities at 15 R_s of 500, 1000, and 1500 km/s. The thick lines show realizations of the word equation (3) in which initial parameters were chosen to give a peak speed near 1000 km/s. (For a discussion of a specification of initial CME and ambient coronal and solar wind conditions, see Chen (1996, 1997)). For the fixed C_D case (dashed line) the drag coefficient was held fixed at unity; for the variable C_D case (solid line) it was set equal to tanh β, where β is the ratio of plasma pressure to magnetic pressure in

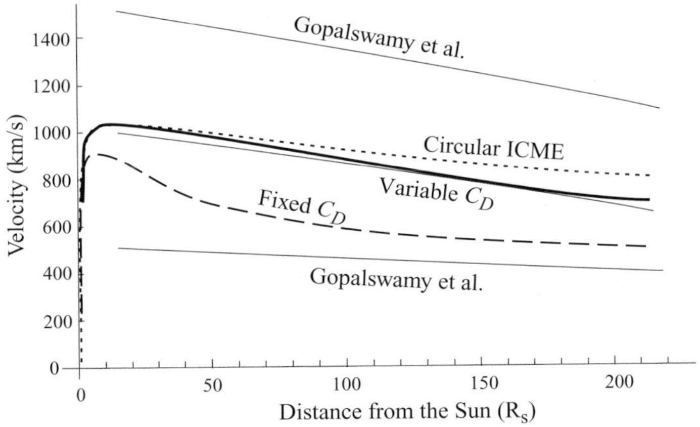

Figure 6. Velocity-versus-distance profiles showing three examples of the Gopalswamy *et al.* template from Equation (1) and three realizations of Equation (3) which differ from one another as labeled.

the ambient solar wind. tanh β is small near the Sun where the magnetic pressure greatly exceeds the plasma pressure, and it approaches unity away from the Sun, where the two pressures are comparable. (Other particulars of the realization of (3) that produced the curves in Figure 6 are given below.) The result demonstrates that the Cargill *et al.* suggestion of a small drag coefficient near the Sun indeed eliminates the concave upward shape of the curve seen in the fixed C_D case. Moreover, the variable C_D case follows the Gopalswamy *et al.* template (almost hidden by the thick line) reasonably well.

3.1.5. *Ratio of Major-to-Minor Axes and Other Values at 1 AU*

In Figure 6 the dotted line labeled "Circular ICME" illustrates by comparison with the "Variable C_D" curve two choices between the ratio of major to minor axes of the ICME cross section. It shows the case of no distortion of the initial circular cross section as the ICME moves outward; that is, the axis ratio is set to unity, although the radius increases according to Equation (3). The "Variable C_D" curve shows the case in which the axis in the direction of motion is determined by Equation (3) but the axis perpendicular to the direction of motion expands kinematically so that the angle subtended remains constant (Riley and Crooker, 2004). The result is a flattened, "pancake" shape, with the perpendicular axis 2.5 times longer than the axis in the direction of motion. Since the "Variable C_D" curve fits the Gopalswamy *et al.* template better than the "Circular ICME" curve, the comparison favors a ratio that is significantly bigger than unity. But it must be said that this result is not definitive since the "Circular ICME" case can be made to fit the Gopalswamy *et al.* template by doubling the drag term (2 tanh β); but then one needs an explanation for a larger-than-expected drag far from the Sun. Further support for a significantly-bigger-than-unity ratio of axes comes from comparing the mass density at 1 AU computed for the "Variable C_D" case (17 cm^{-3}) against the average value $\sim$11 cm^{-3}. This is based on an initial radius of $0.1R_s$ and density of 3.5×10^8 cm^{-3}. The value for the "Circular ICME" case is 43 cm^{-3}, which is too large. But this discrimination against a circular cross section is also somewhat soft because it could be changed by choosing a different initial density. Based on an initial magnetic field strength of 5 G, the computed value at 1 AU for both circular and elliptical cross section cases is 12 nT, which is essentially the same as the observed average value. The computed ICME speed at 1 AU is 677 km/s, and the computed expansion speed is 63 km/s, which is somewhat less than the 108 km/s value that the Owens *et al.* formula gives for a leading edge velocity of 677 km/s. Nonetheless, the computed front-to-back dimension of the ICME at 1 AU in the direction of motion is 0.25 AU, which compares favorably with observed typical values between 0.2 and 0.25 AU.

3.1.6. *Acceleration and Virtual Mass*

Recall that observed accelerations of CMEs in the inner corona vary from a few m/s^2 up to $\sim$1000 m/s^2. Figure 7 shows the computed acceleration in the inner corona using the same realizations of Equation (3) that generated the curves in Figure 6.

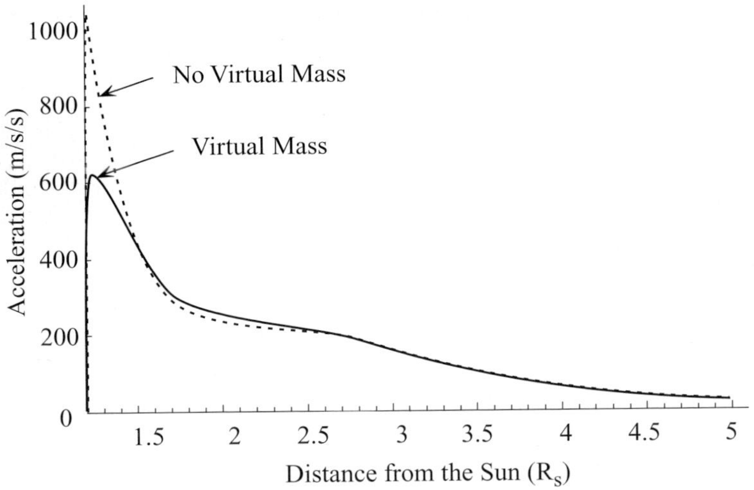

Figure 7. Computed accelerations in the inner corona for the cases of virtual mass and no virtual mass using the equations that generated the "Variable C_D" curve in Figure 6.

The figure illustrates that the simple generalized-buoyancy-versus-gravity-and-drag model that Equation (3) expresses can produce accelerations in the observed range. It also shows that adding virtual mass to the acceleration equation brings the peak acceleration into the observed range, but that after the initial peak, virtual mass has little effect on acceleration. This is because beyond about 1.5 R_s, ambient density, which makes up virtual mass, is relatively small. The initial peak in acceleration is perhaps non-physical since it results from an imposed pressure imbalance between the CME and the ambient medium. Thus, the value of about 200 m/s^2 that persists after the peak is the more appropriate value to use, and it happens to be the observed typical value. Figure 7 also demonstrates that generalized buoyancy is a strong force.

3.1.7. *Comparison with MHD Simulation*

Figure 8a shows a comparison between velocity-versus-distance profiles (out to 50 R_s) obtained from an MHD simulation of a CME (Riley *et al.*, 2003) and an analytic computation based on Equation (3) with initial parameters chosen to match the simulation. The two curves have the same general shapes. Differences in the ambient solar wind speeds cause the MHD simulation velocity to continue to rise at 50 R_s, whereas the analytical velocity remains flat. More importantly the maximum acceleration in the simulation ($\approx$32 m/s^2) is less than that in the analytical case ($\approx$200 m/s^2). Despite these differences, the comparison indicates that the same basic transport physics (expansion, generalized buoyancy, gravity, and variable drag) might operate in both cases.

Figure 8b shows that the MHD simulation generates an ICME cross section at 1 AU that is about twice as flat as that used in the analytical case. The central dimple

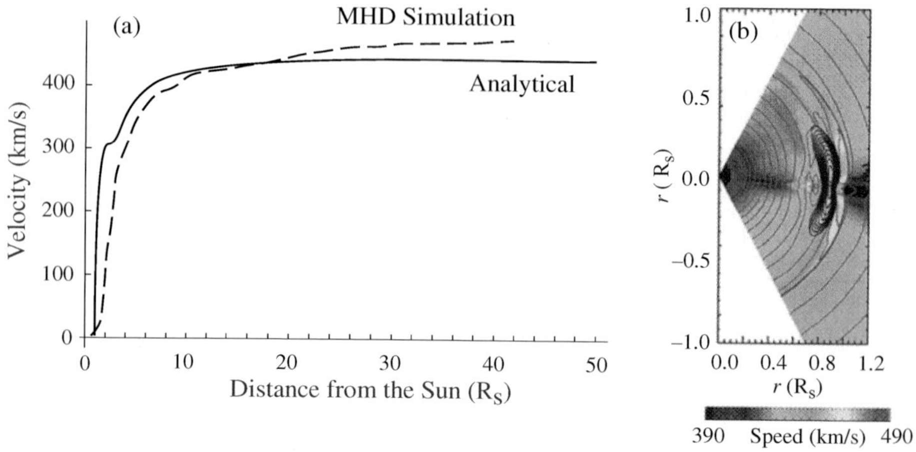

Figure 8. A comparison between analytical results and an MHD simulation of a CME. Panel (a) is the velocity-versus-distance, and panel (b) is the ICME shape at 1 AU.

in the ICME shape results from a higher equatorial density in the ambient solar wind, which might also account for its smaller equatorial thickness (~0.14 AU) compared to the analytical case (0.24 AU).

3.1.8. *Conclusions*

The transport properties of fast, impulsive CMEs can be understood (as modeled analytically) by expansion under magnetic over-pressure leading to expulsion under generalized buoyancy, which can be much stronger than gravity. Drag appears to be weak near the Sun (possibly because of magnetic suppression of a low-pressure wake) but more-or-less normal farther out. Comparison between analytical results and an MHD simulation shows qualitative agreement on essential points.

3.2. Numerical MHD Modeling and Propagation

Because of their large-scale nature (Hundhausen, 1999), CMEs are amenable to treatment using MHD theory. Numerical models have been developed that are able to deduce their global features (see, *e.g.*, Detman *et al.*, 1991; Gosling *et al.*, 1998; Manchester *et al.*, 2004; Odstrcil and Pizzo, 1999a,b,c; Riley *et al.*, 1997; Vandas *et al.*, 1995, 1996; Wu *et al.*, 1995, 1997, 1999). One of these models (Cargill *et al.*, 1996, 2000) exploits the fact that an interplanetary CME can often be regarded, from a dynamic point of view, as being detached from its anchorage back at the Sun. Thus, as in the analytical model in Section 3.1, many aspects of CMEs can be modeled using an azimuthally-symmetric configuration corresponding to a toroidal magnetic flux rope with its axis of rotation running through the poles of the Sun.

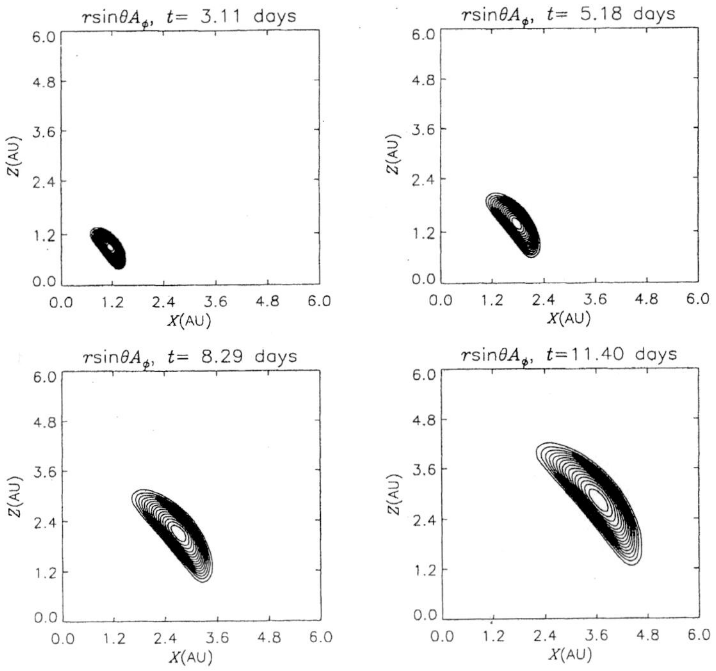

Figure 9. The magnetic field lines of the flux rope projected in the $r - \theta$ plane at four times during its progress from the Sun to 5 AU.

Here, we use the algorithms of Zalesak (1979) and DeVore (1991) to solve the MHD equations to an accuracy of fourth order in space and second order in time. The divergence of the magnetic field is kept at zero by using the vector potential to prescribe the magnetic field. The model is designed to investigate the evolution of magnetic flux ropes as they interact with the surrounding environment. Figure 9 shows the projection of the magnetic field lines of the flux rope in such a simulation. Contour lines of the vector potential are projected onto the x-z plane with x the in-ecliptic coordinate and z the coordinate in the direction of the rotational axis of the Sun. The flux rope has an internal overpressure initially, and as it propagates outwards, it loses its initial circular shape and becomes shell-like with a much larger extent in the meridional than in the radial direction. This is due to the smaller pressure gradient in the meridional direction.

Figure 10 shows the simulated plasma properties of the outward moving magnetic flux rope in a radial cut. The meridional component of the magnetic field B_θ shows a sine-like signal, whereas the azimuthal component of the magnetic field B_ϕ has a bell-like shape (with a molded hat). The latter corresponds to the winding of spiral field lines around the axis of the rope in the ϕ direction, where the height of a single winding increases when the spiral is closer to the axis. The radial velocity in the third panel also shows a sine-like signal, which corresponds to a

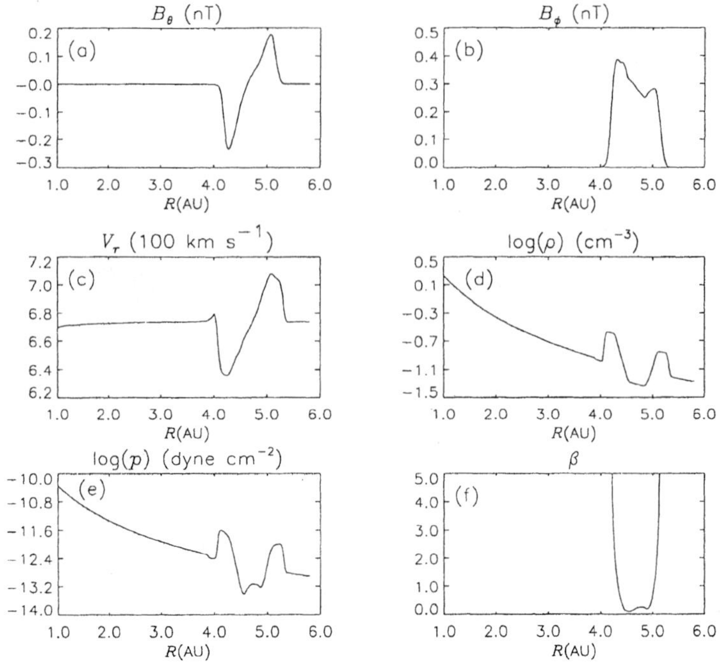

Figure 10. B_θ, B_ϕ, the radial velocity, plasma density and pressure, and plasma β at 11.4 days. Only the outer 200 cells of the simulation are shown.

forward and a reverse shock pair. In the fourth panel we see a density-signal with two bumps at the shocks. The density between those bumps is lower than the density outside the bumps, which is due to the overexpansion. The pressure-signal in the fifth panel shows a similar shape. Finally, the sixth panel shows the plasma β. The region within the CME is a region with low β, and so it is dominated by magnetic forces.

The interaction of CMEs can be studied with MHD simulations. During the active part of the solar cycle there are, on average, about four CMEs per day, so interactions between CMEs may be relatively common. A well-observed example of an interaction of this type occurred on January 10, 2000. Between 16:30 and 19:30 UT, the Radio and Plasma Wave Experiment (WAVES) on the Wind spacecraft (Bougeret *et al.*, 1995) detected an extremely narrow-band radio type II burst, which was flanked by intense radio type III bursts. That event was associated with a slow, dense CME being rammed from behind by a faster, less-dense CME, as observed by LASCO (Brueckner *et al.*, 1995) on SOHO (Gopalswamy *et al.*, 2001, 2002; see also Figure 16 in Forsyth *et al.* (2006), this volume). Since the radio signal outbursts are due to energetic electrons which may originate at shocks, it is useful to study the evolution and structure of the shocks that might form during the collision of the two CMEs. Figure 11 (Schmidt and Cargill, 2004) shows results for a

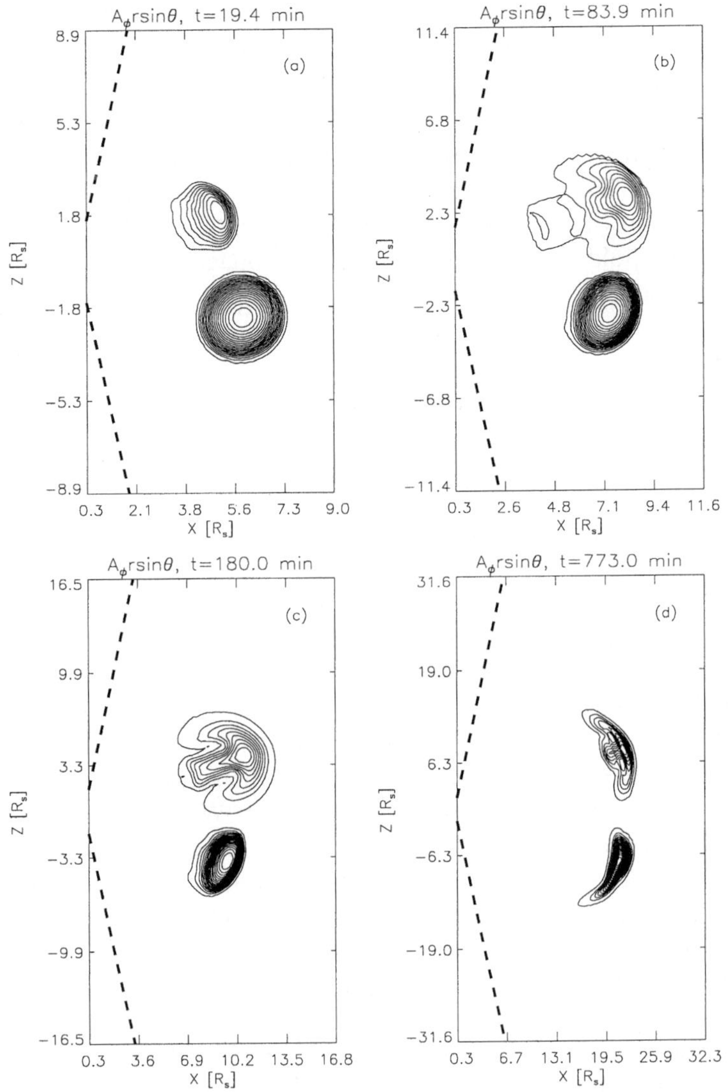

Figure 11. Contour plots of the vector potential at four times during the evolution of two colliding CMEs whose meridional angle is separated by 40° and that interact with each other by shock interaction (Adapted from Schmidt and Cargill, 2004).

simulation of the January 10, 2000 event. The simulation box, indicated by the thick dashed lines, comprises the field of view of the C2 and C3 coronagraphs, and the geometrical dimensions and velocities of the CMEs are taken from the coronagraph observations. After the fast, less-dense (upper) CME overtakes the slow, dense (lower) CME, the lower CME is deflected into the Southern hemisphere by its interaction with the forward shock in front of the fast CME. This shock impacts the slow CME, causing it to flatten at its northern edge. The shock penetrates into the denser material of the slow CME, steepening significantly because of the reduced Alfvén speed there. We find that the steep shock that is generated within the slow CME is quite persistent. It could possibly serve as a site for the diffusive acceleration of particles if there is an adequate seed population of particles nearby. Additional examples of CME interactions that can be studied with MHD simulations can be found in Section 4 of Forsyth *et al.* (2006), this volume. These are reconnection with the ambient field and distortion caused by propagation into solar wind regions with different speeds.

4. Flux Rope Modeling and Fitting

C. CID, P. RILEY

On average, about one third of ICMEs identified in the solar wind contain a rotation in the magnetic field vector (Gosling, 1990). They are commonly called magnetic clouds (MCs, for more on clouds, see Wimmer-Schweingruber *et al.*, 2006, this volume). Their features in the solar wind, strong magnetic field, large rotation of the field vector and a low proton temperature indicate that they have a very well-defined magnetic field topology.

Richardson and Cane (2004) have found that the fraction of ICMEs that are also magnetic clouds varies during the course of the solar cycle from 15% at solar maximum to as much as 100% at solar minimum. Why only some ICMEs contain – or can be identified as – magnetic clouds remains unknown. Perhaps it is an observational selection effect: Whether one observes the necessary signatures to classify the event as a magnetic cloud may depend sensitively on the trajectory of the spacecraft through the structure. On the other hand, it may represent an evolutionary phenomena: simpler magnetic clouds interacting with one another may produce more complex ICMEs that do not retain the classic field rotations and enhancements and/or temperature depressions. Finally, it may suggest distinct birth mechanisms: non-cloud ICMEs are produced by one process, while clouds are produced by another. Numerical simulations currently favor the "selection effect" interpretation (Riley *et al.*, 2003). Whatever the relationship between ICMEs and magnetic clouds turns out to be, clouds are simpler to describe, parameterize, and model. As such, it makes sense to focus our efforts on understanding them first.

Several theoretical models have been proposed for the topology of magnetic clouds, but the procedure to check theoretical predictions from these models with experimental data is complicated. Although magnetic clouds are three-dimensional objects, the magnetic field vector cannot be measured at any desired point inside this structure. Experimental data are limited to a line tracing the trajectory of a spacecraft through the cloud as it travels away from the Sun. In this scenario, the models have to take into account not only the magnetic field topology but also the position inside the magnetic cloud where those experimental data were obtained.

4.1. Approaches to the Problem

Magnetic field configurations for clouds were first proposed in the early 1980s. Burlaga *et al.* (1981) and Klein and Burlaga (1982) considered the possibility of magnetic field lines as a family of circles centered about the axis of the magnetic cloud. Suess (1988) proposed other pinch configurations, but all of them were incompatible with observations.

Goldstein (1983) considered magnetic clouds as force-free configurations, that is, the electric current is parallel to the magnetic field. Then, for static conditions the equation $\nabla \times \mathbf{B} = \alpha(r)\mathbf{B}$ is obtained. Several solutions to this equation were proposed. Burlaga (1988) considered $\alpha(r) = 1$, obtaining the cylindrically symmetric solution proposed by Lundquist (1950). With this solution, magnetic clouds are described as magnetic flux ropes, that is, cylindrical configurations with a two-component magnetic field, one along the axis of symmetry and another in the azimuthal direction. Lepping *et al.* (1990) fitted several examples of magnetic clouds. The procedure starts with a minimum variance analysis, which provides a starting value for the orientation of the axis. Then normalized magnetic field components are fitted with six free parameters related to size, boundaries, latitude and longitude of the axis, and the impact parameter (minimum distance between the spacecraft and the cloud's axis). Finally, the magnetic field strength is fitted, involving one more free-parameter, the magnetic field strength at the axis.

Experimental data show that the maximum of the magnetic field profile is often displaced towards the leading edge of the cloud. However, for the models described above, this maximum is located at the closest distance to the axis, that is, in the middle of the magnetic cloud time interval. More recently, other configurations have been proposed, such as spheromak solutions of the force-free equation (Ivanov and Kharshiladze, 1985; Vandas *et al.*, 1991, 1992, 1993) or toroidal solutions (Ivanov *et al.*, 1989; Romashets and Ivanov, 1991). Moreover, the static assumption was removed, and the effects of expansion and interaction with the ambient plasma were included (Osherovich *et al.*, 1993; Farrugia, 1995; Marubashi, 1997).

Pressure gradients have been observed inside MCs, indicating that they are not force-free structures. To address this, non force-free methods have been introduced. Mulligan and Russell (2001) proposed a model using exponential distributions for magnetic field strengths that can create both cylindrically symmetric and non-symmetric magnetic topologies. Ten parameters are used to describe a cylindrically symmetric flux rope. Seven are related solely to the magnetic field topology and the other three to the orientation of the symmetry axis and to the impact parameter. For the non-symmetric topologies, the number of free parameters was increased to eleven.

Another analytical approach to modeling non-force-free magnetic clouds is to assume a current density vector and then solve Maxwell's equations for the magnetic field vector. Hidalgo *et al.* (2000) considered a cylindrical geometry with the axial and poloidal components of the current density constant and a null radial component. An improved version of this model shows the first fitting of thermal pressure inside a MC (Cid *et al.*, 2002; Hidalgo *et al.*, 2002a). Having taken into account the interaction of the cloud with the ambient solar wind, Maxwell's equations are solved in elliptical coordinates (Hidalgo *et al.*, 2002b). Projections of magnetic field lines on a cross section perpendicular to the axis are then ellipses, centered on the cloud axis. The expansion of magnetic clouds has been included by Hidalgo *et al.*, (2003), involving nine free parameters, four of them related to the orientation of the cloud and the spacecraft path.

The procedure followed to fit experimental data to these cloud models is to calculate the minimum of the sum of the residuals between the data and the model: $\chi^2 = \sum(B^2_{x,\mathrm{exp}} - B^2_{x,\mathrm{th}} + B^2_{y,\mathrm{exp}} - B^2_{y,\mathrm{th}} + B^2_{z,\mathrm{exp}} - B^2_{z,\mathrm{th}})$ where $B_{i,\mathrm{exp}}$, $(i = x, y, z)$ are the Cartesian components of the magnetic field observations in the cloud interval and $B_{i,\mathrm{th}}$, $(i = x, y, z)$ are the components of the theoretical model rotated into the frame of the observations.

Magnetic cloud properties have also been determined using the Grad-Shafranov reconstruction technique (Hu and Sonnerup, 2001, 2002). This method assumes that the cloud is an asymmetric cylindrical structure in approximate magnetostatic equilibrium. The structure perpendicular to its axis can be recovered using the Grad-Shafranov equation. With this technique, the boundaries of the cloud need not first be identified in the data.

Within the constraints of the caveats noted above, flux rope fitting (FRF) techniques can be an invaluable tool for extracting information about the properties of magnetic clouds. However, it has proven difficult to assess their accuracy from single-spacecraft data. In contrast, large-scale MHD simulations of CME evolution can provide both a global view as well as a localized time series at specific points in space. Riley *et al.* (2004) applied 5 different fitting techniques to 2 hypothetical time series derived from MHD simulation results (Figure 12). Independent teams performed the analyses of the events in "blind tests," for which no information, other than the time series, was provided. From the results, they inferred the following: (1) Accuracy decreases markedly with increasingly glancing encounters; (2) Correct identification of the boundaries of the flux rope can be a significant limiter; and (3) Results from techniques that infer global morphology must be viewed with caution. In spite of these limitations, FRF techniques appear to be a useful tool for describing in-situ observations of flux rope CMEs.

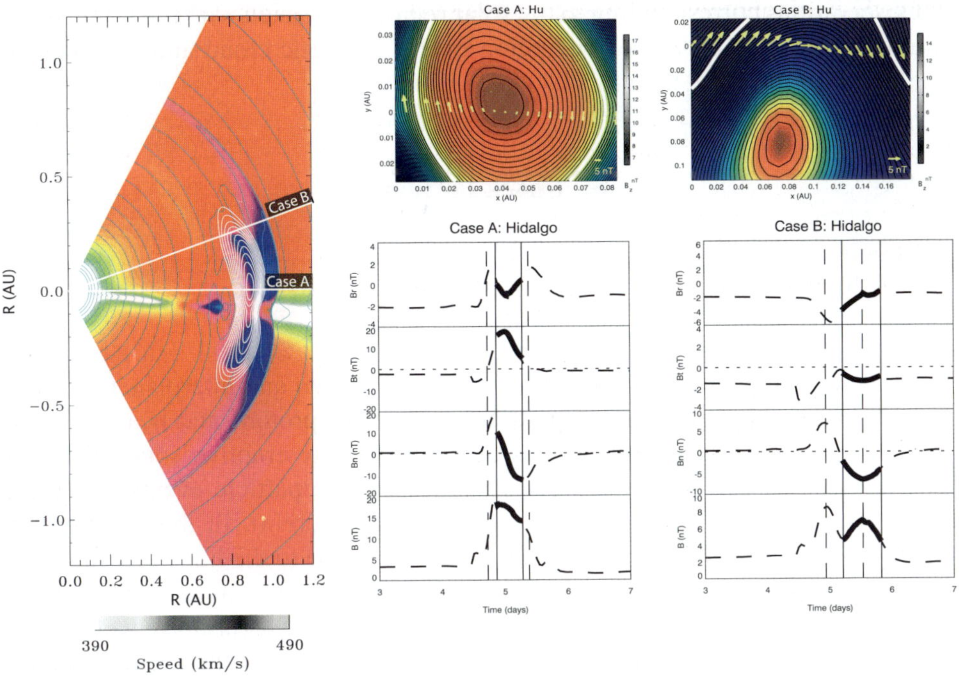

Figure 12. The left panel summarizes simulated arrival of flux rope CME to 1 AU. The two cases, labeled A and B were obtained by placing hypothetical spacecraft at these latitudes at a distance of 1 AU from the Sun. The fits obtained from the GSR technique are shown in the upper right and the fits obtained by the Current Density fitting technique of Hidalgo *et al.* are shown in the lower right.

4.2. Future Directions

A major unresolved issue with magnetic cloud models is that there is no independent way to assess the errors of the fit. One can and does, of course, compute deviations from the observed data with the derived profiles. In fact, minimizing this deviation defines the best fit to the data. But even a "reasonable" fit does not guarantee that the derived parameters and, moreover, the assumed field topology are "correct". The availability of data along at least two different lines inside the same magnetic cloud could help in this regard. At present, the models can reproduce the data measured at one spacecraft; however, the orientations inferred at other locations often do not appear to be consistent.

Using multipoint measurements of magnetic clouds requires extending the model to optimize magnetic cloud data from multiple spacecraft simultaneously. Noting that the theoretical models described above rely on a cylindrical approach, they could only be applied for observations for which the spacecraft are radially aligned. In this case, special care should be taken to obtain a global (non-local) magnetic field topology for the magnetic cloud.

Observations corresponding to the solar region associated with the CME should also be studied to verify the orientation of the magnetic cloud and the magnetic topology inferred from the model.

5. Shock Formation

G. MANN

5.1. OVERVIEW

Shock waves play an important role in the solar corona and interplanetary space, since they are able to accelerate particles (electrons, protons and heavy ions) up to high energies. Thus, shock waves accompanied by eruptive solar events are generally considered as one source of solar energetic particle events (SEP) (see Klecker *et al.*, 2006, this volume, and references therein). In the solar corona, shock waves can be produced either as blast waves due to the huge pressure pulse accompanying the flare and/or as piston driven shocks, *i.e.*, as a bow shock of a rising CME. It should be emphasized that a shock is a discontinuity well-defined in magnetohydrodynamics and, consequently, independent of the exciting agent. It represents a discontinuity with a transmitted mass flow, which is decelerated from a super-Alfvénic to a sub-Alfvénic speed. Thus, the shock is a dissipative structure, in which kinetic energy of a directed plasma flow is partly transferred to heating of the plasma. Since the necessary dissipation doesn't take place by means of particle collisions, these shocks are usually called collisionless shocks. The conservation laws lead to the well-known Rankine-Hugoniot relationships relating the quantities in the upstream region to those in the downstream region (Edmiston and Kennel, 1984). They result in an upper limit on the density and magnetic field jump across the shock, *i.e.*, $B_2/B_1 \leq N_2/N_1 \leq 4$, where B_2 and N_2 (B_1 and N_1) denote the magnitude of the magnetic field and the particle number density in the downstream (upstream) region.

5.2. SHOCK STRUCTURE

In-situ measurements by various spacecraft missions have revealed sub-structures of collisionless shocks in space plasmas. Generally, collisionless shocks can be divided into super- and sub-critical shocks as well as quasi-parallel and quasi-perpendicular shocks (see, *e.g.*, Kennel *et al.* (1985) and references therein). The critical fast Mach number M_f^* is defined by equating the normal component of the downstream flow velocity in the shock frame to the ordinary sound speed (Edmiston and Kennel, 1984). In contrast to sub-critical shocks ($M_f < M_f^*$; M_f, fast magnetosonic Mach number) resistivity in a super-critical shock ($M_f > M_f^*$) cannot provide all the dissipation necessary for a shock transition according to the Rankine-Hugoniot relationships. M_f^* is strongly dependent on the upstream plasma β and the angle $\theta_{B,n}$. For low plasma β, as usually found under coronal circumstances, M_f^* varies between 1.5 and 2.7. Its maximum is 2.76 for $\theta_{B,n} = 90°$ and vanishing plasma-β (Edmiston and Kennel, 1984). Thus, other processes like wave-particle interactions provide the dissipation required for fast magnetosonic, super-critical shock formation. This is the reason why such shocks are able to accelerate particles and are accompanied by large amplitude magnetic field fluctuations (Kennel *et al.*, 1985). If the angle $\theta_{B,n}$ between the upstream magnetic field and the shock normal is greater or smaller than 45°, the shock is called quasi-perpendicular or quasi-parallel, respectively. In the case of a quasi-perpendicular shock geometry, the shock transition has a typical scale length of a few ion inertial lengths, c/ω_{pi}, (corresponding to about 20 m in the corona at the altitude where the electron plasma frequency is 100 MHz and about 100 km in interplanetary space at 1 AU), where $\omega_{pi} = (4\pi e^2 N_p/m_p)^{1/2}$ is the proton plasma frequency, c is the speed of light, e is the elementary charge, N_p is the proton number density, and m_p is the proton mass. Immediately at the shock transition the magnetic field shows a so-called "overshoot," *i.e.*, the magnetic field is locally compressed to a maximum $B_{\max}$, which is higher than the downstream magnetic field B_2 ($<B_{\max}$) according to the Rankine-Hugoniot relationships. In the case of a quasi-parallel super-critical shock, its transition is extended up to a few hundreds of ion inertial lengths. It is accompanied by intermittent large-amplitude magnetic field fluctuations, the so-called SLAMS (Schwartz *et al.*, 1992; Mann *et al.*, 1994), in which the super-Alfvénic flow is partially decelerated.

5.3. Conditions Leading to Shock Formation

Generally, shocks are formed anytime a compressive wave is generated, since the nonlinear dynamics associated with the Reynolds stresses in the equation of motion cause such a wave to steepen with time. However, dissipation may damp the wave before a shock can form, so physically significant shocks tend to occur in regions where the velocity of the exciting agent exceeds the local fast magnetosonic speed v_f, which is related to the Alfvén speed v_A and sound speed c_s by $v_A \le v_f \le (v_A^2 + c_s^2)^{1/2}$. Consequently, the spatial behaviour of the Alfvén speed in the solar corona and interplanetary space is of interest, in order to evaluate where shock formation can take place. The origin of shock waves and their propagation through the corona and interplanetary space have been investigated by several authors (Gopalswamy *et al.*, 1997, 2000; Aurass *et al.*, 1998; Klassen *et al.*, 1999; Klein *et al.*, 1999) by evaluating dynamic radio spectra of type II radio bursts as signatures of shocks, radioheliographic observations, and X-ray images.

The Alfvén speed is defined by

$$v_A = \frac{B}{\sqrt{4\pi \mu \mathrm{m}_p N}}$$

with the magnetic field B and the full particle number density N. The symbol μ denotes the mean molecular weight, which has a value of 0.6 in the corona (Priest 1982). The electron number density N_e is related to the full particle number density N by $N = 1.92N_e$. Since v_A depends on the magnetic field B and the full particle number density N, a model of the magnetic field and the density in the corona and interplanetary space is needed for studying the radial behaviour of the Alfvén speed.

In the solar corona shock waves are established near but out of active regions (Aurass *et al.*, 1998, Klassen *et al.*, 1999; Klein *et al.* 1999; Gopalswamy *et al.*, 2000). Therefore, a one-fold Newkirk (1961) model

$$N_e(R) = N_0\, 10^{4.32 R_S/R}$$

($N_0 = 4.2 \times 10^4$ cm^{-3}; R_S, radius of the Sun) is chosen as an appropriate density model, since Koutchmy (1994) showed by white-light scattering observations that a this model fits well the conditions above quiet equatorial regions. Note that it corresponds to a barometric height formula with a temperature of 1.4×10^6 K, which is a typical value in the solar corona outside active regions. It is well known that the Newkirk model is a hydrostatic model and does not take into account the solar wind. To do so, Mann *et al.* (1999a) constructed a heliospheric density model as a special solution of Parker's (1958) wind equation. This model agrees very well with the observations from the corona out to 5 AU (see also Leblanc *et al.* (1994). Thus, the Newkirk model and the Mann *et al.* model are used for regions $R \leq 1.8R_S$ and $R \geq 1.8R_S$, respectively, with continuity at $R = 1.8R_S$. The radial behaviour of such a density model is presented in Figure 13.

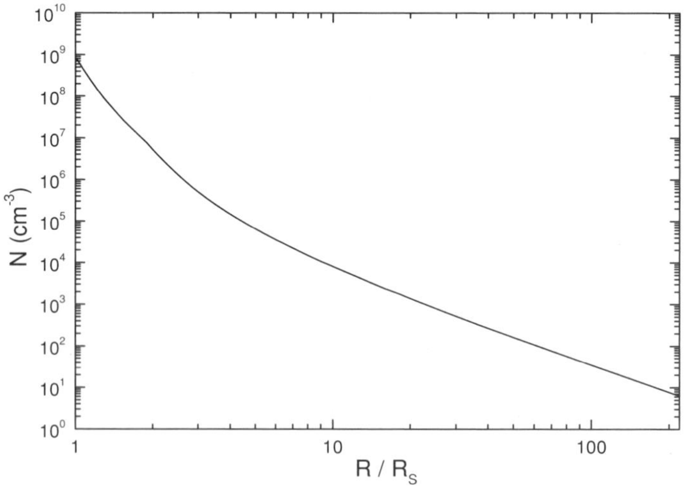

Figure 13. Radial behaviour of the electron number density according to the one-fold Newkirk (1961) model for the solar corona and the heliospheric density model by Mann *et al.* (1999a) for interplanetary space.

The model magnetic field $\mathbf{B}$ is composed of that of an active region $\mathbf{B}_{ar}$ and that of the quiet Sun $\mathbf{B}_{qS}$, *i.e.* $\mathbf{B} = \mathbf{B}_{ar} + \mathbf{B}_{qS}$. Here, an active region is modeled by a magnetic dipole with moment $\mathbf{M}$ and length λ. It is located at a depth $\lambda/2$ below the photosphere along the radial axis away from the center of the Sun. Then, the magnetic field of such a dipole is given by

$$\mathbf{B}_{ar} = \frac{3(\mathbf{M} \cdot \mathbf{r})\mathbf{r}}{r^5} - \frac{\mathbf{M}}{r^3} \tag{5}$$

where $\mathbf{r}$ denotes the distance from the center of the dipole. If B_0 is the magnitude of the magnetic field on the axis of the dipole, the magnetic moment is related to B_0 by $M = B_0\lambda^3/16$. Here, $B_0 = 0.8$ kG is taken as a typical value of the magnetic field in an active region (see Priest, 1982). The magnetic field of the quiet Sun is assumed to be radially directed so that its radial behaviour is given by

$$B_{qS} = B_S \left(\frac{R_S}{R}\right)^2 \tag{6}$$

This approach is appropriate outside the heliospheric current sheet (Banaszkiewicz *et al.*, 1998). The statistical analysis of coronal transient (or EIT) waves (Klassen *et al.*, 2000) provided a mean magnetic field of 2.2 G for the quiet Sun at the photosphere (Mann *et al.*, 1999b).

For this model of the Alfvén speed, the solar magnetic dipole defines a framework of cylindrical coordinates with an azimuthal symmetry, where the dipole is directed along the z-axis. The field magnitude, B, the full particle number density, N, and, subsequently, the Alfvén speed, v_A can be found at each point in the corona and interplanetary space. Figure 14 shows the radial behaviour of the Alfvén speed outward from the photosphere along a line inclined at 45° from the z-axis. The solid and dashed lines show cases in which the magnetic dipole is directed parallel (+ sign) and anti-parallel (− sign) with respect to that of the quiet Sun, respectively. For completeness, the dotted line represents the Alfvén speed due only to the quiet Sun field, *i.e.*, without any active region. Generally, a local maximum of the Alfvén speed of 740 km/s is found at 3.8 R_S. Furthermore, a local minimum of v_A is established in the middle of the corona. This minimum is strongly developed in the case of anti-parallel magnetic field orientation. A much more detailed discussion of this approach is given by Mann *et al.* (2003).

Of course, special values of the parameters (*e.g.*, B_0, B_S, and λ) have been used in this study. In reality they can change from case to case on the Sun. Nevertheless, the occurrence of a local minimum in the Alfvén speed near (but outside of) active regions in the middle of the corona and a local maximum of $\approx$800 km/s at distances 3–5R_S from the center of the Sun, *i.e.*, in the near-Sun interplanetary space, is the generally valid result of this study. This behaviour of the Alfvén speed has important consequences for the formation and development of shock waves in the corona and interplanetary space and their ability to accelerate particles. Inspecting the results by Edmiston and Kennel (1984), the first critical Mach number M_f^* is about 2 under

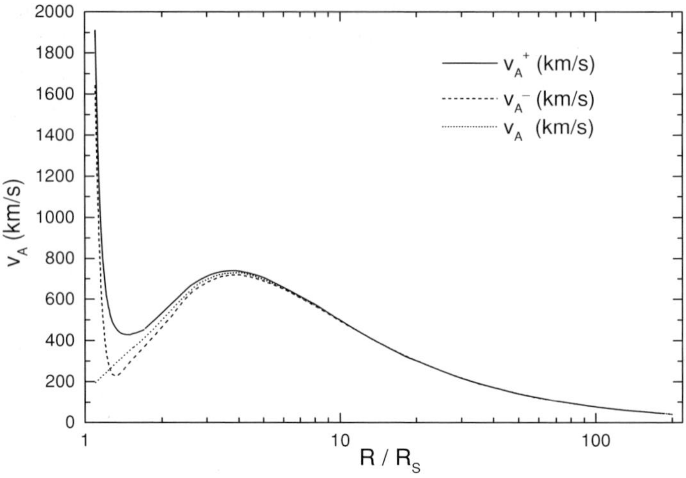

Figure 14. Alfvén speed along a straight line with $\delta = 45°$ away from an active region as a function of the radial distance R. Further explanations are given in the text.

coronal circumstances. Thus, an agent exciting a super-critical, fast magnetosonic shock wave should have a velocity exceeding at least twice the Alfvén speed (see, also, Cairns *et al.* (2003)). Super-critical shocks are able to produce energetic particles, which are responsible for solar energetic particle events (Kahler, 1994; Reames *et al.*, 1996; Klein and Trottet, 2001; Vainio and Khan, 2004). Figure 14 implies that such shocks can be generated in the middle of the corona $\approx 1.2 - 2R_S$ and in interplanetary space beyond $\approx 8R_S$ (Gopalswamy *et al.*, 2001; Mann *et al.*, 2003).

6. Particle Acceleration and Transport

J. KÓTA, M. A. LEE, R. VAINIO

Particle acceleration is a fundamental feature of solar activity. It occurs at sites of magnetic reconnection in solar flares and at shocks driven by CMEs. These two sites appear to be responsible for the two classes of solar energetic particle (SEP) events observed in interplanetary space: impulsive and gradual events, respectively (see Mikić and Lee, 2006, this volume). Impulsive events have ion charge-state distributions characteristic of heated flare plasma. The particles in impulsive events are thought to be accelerated either by reconnection electric fields (Litvinenko, 1996), turbulence generated at the reconnection site (Emslie *et al.*, 2004), or at coronal shocks produced by reconnection jets or heating (Tsuneta and Naito, 1998). Impulsive events are not directly related to a CME and are not considered further in this section. However, CMEs are generally accompanied by flares arising from changes in magnetic topology in the wake of the CME. Thus, gradual events which are observed on magnetic field lines favorably connected to the flare site may very well include a flare/impulsive component. The existence and/or importance of this component is currently controversial (Cane *et al.*, 2002, 2003; Tylka *et al.*, 2005).

6.1. Ion Shock Acceleration

A fast CMEs ($V_{\mathrm{CME}} \geq 10^3$ km/s) certainly drives a shock wave ahead of it. The strength of the shock may initially decrease but will increase beyond $\approx 4R_S$ as the Alfvén speed decreases, as described in Section 5.3 and Figure 14. This CME-driven shock is an ideal site for ion acceleration all the way to Earth orbit and beyond.

6.1.1. *Focused Transport Equation*

The focused transport equation, which is restricted to field-aligned transport of particles, is the starting point for most treatments of SEP transport and acceleration. The full version of the equation, appropriate to oblique or quasi-perpendicular shocks, may be written as (Skilling, 1971; Ruffolo, 1995; Isenberg, 1997; Kóta and Jokipii, 1997):

$$\begin{aligned}
&\frac{\partial f}{\partial t} + (V_i + w\,\mu b_i)\frac{\partial f}{\partial x_i} \\
&\quad + \left(\frac{1-\mu^2}{2}\right)\left[\frac{w}{L} - \frac{2}{w}b_i\,\frac{DV_i}{Dt} + \mu(\delta_{ij} - 3b_i b_j)\frac{\partial V_i}{\partial x_j}\right]\frac{\partial f}{\partial \mu} \\
&\quad + \left[-\frac{\mu b_i}{w}\,\frac{DV_i}{Dt} + \mu^2 b_i b_j \frac{\partial V_i}{\partial x_j} + \frac{1-\mu^2}{2}(\delta_{ij} - b_i b_j)\frac{\partial V_i}{\partial x_j}\right] p\frac{\partial f}{\partial p} \\
&= \frac{\partial}{\partial \mu}\left(\frac{D_\mu}{2}\frac{\partial f}{\partial \mu}\right) + q
\end{aligned} \tag{7}$$

where the particle speed, w, momentum, p, and the cosine of the pitch angle, μ are measured in the frame co-moving with the solar wind plasma at velocity, V_i. The symbol b_i represents the unit vector in the direction of the magnetic field, B_i, and L is the adiabatic focusing length ($1/L = b_i B^{-1}\partial B/\partial x_i$). The notation DV_i/Dt refers to the total derivative, *i.e.*, the acceleration of the fluid. For high particle speed ($w \gg V$), the DV_i/Dt terms can be neglected, and a relativistic correction appears for relativistic speed (Ruffolo, 1995).

This equation applies under broad conditions and remains valid for highly anisotropic pitch-angle distributions. Scattering may be either strong or weak, up to the scatter-free limit. The particle speed may be either higher or lower than the fluid speed. Terms of the order of $(V/w)^2$ can be neglected at high energies but may be of importance in the injection process. Particles in this description are assumed to remain tied to the field lines, reducing the spatial variation to one dimension, so as to keep the computation at a tractable level. The drift motion of particles across the magnetic field is neglected, which should not be a severe omission if particle gradients across the field are small. The role of cross-field diffusion is still not fully understood and requires further theoretical efforts. Another, equivalent formulation of the same equation, written in conservation form, was considered by Ruffolo (1995).

A major difference from the spatial diffusion formulation of shock acceleration discussed in Section 6.1.2 below is that the acceleration rate depends on pitch-angle if the compression of the fluid is not isotropic. Particles moving along the field ($\mu \approx \pm 1$) sense compression parallel to the field, while particles moving at near 90° pitch angle ($\mu \approx 0$) sense compression perpendicular to the field. The compression rates in the two respective directions can be expressed in terms of the change in (n/B) and B, respectively (Kóta and Jokipii, 1997), and turn out to be different for parallel and perpendicular shocks. Here n is the plasma density. At a parallel shock, n (and hence n/B) changes, while B remains unchanged. A perpendicular shock, on the other hand, changes B but not (n/B).

The equation of focused transport has been solved by direct numerical methods but, also, may be solved by employing the Monte Carlo method (Kocharov *et al.*, 1998; Vainio *et al.*, 2000), where individual test particles are simulated under the guiding-center approximation in a given background field geometry. Particle scattering off turbulence and/or waves is simulated via a random generator by performing small-angle scatterings to the particle propagation direction in the frame co-moving with the scattering centers. This method is more time consuming than a finite difference scheme, but the advantage of the method is that it is flexible: once the scattering center velocities and background field geometries are specified, the method automatically yields the adiabatic and Fermi-mechanism particle energy changes. Thus, it has already been employed in studies of coronal (Vainio *et al.*, 2000) and interplanetary (Vainio, 1997) shock acceleration.

6.1.2. *Diffusive Shock Acceleration*

If scattering by turbulence is sufficiently strong to keep the particle distribution close to isotropic, the focused transport equation can be approximated by Parker's equation

$$\frac{\partial f_0}{\partial t} + V\frac{\partial f_0}{\partial s} - \frac{1}{A}\frac{\partial (AV)}{\partial s}\frac{p}{3}\frac{\partial f_0}{\partial p} = \frac{1}{A}\frac{\partial}{\partial s}\left(A\kappa_{\parallel}\frac{\partial f_0}{\partial s}\right) \tag{8}$$

where $f_0(s, p, t)$ is the isotropic part of the distribution function, s is the coordinate measured along the field lines, V is the speed of the scattering centers along the field lines, $A(s)$ is the flux-tube cross-sectional area, $\kappa_{\parallel} = \frac{1}{3}v\lambda_{\parallel}$ is the diffusion coefficient parallel to the field lines, and $\lambda_{\parallel}$ is the parallel scattering mean-free path. In this description, guiding-center drifts and perpendicular diffusion are also neglected. In a radial geometry, $s = r$ and $A = r^2$, and with a time-dependent radial solar wind speed, $V = V(r, t)$, this equation allows a simplified description of the

CME-driven shock as a propagating spherical discontinuity separating the ambient and the shocked plasma. With simplifying (and not very realistic) assumptions for κ_{rr} and V, it can be solved analytically (Lee and Ryan, 1986).

Let us consider a spherical shock wave propagating outwards from the Sun with a speed of V_s. If the mean free path is so short that the diffusion length, $\kappa_\parallel/V$, is shorter than the scale length of the system,

$$\kappa_\parallel/V \ll r, \tag{9}$$

the particle distribution around the shock position, $r = r_s(t)$, can be described in a one-dimensional geometry, *i.e.*,

$$\frac{\partial f}{\partial t} + u\frac{\partial f}{\partial x} - \frac{p}{3}\frac{\partial u}{\partial x}\frac{\partial f}{\partial p} = \frac{\partial}{\partial x}\left(\kappa_{xx}\frac{\partial f}{\partial x}\right), \tag{10}$$

where $\kappa_{xx} = \kappa_\parallel \cos^2\psi$ and $\psi = \sphericalangle(\mathbf{r}, \mathbf{B})$, $x = r_s - r$ is the local, co-moving coordinate along the shock normal pointing towards the shocked plasma, $u = \dot{r}_s - V$ and $\partial u/\partial x$ is negligible away from the shock. Thus, in a quasi-steady state, the distribution function can be obtained as a solution of a one-dimensional diffusion convection equation on both sides of the shock, and matching these solutions gives the canonical power-law spectrum of diffusive shock acceleration (see, *e.g.*, Vainio, 1999, for a review)

$$f(0, p) = \sigma\left[\frac{Q_0}{4\pi u_1 p_0^3} + \int_{p_0}^{p}\frac{dp'}{p'}\left(\frac{p'}{p_0}\right)^{\sigma} f_{-\infty}(p')\right]\left(\frac{p_0}{p}\right)^{\sigma}, \tag{11}$$

where u_1 is the upstream scattering-center speed along the shock normal, $\sigma = 3\rho/(\rho - 1)$ and ρ is the scattering-center compression ratio at the shock.[1] The first term describes particles injected into the acceleration process from low energies [from below the injection momentum p_0] at rate Q_0 [$\mathrm{cm^{-2}s^{-1}}$], and the second term describes energetic particles of the ambient medium (distributed in momentum space as $f_{-\infty}(p)$) overtaken and re-accelerated by the shock.

The spectrum calculated at the shock holds in the downstream medium in the absence of adiabatic energy losses as well. In its derivation, however, one assumes that the system has an extent much larger than the diffusion length. This means that only a minor fraction of upstream ions may escape to the ambient medium. In the upstream region, the particle distribution decays exponentially with distance from the shock, *i.e.*,

$$f(x, p) = f(0, p)\exp\left(-\int_x^0 \frac{u_1\,dx}{\kappa_{xx}(x, p)}\right); \quad x < 0. \tag{12}$$

[1] Note that ρ does not necessarily equal the gas compression ratio of the shock. Vainio and Schlickeiser (1999) showed that for a low-Mach-number parallel shock, ρ can become very large if the downstream scattering centers are Alfvén waves transmitted from upstream.

If particles escape at some distance $L_1(p)$ ahead of the shock, *e.g.*, due to adiabatic focusing, and if the number of upstream diffusion lengths,

$$\eta(p) = \int_{-L_1(p)}^{0} \frac{u_1\, dx}{\kappa_{xx}(x, p)}, \tag{13}$$

is not large, the spectral form (for freshly injected ions) is (Vainio *et al.*, 2000)

$$f(0, p) = \frac{\sigma Q_0}{4\pi u_1 p_0^3} \left(\frac{p_0}{p}\right)^{\sigma} \exp\left\{-\sigma \int_{p_0}^{p} \frac{dp'}{p'(e^{\eta(p')} - 1)}\right\}. \tag{14}$$

The rate of particle escape ahead of the shock into the flux tube is (Vainio *et al.*, 2000)

$$\frac{d^2 N_{\mathrm{esc}}}{dp\, dt} = \frac{u_1 A_s\, 4\pi p^2 f(0, p)}{e^{\eta(p)} - 1}, \tag{15}$$

where $A_s = A(r_s)$ is the flux-tube cross-sectional area at the shock. This spectral form is relevant to SEPs observed ahead of the CME-driven shock, if a quasi-stationary state in the vicinity of the shock can be assumed.

The steady-state spectrum holds up to a cut-off energy, E_c, determined by a balance between energy-losses and acceleration rate, by particle escape, or by the available acceleration time. The acceleration rate is determined by the rate of particle scattering near the shock, and the acceleration time scale at momentum p is

$$\tau(p) \equiv \frac{p}{\dot{p}} = \frac{\sigma}{u_1} \left(\frac{\kappa_{xx}(0-, p)}{u_1} + \frac{\kappa_{xx}(0+, p)}{u_2}\right). \tag{16}$$

In the downstream region, the turbulence is amplified from the upstream levels, and the mean free path is typically at least an order of magnitude shorter there than in the upstream region (Vainio and Schlickeiser, 1999). Thus, by neglecting the downstream contribution and equating the acceleration time scale with the dynamical time scale of the shock, $r_s/\dot{r}_s$, one gets an equation determining the cut-off momentum, p_c, as

$$\frac{\sigma \kappa_{xx}(0-, p_c)}{u_1^2} \approx \frac{r_s}{\dot{r}_s}. \tag{17}$$

The time scale for adiabatic deceleration is $\tau_{\mathrm{ad}} \sim (r/2V)$, producing a very similar cut-off momentum. Also particle escape by adiabatic focusing produces, apart from numerical constants, the same value for the cut-off momentum (Vainio *et al.*, 2000). Note, finally, that below the cut-off energy, the one-dimensional approximation is valid (*cf.*, Equattion (9)).

6.1.3. *Wave Excitation*

The SEP mean free path at 1 AU is observationally in the range 0.1–1 AU (Palmer, 1982; Dröge, 2000). If such values of $\lambda_{\|}$ were prevalent in the inner heliosphere and solar corona, as well, there would be no possibility for diffusive acceleration at

CME-driven shocks beyond 1-MeV energies. The accelerated particle distribution upstream of the shock front, however, is known to be unstable against generating hydromagnetic waves (Lee, 1983; Gordon *et al.*, 1999). The upstream particle scattering is, therefore, determined by the distribution of the particles themselves.

In steady state, the magnetic energy density of the unstable hydromagnetic waves is $W_B = \frac{1}{3} V_A \cos\psi (u_1 - V_A \cos\psi)^{-1} W_P$ (Gordon *et al.*, 1999), where V_A is the Alfvén speed and W_P is the energy density of the accelerated particles. By taking a sharpened resonance condition for the wave–particle interactions, *i.e.*, $k = m|\Omega_0|/p$, the wavenumber spectrum of the unstable waves is

$$\frac{k\,I(x,k)}{8\pi} \approx \frac{4\pi}{3} \frac{V_A \cos\psi}{u_1 - V_A \cos\psi} E\, p^3\, f(x,p), \tag{18}$$

where E is the particle energy. This gives an estimate for the mean free path of the energetic particles as

$$\lambda_\parallel(x,p) \approx \frac{v}{\Omega} \frac{2B^2}{\pi k I(k)} = \frac{v}{\Omega} \frac{2}{\pi} \left[\frac{u_1 - V_A \cos\psi}{V_A \cos\psi} \frac{B^2/8\pi}{(4\pi/3) E\, p^3\, f(x,p)} \right]. \tag{19}$$

where Ω is the ion gyro-frequency. To avoid non-linear wave amplitudes, the factor in brackets should be replaced by unity if its value gets smaller than this. For freshly injected particles ($f_{-\infty}(p) = 0$) with $p_0 = m u_1$ and $Q_0 = \varepsilon u_1 n_1$ (where ε is the fraction of the injected particles), we can use the spectrum at the shock, Equation (11), to estimate the upstream mean free path at the shock as

$$\lambda_\parallel(0-,p) \sim \frac{pc}{eB} \frac{2}{\pi} \left[\frac{3}{\sigma} \frac{M_A - \cos\psi}{\varepsilon M_A^2 \cos\psi} \frac{\frac{1}{2} m u_1^2}{E} \left(\frac{p}{m u_1} \right)^{\sigma - 3} \right], \tag{20}$$

and the bracketed quantity is seen to decrease with non-relativistic (relativistic) energy for shocks with $\sigma < 5$ (4), *i.e.*, $\rho > \frac{5}{2}$ (4). Particles accelerated by strong shocks are, thus, very efficient in generating upstream waves.

Using Equations (17) and (20) and $\frac{1}{2} v p = E$ gives the cut-off momentum of the shock accelerated particles as

$$\left(\frac{p_c}{m u_1} \right)^{\sigma - 3} \approx \frac{r_s}{\dot r_s} \frac{\pi e B_s}{2mc \cos\psi} \frac{\varepsilon M_A^2}{M_A - \cos\psi} = \frac{1.4 \times 10^7}{V_{s8} \cos\psi} \frac{\varepsilon M_A^2}{M_A - \cos\psi} \frac{r_\odot}{r_s}, \tag{21}$$

where V_{s8} is the shock speed in $1000\ \mathrm{km\,s^{-1}} = 10^8\ \mathrm{cm\,s^{-1}}$. Thus, shocks in the outer solar corona with $V_{s8} \approx 1$, $u_1 \approx 1000\ \mathrm{km\,s^{-1}}$, $r_s \approx 5\, r_\odot$, $M_A \approx 2$, $\cos\psi \approx 1$, $\sigma \approx 5$, and $\varepsilon \approx 0.01$ can accelerate particles up to relativistic energies ($p_c \approx 1\ \mathrm{GeV}/c$).

Apart from some numerical factors, Equation (21) is the same as that used by Rice *et al.* (2003) in a study of CME shock acceleration. These authors performed MHD simulations of shock propagation through the inner heliosphere. They computed the energy spectrum at the shock by using an injection energy equal to 50% of the downstream thermal energy and the maximum momentum calculated from the dynamical time scale, similar to the above analysis. These spectra were then allowed to convect to the downstream region by applying adiabatic deceleration, and to diffuse upstream of the shock front. These authors arbitrarily took $\eta(p) = 4$ at all momenta and, thus, calculated the flux of particles escaping to the ambient medium. In a parallel study, Li *et al.* (2003) used this particle injection for a focused transport simulation in a relatively undisturbed solar wind. These computations could qualitatively reproduce the essential features of CME-related SEP events.

The mean free path obtained above corresponds to the quasi-stationary state. Vainio (2003) studied the initial phase of wave generation ahead of the shock wave and showed that the escaping particle flux in case of a very strong moving source, like a strong CME-driven shock, would have a plateau-type power-law spectrum with a hard spectrum $dN_{\rm esc}/dp \propto p^{-1}$, indicating that $e^{\eta} - 1 \propto p^{3-\sigma}$, *cf.*, Equation (15). Without any free parameters, the particle escape model of Vainio (2003) could reproduce the SEP source profile deduced for the large 19 Oct 1989 SEP event at energies around 10 MeV. The model also predicted that wave generation in the solar corona during more typical small gradual events with 1-MeV-proton peak intensities below 10 $({\rm cm}^2\ {\rm s\ sr\ MeV})^{-1}$, often associated with slower ($<$1000 km s^{-1}) CMEs, would be negligible. For such events, diffusive shock acceleration in the low corona ($r < 2 r_{\odot}$) by refracting shock waves was studied by Vainio and Khan (2004). They showed that high-frequency Alfvén waves emitted from the solar surface and related to coronal ion-cyclotron heating would be able to provide enough scattering to accelerate protons up to tens of MeVs. The spectra of escaping particles would be consistent with those observed during typical small gradual events. Note that in the case of external upstream turbulence, η is usually a rapidly decreasing function of momentum, and the spectrum of the particles escaping to the upstream region is severely modulated (*cf.*, Equation (15)). Thus, in this case models connecting the observer to the downstream region are favored.

6.2. Ion Injection at Shocks

The distribution function described by Equation (11) depends on the injection rate Q_0, which cannot be determined from Equation (8), or even Equation (7). Equation (8) is restricted to energies much greater than u_1; Equation (7) presumes diffusive transport in μ which may not be valid for the scattering of particles with $v \approx u_1$ by the electromagnetic fluctuations at the shock front. Similarly the behavior of particles with $v \approx u_1$ in the advected distribution, $f_{-\infty}(p)$, is not well-described by Equation (11). The first-order Fermi mechanism may or may not operate for these low-energy particles, or it may operate for a small fraction of the advected particles. At the quasi-parallel Earth's bow shock or interplanetary traveling shocks, $\approx 10^{-2}$ of the incident solar wind ions are typically "injected" into the process of diffusive shock acceleration (Lee, 1982; Gordon *et al.*, 1999).

The behavior of the low-energy ions at a shock can best be revealed by numerical simulations (Scholer *et al.*, 2000), although injection is likely to be sensitive to the assumed dimensionality of the simulation. Monte Carlo calculations using prescribed scattering mean free paths are conceptually instructive but are overly simplified (Ellison *et al.*, 1990). Malkov (1998) has presented an analytical theory of injection at a parallel shock based on a simplified version of Equation (7) with diffusive transport in μ.

A question of considerable importance is the energy threshold for diffusive shock acceleration as a function of ψ; that is, above what energy does first order Fermi acceleration occur as described by Equation (8)? At lower energies particle mobility normal to the shock is too small to allow particles to scatter back and forth across the shock, and they are swept downstream. A reasonable choice for the threshold speed is $u_1/\cos\psi$ in the upstream plasma frame. A particle with such a speed is kinematically able to escape the shock in the upstream direction by moving along the ambient field with $|\mu| = 1$. For quasi-perpendicular shocks, perpendicular diffusion can substantially reduce this threshold speed (Giacalone, 2005), but the actual reduced value is controversial. The threshold speed determines the mix of solar wind and ambient suprathermal/energetic particles accelerated at the shock, which in turn determines SEP composition (Tylka *et al.*, 2005). That the threshold energy increases markedly with ψ is shown by the decrease in ion intensity as ψ increases and the overall irregularity of ion intensity at quasi-perpendicular shocks (van Nes *et al.*, 1984).

6.3. Electron Shock Acceleration

In principle, electrons satisfy the same Equations (7) and (8) as do ions, and are subject to the process of diffusive shock acceleration. However, by virtue of their large speed and no established instability to enhance their pitch-angle scattering rate, $\kappa_{\|}$, is large, and the process is correspondingly slow. As a result, energetic electron enhancements at shocks are not large. An exception is at nearly perpendicular shocks where a single reflection of an electron, by mirroring in the region of increased magnetic field strength at the shock (the "shock drift" mechanism), can lead to substantial energy gain. Such electrons are observed at Earth's bow shock adjacent to the magnetic field line tangent to the shock surface (Wu, 1984). This mechanism is responsible for the unstable electron beams which excite type II radio bursts (see Pick *et al.*, 2006, this volume). The origin of the solar energetic electron beams in interplanetary space is controversial but it may be coronal shocks (Haggerty and Roelof, 2001; Ellison and Ramaty, 1985).

6.4. Temporal and Spatial Variations

An important aspect of CME-driven shock acceleration is the inherent temporal and spatial variation of the resulting SEP events. The driven shock first forms, then strengthens as the coronal Alfvén speed decreases, and finally weakens as the shock expands into the heliosphere. The shock is also generally stronger near the nose and weaker on the flanks, which contributes to a dependence of particle intensity on magnetic field line geometry. As an observer's field line sweeps across the shock surface with increasing time, shock strength and ψ change. These variations were recognized by Cane *et al.* (1988) and invoked to argue for a shock origin of SEP events. In addition, ions escaping the shock upstream undergo a transition from scatter-dominated transport adjacent to the shock to nearly scatter-free transport further from the shock. This transition leads to a marked temporal variation in the ion intensity, energy spectrum and anisotropy as an event approaches the observer. In principle these variations provide information on shock geometry and ion transport. Current models attempt to include some of these features (Sokolov *et al.*, 2004).

6.5. Challenges

The origin of SEPs at CME-driven shocks can account for most of the observed features of these events. Also the basic theoretical concepts are quite well understood. Nevertheless, there are several challenges to developing a complete theory:

(i) The intensity of upstream escaping ions and the high-energy cutoff are very sensitive to the form of the upstream excited wave spectrum. The wave spectrum needs to be calculated carefully with the correct resonance condition.

(ii) The transition from scatter-dominated ion transport adjacent to the shock to nearly scatter-free transport further away must be handled properly (Lee, 2005).

(iii) Ion injection from both the solar wind and the ambient suprathermal/energetic ion population must be included with appropriate thresholds and injection rates.

(iv) The role of perpendicular diffusion at quasi-perpendicular shocks needs to be assessed including possible instabilities and its influence on injection thresholds.

(v) The compression ratio, which plays a central role in the theory of diffusive shock acceleration, is the average-wave-frame compression ratio, which can be quite different from the plasma-frame compression ratio for these generally weak shocks. This issue must be addressed for oblique shocks including the determination of the downstream average-wave-frame.

(vi) The transport of SEPs downstream of the shock similarly transitions from scatter-dominated to nearly scatter-free as the turbulence decays, ions fill the inner heliosphere, and they cool adiabatically (Reames *et al.*, 1996; Lee, 2005). This transition needs to be included quantitatively with appropriate turbulence decay rates.

(vii) Several current models include time-dependence under an adiabatic approximation (Zank *et al.*, 2000; Lee, 2005), which is not valid at early times and high energies. The approximation needs to be relaxed.

7. Concluding Remarks

Future progress in understanding the origin and evolution of CMEs depends critically on obtaining new observations. More information is needed about the structure of the magnetic field in the corona before and after eruption. The magnetic field in the lower corona plays a fundamental role in the genesis of CMEs. Equally important are new observations of the magnetic field in the solar wind. Knowledge of how this field is distributed in space as a function of time would resolve several outstanding issues about CMEs and ICMES. Thus, ongoing efforts to improve vector magnetic field observations of the solar surface and to obtain multi-spacecraft observations of the interplanetary field need to be sustained. High resolution images of plasma structures in the corona and chromosphere also provide information about the coronal magnetic field because the low plasma β conditions there cause gradients in the plasma properties to be strongest in the direction perpendicular to the magnetic field. Virtually all of the evidence for the existence of neutral points and current sheets in the corona is based on such images.

More accurate observations are needed of the acceleration profile of CMEs and associated structures at altitudes below one solar radius where the strongest acceleration occurs. Because acceleration is proportional to the force acting on the plasma, such observations provide a strong constraint on theories of CME initiation. For example, initiation mechanisms based on ideal-MHD processes predict that the peak acceleration will occur at a time on the order of the Alfvén time scale of the erupting region, which is typically no more than a few minutes. Thus, models of this type may not be able to account for slowly accelerating CMEs which take several hours to reach their peak value (Zhang *et al.*, 2004). The various models also predict different reconnection signatures. While all the models, predict reconnection to occur after CME onset, some also require it before onset (*e.g.* Moore and Roumeliotis, 1992; Antiochos *et al.*, 1999).

Multi-spacecraft, in-situ observations of ICMEs are needed to resolve issues about the large scale structure of the magnetic field ejected during a CME. Such observations have already shown that the erupted field is in the form of a flux rope, but the distribution of the current density within the flux rope, and the overall structure along the axial field are not well known. Furthermore, the interaction of magnetic field of the ICME with the ambient field of the corona is not well understood. The long term fate of ICMEs in the outer heliosphere is also an unsolved question. Beyond about 10 AU, it is difficult to identify individual ICMEs, and it is thought that they merge together with CIRs to form GMIRs. Compositional measurements with multiple spacecraft would also be a great help in sorting out the distinction between impulsive and gradual particle events.

Acknowledgments

This work was supported at the University of New Hampshire in part by grants ATM-0327512, ATM-0422764, and ATM-0518218 from the US National Science Foundation; NASA grants NNH05-AA131 and NNG05-GL40G, and the US Dept. of Defense MURI program on Space Weather. Work at SAIC was supported by the NASA LWS, SR&T, and SECTP programs, and by NSF through the SHINE program and the Center for Integrated Space Weather Modeling.

References

Aly, J. J.: 1984, *Astrophys. J.* **283**, 349–362.

Aly, J. J.: 1991, *Astrophys. J.* **375**, L61–L64.

Amari, T., Luciani, J. F., Mikić, Z., and Linker, J. A.: 2000, *Astrophys. J.* **529**, L49–L52.

Amari, T., Luciani, J. F., Aly, J. J., Mikić, Z., and Linker, J. A.: 2003, *Astrophys. J.* **585**, 1073–1086.

Amari, T., Luciani, J. F., Aly, J. J., Mikić, Z., and Linker, J. A.: 2003, *Astrophys. J.* **595**, 1231–1250.

Antiochos, S., DeVore, C., and Klimchuk, J.: 1999, *Astrophys. J.* **510**, 485–493.

Antiochos, S. K., and DeVore, C. R.: 1999, in: *Magnetic Helicity in Space and Laboratory Plasmas: Geophysical Monograph 111*, M. R., Brown, R. C., Canfield and A. A. Pevtsov (eds.), Amer. Geophys. Union, Washington, DC, p. 187.

Antiochos, S. K.: 1998, *Astrophys. J.* **502**, L181–L184.

Anzer, U.: 1978, *Solar Phys.* **57**, 111–118.

Aulanier, G., DeLuca, E. E., Antiochos, S. K., McMullan, R. A., and Golub, L.: 2000, *Astrophys. J.* **540**, 1126–1142.

Aurass, H., Hofmann, A., and Urbarz, H.-W.: 1998, *Astron. Astrophys.* **334**, 289–302.

Banaszkiewicz, M., Axford, W. I., and McKenzie, J. F.: 1998, *Astron. Astrophys.* **337**, 940–948.

Birn, J., Forbes, T. G., and Schindler, K.: 2003, *Astrophys. J.* **588**, 578–585.

Biskamp, D., and Welter, H.: 1989, *Solar Phys.* **120**, 49–77.

Bougeret, J.-L., Kaiser, M. L., Kellogg, P. J., Manning, R., Goetz, K., Monson, S. J., *et al.*: 1995, *Space Sci. Rev.* **71**, 231–263.

Brueckner, G. E., Howard, R. A., Koomen, M. J., Korendyke, C. M., Michels, D. J., Moses, J. D., *et al.*: 1995, *Solar Phys.* **162**, 357–402.

Burlaga, L.: 1981, *J. Geophys. Res.* **86**, 6673–6684.

Burlaga, L.: 1988, *J. Geophys. Res.* **93**, 7217–7224.

Cairns, I. H., Knock, S. A., Robinson, P. A., and Kuncic, Z.: 2003, *Space Sc. Rev.* **107**, 27–34.

Cane, H. V., Reames, D. V., and von Rosenvinge, T. T.: 1988, *J. Geophys. Res.* **93**, 9555–9567.

Cane, H. V., Erickson, W. C., and Prestage, N. P.: 2002, *J. Geophys. Res.* **107**, Issue A10 CiteID 1315 DOI 10.1029/2001JA000320.

Cane, H. V., von Rosenvinge, T. T., Cohen, C. M. S., and Mewaldt, R. A.: 2003, *Geophys. Res. Lett.* **30**, Issue A12 CiteID 8017, DOI 10.1029/2002GL016580.

Cargill, P. J., Chen, J., Spicer, D. S., and Zalesak, S. T.: 1996, *J. Geophys. Res.* **101**, 4855–4870.

Cargill, P. J., Schmidt, J. M., Spicer, D. S., and Zalesak, S. T.: 2000, *J. Geophys. Res.* **105**, 7509–7520.

Chen, J., Santoro, R. A., Krall, J., Howard, R. A., Duffin, R., Moses, J. D., *et al.*: 2000, *Astrophys. J.* **533**, 481–500.
Chen, P. F., Shibata, K., and Yokoyama, T.: 2001, *Earth, Planets and Space* **53**, 611–614.
Chen, J.: 1989, *Astrophys. J.* **338**, 453–470.
Chen, J.: 1996, *J. Geophys. Res.* **101**, 27499–27519.
Chen, J.: 1997, *Coronal Mass Ejections. Geophys. Monograph 99*, N., Crooker, J. A., Joselyn and J. Feynman (eds.), Amer. Geophys. Union, Washington, DC, pp. 65–81.
Chen, J., Howard, R. A., Brueckner, G. E., Santoro, R., Krall, J., Paswaters, S. E., St. Cyr, O. C., Schwenn, R., Lamy, P., and Simnett, G. M.: 1997, *Astrophys. J.* **490**, L191–L194.
Cheng, C. Z., Ren, Y., Choe, G. S., and Moon, Y.-J.: 2003, *Astrophys. J.* **596**, 1341–1346.
Cid, C., Hidalgo, M. A., Nieves-Chinchilla, T., Sequeiros, J., and Viñas, A. F.: 2003, *Solar Phys.* **207**, 187–198.
Detman, T. R., Dryer, M., Yeh, T., Hun, S. M., Wu, S. T., and McComas, D. J.: 1991, *J. Geophys. Res.* **96**, 9531–9540.
DeVore, C. R.: 1991, *J. Comput. Phys.* **92**, 142–160.
Dröge, W.: 2000, *Astrophys. J.* **537**, 1073–1079.
Dryer, M., Wu, S. T., Steinolfson, R. S., and Wilson, R. M.: 1979, *Astrophys. J.* **227**, 1059–1071.
Edmiston, J. P., and Kennel, C. F.: 1984, *J. Plasma Phys.* **32**, 429–443.
Ellison, D. C., Möbius, E., and Paschmann, G.: 1990, *Astrophys. J.* **352**, 376–394.
Ellison, D. C., and Ramaty, R.: 1985, *Astrophys. J.* **298**, 400–408.
Emslie, A. G., Miller, J. A., and Brown, J. C.: 2004, *Astrophys. J.* **602**, L69–L72.
Fan, Y.: 2005, *Astrophys. J.* **630**, 543–551.
Farrugia, C. J., Osherovich, V. A., and Burlaga, L. F.: 1995, *J. Geophys. Res.* **100**, 12293–12306.
Forbes, T. G.: 1990, *J. Geophys. Res.* **95**, 11919–11931.
Forbes, T. G., and Isenberg, P. A.: 1991, *Astrophys. J.* **373**, 294–307.
Forbes, T. G., Priest, E. R., and Isenberg, P. A.: 1994, *Solar Phys.* **150**, 245–266.
Forbes, T. G., and Priest, E. R.: 1995, *Astrophys. J.* **446**, 377–389.
Forbes, T. G.: 2000, *J. Geophys. Res.* **105**, 23153–23165.
Forsyth, R. J., Bothmer, V., *et al.*: 2006, *Space Sci. Revs.*, this volume.
Gary, G. A., and Moore, R. L.: 2004, *J. Geophys. Res.* **611**, 545–556.
Giacalone, J.: 2005, *Astrophys. J.* **628**, L37–L40.
Gibson, S. E., and Low, B. C.: 1998, *Astrophys. J.* **493**, 460–473.
Goldstein H.: 1983, M. Neugebauer (ed.), *Solar Wind Five, NASA Conf. Publ.* **CP-2280**, Washington, DC, pp. 731–733.
Gopalswamy, N., Lara, A., Lepping, R. P., Kaiser, M. L., Berdichevsky, D., and St. Cyr, O. C.: 2000, *Geophys. Res. Lett.* **27**, 145–148.
Gopalswamy, N., Yashiro, S., Kaiser, M. L., Howard, R. A., and Bougeret, J. L.: 2001, *Astrophys. J.* **548**, L91–L94.
Gopalswamy, N., Yashiro, S., Michalek, G., Kaiser, M. L., Howard, R. A., Reames, D. V., *et al.*: 2002, *Astrophys. J.* **572**, L103–L106.
Gopalswamy, N., Kundu, M. R., Manoharan, P. K., Rauolt, A., Nitta, A., and Zarka, P.: 1997, *Astrophys. J.* **486**, 1036–1048.
Gopalswamy, N., Kaiser, M. L., Sato, J., and Pick, M.: 2000, R., Ramaty and N.,, Mandzhavidze (eds.), *High Energy Physics: Anticipating HESSI, PASP Conf. Ser.* **206**, San Francisco, pp. 351–360.
Gopalswamy, N., Lara, A., Kaiser, M. L., and Bougeret, J.-L.: 2001, *J. Geophys. Res.* **106**, 25261–25273.
Gordon, B. E., Lee, M. A., Möbius, E., and Trattner, K. J.: 1999, *J. Geophys. Res.* **104**, 28263–28277.
Gosling, J. T.: 1990, in: C. T., Russell, E. R., Priest and L. C. Lee (eds.), *Physics of Magnetic Flux Ropes Geophys. Monograph 58*, Amer. Geophys. Union, Washington, DC, pp. 343–364.

Gosling, J. T., Riley, P., McComas, D. J., and Pizzo, V. J.: 1998, *J. Geophys. Res.* **103**, 1941–1941.
Haggerty, D. K., and Roelof, E. C.: 2001, *Proc. 27th Internat. Cosmic Ray Conf. (Hamburg)* **8**, 3238–3241.
Harrison, R.: 1986, *Astron. Astrophys.* **162**, 283–291.
Hidalgo, M. A.: 2003, *J. Geophys. Res.* **108**, Issue A8 CiteID 1320 DOI 10.1029/2002JA009818.
Hidalgo, M. A., Cid, C., Medina, J., and Viñas, A. F.: 2000, *Solar Phys.* **194**, 165–174. 10.1029/2001GL013875.
Hidalgo, M. A., Cid, C., Viñas, A. F., and Sequeiros, J.: 2002a, *J. Geophys. Res.* **107**, Issue A1 CiteID 1002 DOI 10.1029/2001JA900100.
Hidalgo, M. A., Nieves-Chinchilla, T., and Cid, C.: 2002b, *Geophys. Res. Lett.* **29**, Issue 13 CiteID 1637 DOI
Hu, Q., and Sonnerup, B. U. Ö.: 2001, *Geophys. Res. Lett.* **28**, 467–470.
Hu, Q., and Sonnerup, B. U. Ö.: 2002, *J. Geophys. Res.* **107**, Issue A7 CiteID 1142, DOI 10.1029/2001JA000293.
Hundhausen, A. J.: 1999, in:Strong, K. T., Saba, J. L. R., Haisch, B. M., and Schmelz, J. T. (eds.), *The Many Faces of the Sun: A Summary of Results from NASA's Solar Maximum Mission*, Springer, New York, p. 143.
Illing, R. M. E., and Hundhausen, A. J.: 1985, *J. Geophys. Res.* **90**, 275–282.
Illing, R. M. E., and Hundhausen, A. J.: 1986, *J. Geophys. Res.* **91**, 10951–10960.
Isenberg, P. A.: 1997, *J. Geophys. Res.* **102**, 4719–4724.
Ivanov, K. G., and Kharshiladze, A. F.: 1985, *Solar Phys.* **98**, 379–386.
Ivanov, K. G., Kharshiladze, A. F., Eroshenko, E. G., and Styazhkin, V. A.: 1989, *Solar Phys.* **120**, 407–419.
Kahler, S. W.: 1994, *Astrophys. J.* **428**, 837–849.
Kennel, C. F., Edmiston, J. P., and Hada, T.: 1985, in: R. G., and Tsurutani, B. T. (eds.), *Collisionless Shocks in the Heliosphere: A Tutorial Review*, Stone, Geophysical Monograph **34**, Amer. Geophys. Union, Washington, DC, pp. 1–36.
Klassen, A., Aurass, H., Klein, K.-L., Hofmann, A., and Mann, G.: 1998, *Astron. Astrophys.* **343**, 287–296.
Klassen, A., Aurass, H., Mann, G., and Thompson, B. J.: 2000, *Astron Astrophys. Suppl.* **141**, 357–369.
Klein, K.-L., Khan, J. I., Vilmer, N., Delouis, J.-M., and Aurass, H.: 1999, *Astron. Astrophys.* **346**, L53–L57.
Klein, K.-L., and Trottet, G.: 2001, *Space Sci. Rev.* **95**, 215–225.
Klein, L. W., and Burlaga, L. F.: 1982, *J. Geophys. Res.* **87**, 613–624.
Klecker, B., Kunow, H., *et al.*: 2006, *Space Sci. Revs.*, this volume.
Kliem, B., Titov, V., and Török, T.: 2004, *Astron. Astrophys.* **413**, L23–L26.
Klimchuk, J. A.: 2001, in: Song, P., Siscoe, G., and Singer, H. (eds.), *Space Weather* (Geophysical Monograph 125), Amer. Geophys. Union, Washington, D.C., pp. 143–157.
Kocharov, L., Vainio, R., Kovaltsov, G. A., and Torsti, J.: 1998, *Solar Phys.* **182**, 195–215.
Kosovichev, A. G., and Zharkova, V. V.: 2001, *Astrophys. J.* **550**, L105–L108.
Kóta, J., and Jokipii, J. R.: 1997, *Proc. 25th Internat. Cosmic Ray Conf. (Durban)* **1**, 213–216.
Koutchmy, S.: 1994, *Adv. Space Res.* **14**(4), 29–39.
Krall, J., Chen, J., and Santoro, R.: 2000, *Astrophys. J.* **539**, 964–982.
Krall, J., Chen, J., Duffin, R. T., Howard, R. A., and Thompson, B. J.: 2000, *Astrophys. J.* **562**, 1045–1057.
Kumar, A., and Rust, D. M.: 1996, *J. Geophys. Res.* **101**, 15667–15684.
Kuperus, M., Raadu, M. A.: 1974, *Astron. Astrophys.* **31**, 189–193.
Kusano, K., Maeshiro, T., Yokoyama, T., and Sakurai, T.: 2004, *Astrophys. J.* **610**, 537–549.
Le Méhauté, B.: 1976, *An Introduction to Hydrodynamics and Water Waves*, Springer-Verlag, New York.

Leblanc, Y., Dulk, G. A., and Bougeret, J.-L.: 1998, *Solar Phys.* **183**, 165–176.
Lee, M. A., and Ryan, J.: 1986, *Astrophys. J.* **303**, 829–842.
Lee, M. A.: 1982, *J. Geophys. Res.* **87**, 5063–5080.
Lee, M. A.: 1983, *J. Geophys. Res.* **88**, 6109–6119.
Lee, M. A.: 2005, *Astrophys. J. Suppl.* **158**, 38–67.
Lepping, R. P., Burlaga, L. F., and Jones, J. A.: 1990, *J. Geophys. Res.* **95**, 11957–11965.
Lepping, R. P., Berdichevsky, D. B., Szabo, A., Arqueros, C., and Lazarus, A. J.: 2003, *Solar Phys.* **212**, 425–444.
Leroy, J. L., Bommier, V., and Sahal-Brechot, S.: 1983, *Solar Phys.* **83**, 135–142.
Leroy, J. L., Bommier, V., and Sahal-Brechot, S.: 1984, *Astron. Astrophys.* **131**, 33–44.
Li, G., Zank, G. P., and Rice, W. K. M.: 2003, *J. Geophys. Res.* **108**, Issue A2 CiteID 1082, DOI 10.1029/2002JA009666.
Lin, J., Forbes, T. G., Isenberg, P. A., and Démoulin, P.: 1998, *Astrophys. J.* **504**, 1006–1019.
Lin, J., Soon, W., and Baliunas, S.: 2003, *New Astron. Revs.*, **47**, 53–84.
Lin, J., Raymond, J. C., and van Ballegooijen, A. A.: 2004, *Astrophys. J.* **602**, 422–435.
Linker, J., and Mikić, Z.: 1995, *Astrophys. J.* **103**, L45–L48.
Linker, J. A., and Mikić, Z.: 1999, in: Crooker, N., Joselyn, J., and Feynman, J. (eds.), *Coronal Mass Ejections*, *Geophys. Monograph*, **99**, Amer. Geophys. Union, Washington, pp. 269–277.
Linker, J. A., Mikić, Z., Biesecker, D. A., Forsyth, R. J., Gibson, S. E., Lazarus, A. J., et al.: 1999, *J. Geophys. Res.* **104**, 9809–9830.
Linker, J. A., Lionello, R., Mikić, Z., and Amari, T.: 2001, *J. Geophys. Res.* **106**, 25165–25176.
Linker, J. A., Mikić, Z., Lionello, R., Riley, P., Amari, T., and Odstrcil, D.: 2003, *Phys. Plasmas* **10**, 1971–1978.
Lionello, R., Mikić, Z., Linker, J. A., and Amari, T.: 2002, *Astrophys. J.* **581**, 718–725.
Lites, B. W., Low, B. C., Martinez Pillet, V., Seagraves, P., Skumanich, A., Frank, Z. A., *et al.*: 1995, *Astrophys. J.* **446**, 877–894.
Litvinenko, Y. E.: 1996, *Astrophys. J.* **462**, 997–1004.
Litvinenko, Y. E., and Martin, S. F.: 1999, *Solar Phys.* **190**, 45–58.
Livi, S. H. B., Wang, J., Martin, S. F.: 1985, *Aust. J. Phys.* **38**, 855–873.
Livi, S. H. B., Martin, S. F., Wang, H., and Guoxiang, A.: 1989, *Solar Phys.* **121**, 197–214.
Low, B. C.: 1981, *Astrophys. J.* **251**, 352–363.
Low, B. C.: 2001, *J. Geophys. Res.* **106**, 25141–25164.
Low, B. C., and Zhang, M.: 2002, *Astrophys. J.* **564**, L53–L56.
Lundquist, S.: 1950, *Ark. Fys.* **2**, 361–365.
Lynch, B., MacNeice, P., Antiochos, S. K., Zurbuchen, T., and Fisk, L.: 2004, *Astrophys. J.* **617**, 589–599.
MacNeice, P. J., Antiochos, S. K., Phillips, A., Spicer, D. S., DeVore, C. R., and Olson, K.: 2004, *Astrophys. J.* **614**, 1028–1041.
Malkov, M. A.: 1998, *Phys. Rev. E* **58**, 4911–4928.
Manchester, W. B., Gombosi, T. I., Roussev, I., Ridley, A., De Zeeuw, D. L., Sokolov, I. V., et al.: 2004, *J. Geophys. Res.* **109** Issue A2 CiteID A02107, DOI 10.1029/2003JA010150.
Mann, G., Lühr, H., and Baumjohann, W.: 1994, *J. Geophys. Res.* **99**, 315–333.
Mann, G., Jansen, F., MacDowall, R. J., Kaiser, M. I., and Stone, R. G.: 1999a, *Astron. Astrophys.* **348**, 614–620.
Mann, G., Aurass, H., Klassen, A., Estel, C., and Thompson, B. J.: 1999b, *Proc. 8th SOHO Workshop* ESA **SP-446**, 477–481.
Mann, G., Klassen, A., Aurass, H., and Classen, H.-T.: 2003, *Astron. Astrophys.* **400**, 329–336.
Manoharan, P. K., and Kundu, M.: 2003, *Astrophys. J.* **592**, 597–606.

Martin, S. F., and Echols, C. R.: 1994, in: Rutten, R. J., and Schrijver, C. J. (eds.), *Solar Surface Magnetism, Proc. NATO Adv. Res. Workshop, Soesterberg, The Netherlands, November 1–5, 1993*, Kluwer Academic Publishers, Dordrecht, pp. 339–346.
Martin, S. F., Livi, S. H. B., Wang, J., and Shi, Z.: 1984, in: Hagyard, M. J. (ed.), *Measurements of Solar Vector Magnetic Fields*, NASA Conf. Pub. 2374, pp. 403–436.
Martin, S. F., Livi, S. H. B., and Wang, J.: 1985, *Australian J. Phys.* **38**, 929–959.
Martin, S. F., Bilimoria, R., and Tracadas, P. W.: 1994, in: Rutten, R. J., and Schrijver, C. J. (eds.), *Solar Surface Magnetism, Proc. NATO Adv. Res. Workshop, Soesterberg, The Netherlands, November 1–5, 1993*, Kluwer Academic Publishers, Dordrecht, p. 303–338.
Marubashi, K.: 1997, *Coronal Mass Ejections. Geophys. Monograph 99*, Crooker, N., Joselyn, J. A., and Feynman, J. (eds.), Amer. Geophys. Union, Washington, DC, pp. 147–156.
McComas, D. J., Gosling, J. T., Hammond, C. M., Moldwin, M. B., Philips, J. L., and Forsyth, R. J.: 1994, *Geophys. Res. Lett.* **21**, 1751–1754.
Mikić, Z., Barnes, D., and Schnack, D.: 1988, *Astrophys. J.* **328**, 830–847.
Mikić, Z., and Lee, M. A.: 2006, *Space Sci. Revs.*, this volume.
Moldwin, M. B., Phillips, J. L., Gosling, J. T., Scime, E. E., McComas, D. J., Bame, S., et al.: 1995, *J. Geophys. Res.* **100**, 19903–19910.
Moore, R. L., and Roumeliotis, G.: 1992, in: Švestka, Z., Jackson, B., and Machado, M. (eds.), *Eruptive Solar Flares. Proc. IAU Colloquium 133, Iguazu, Argentina, August 2–6, 1991*, New York, Springer-Verlag, pp. 69–78.
Mulligan, T., and Russell, C. T.: 2001, *J. Geophys. Res.* **106**, 10581–10596.
Newkirk, G. A.: 1961, *Astrophys. J.* **133**, 983–994.
Odstrcil, D., and Pizzo, V. J.: 1999a, *J. Geophys. Res.* **104**, 483–492.
Odstrcil, D., and Pizzo, V. J.: 1999b, *J. Geophys. Res.* **104**, 493–504.
Odstrcil, D., and Pizzo, V. J.: 1999c, *J. Geophys. Res.* **104**, 28225–28239.
Odstrcil, D., Linker, J., Lionello, R., Mikić, Z., Riley, P., Pizzo, V., *et al.*: 2002, *J. Geophys. Res.* **107**, Issue A12 CiteID 1493, DOI 10.1029/2002JA009334.
Osherovich, V. A., Farrugia, C. J., Burlaga, L. F., Lepping, R. P., Fainberg, J., and Stone, R. G.: 1993, *J. Geophys. Res.* **98**, 15331–15342.
Owens, M. J., Crooker, N. U., Siscoe, G., Pagle, C., and Cargill, P. J.: 2005, *J. Geophys. Res.* **110**, Issue A1 CiteID A01105 10.1029/2004JA010814.
Palmer, I. D.: 1982, *Rev. Geophys. Space Phys.* **20**, 335–351.
Parker, E. N.: 1958, *Astrophys. J.* **128**, 664–672.
Pick, M., Forbes, T. G., Mann, G., *et al.*: 2006, *Space Sci. Revs.*, this volume.
Priest, E. R.: 1982, *Solar Magnetohydrodynamics*, Reidel, Dordrecht.
Reames, D. V., Barbier, L. M., and Ng, C. K.: 1996, *Astrophys. J.* **466**, 473–486.
Reiner, M. J., Kaiser, M. L., and Bougeret, J.-L.: 2003, in: Velli, M. *et al.* (eds.), *Solar Wind Ten*, Amer. Inst. Phys., New York, pp. 152–155.
Rice, W. K. M., Zank, G. P., and Li, G.: 2003, *J. Geophys. Res.* **108** Issue A10 CiteID 1369 DOI 10.1029/2002JA009756.
Richardson, I. G., Cane, H. V.: 2004, *Geophys. Res. Lett.* **31** Issue 18 CiteID L18804 DOI 10.1029/2004GL020958.
Riley, P., and Crooker, N. U.: 2004, *Astrophys. J.* **600**, 1035–1042.
Riley, P., Gosling, J. T., and Pizzo, V. J.: 1997, *J. Geophys. Res.* **102**, 14677–14686.
Riley, P., Linker, J. A., Mikić, Z., Odstrcil, D., Zurbuchen, T., Lario, D., *et al.*: 2003, *J. Geophys. Res.* **108** Issue A7 CiteID 1272 DOI 10.1029/2002JA009760.
Riley, P., Linker, J. A., Lionello, R., Mikic, Z., Odstrcil, D., Hidalgo, M. A., et al.: 2004, . *Atmos. Solar-Terr. Phys.* **66**, 1321–1331.
Rogers, B. N., Drake, J. F., and Shay, M. A.: 2000, *Geophys. Res. Lett.* **27**, 3157–3160.
Romashets, E. P., and Ivanov, K. G.: 1991, *Geomag. Aeron.* **31**, 583–586.

Roussev, I. I., Forbes, T. G., Gombosi, T., Sokolov, I. V., De Zeeuw, D., and Birn, J.: 2003, *Astrophys. J.* **588**, L45–L48.
Roussev, I. I., Sokolov, I. V., Forbes, T. G., Gombosi, T. I., Lee, M. A., and Sakai, J. I.: 2004, *Astrophys. J.* **605**, L73–L76.
Ruffolo, D.: 1995, *Astrophys. J.* **442**, 861–874.
Russell, C. T., and Mulligan, T.: 2002, *Adv. Sp. Res.* **29**, 301–306.
Rust, D. M., and Kumar, A.: 1994, *Solar Phys.* **155**, 69–97.
Schmidt, J. M., and Cargill, P. J.: 2001, *J. Geophys. Res.* **106**, 8283–8290.
Schmidt, J. M., and Cargill, P. J.: 2003, *J. Geophys. Res.* **108**, Issue A1 CiteID 1023, DOI 10.1029/2002JA009325.
Schmidt, J. M., and Cargill, P. J.: 2004, *Annales Geophys.* **22**, 2245–2254.
Schmidt, J. M.: 2000, *Solar Phys.* **197**, 135–148.
Scholer, M., Kucharek, H., and Giacalone, J.: 2000, *J. Geophys. Res.* **105**, 18285–18294.
Schwartz, S. J., Burgess, D., Williamson, W. P., Kessel, R. L., Dunlop, M., and Lühr, H.: 1992, *J. Geophys. Res.* **97**, 4209–4227.
Shafranov, V.D.: 1966, *Revs. of Plasma Phys.* **2**, 103–151.
Sheeley, Jr., N. R., Walters, J. H., Wang, Y.-M., and Howard, R. A.: 1999, *J. Geophys. Res.* **104**, 24739–24767.
Schwenn, R., Raymond, J. C., *et al.*: 2006, *Space Sci. Revs.*, this volume.
Srivastava, N., Schwenn, R., Inhester, B., Martin, S. F., and Hanaoka, Y., *Astrophys. J.* **534**, 468–481.
Skilling, J.: 1971, *Astrophys. J.* **170**, 265–273.
Sokolov, I. V., Roussev, I. I., Gombosi, T. I., Lee, M. A., Kóta, J., Forbes, T. G., *et al.*: 2004, *Astrophys. J.* **616**, L171–L174.
Somov, B. V., Kosugi, T., Hudson, H. S., Sakao, T., Masuda, S.: 2002, *Astrophys. J.* **579**, 863–873.
Steinolfson, R. S., and Nakagawa, Y.: 1977, *Astrophys. J.* **215**, 345–355.
Sterling, A. C., and Moore, R. L.: 2001, *J. Geophys. Res.* **106**, 25227–25238.
Sterling, A. C., and Moore, R. L.: 2004, *Astrophys. J.* **602**, 1024–1036.
Sterling, A. C., Moore, R. L., Qiu, J., and Wang, H.: 2001, *Astrophys. J.* **561**, 1116–1126.
Sturrock, P.: 1991, *Astrophys. J.* **380**, 655–659.
Sturrock, P. A., Weber, M., Wheatland, M. S., and Wolfson, R.: 2001, *Astrophys. J.* **548**, 492–496.
Subramanian, P., Ananthakrishnan, S., Janardhan, P., Kundu, M. R., White, S. M., and Garaimov, V. I.: 2003, *Solar Phys.* **218**, 247–259.
Suess, S. T.: 1988, *J. Geophys. Res.* **93**, 5437–5445.
Švestka, Z., and Cliver, E.: 1992, in: Švestka, Z., Jackson, B., and Machado, M. (eds.), *Eruptive Solar Flares*, New York: Springer-Verlag, pp. 1–14.
Titov, V., and Démoulin, P.: 1999, *Astron. Astrophys.* **351**, 701–720.
Tokman, M., and Bellan, P.: 2002, *Astrophys. J.* **567**, 1202–1210.
Tsuneta, S., and Naito, T.: 1998, *Astrophys. J.* **495**, L67–L70.
Tylka, A. J., Cohen, C. M. S., Dietrich, W. F., Lee, M. A., Maclennan, C. G., Mewaldt, R. A., *et al.*: 2005, *Astrophys. J.* **625**, 474–495.
Vainio, R., and Khan, J. I.: 2004, *Astrophys. J.* **600**, 451–457.
Vandas, M., Fischer, S., and Geranios, A.: 1991, *Planet. Space Sci.* **39**, 1147–1154.
Vandas, M., Fischer, S., Pelant, P., and Geranios, A.: 1992, in: Marsch, E., and Schwenn, R. (eds.), *Solar Wind Seven*, *Pergamon*, New York, pp. 671–674.
Vandas, M., Fischer, S., Pelant, P., and Geranios, A.: 1993, *J. Geophys. Res.*, **98** 11467–11475.
Vainio, R., and Schlickeiser, R.: 1999, *Astron. Astrophys.* **343**, 303–311.
Vainio, R., Kocharov, L., and Laitinen, T.: 2000, *Astrophys. J.* **528**, 1015–1025.
Vainio, R.: 1997, *Proc. 25th Internat. Cosmic Ray Conf. (Durban)* **1**, 253–256.

Vainio, R.: 1999, in: Ostrowski, M., and Schlickeiser, R. (eds.), *Plasma Turbulence and Energetic Particles in Astrophysics*, Obserwatorium Astronomiczne, Uniwersytet Jagielloński), Cracow, pp. 232–245.
Vainio, R.: 2003, *Astron. Astrophys.* **406**, 735–740.
van Ballegooijen, A. A.: 2004, *Astrophys. J.* **612**, 519–529.
van Ballegooijen, A.A., and Martens, P. C. H.: 1989, *Astrophys. J.* **343**, 971–984.
van der Linden, R. A. M., Hood, A. W., and Goedbloed, J. P.: 1994, *Solar Phys.* **154**, 69–96.
van Nes, P., Reinhard, R., Sanderson, T. R., Wenzel, K.-P., and Zwickl, R. D.: 1984, *J. Geophys. Res.* **89**, 2122–2132.
van Tend, W., and Kuperus, M.: 1978, *Solar Phys.* **59**, 115–127.
Vandas, M., Fischer, S., Dryer, M., Smith, Z., and Detman, T.: 1995, *J. Geophys. Res.* **100**, 12285–12292.
Vandas, M., Fischer, S., Dryer, M., Smith, Z., and Detman, T.: 1996, *J. Geophys. Res.* **101**, 2505–2510.
Wagner, W., Hildner, E., House, L., Sawyer, C., Sheridan, K., and Dulk, G.: 1981, *Astrophys. J.* **244**, L123–L126.
Wang, H., Qiu, J., Jing, J., and Zhang, H.: 2003, *Astrophys. J.* **593**, 564–570.
Wang, Y.-M.: 2001, *Astrophys. J.* **560**, 456–465.
Wang, Y.-M., Nash, A. G., and Sheeley, Jr. N. R.: 1989, *Science* **245**, 712–718.
Wang, Y.-M., and Sheeley Jr., N. R.: 1990, *Astrophys. J.* **365**, 372–386.
Welsch, B. T., Abbett, W. P., Fisher, G. H., and Regnier, S.: 2004, *Astrophys. J.* **610**, 1148–1156.
Wimmer-Schweingruber, R. F., Crooker, N. U., *et al.*: 2006, *Space Sci. Revs.*, this volume.
Wood, B. E., Karovska, M., Chen, J., Brueckner, G. E., Cook, J. W., and Howard, R. A.: 1999, *Astrophys. J.* **512**, 484–495.
Wu, C. S.: 1984, *J. Geophys. Res.* **89**, 8857–8862.
Wu, S. T., Guo, W. P., and Wang, J. F.: 1995, *Solar Phys.* **157**, 325–248.
Wu, S. T., Guo, W. P., and Dryer, M.: 1997, *Solar Phys.* **170**, 265–282.
Wu, S. T., Guo, W. P., Michels, D. J., and Burlaga, L. F.: 1999, *J. Geophys. Res.* **104**, 14789–14802.
Wu, S. T., Andrews, M. D., and Plunkett, S. P.: 2001, *Space Sci. Revs.* **95**, 191–213.
Zalesak, S. T.: 1979, *J. Comput. Phys.* **3** 335–362.
Zank, G. P., Rice, W. K. M., and Wu, C. C.: 2000, *J. Geophys. Res.* **105**, 25079–25096.
Zhang, J.: 2005, in: Dere, K., Wang, J., and Yan, Y. (eds.), *Coronal and Stellar Mass Ejections, Proc. IAU Symp. 226*, Cambridge Univ. Press., Cambridge, UK, pp. 65–70.
Zhang, J., Dere, K., Howard, R., and Vourlidas, A.: 2004, *Astrophys. J.* **604**, 420–432.

THE PRE-CME SUN

Report of Working Group E

N. GOPALSWAMY[1], Z. MIKIĆ[2], D. MAIA[3], D. ALEXANDER[4], H. CREMADES[5], P. KAUFMANN[6], D. TRIPATHI[5] and Y.-M. WANG[7]
[1]*NASA Goddard Space Flight Center, Greenbelt, Maryland, USA*
[2]*SAIC, 10260 Campus Point Drive, San Diego, CA 92121-1578 USA*
[3]*CICGE, Faculdade Ciências Universidade Porto, Portugal*
[4]*Department of Physics and Astronomy, Rice University, 6100 Main St, Houston, TX 77005, USA*
[5]*Max-Planck-Institute for Solar System Research, 37191 Katlenburg-Lindau, Germany*
[6]*CRAAM, Mackenzie Presbyterian University, S. Paulo, Brazil*
[7]*Naval Research Laboratory, Washington, DC, USA*
*(*Author for correspondence: E-mail: gopals@ssedmail.gsfc.nasa.gov)*

(Received 3 February 2006; Accepted in final form 12 June 2006)

Abstract. The coronal mass ejection (CME) phenomenon occurs in closed magnetic field regions on the Sun such as active regions, filament regions, transequatorial interconnection regions, and complexes involving a combination of these. This chapter describes the current knowledge on these closed field structures and how they lead to CMEs. After describing the specific magnetic structures observed in the CME source region, we compare the substructures of CMEs to what is observed before eruption. Evolution of the closed magnetic structures in response to various photospheric motions over different time scales (convection, differential rotation, meridional circulation) somehow leads to the eruption. We describe this pre-eruption evolution and attempt to link them to the observed features of CMEs. Small-scale energetic signatures in the form of electron acceleration (signified by nonthermal radio bursts at metric wavelengths) and plasma heating (observed as compact soft X-ray brightening) may be indicative of impending CMEs. We survey these pre-eruptive energy releases using observations taken before and during the eruption of several CMEs. Finally, we discuss how the observations can be converted into useful inputs to numerical models that can describe the CME initiation.

Keywords: coronal mass ejections, flares, radio bursts, filaments, prominences, streamers, solar magnetism, active regions, closed and open magnetic fields, pre-eruption signatures, energy storage and release, helicity

1. Introduction to the Pre-CME Sun

Understanding the pre-eruption state of the solar sources of CMEs is a key aspect in unraveling the mystery behind CME initiation. There is also an important practical application: if it is possible to identify some parameter that may indicate a high probability for eruption, then one should be able to predict the imminent occurrence of a CME and hence the accompanying interplanetary consequences. In order to understand the pre-eruption state, we need to consider three aspects of an eruption region on the Sun: (i) the magnetic structure, (ii) the evolution, and (iii) the energetic

Space Science Reviews (2006) 123: 303–339
DOI: 10.1007/s11214-006-9020-2

signatures. The very basic requirement in an eruption region is that it contains closed magnetic field structure identified in longitudinal magnetograms by pairs of positive and negative magnetic polarity patches. A simple closed field structure may not have enough free energy to be ejected as a CME, so the closed field structure has to be distorted by pre-eruption evolution that adds free energy, generally referred to as energy build-up. Shearing motions, flux emergence and cancellation, and sunspot rotations in the source region are some of the processes that occur in the energy build-up phase and can result in small-scale energy releases distinct from the eruption itself. Such small-scale heating and non-thermal radio emission (indicative of particle acceleration) are considered to be signatures of the energy build-up and release.

It must be noted that extensive literature exists on preflare activities (see e.g. Priest *et al.*, 1986; Simnett, 1999). While not all flares are associated with CMEs, the precursor activity in eruptive flares is expected to be similar to that for CMEs. All the aspects of the structure, evolution and energetics of the preflare source region are applicable to the pre-CME source region.

The free magnetic energy is typically found to be twice the potential energy of the eruption region (Mackay *et al.*, 1997; Forbes, 2000; Metcalf *et al.*, 1995), although recent results indicate the possibility of much higher factors (Metcalf *et al.*, 2004). In this chapter we focus on the energy build-up and describe the three basic aspects of eruption regions.

After an introduction to the pre-CME Sun (Section 1), the pre-eruption structure (Section 2), pre-eruption evolution (Section 3), and pre-eruption energetic signatures (Section 4) are presented. Global issues related to the pre-CME Sun are discussed in Section 5. The final Section 6 contains discussion and summary with a brief note on future perspectives.

2. Pre-Eruption Structure

The characteristics and dynamics of CMEs are inherently related to the properties and behavior of their associated near-surface features. Consequently, what we observe in a CME source region prior to the eruption may yield important insight into the eruption process itself. Quiescent features in the various layers of the solar atmosphere need to be pieced together to fully understand the pre-eruption structure of a CME. The type of magnetic field configurations we infer from observations and how they relate to the observed structure of CMEs is presented in this section.

2.1. Closed and Open Field Regions on the Sun

The basic magnetic structure of the solar atmosphere consists of open and closed field. The closed field regions are typically comprised of active regions, quiescent filament regions, and transequatorial interconnecting regions. Filaments are regarded as one of the fundamental solar sub-structures relevant to coronal mass ejections: solar source regions occasionally include both filaments and active regions, with the filaments lying outside, but close to the active regions. The primary difference between the filament regions and active regions is simply characterized by the strength and structure of the magnetic field involved. Active regions typically have much higher field strength in a compact volume, while filament regions have lower field strength, typically elongated along the filament axis (although there are exceptions). The post-eruption structures are also accordingly different in the two cases. Transequatorial structures generally appear during solar minimum, when the active regions are much closer to the equator. An eruption involving a transequatorial region occurs over a larger volume than one involving a single active region.

Figure 1 shows the Sun during October 17–18, 1999 at four different wavelengths, corresponding to four different heights in the atmosphere: a longitudinal

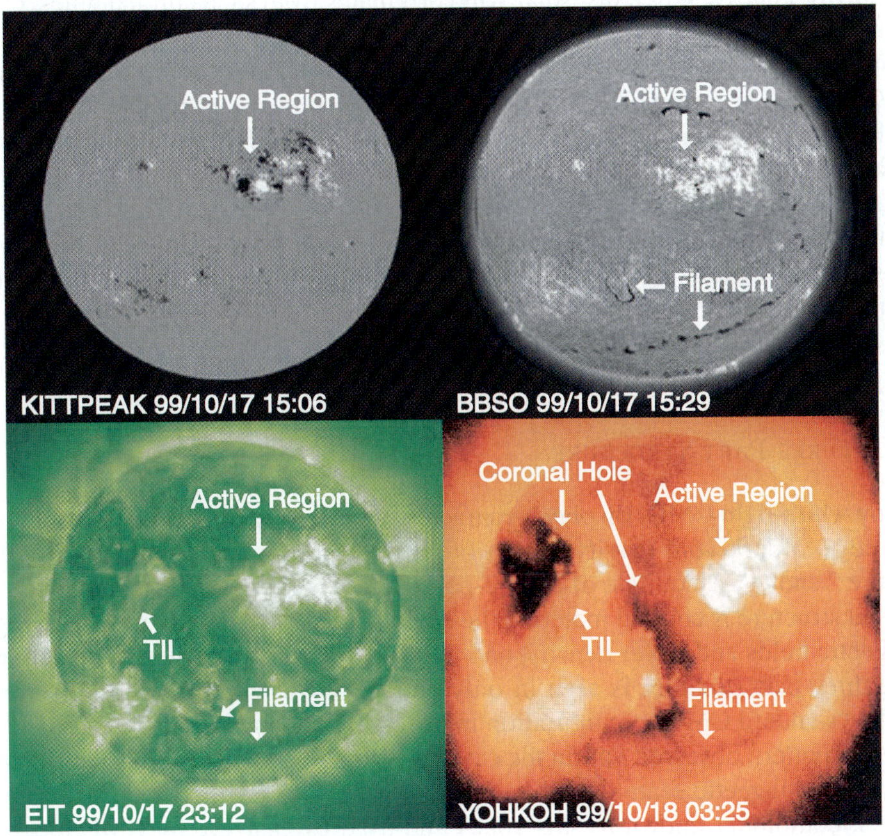

Figure 1. Large-scale solar magnetic regions such as active regions and filament regions that have closed field structure (CME-producing regions) and coronal holes with open field structure that do not produce CMEs. TIL–transequatorial interconnecting region.

magnetogram from SOHO/MDI, an Hα picture from the Big Bear Solar Observatory, an EUV image at 195 Å from SOHO/EIT, and a soft X-ray image from Yohkoh. The active region complex contains five numbered regions, the largest one being AR 8731. The active region complex appears brightest in all images. The active region loops are hotter and denser and hence have bright coronal emission. The elongated dark features are the filaments, which delineate opposite polarity patches both in the active region and in the quiet regions. As the active region developed, the U-shaped filament in the Hα image erupted, as captured in the SOHO/EIT image: the filament appears bright in EUV probably because it was heated during eruption. A prominent transequatorial structure can be seen in the SXT and in EIT images (marked TIL), connecting AR 8735 in the north (N18E22) and AR 8736 (S20E35) in the south. In principle, all of these closed field structures could erupt provided they have enough free energy. One can also see coronal holes, which appear as dark patches in the SXT and EIT images. The magnetic field inside the largest coronal hole in the northwest quadrant is enhanced and of unipolar (positive) polarity. While fast solar wind originates from such coronal holes, CMEs do not.

White-light observations of CMEs show a wealth of different morphologies, ranging from amorphous blobs, to simple narrow jet-like features, up to highly structured and complicated entities. While CMEs exclusively originate from closed magnetic field regions, there are situations in which there is an interaction between closed and open field regions. We shall first consider the structure of CMEs that originate from closed field regions and then briefly discuss narrow CMEs resulting from the interaction between closed and open field structures.

2.2. Three-Part Structure

Many white-light CMEs display a characteristic three-part structure: a bright leading edge, a dark void (cavity) and a bright core (Illing and Hundhausen, 1985). Figure 2 shows an example of a three-part CME recorded by SOHO/LASCO. The frontal structure is coronal material, the cavity also is coronal, but may have higher magnetic fields and lower density, and the bright core is the eruptive prominence. The three-part structure is seen in only about 30% of CMEs, yet this is viewed as the "standard CME" configuration in observational and theoretical studies. The circular pattern observed within the cavities of many CMEs suggests the existence of helical magnetic features (Chen *et al.*, 1997; Dere *et al.*, 1999; Wood *et al.*, 1999), commonly known as flux ropes and used extensively by CME modelers. *In-situ* measurements of magnetic clouds linked to filament eruptions (Marubashi, 1986; Gopalswamy *et al.*, 1998), and the discovery of filament material inside the magnetic cloud (Burlaga *et al.*, 1998) support this association.

Figure 2. SOHO/LASCO image (with an EIT 195 image superposed) obtained on 2001 December 20 showing the three-part structure of a CME above the southwest limb.

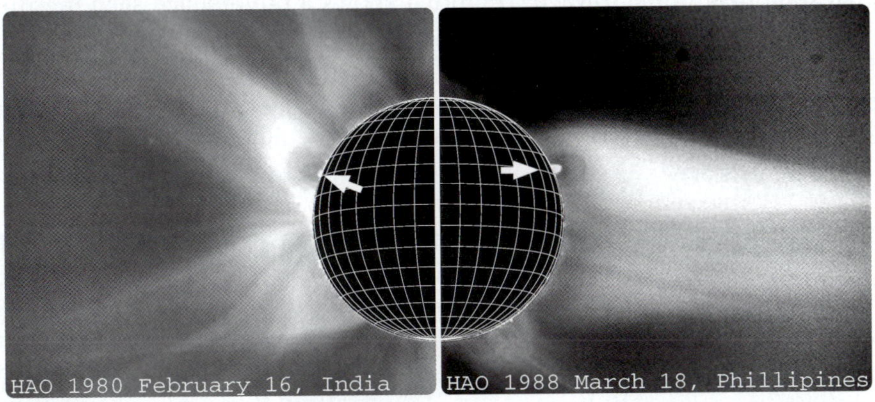

Figure 3. White-light eclipse images showing bright streamers overlying cavities with filaments inside (indicated by the arrows). Courtesy High Altitude Observatory.

The three-part structure is commonly observed in eclipse pictures (Saito and Tandberg-Hanssen, 1973). Figure 3 shows two eclipse images where the three-part structure is clearly evident. The subsequent eruption of this structure then becomes the three-part CME. The prominence and cavity become the bright core and flux rope of the CME, while the streamer deforms to become the frontal structure. The direct connection between pre-eruption and CME components can be shown by means of height-time plots, which reveal the continuity of the three different CME parts when the fields of view of different instruments overlap (Illing and Hundhausen, 1985; Srivastava *et al.*, 1999; Plunkett *et al.*, 2000; Gopalswamy *et al.*, 2003b).

2.2.1. *The Bright Core*

The bright core is the feature of CMEs whose near-surface counterpart is best known. The white-light core of CMEs has been shown to be the eruptive filament observed in Hα (House *et al.*, 1981) or in micro-waves (Gopalswamy, 1999) close to the surface. Solar filaments, called "prominences" when viewed above the limb, have an observational heritage significantly older than that of CMEs. Prominence eruptions, discovered in the 1800s, were in fact one of the earliest forms of mass ejection to receive attention (for a historical introduction see Tandberg-Hanssen, 1995). Filaments are lanes of cool (7000 K) plasma embedded in the corona that lie, without exception, above the photospheric magnetic neutral lines (Babcock and Babcock, 1955). They are located within the so-called filament channels, where the orientation of chromospheric fibrils along the filament axis suggests a strongly sheared magnetic field (Martin, 1998a). Filaments have traditionally been observed in the Hα line, but are also visible in the EUV or soft X-rays as dark structures (Vaiana *et al.*, 1973). Prominences are optically thick in microwaves, so they appear as dark elongated features on the disk and bright features above the limb as in Hα (Gopalswamy, 1999). In fact prominences are the brightest features outside the solar disk in microwaves. Figure 4 shows a prominence seen in different wavelengths.

Although bright cores are commonly observed in CMEs, their presence is not necessary: many structured CMEs are not three-part but two-part, being composed only of a bright leading edge and a cavity. Even for those CMEs associated with

Figure 4. Prominence seen in different wavelengths. (left) The quiescent prominence is seen in EUV as a series of dark features, each looking like an inverted "Y". The bright (middle) prominence seen in Hα in emission. Note the good correspondence between each of the features seen in EUV and Hα. (right) In microwaves the prominence is seen in emission, and there is good correspondence with other wavelengths.

prominence eruptions, only 60% to 70% show a bright core (Hori and Culhane, 2002; Gopalswamy *et al.*, 2003b). Not all filaments that erupt lead to CMEs, and there are some strikingly different results on the fraction of prominence eruptions that lead to CMEs. Early work by Webb *et al.* (1976) found that all filament disappearances they examined had a "transient coronal manifestation". This result was confirmed in more recent work by Gilbert *et al.* (2000), Hori and Culhane (2002), and Gopalswamy *et al.* (2003b), with association rates that varied between 70% to 90%. On the other hand, Yang and Wang (2002) have found that, in a total of 431 prominence eruptions, only 10% to 30% were associated with CMEs. The discrepancy may be due in part to selection criteria, mainly in the maximum height attained by the prominence. In fact, Munro *et al.* (1979) observed that while all prominence eruptions seen beyond $1.2R_\odot$ were associated with CMEs, only about 60% of those seen beyond 1.1 but below $1.2R_\odot$ were associated with CMEs. They noted also that about two thirds of reported eruptive prominences are in fact not seen above $1.1R_\odot$; and these show a rate of association with CMEs of only about 10%. The relation between prominence height and CME association has recently been confirmed using the height-time history of eruptive prominences and their associated CMEs (Gopalswamy *et al.*, 2003b).

2.2.2. *The Cavity*

The cavity is a localized lack of emission (see Figure 3), in white-light images. Engvold (1988) found this brightness depression to be due to a drop of 50% to 75% in the local electron density in the immediate vicinity of the filament and the overlying coronal arcades. It is tempting to see this cavity as the outer part of the magnetic structure (flux rope) sustaining the filament (Low and Hundhausen, 1995), but other possibilities have been proposed. Martin (1998a) defines the cavity as the volume between features of opposite chirality. The void exists because structures with opposite chirality cannot conjoin.

Cavities are also seen in EUV and soft X-ray images (Vaiana *et al.*, 1973; Hudson *et al.*, 1999). An example, observed by EIT in 195Å, on 1997 December 3, is shown in Figure 5. On this day three filament channels were seen extending close to the limb, with dark thin filaments clearly visible inside each of them. Overlying each of these filaments a cavity is seen reaching an altitude about $0.15R_\odot$ over the solar limb.

EUV and X-ray observations show cavities only at the limb. At radio wavelengths, cavities can be observed on the disk as brightness depressions (Kundu and McCullough, 1972; Gopalswamy *et al.*, 1991; Marqué, 2004). The brightness depressions at 20 cm wavelength were found to have a larger extent than the filament itself at higher temperature (Gopalswamy *et al.*, 1991). Decimetric radio images from the Nançay Radioheliograph show cavities overlying filaments observed in Hα. The observed radio depression in the a cavity implies a density drop of 25% to 50% from the mean coronal density (75% for one event). The radio cavities match the extent of filament channels, and are continuous features even when the associated Hα filament is not. When seen erupting, the radio depression overlies

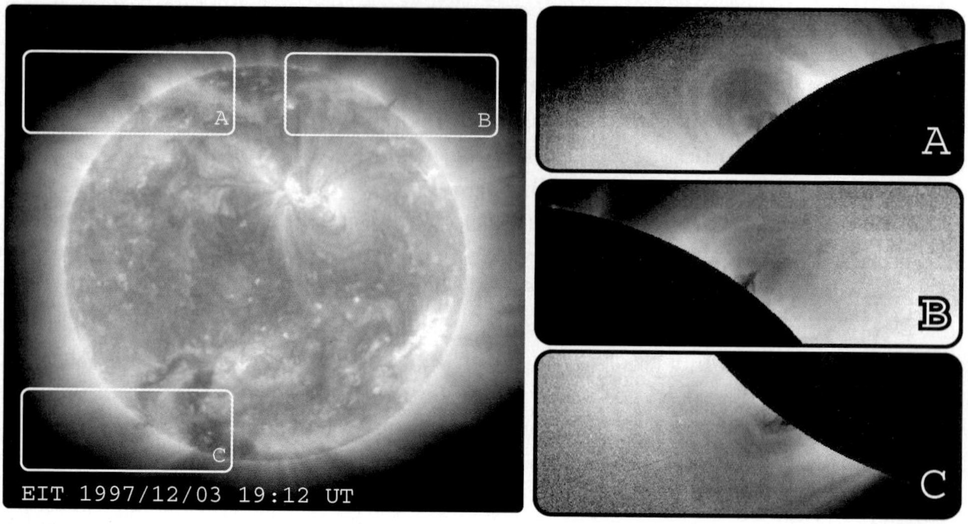

Figure 5. Cavities overlying EUV filaments as seen by EIT on 1997 December 3, in 195 Å bandpass.

the erupting filament (Marqué *et al.*, 2002), suggesting that the radio depression is linked to the filament, in agreement with flux-rope models of filaments (Low and Hundhausen, 1995).

2.2.3. *The Frontal Structure*

The frontal structure of CMEs (the bright leading edge) represents density enhancements from 10 to 100 times the background corona (below $6R_{\odot}$). The average mass of CMEs is about 2×10^{15} g, with a significant fraction of CMEs (15%) having masses below 10^{14} g (Vourlidas *et al.*, 2002). The question is where does all that mass come from and what kind of features in the low corona account for the frontal structure? It has been suggested that the mass in the overlying pre-eruptive arcade could account for most of the mass in CMEs (Hudson *et al.*, 1996; Gopalswamy and Hanaoka, 1998). Some observations indicate that prominences also carry approximately the same mass as the frontal structure (Gopalswamy and Hanaoka, 1998). As the CME proceeds into the corona, it will likely sweep up additional material so that the mass would increase, at least in the beginning. As the CME moves out, the frontal structure near the Sun is likely to evolve into the sheath material behind interplanetary shocks observed *in situ* (Gopalswamy, 2003). Transequatorial loop systems (TLS) have masses around 10^{15} g (Khan and Hudson, 2000). These may become part of the larger CMEs (Delannée and Aulanier, 1999; Wills-Davey and Thompson, 1999; Khan and Hudson, 2000) and as such can account for a substantial fraction of the CME mass. Glover *et al.* (2003) have found that TLS are likely to contribute mass to a CME if at least one of the following happens: (1) two active regions are involved, (2) the TLS is extended, (3) the longitudinal asymmetry is

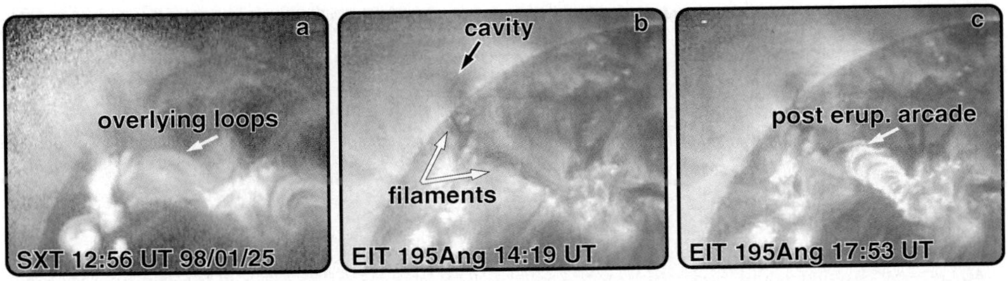

Figure 6. Pre-event versus post-event configuration of a CME source region. (a) and (b) The pre-event shows a system of sheared loops overlying a filament. (c) After the CME lift-off magnetic restructuring gives rise to an arcade of loops perpendicular to the neutral line.

high. Unfortunately no comprehensive study of TLS association with CMEs has been made to date.

2.2.4. *Relation to Post-Eruption Arcade*

Post-eruptive arcades (PEAs) are defined as the formation of transient large-scale loop systems straddling the neutral line, following an eruption. Figure 6 shows an example of such a feature. The PEAs are typically observed over a time period of several hours on the solar disk in EUV and X-rays (Webb *et al.*, 1978; Zarro *et al.*, 1999; Sterling *et al.*, 2000; Gopalswamy, 2003; Tripathi *et al.*, 2004). In Hα, they are referred to as post-flare loops. Occasionally, the PEAs are observed in microwaves as enhanced thermal free-free emission (Hanaoka *et al.*, 1994). The foot points of a series of PEA loops form the ribbons visible in Hα (the two-ribbon flares). EUV PEAs observed by SOHO/EIT (195 Å) from 1997 to 2002 had an average duration of $\sim$7 hours and an average length of $\sim$15 degrees (Tripathi *et al.*, 2004). Kopp and Pneuman (1976) have interpreted the formation of loop systems as a consequence of magnetic reconnection in the aftermath of the CME eruption. One of the interesting questions is how these PEAs evolve to become the frontal structures of future CMEs.

2.3. Filament Eruption Near Coronal Holes

Filaments located in the vicinity of coronal holes have shown some proclivity for eruption. This was first pointed out by Webb *et al.* (1978) using Hα and Skylab X-ray data. Webb *et al.* (1978) studied the relation between X-ray PEAs and the long and short-term variability in coronal hole boundaries and found that they occurred over neutral lines which form the borders of evolving coronal holes. The associated filaments were observed to rotate rigidly (compared to the photospheric plasma) similar to the coronal holes (Adams, 1976). A similar study was conducted using Yohkoh data by Bhatnagar (1996). Webb *et al.* (1976) also found that the eruptions are associated with coronal holes growing in area while Bravo (1996) suggested that

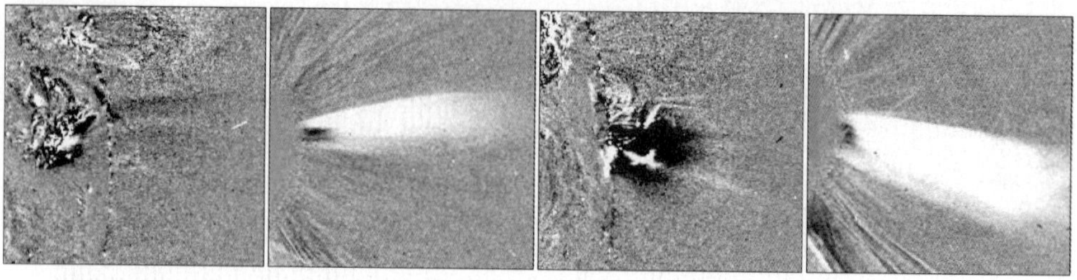

Figure 7. Narrow, jet-like CMEs originating from an active region inside a coronal hole, 1999 November 12 and 14. (a) and (c) show the two events as observed near the solar surface with EIT (each panel represents the difference between two Fe XII λ195 images taken 12 min apart). (b) and (d) show the same events as observed above $r \sim 2R_\odot$ over the west limb with the LASCO C2 coronagraph. Both CMEs had velocities of the order of 1000 km/s.

the coronal hole area may increase due to the eruption of nearby filaments. Thus the location of filaments near coronal holes can be considered as a pre-eruption state with a propensity for eruption. These studies need to be repeated with the uniform and extensive data set available from SOHO. For example, SOHO/EIT data can provide information on coronal holes and filaments. Filament eruptions are readily observed in the EUV images and can be compared with SOHO/LASCO data for the associated CMEs. Eruptions of filaments not located close to coronal holes can be used as a control sample.

2.4. Jets, Narrow CMEs, and Their Relation to Coronal Holes

There is a particular class of mass ejection that appears to be triggered by the interaction between closed and open magnetic fields and that is closely associated with coronal holes. During the 1996–1997 activity minimum, the LASCO C2 coronagraph detected faint linear structures propagating outward at high speeds from the Sun's polar regions. These white-light jets, which occurred at an average rate of 3–4 per day and which were also seen in O VI and Lyα with UVCS (Dobrzycka *et al.*, 2002), had angular widths of $1^\circ - 4^\circ$ and leading-edge velocities of ~400–1100 km/s. They were shown to be the outward extensions of EIT jets that originated from flaring EUV bright points inside the polar coronal holes (Wang *et al.*, 1998). During the 1999–2001 activity maximum, both white-light jets and narrow CMEs having widths $\lesssim 15^\circ$ and velocities of up to 1000 km/s were observed to originate from active regions located inside or along the boundaries of low-latitude coronal holes (Wang and Sheeley, 2002). A single active region often gave rise to a succession of such ejections over its lifetime, which were detected by LASCO as the coronal hole rotated past the solar limb (see Figure 7). The basic mechanism underlying all of these jet-like events is presumed to be "interchange" magnetic reconnection between the bipole and coronal hole fields, whereby material

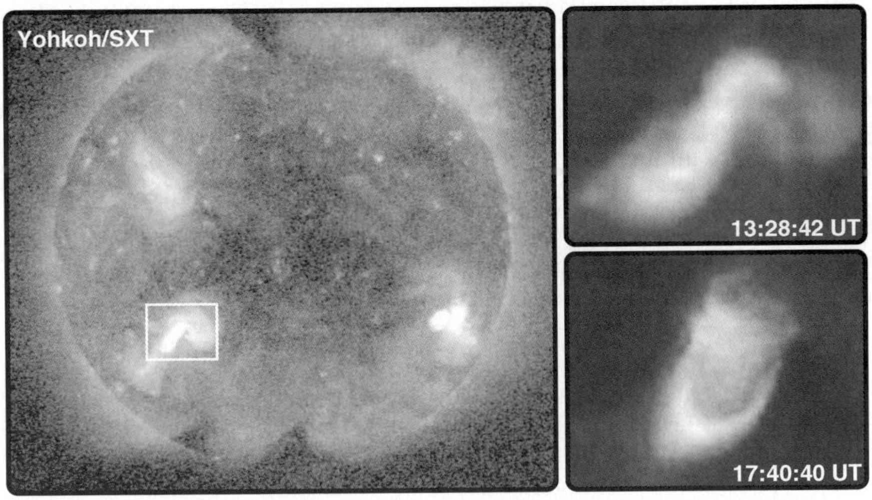

Figure 8. Soft X-ray sigmoid observed by the Yohkoh/SXT. This region erupted with an associated CME: the reconfiguration of the coronal structure in the aftermath of the eruption is evident. Courtesy Yohkoh Team.

is transferred from closed to open fields lines to produce a fast, highly collimated ejection.

2.4.1. *Sigmoids*

An observational manifestation of the connection between coronal structure and CME production is the soft X-ray sigmoid (Rust and Kumar, 1996; Pevtsov and Canfield, 1999; Canfield *et al.*, 2000). An example of a sigmoid that gives rise to an eruption is shown in Figure 8. Active regions exhibiting sinuous-S or reverse-S shapes have been shown to exhibit a greater tendency to erupt (Canfield *et al.*, 1999). Leamon *et al.* (2002), further demonstrated that when active region sigmoids do erupt, they tend to produce at least moderate geomagnetic storms. Relating the sigmoids at X-ray (and other) wavelengths to magnetic structures and current systems in the solar atmosphere is the key to understanding their relationship to CMEs. The sigmoidal shape suggests that the twist in the magnetic structure may have something to do with the eruption. For instance, the stability of a flux tube seems to be related to the amount of twist it contains (e.g. Priest, 1984; Rust and Kumar, 1994b) with the result that a flux tube becomes unstable above a critical twist of order 2π. However, Leamon *et al.* (2003) have argued that the observed sigmoids fall well short of this critical twist, albeit using an approach which measures an active-region averaged value for the force-free parameter, α, as a measure of the twist in the sigmoid.

The transition from a sigmoid to an arcade in the X-ray observations during the initiation of a CME is strongly suggestive that an instability, and associated topological reconfiguration, has occurred. However, sigmoids may also disappear

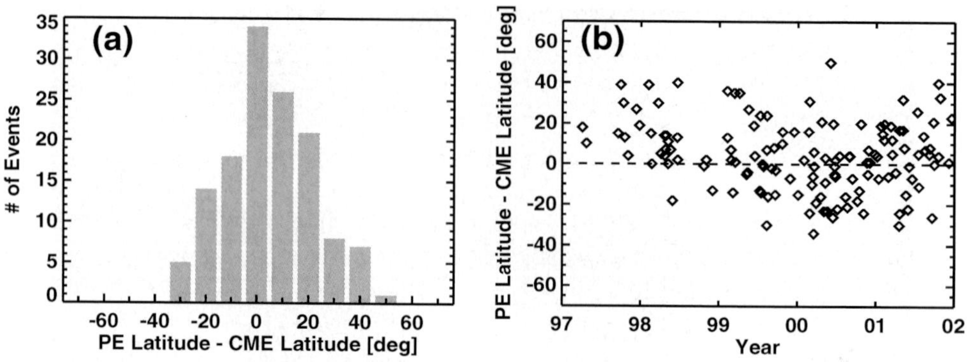

Figure 9. The latitude difference between CMEs and prominence eruptions (PEs) for the period 1996–2001: (a) the distribution of the latitude difference, (b) the latitude difference as a function of time. The reason for the positive offset of PEs in (a) is obvious in (b) because the offset is positive during solar minimum. Towards and during solar maximum, there is no net offset. During solar minimum the dipole field of the Sun is strong and guides the CMEs towards the equator. During the rising phase and maximum the dipole field is considerably weakened and hence the PEs and CMEs have roughly the same position angle. Adapted from Gopalswamy *et al.* (2003b).

and reappear without a detectable CME (see Gibson *et al.*, 2002). We need to understand more about the formation and evolution of sigmoid structures in active regions and to explore the conditions that drive them to eruption. Key components would be the role played by magnetic helicity and how this helicity reaches the corona, via photospheric motions (shearing) or direct emergence of twisted flux.

2.5. Guiding Streamers

During solar minimum, the CME latitude shows a single broad peak centered on the equator, while the source region latitude exhibits a two-peak behavior, with maxima at around $\pm 30°$ (Plunkett *et al.*, 2002; Gopalswamy *et al.*, 2003b; Cremades and Bothmer, 2004). This effect is clearly evident in Figure 9, showing the distribution of the offset between the latitudes of prominence eruptions (PEs) and CMEs (derived from the central position angles). The offset is clearly positive during solar minimum years (the PE latitude is poleward of the CME latitude). The reason for this offset can be explained by the strong influence of the large-scale global solar magnetic field on the erupting structure, as evidenced by the nonradial motion of prominences (Gopalswamy *et al.*, 2000; Filippov *et al.*, 2001). The typical scenario is that the prominence originates from the active region belt, and is channeled into the streamer centered at the heliospheric current sheet, typical of minimum solar activity. Since the majority of the CMEs observed in the field of view of SOHO/LASCO C2, maintain their direction of propagation, the deflection must have already taken place at lower altitudes, i.e. below 2 solar radii. The marked trend in CME position angle toward the equator is not seen at solar maximum: CMEs get deflected in all directions, reflecting the varied influence of the solar dipolar field on CMEs during solar minimum and maximum.

3. Pre-Eruption Evolution

Pre-eruption evolution can be loosely defined as the process by which closed field structures on the Sun are energized (or become non-potential). One of the simplest ways in which this can happen is by the deformation of the closed field lines as a result of the change in footpoint separation or twisting of one or both footpoints. The deformation resulting from the footpoint translations is referred to as magnetic shear and statistically, regions with sheared magnetic field lines are prone to eruption. Vector magnetograms provide direct evidence for sheared magnetic fields near the photospheric neutral lines (e.g. Krall *et al.*, 1982; Heyvaerts and Hagyard, 1991) and rotation of sunspots causing magnetic shear is often observed. Measurement of shear in active regions is possible from vector magnetograms, and hence provides a simple means of predicting the possibility of large eruptions. Theoretically, as the shear reaches a critical value, the system goes out of equilibrium and the sheared structure erupts (Hagyard *et al.*, 1984). Other processes such as flux emergence and vertical flows may also trigger the eruption of sheared structures (Ambastha, 1998). Flux emergence has long been associated with filament eruption (Canfield *et al.*, 1974), a good indicator of CMEs (Gopalswamy, 2003 and references therein). However, flux emergence seems to be ubiquitous in active regions and its utility as a pre-eruption signature requires more study. Filament activation observed for about 10–30 min before eruption provides further insight into the eruption process itself. The filament activation may represent changes in the background magnetic field or the increase in filament current (see, e.g., Kuperus and van Tend, 1981) which may result from changes caused by shear or flux emergence.

3.1. Shear, Differential Rotation

In the solar corona, the magnetic helicity and excess energy must be supplied by the photospheric or sub-photospheric activity. While theoretical studies of helicity injection have concentrated on the emergence of twist and/or writhe in confined flux tubes (e.g. Magara and Longcope, 2001), observationally most of the emphasis has been on the shearing of the footpoints of coronal structures (Canfield and Pevtsov, 1998; Kusano *et al.*, 2002; Moon *et al.*, 2002) with some effort to determine the flux emergence component (Nindos and Zhang, 2002; Kusano *et al.*, 2002).

Recent results have been confusing about the role the two components play in the production of flares and CMEs, through the build-up of helicity in the corona. On the one hand, DeVore (2000) has argued that a significant quantity of magnetic helicity is injected by the action of differential rotation over the lifetime of an active region; enough to explain the total ejected helicity detected in interplanetary magnetic clouds, often treated as the interplanetary counterparts of CMEs (Gosling *et al.*, 1995). This assertion has been contested by Démoulin *et al.* (2002) and Green *et al.* (2002) who argue that the helicity injected by differential rotation is 5 to 50 times smaller than that inferred to be carried away in CMEs, leaving these authors to conclude that the bulk of the helicity injection is provided by the twist in the sub-photospheric part of the magnetic flux tubes forming active regions. Proving this observationally is difficult since the measurement of the helicity injection from flux emergence not only requires vector magnetic field information but also knowledge of the vertical velocity component of the emergence (cf. Kusano *et al.*, 2002).

In the debate over the role of differential rotation, the strong local shearing often observed near the magnetic neutral line(s) of flare-productive active regions (e.g Harvey and Harvey, 1976) is frequently neglected. Such strong local shear may contribute significantly to the helicity injection into large but otherwise local structures associated with the active region. Recent studies by Kusano *et al.* (2002), Nindos and Zhang (2002) and Moon *et al.* (2002) have demonstrated that multiple flaring is often associated with regions of strongly sheared footpoint motions and local helicity injection. In the case discussed by Kusano *et al.* (2002) the shearing motions can contribute as much, if not more, helicity as the flux emergence.

3.2. Photospheric Flows

The evolution of the magnetic field is crucial in driving the solar corona to erupt in a CME. Unfortunately, routine observations of the 3D magnetic field can only be performed in the solar photosphere, where the conditions are demonstrably non-force-free (Metcalf *et al.*, 1995). The desired coronal field is then obtained using extrapolation techniques with the photospheric observations serving as a boundary condition and assuming, typically, that the field is force-free (e.g. Démoulin *et al.*, 1997). More recent approaches, made viable by continuously improving computational resources, adopt fully 3D MHD modeling of the of the coronal magnetic field incorporating the observed field and plasma data at the photospheric boundary (Abbett and Fisher, 2003). These data-driven boundary conditions then require detailed observations of the evolution not only of the magnetic field but also of the plasma, i.e. in order to fully understand the physical processes governing CME initiation we need to understand the role played by plasma flows in the photosphere.

The coupling of the photospheric dynamics and the coronal evolution is, at best, unclear. Evidence of flows associated with eruptive phenomena comes in the form of divergence of magnetic flux as new flux emerges, shearing motions in the photosphere, and the collision of opposite magnetic polarities leading to flux cancellation. Early studies on large flares indicated that shearing motions in the photosphere provided the mechanism by which energy was stored in the corona prior to eruption (e.g. Hagyard *et al.*, 1984). However, the relationship between sheared field and eruptive phenomena has not been conclusive, with some events occurring far from sheared neutral line and some eruptive sites exhibiting increased shear after the event (see review by Wang, 1993). Recently, observations of rotating sunspots have been instrumental in relating photospheric motions to helicity injection and subsequent CME initiation (Brown *et al.*, 2003). Numerical simulations generally incorporate photospheric motions in an ad hoc way, although efforts to include observations are progressing (see below). CME initiation models have been developed where the corona is driven to eruption by the emergence of twisted flux from below the surface (e.g. Fan, 2001), by subjecting the footpoints of coronal field to vortical motions (e.g. Roussev *et al.*, 2004a), and by including flux cancellation via converging flows (Amari *et al.*, 2003a).

Ideally, one would need to categorize all three components of the photospheric velocity. This is very difficult in practice although some progress is being made with a variety of techniques. An older and commonly used method is that of local correlation tracking (LCT: November and Simon, 1988) where displacements of identifiable features observed in the white-light continuum are correlated to determine the bulk transverse plasma flow. The problems with this approach are evident. One must not only be able to identify coherent features in the data but these features must last sufficiently long to be observed in several images: LCT results on SOHO/MDI data can require several hours of continuous data (see Berger *et al.*, 1998). In the source regions of flares and CMEs the field is evolving rapidly, with emerging and canceling flux limiting the accuracy of the LCT approach. Welsch *et al.* (2004) modified the standard LCT approach which incorporates the inductive equation to better determine the transverse velocity components. However, the LCT velocity may not be consistent with the induction equation and, consequently, Longcope (2004) has developed a parametric approach to solving the inductive equation independently from the LCT velocity results.

The detailed coupling between the observational data and the theoretical modeling is yielding important new results that will allow us to determine which properties of the field and velocity evolution are critical in driving the corona to eruption.

3.3. Emerging Flux

The connection between emerging flux and the CME can be readily established by comparing magnetograms and filament observations because the filament starts erupting very close to the site of emerging flux. The current paradigm is that the emerging flux leads to reconnection with the pre-existing flux. This means that the orientation of the emerging field lines should be favorable for reconnection. It is well known that flux emergence rate is an order of magnitude higher in active regions than in quiet regions (Liggett and Zirin, 1984). Since flux emergence is ubiquitous, it is not often easy to directly link the flux emergence to filament eruptions. However, in retrospect, it is generally possible to identify magnetic field changes on either side of the neutral line (Lara *et al.*, 2000).

Figure 10 shows a case of well-observed flux emergence on 2000 September 12. The flux emerged on the positive polarity side of the photospheric neutral line, with a large overlying filament. The region itself is an extended bipolar region with no sunspots (AR 9163 located at S17W09). The positive polarity of the emerging flux is not seen distinctly because it emerges into the existing positive polarity. The negative polarity region appears as a tiny circular region around 06:23 UT, rapidly spreading to become an elongated region by 11:15 UT. The filament, visible in the EIT 195 Å images, showed signs of activation coincident with the flux emergence, displayed rapid motion along the axis and then lifted off by 11:15 UT. The filament eruption was followed by an M1.0 flare starting at 11:32 UT. The associated CME first appeared above the limb at 11:54 UT. The CME had a relatively large acceleration (58.2 m/s^2), typical of CMEs associated with filament eruption. The first break-off of the filament occurred from the vicinity of the emerging flux. The emerging flux region appears as a dark spot in the EUV image, because it was hot and hence outside of the EUV pass band. This is verified from the Yohkoh/SXT images, which show the same spot as a bright feature. Yohkoh SXT images also show that the overlying field lines are sheared and hence highly non-potential. After the eruption, the post-flare X-ray loops appeared potential, roughly perpendicular to the neutral line. There was also a tiny cusp-shaped feature at the site of flux emergence visible in the Yohkoh/SXT image, suggesting reconnection between the newly emerging and preexisting fields. The cadence of the Hα pictures was not sufficiently high to see the details of the emerging flux manifestation at the chromospheric level, but a clear arch filament system is evident (See Bruzek, 1967 for details on the arch filament system).

The case study presented above is considered to be a case of flux emergence favorable for reconnection. Feynman and Martin (1995a) investigated a large number of filament eruptions and concluded that two-thirds of all the filament eruptions they studied occurred after substantial amounts of flux emerged in the vicinity of the filament. They also found that about 84% of filaments not associated with flux emergence did not erupt. They further note that the filament invariably erupted when the emerging flux had an orientation favorable for reconnection with the large-scale coronal structures overlying the filament. Green *et al.* (2003) considered the CME and flare productivity of 4 young active regions during their disk passage and found that the majority of CMEs occurred during or after the flux emergence. Numerical simulations by Chen and Shibata (2000) support this interpretation (see Figure 11).

There are exceptions to the scenario presented above. Wang and Sheeley (1999) found 3 cases of flux emergence associated with filament eruption; they also found filament eruptions with no evidence for flux emergence. Subramanian and Dere (2001) studied the solar sources of 32 CMEs observed by SOHO; 41% of the CMEs originated from active regions with no obvious signs of filament eruption. Of the remaining CMEs, 45% from active regions and 15% from quiet regions were associated with filament eruptions. CMEs without filament eruptions originated from young active regions (lifetime $<$3 months), while ARs with filament eruptions

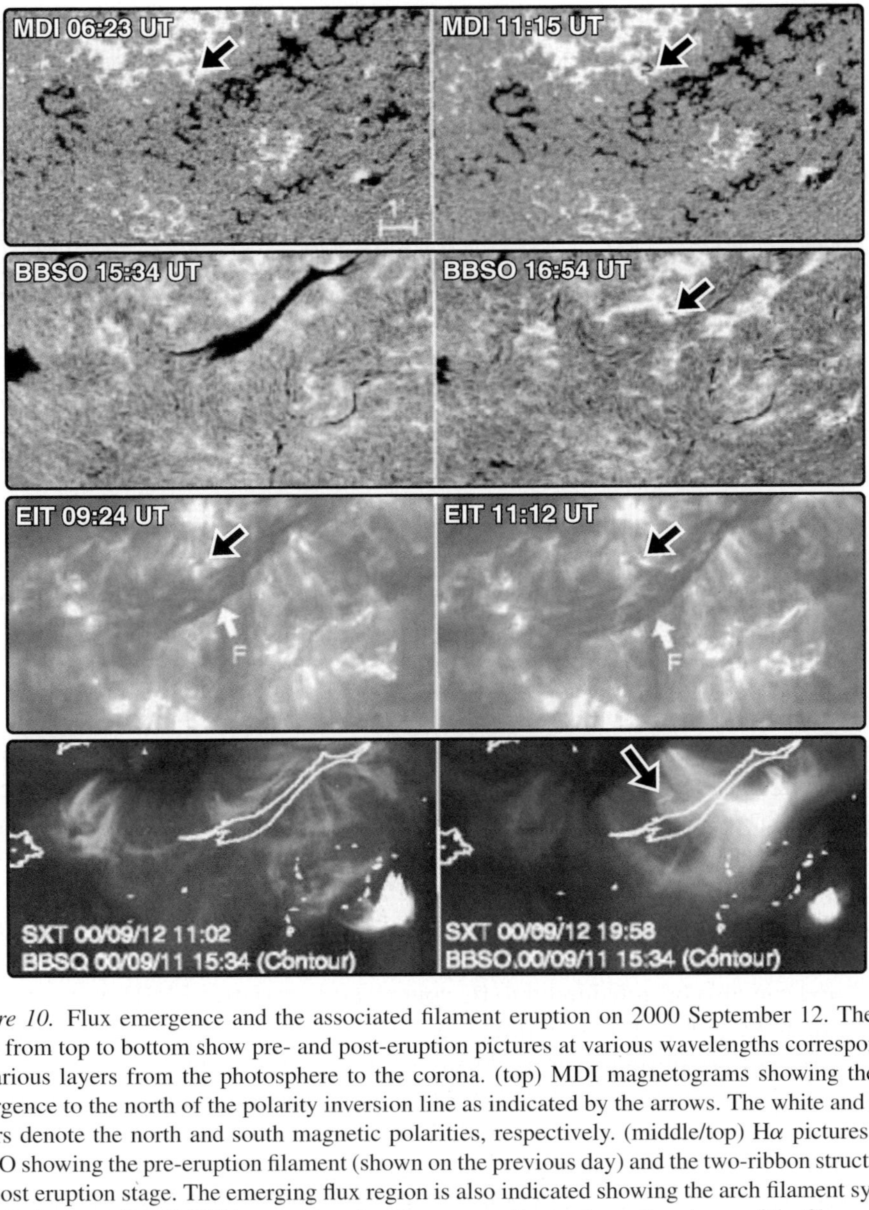

Figure 10. Flux emergence and the associated filament eruption on 2000 September 12. The four rows from top to bottom show pre- and post-eruption pictures at various wavelengths corresponding to various layers from the photosphere to the corona. (top) MDI magnetograms showing the flux emergence to the north of the polarity inversion line as indicated by the arrows. The white and black colors denote the north and south magnetic polarities, respectively. (middle/top) Hα pictures from BBSO showing the pre-eruption filament (shown on the previous day) and the two-ribbon structure in the post eruption stage. The emerging flux region is also indicated showing the arch filament system. (middle/bottom) SOHO/EIT images showing the pre-eruption and erupting stages of the filament (F). Changes in the emerging flux region are also seen. (bottom) Yohkoh/SXT images with the superposed outline of the filament (white contour). The pre-eruption X-ray loops are sheared parallel to the filament and some overlie the filament. In the post-eruption phase the SXT loops are perpendicular to the pre-eruption filament axis.

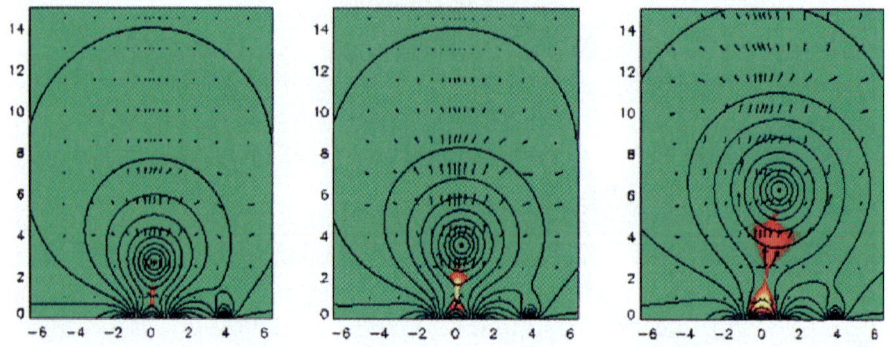

Figure 11. Results of a 2D MHD simulation showing CME eruption following flux emergence at [4,0] to the right of the neutral line (perpendicular to the plane of the paper at the location [0,0]). The solid line represents the magnetic field lines; the arrows indicate the direction and magnitude of flow. The color represents the temperature. (from Chen and Shibata, 2000).

were older (6–7 months). Reconnection-favoring flux emergence seems to be the key aspect in this type of eruption. Lack of flux emergence has to be understood in terms of other types of evolution such as flux cancellation and shearing.

3.4. Helicity Evolution

In recent years the role of helicity injection, or helicity-charging (Rust and Kumar, 1994a,b), has been a focal point in the discussion of eruptive events. In particular, it has been argued that CMEs are the means by which the solar corona expels magnetic helicity accumulated over hours and days by the combination of local shearing motions, differential rotation and the emergence of twisted flux systems (Low, 1996; DeVore, 2000; Démoulin *et al.*, 2002). The attractiveness of magnetic helicity for such studies lies in the fact that it is a globally conserved quantity in ideal MHD and can also be considered to be conserved in resistive MHD on time scales shorter than the global diffusion time scale (Berger, 1984). This property opens up a wide array of possibilities for exploring the CME process both theoretically and observationally (see articles in Brown *et al.*, 1999). An observational manifestation of the connection between helicity and CME production is the soft X-ray sigmoid discussed earlier in this chapter.

Since the solar corona is an open volume with the photosphere as a boundary with normal flux, the magnetic helicity can be transported across the boundary by velocity fields in the photosphere. The Poynting theorem for magnetic helicity, H, in an open volume is given by Berger and Field (1984) as

$$\frac{dH}{dt} = 2\oint (B \cdot A_p) v_z dS - \oint (v \cdot A_p)\, B_z dS$$

where B is the vector magnetic field and v is the velocity vector field (the subscript denoting the vertical component of each of these quantities), A_p is the unique vector potential of the potential field satisfying the conditions: $\nabla \times A_p \cdot z = B_z$, $\nabla \times A_p = 0$; $A_p \cdot z = 0$. With this definition, H is then the "relative helicity" (Berger and Field, 1984). Thus, the magnetic helicity in an open volume can change either by the passage of field lines through the surface (first term on RHS) or by the horizontal motions of the field lines (second term on RHS).

It is important to consider all potential contributions to the helicity injected into the sigmoid volume. Several studies point to the importance of intrinsic twist in the structures emerging from below the photosphere. Leka *et al.* (1996) showed that most field emerges twisted in a detailed active region study. Moreover, CME studies imply that helicity injection by differential rotation is 2–3 orders of magnitude lower than that required to explain helicity seen in magnetic clouds (Green *et al.*, 2002) and theoretical flux-rope models imply twisting of buoyant flux tubes by convective motions prior to emergence (Fan, 2001). On the other hand, measurement of helicity carried away by CMEs relies on uncertain magnetic cloud measurements and conclusions differ (DeVore, 2000). Strong localized photospheric motions can inject sufficient helicity to drive large flares (Moon *et al.*, 2002).

The emerging flux tube simulations of Magara and Longcope (2001), for example, suggest that neutral-line shear and sigmoidal field arise as a natural by-product of flux emergence with the observed sigmoidal field being the consequence of the emerging inner toroidal field lines of the flux tube. One would also expect a separation and rotation of the opposing polarity regions toward a more axial field orientation. In those cases where flux emergence is found to be important we would need to derive, or estimate, the vertical component of the photospheric flows. This cannot be done directly from observations. However, Kusano *et al.* (2002) have developed an approach which extracts an estimate of the vertical velocity from an inversion of the induction equation, utilizing the observed vector field components and transverse velocity field. Future helicity studies will require detailed measurements of the vector magnetic field and photospheric flow patterns before, during and after significant eruptions. Various approaches to this problem are currently underway at a number of institutions.

3.5. Flux Cancellation

As described previously, the emergence of new magnetic flux, especially in the vicinity of pre-existing magnetic fields, is connected with solar eruptions (e.g., Gaizauskas *et al.*, Zwaan, Feynman and Martin, 1983, 1985, 1995b). Flux cancellation is another mechanism that can trigger CMEs, as discussed by Mikić and Lee (2006, this volume) and by Forbes *et al.* (2006, this volume). The process of flux cancellation has been defined observationally as the mutual disappearance of magnetic fields of opposite polarity at the neutral line separating them (Martin *et al.*, 1985). Flux cancellation has also been identified as a key element in the formation of prominences and filaments (Gaizauskas *et al.*, 1997; Martin, 1998b; Litvinenko and Martin, 1999; van Ballegooijen *et al.*, 2000; Martens and Zwaan, 2001).

Because of the close relationship between flux cancellation and solar eruptions, the role of flux cancellation in prominence formation, eruption, and CME initiation has been studied extensively using numerical models (Linker *et al.*, 2001, 2003a,b; Roussev *et al.*, 2004b; Amari *et al.*, 1999, 2000, 2003b,c; Lionello *et al.*, 2002; Welsch *et al.*, 2005). These models show that a magnetic configuration subject to flux cancellation can initially exhibit stable behavior with a magnetic field topology that can support prominence material. If flux cancellation is continued beyond a critical value, the configuration erupts. When the configuration is close to the critical state, even a small amount of flux cancellation can trigger a violent eruption. Hence, the triggering event may appear to be unremarkable in photospheric magnetograms.

The halo CME observed on May 12, 1997 illustrates the role of flux cancellation in triggering CMEs. This event occurred during solar minimum when the Sun had a very simple structure. The CME originated from an isolated active region near disk center (AR 8038 at N21W08). The CME was accompanied by a C1.3 flare at 04:42UT. Observations of this event have been analyzed in detail (Plunkett *et al.*, 1998; Thompson *et al.*, 1998; Webb *et al.*, 2000; Gopalswamy and Kaiser, 2002). It has been selected as a SHINE Campaign Event for detailed theoretical study (Gopalswamy, 2005). MDI magnetograms preceding the CME indicate that there is a substantial amount of flux cancellation near the main neutral line in the active region, as shown in Figure 12. Whether this triggered the eruption is a question that is presently being investigated.

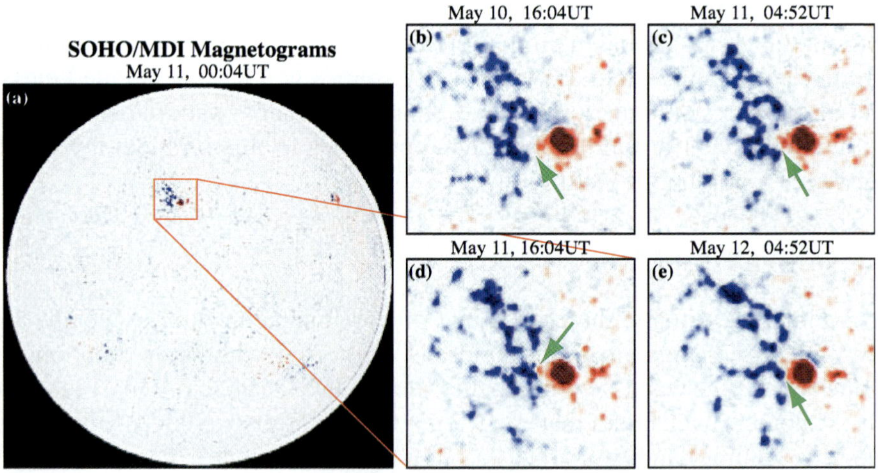

Figure 12. A sequence of MDI magnetograms preceding the halo CME on May 12, 1997. A C1.3 flare occurred at 04:42UT. (a) A full-disk MDI magnetogram showing AR 8038; (b)–(e) detailed magnetograms of the active region approximately 36 hours, 24 hours, and 12 hours prior to eruption, and at the time of eruption. The green arrow indicates an area with substantial convergence of opposite polarity flux at the neutral line, which may have triggered the eruption.

3.6. Successive Eruptions

Some active regions show rapid succession of flares and CMEs over a time scale of minutes to hours. This time scale is too small compared to the typical time scale for energy build-up in active regions. The rapid succession of flares and CMEs therefore represents a fragmented energy release. i.e., all the free energy available in an active region is not released in a single eruption. Many of the SEP-producing active regions are CME-prolific (Gopalswamy *et al.*, 2004b). It becomes difficult to distinguish between successive eruptions and a precursor-CME combination. This raises the question whether a preceding eruption makes it easier for the next eruption.

Examination of the angular difference distribution of successive CME pairs indicates an overabundance of these pairs (relative to background levels) when the position angle difference is within 10°; such position-angle coincidence is indicative of successive CMEs from the same source region (Moon *et al.*, 2003). Figure 13 shows the heliographic locations of eruptions from AR 9236 and the distribution of flare and CME recurrence times (Gopalswamy *et al.*, 2005). Only flares of C-class and above were considered. The average CME recurrence time was 4.6 h with 17 CMEs occurring over a five-day period. Over the same period, there were 33 C, M, and X-class flares with an average recurrence time of 3.6 h. Note that the smallest bins in the CME and flare recurrence distributions contain most of the events. For the CMEs in this bin, it is difficult to identify precursors. The time separation between the two big CMEs was about 10 h. But whether the preceding CMEs have any role in creating the conditions for a subsequent eruption is an open question. For instance, the occurrence of a preceding event might contribute to further destabilization of the source region leading to further eruptions. The active region in Figure 13 was also undergoing dynamic restructuring due to flux emergence in the core of the active region (Nitta and Hudson, 2001), so this region presents one of the best examples to study CME-rich active regions. On the other hand occurrence of a preceding event might contribute to further destabilization of the whole region.

3.7. Streamer Distension

Evolution of coronal streamers as a precursor to CMEs has been considered in the past. Slow addition of mass to the closed field regions of the streamer may result in its subsequent eruption (Wolfson *et al.*, 1987) seen as "bugles" (Howard *et al.*, 1985; Hundhausen, 1993). Another aspect of this configuration is that the mass in the streamer restrains the cavity, and hence allows for the build-up of sufficient energy to produce eruptions (Wolfson and Saran, 1998). There is also a strong correlation between the direction of propagation of CMEs and the location of streamers, often regardless of the location of the associated eruptive activity at the Sun. As such, streamers may hold clues to understanding the initiation and progression of

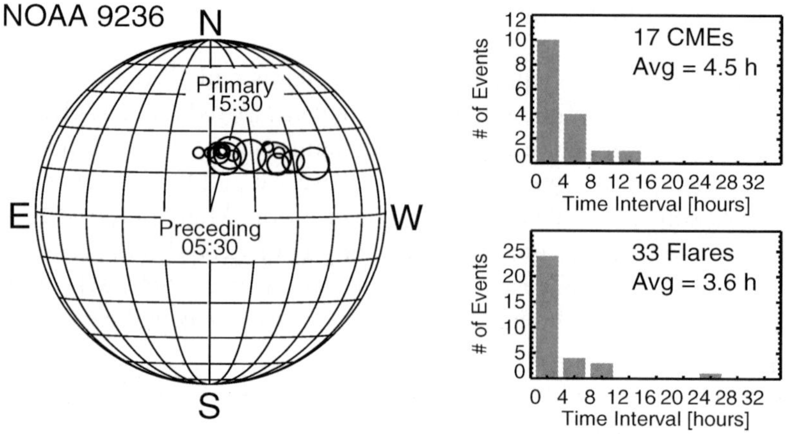

Figure 13. (left) Heliographic coordinates of flares from AR 9236 (November 22–27 2000). The locations and onset times of two halo CMEs on November 24 are also indicated. The size of the circles indicate the flare size (small: C-class, medium: M-class and large: X-class flares). (right) Distributions of time intervals between successive CME (upper) and flares (lower). The average recurrence times are 4.5 h (CMEs) and 3.6 h (flares). A SEP associated halo CME (Primary) at 15:30 UT, was preceded within 10 h by another halo at 05:30 UT on November 25.

CMEs low in the corona. In another kind of streamer evolution, the eruption of the associated prominence is partial and results in some noticeable streamer changes. Gopalswamy *et al.* (2003b) found that about 21% of the slow prominence eruptions were associated with changes in the overlying white-light streamers, instead of CMEs. In these "transverse" prominence eruptions, the prominence material did not reach heights beyond about $0.16R_{\odot}$ above the surface. An interesting result was that the majority of these streamers erupted in the next 4 to 42 hours as CMEs (Gopalswamy *et al.*, 2004a).

Very slow evolution of streamers in brightness and size has been linked to changes in photospheric magnetic flux. In particular, increase in brightness is known to be correlated with decrease in photospheric flux, which was interpreted as opening of flux tubes under the streamer (Poland and MacQueen, 1981). In this sense, we think that the small-scale prominence eruptions produce a similar effect of opening some field lines. Unfortunately, we cannot verify this because it is difficult to observe the photospheric magnetic field changes for limb events. Prolonged coronal dimming associated with some eruptions may also correspond to these field openings (Gopalswamy, 1999). We need a reverse study to look at a large number of streamers as they evolve to arrive at a firm conclusion regarding the connection between the small-scale prominence eruptions and activations on the one hand and the stability of the streamers on the other.

4. Pre-Eruption Energetic Signatures

The energetic signatures preceding the launch of CMEs are transient in nature, unlike the evolutionary signatures discussed in the previous section. The pre-eruption signatures can be thermal or non-thermal. Thermal signatures typically appear as a weak enhancement in the soft X-ray light curve. Imaging observations are essential for checking if these weak enhancements originate from the same source region as the main CME and its associated flare. Non-thermal signatures are primarily observed at radio wavelengths. The radio emission is produced by non-thermal electrons accelerated in the eruption region. The thermal and non-thermal signatures may not be independent because both heating and particle acceleration can occur in the same process such as reconnection. Hard X-ray emission may also result from the non-thermal electrons if they are stopped by dense plasma in the vicinity of the acceleration region. As we noted before, it is sometimes difficult to determine whether the precursor is a separate event or a true precursor, especially when non-thermal emission is involved.

4.1. Thermal Signatures

In addition to the soft X-ray sigmoid signatures discussed above, the pre-eruptive atmosphere shows signatures of an impending eruption in a number of different wavelengths. We have already discussed one example in Section 3.3. Gopalswamy (1999) discussed another filament eruption event associated with a weak halo CME on 1998 January 21. The magnetic field underlying the quiescent filament was very weak. The CME first appeared in the LASCO/C2 field of view at 06:57 UT. Figure 14 shows a set of X-ray structures (marked X) that brightened near the initial location of the filament that subsequently erupted (marked F). There was also a compact brightening in X-rays (not shown) located at the apex of a triangular X-ray structure, which was not observed in EUV images, suggesting that it was indeed a hot structure (Gopalswamy, 2003). The filament completely disappeared about 3 hours after the X-ray brightening. The CME appeared above the southern limb after the filament disappearance.

Recently, Sterling and Moore (2001a) identified EUV structures in the SOHO/EIT data, dubbed EIT crinkles. These structures take the form of small-scale transient sources of emission at a significant distance from the main flaring core. These structures first propagated away from and then faded in towards the central core region. Concurrent with this EUV signature, the soft X-ray emission exhibited faint transient brightening features connecting the main flare core with the location of the EIT crinkles. This indicated that the EUV emission delineates the low temperature foot points of newly heated coronal loops. Based on this evidence Sterling and Moore (2001b) argued that a model akin to the breakout model of Antiochos (1998) and Antiochos *et al.* (1999) best describes the eruptive process in these events.

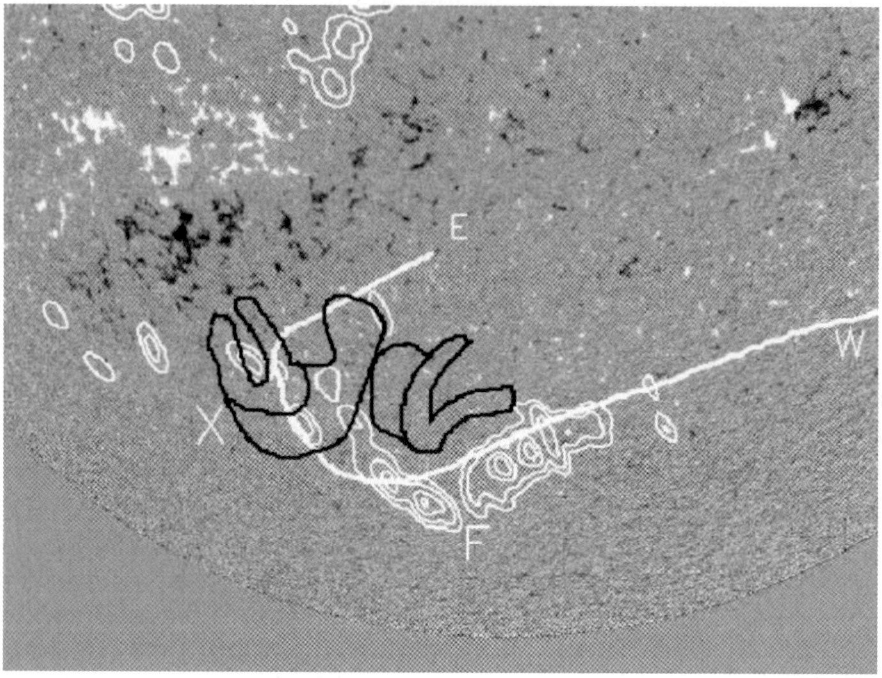

Figure 14. Composite picture of the filament eruption associated with a halo CME on 1998 January 21. The erupting filament (F), observed by the Nobeyama Radioheliograph is shown in white contours. The dark contours (X) define the outline of the initial soft X-ray brightening as observed by Yohkoh/SXT. The elongated neutral line (E) inferred from the MDI magnetogram is also shown. (Adapted from Gopalswamy, 1999).

4.2. Non-Thermal Signatures

Jackson and Sheridan (1979) found a broad maximum in the number of isolated metric-wave Type III bursts about 5 h prior to large H_α solar flares. The maximum was more pronounced if only the type III bursts co-located with the flare are considered. A similar result was obtained when they considered type III bursts prior to CME onsets (Jackson *et al.*, 1978). From these works, they concluded that the Type III burst activity was a good indicator of coronal energy input and storage near the time of a major eruption. Type III bursts are caused by mildly relativistic electrons, so they are non-thermal in nature and might represent small-scale energy releases. Kundu *et al.* (1987) used a single event with radio imaging to show that precursor type III bursts occur at the same location as the main flare and suggest that small-scale reconnection occurred continuously starting about 1 h before the flare.

It is possible to investigate the behavior of filaments and their cavities using weak non-thermal radio bursts which precede filament eruptions. Figure 15 presents an event observed on 1999 September 2 where a small filament in the vicinity of an

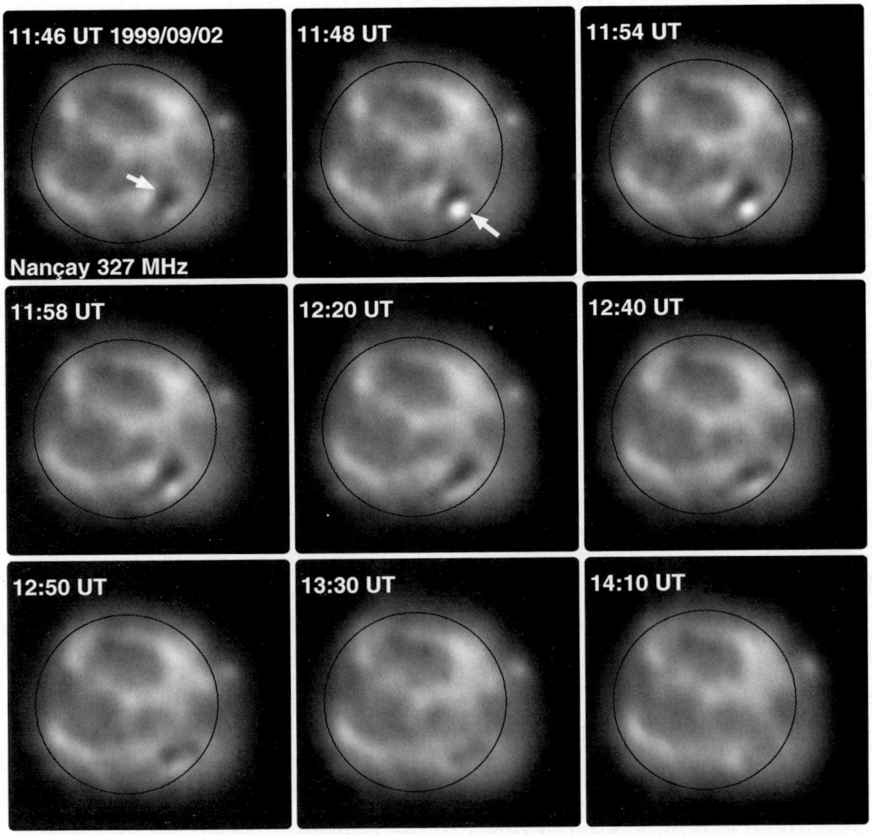

Figure 15. Nançay Radioheliograph observations at 327 MHz showing the slow evolution of a filament cavity, indicated by an arrow on the 11:46 UT image, linked with changes seen in white light streamers. The onset of the changes seen in the radio cavity (slow distension and fading) coincide with bright non-thermal bursts, indicated by an arrow on the 11:48 UT image, seen in the vicinity of the filament cavity, starting at about 11:48 UT and lasting roughly 10 min at 327 MHz. (Adapted fom Marqué, 2004.)

active region erupted (Marqué, 2004). The earliest signatures of the event were faint non-thermal bursts occurring on the side of the cavity and lasting for about 10 min. Following the radio bursts the cavity distended and slowly faded away, implying a slow eruption.

Repeated acceleration of electrons at multiple locations underlying a coronal streamer was reported by Klein *et al.* (1997), as inferred from type III bursts. A few hours after the start of these type III bursts, the streamer erupted. The type III bursts (reverse and normal drift) were also coincident with soft X-ray brightenings. These authors suggest that the type III bursts are indicative of pre-eruption instabilities in the large-scale structure that eventually erupts. See Pick *et al.* (2006, this volume) for a discussion of radio signatures occurring in the eruptive phase of CMEs.

Non-thermal signatures in hard X-rays and decimetric spikes preceding the main eruption by about 2 h was reported during the September 24, 2001 CME (Fárník *et al.*, 2003). There were also accompanying thermal signatures in the form of GOES soft X-ray enhancements. About 1 h before the main eruption, an unusual drifting radio continuum was observed together with two radio sources (at 327 and 164 MHz) in positions corresponding to expanding loops seen in Yohkoh/SXT and SOHO/EIT images, accompanied by a filament disappearance during the same period. Expanding loops and filament disappearance are characteristics of a preceding CME, but it is unlikely that such a preceding CME would be observed in white light because of the disk center location of the eruption and a major CME from the same region within an hour. Events like this demonstrate the complexity in understanding the precursors, because one might have to consider smaller preceding CMEs as part of the pre-event evolution.

4.3. Submillimeter Pulsations

Rapid (subsecond) pulsating bursts at submillimeter wavelengths have been found to be closely associated with CME launch times (Kaufmann *et al.*, 2003). Figure 16 shows an example of a submillimeter pulsating burst on 2001 December 13. The wavelet decomposition scalogram at 212 GHz shows the submillimeter activity preceding the CME launch by several minutes.

The submillimeter pulsating bursts may or may not have an associated impulsive bulk emission. A sample of 20 such pulsating events were analyzed in more detail, and all of them exhibited similar onset of the pulsating bursts before CME launch times. It is not possible to establish accurate statistics because the definition of the pulse onset times depends on the telescope sensitivity, which is highly variable due to atmospheric transmission conditions for different observations. However, there is no doubt that the onset times of the submillimeter pulsating bursts precede the CME launch times by a few to tens of minutes. The origin of these rapid pulses is unknown. Their association with the pre-CME conditions is significant, which needs to be further investigated by more observations covering a wider range of frequencies, and by theoretical interpretation on the possible physical mechanisms involved.

4.4. Prolonged Transequatorial Dimming

During 1998 April 27–May 9, a series of transequatorial eruptions occurred involving a major active region in the southern hemisphere (AR 8210) and a smaller region in the northern hemisphere. The first major eruption was a fast (1385 km/s) CME on 1998 April 27 appearing at 08:56 UT in the SOHO/LASCO FOV (Gopalswamy *et al.*, 1999). From the height-time plot, the CME onset was estimated to occur within 08:45–08:55 UT. About 2 h before the CME onset, the

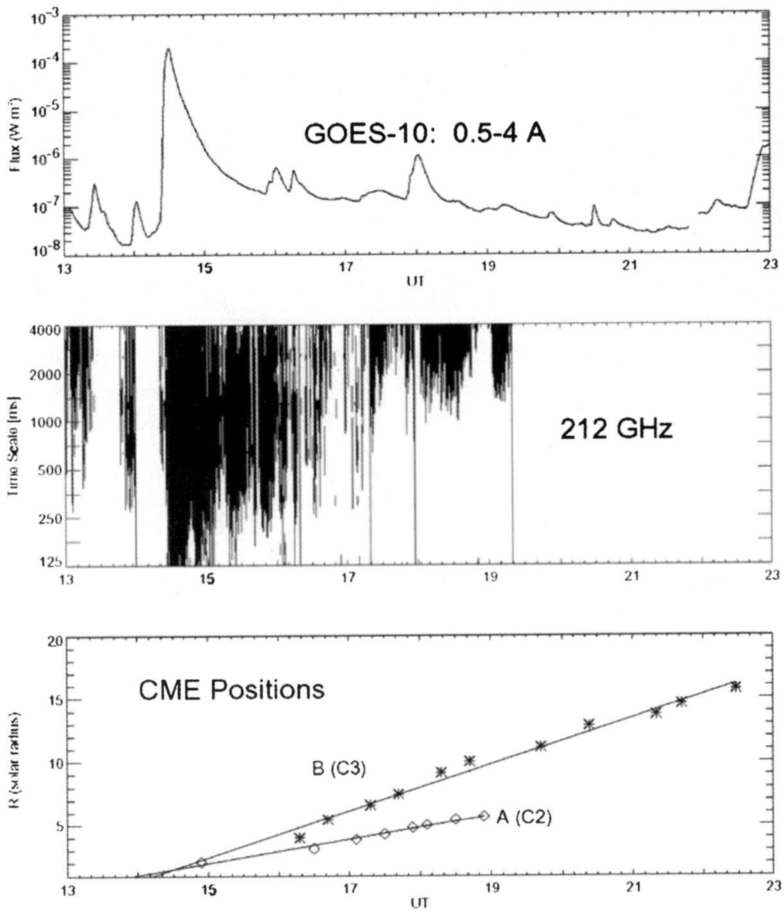

Figure 16. GOES soft ray light curve (top), millisecond pulsations observed at 212 GHz (middle) and the height-time plots of two associated CMEs labeled A (observed by LASCO C2) and B (observed by LASCO C3). The CME onsets coincided with small GOES soft X-ray spike and associated millisecond pulsations.

transequatorial region started showing weak dimming for about an hour (06:42 to 08:06 UT) followed by a deeper dimming (see Figure 17) that reached a maximum at about 08:45 UT, coinciding with the projected onset of the CME.

Figure 18 shows the EUV intensity in the active region (AR) and the transequatorial region (D), compared with the GOES soft X-ray light curve (G). The flare itself started only at 08:55 UT (see the curves AR and G). The CME was already at a height of $2.6R_{\odot}$ at 08:56 UT, just one minute after the flare onset. The CME must have traveled a distance of 1.6 Ro in 15 min before the start of the flare, which gives an average speed of 1237 km/s, consistent with that obtained from the height-time plot. Another interesting aspect of this event is that there was a long wavelength

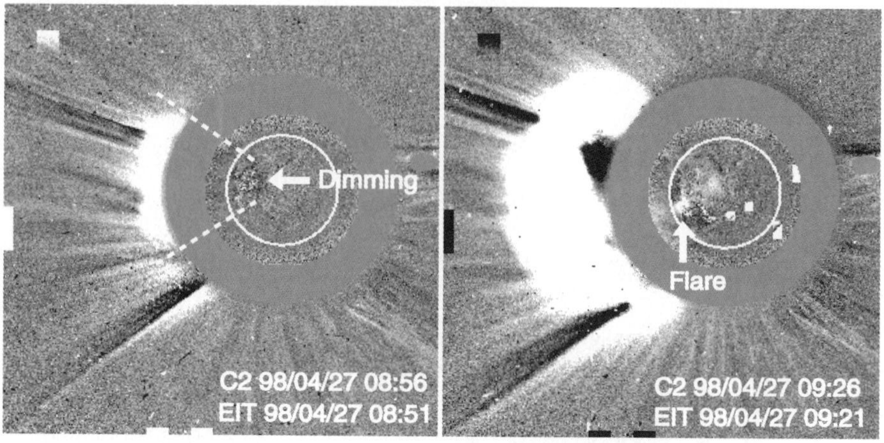

Figure 17. (left) SOHO/LASCO CME on 1998 April 27, which appeared first above the occulting disk at 08:56 UT. The angular extent of the CME can be seen to be consistent with the extent of the transequatorial dimming observed by SOHO/EIT. (right) The CME in the next frame with the associated flare in AR 8210. The projected onset of the CME is estimated to occur within 08:45–08:51 UT from the height-time plots. Note the widespread disturbance in the transequatorial region in the EIT difference image at 09:21 UT.

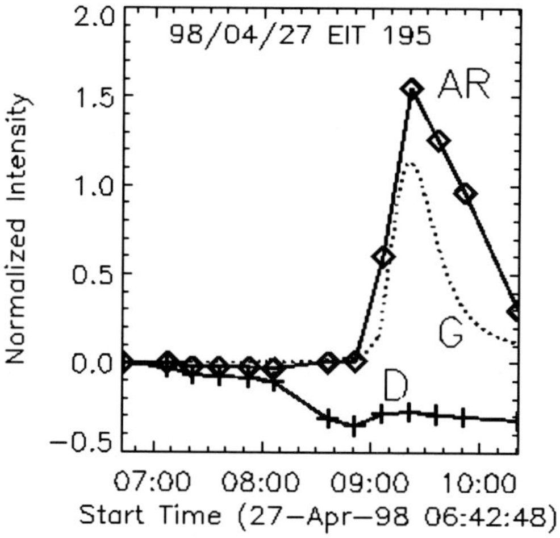

Figure 18. SOHO/EIT 195 Å light curve showing the EUV intensity in the active region (AR) and the transequatorial dimming region (D) compared with the GOES soft X-ray light curve (G) on 1998 April 27. The flare emission starts only at 08:45 UT in AR 8210, whereas the dimming had started two hours earlier. The CME onset was about 40 min before the flare onset.

(<14 MHz) type III burst when the weak dimming started, and a series of type III bursts at the start of the deep dimming, leading up to the intense complex type III bursts at the time of the flare. The dimming and the weak type III bursts may be indicative of the subsequent eruption. However, Khan and Hudson (2000) came to the opposite conclusion for three eruptions that occurred when AR 8210 was on the western hemisphere. They suggested that flares from AR 8210 produced blast waves whose impact caused the eruption of the transequatorial structure.

5. Global Issues

In this section, we discuss two related issues regarding the nature of the pre-eruption configurations of the source regions: (i) the prominence configuration (normal and inverse polarity) that might lead to possible two types of CMEs, and (ii) the high-latitude CMEs associated with polar crown prominences. The high-latitude CMEs are clearly prominence related, while the low-latitude ones can be either flare related or prominence related. While the distinction between the two types may not be significant as far as the initiation mechanism is concerned, it may be important in the context of the solar polarity reversal. The high-latitude CMEs are purely prominence related, whereas the low-latitude CMEs are mostly active-region (and hence flare) related. The high-latitude CMEs are related to solar polarity reversal.

5.1. Normal and Inverse Polarity CMEs

MacQueen and Fisher (1983) had suggested that different acceleration mechanisms may be operating in CMEs associated with prominence eruptions and flares. The flare-related CMEs are characterized by higher constant speed, while the prominence-related CMEs are slower and accelerating (see also St. Cyr *et al.*, 1999). (Tappin and Simnett, 1997) used 149 LASCO CMEs and found that the constant speed CMEs were generally faster. Examples of height-time profiles showing constant speed and accelerating CMEs were also reported by others (Sheeley *et al.*, 1999; Andrews and Howard, 2001; Gopalswamy *et al.*, 2001). Moon *et al.* (2002) also found a clear difference in speeds of flare-related (759 km/s) and prominence-related (513 km/s) CMEs. The flare-related CMEs also showed a tendency for deceleration, but this probably reflects the fact that they are faster (Gopalswamy *et al.*, 2001). The question is whether the speed difference represents a qualitative distinction between the two types of CMEs. Studying the acceleration of CMEs, Chen and Krall (2003) concluded that one mechanism is sufficient to explain flare- and prominence-related CMEs. Furthermore, the distinction between the two is somewhat ambiguous because flares often involve prominence eruptions and quiescent prominence eruptions result in two-ribbon flares. The primary distinction may be the strength of the magnetic fields and the inverse and normal polarity configurations in the pre-eruption stage. In fact, two different pre-eruption prominence configurations have been suggested as the basic reason for the two types of CMEs, accelerating and constant-speed (Low and Zhang, 2002). These authors assume that the flux rope is a basic ingredient in both configurations. They propose that loss of equilibrium of the flux ropes is the mechanism that initiates CMEs (Low, 1996). In the pre-eruptive state, the flux rope is held down by the prominence mass and the magnetic tension of the overlying fields. A CME is produced when the confinement of the flux rope breaks down for a variety of reasons, such as loss of prominence mass (Low and Zhang, 2002). The interaction between the current in the flux rope and the current sheets in the overall configuration determines the nature of the eruption and the dynamics of the flux rope. (Low and Zhang, 2002) claim that the differences in this interaction can account for the predominance of accelerating CMEs from inverse polarity configurations and constant speed CMEs from normal polarity configurations.

5.2. High-Latitude CMEs and Polarity Reversal

As we discussed before, the presence of a closed field structure on the Sun is a basic requirement for CMEs. The number of closed field regions on the Sun depends on the phase of the solar cycle. During solar minimum, most of the closed field regions are confined to the equatorial streamer belt. As the solar cycle progresses, closed field regions spread to all latitudes. This is often reflected in the tilt angle of the heliospheric current sheet. The sunspot number is often used as a measure of solar activity, but it is not a good indicator of CMEs especially during solar maximum. This is reflected in the less-than-perfect correlation between the CME rate and sunspot number. The main reason is that solar activity defined by sunspots is confined to lower latitudes ($<45^\circ$). There is another source of CMEs, especially during the rising and maximum phases of the solar cycle: the polar crown filaments (PCFs). It is well known that the PCFs begin appearing at the poles in the rising phase and disappear somewhere during the maximum of activity (see Cliver *et al.*, 1994). No sunspots are associated with the PCFs, so the PCF-related CMEs (or high-latitude CMEs) are not expected to be correlated with sunspot activity. The rate of high-latitude CMEs is clearly related to the migration of closed field structures to the poles. A close temporal relationship between the cessation of high-latitude CMEs and the polarity reversal at the solar poles has been demonstrated for solar cycles 21 and 23 (Gopalswamy *et al.*, 2003a). Typically, the high-latitude CME rate falls to the background value just before the reversal in each pole. High-latitude CMEs provide a natural explanation for the disappearance of PCFs, which need to be removed before the poles can develop an open field structure of the opposite polarity.

6. Discussion, Summary and Future Perspectives

So far we have discussed observations and interpretations of the three basic aspects of the pre-CME Sun: pre-eruption structure, evolution, and energetic signatures. In this section, we briefly discuss how these observations are being used in numerical modeling of CME initiation. Then we shall present a summary and future perspectives.

6.1. Observational Input to CME Initiation Models

The main observational inputs for modeling CME initiation have been photospheric magnetic field measurements, especially from vector magnetograms. Particularly useful are sequences of magnetic measurements that can be used to model the storage of energy, the build-up of twist, and possibly the crossing of a stability threshold.

The development of CME initiation models hinges principally on our understanding of the detailed structure and evolution of the coronal magnetic field in the pre-eruptive state (e.g., Mikić and Lee, 2006, this volume; Forbes *et al.*, 2006, this volume). Unfortunately, it is difficult to measure coronal magnetic fields in detail, so we have to rely on photospheric measurements, extrapolated into the corona using model fields. Moreover, at present we routinely only measure these magnetic fields using (longitudinal) magnetographs, which only supply one component of the magnetic field (the normal component at disk center). The energization of coronal fields can only be determined from vector magnetic field measurements, and this information has been scant.

We have had to rely on inferences of magnetic structure, principally from X-ray and EUV emission. The high-resolution EUV images from the TRACE mission have been especially useful in this regard, since they tend to emphasize what appear to be traces of individual magnetic field lines. At times, white-light coronagraph images, and especially movies, can help us infer the pre- and post-eruptive structure of CMEs. The twist of embedded filaments is sometimes clearly visible in coronagraph movies of streamer eruptions. Radio emission measurements may be used to infer the strength, and to a more limited extent the direction, of the coronal magnetic field. Of course, these qualitative morphological indications of the magnetic field geometry are not suitable for detailed modeling purposes. Our poor knowledge of CME initiation is largely a consequence of our limited ability to measure the coronal magnetic field.

Fortunately, two upcoming missions will greatly help in this regard. STEREO will image the corona in EUV and white light from two different vantage points, which, in combination with observations from SOHO, TRACE, and the ground, will help to deconvolve projection effects and help us to understand the three-dimensional nature of coronal features. Solar B will for the first time measure vector magnetic fields at high resolution from space, as well as X-ray emission at high resolution. These observations are expected to greatly advance our understanding of CME initiation.

6.2. Summary and Future Perspectives

Appearance of a large-scale closed magnetic field structure on the Sun can be considered as the lowest form of pre-CME evolution necessary for a CME. Thus, the ability of the Sun to emerge closed field structure from below the photosphere into the atmosphere is a basic requirement for a CME. Active regions and filament channels are the basic units of closed field structures, which often occur together and in clusters. During solar maximum, a large number of these structures are present on the Sun with a correspondingly higher CME rate. Also the latitudinal distribution of the magnetic regions, governing the locations of eruptions on the Sun, are different between solar maximum and minimum. For example, polar crown eruptions occur starting a couple of years before the solar polarity reversal and then practically disappear after that. Next, the magnetic source region has to build-up free energy. Identification of the signatures of this energy build-up is a crucial step in deciding whether and when a CME will occur. Flux emergence and cancellation, shear and converging flows are some of the photospheric signatures that indicate energy build-up. During the build-up phase, minor energetic events occur, which are denoted as precursors or pre-eruption energy release. Both thermal (soft X-rays, EUV, Hα) and non-thermal (radio, hard X-rays) precursors have been observed. Sometimes it is difficult to decide whether a pre-eruption energy release is a true precursor or a separate eruption. Even a prior smaller eruption may lead to a bigger subsequent eruption by destabilizing the larger structures in the eruption region. While this simplified picture provides insight, the reality is more complex and we are far from predicting whether an active region is likely to erupt by looking at its evolution prior to eruption. Progress needs to be made in answering a number of open questions regarding the role of flux ropes in CMEs: When are they formed, before or during eruption? If flux ropes form before eruption, do they formed in the atmosphere or do they form in the subsurface layers and then subsequently emerge? Do all CMEs involve flux ropes? An answer in the affirmative would simplify the modeling of CMEs. Unfortunately, there is an observational difficulty. While the filament or coronal cavity overlying quiescent prominences is often thought to be a flux rope, it is hard to observe cavities overlying active region filaments. It is important to know whether active-region filaments have cavities or not. Cavities are well observed only in CMEs associated with filament eruptions. In disk CMEs, however, we often observe only two parts: the frontal structure and the bright core. Is this difficulty in observing cavities a projection effects or because they do not exist? For example, radio observations often show the eruption of cavities from the solar disk, suggesting that they are coherent structures. Is the filament a flux rope as some modelers assume? Why are some CMEs not associated with filaments/filament eruptions? Other questions pertaining to the evolution of the pre-eruptive structures towards eruption also required detailed study: What is the role of flux emergence/cancellation/shear in flux-rope formation and CME initiation? What are the observed characteristics of the source of fast CMEs and how do they differ from those of slow CMEs? What is the location of reconnection with respect to the filament?

We have made enormous progress in understanding the gross properties of CMEs, thanks to the continuous and uniform observations of high quality from the SOHO mission. Yet, the observational coverage of the early life of CMEs is limited, mainly due to the poor cadence of observations. For example, the CME watch in EUV (195 Å) has a cadence of 12 min over a very limited field of view, preventing us from having a close look at CME initiation and early evolution. High cadence images are needed to systematically identify pre-eruption signatures, which is crucial in developing tools to predict the impending onset of CMEs. We still lack quantitative understanding in connecting magnetic complexity in a source region to CME productivity. Since vector magnetograms provide most of the information on the free energy available in active regions, they should be integrated with MHD models. Finally, a better understanding of the helioseismic subsurface imaging of active regions might provide important clues to the build-up of energy in active regions.

Acknowledgements

The open data policy of the SOHO project has contributed enormously in the CME-related studies. We thank S. Yashiro for helping with some of the figures.

References

Abbett, W. P., and Fisher, G. H.: 2003, *Astrophys. J.* **582**, 475–485.
Adams, W. M.: 1976, *Sol. Phys.* **47**, 601–605.
Amari, T., Luciani, J. F., Aly, J. J., Mikić, Z., and Linker, J.: 2003a, *Astrophys. J.* **585**, 1073–1086.
Amari, T., Luciani, J. F., Aly, J. J., Mikić, Z., and Linker, J.: 2003b, *Astrophys. J.* **585**, 1073–1086.
Amari, T., Luciani, J. F., Aly, J. J., Mikić, Z., and Linker, J.: 2003c, *Astrophys. J.* **595**, 1231–1250.
Amari, T., Luciani, J. F., Mikić, Z., and Linker, J.: 1999, *Astrophys. J.* **518**, L57–L60.
Amari, T., Luciani, J. F., Mikić, Z., and Linker, J.: 2000, *Astrophys. J.* **529**, L49–L52.
Ambastha, A.: 1998, *ASP Conf. Ser. 140: Synoptic Solar Physics*. pp. 113–119.
Andrews, M. D., and Howard, R. A.: 2001, *Space Science Reviews* **95**, 147–163.
Antiochos, S. K.: 1998, *Astrophys. J.* **502**, L181.
Antiochos, S. K., DeVore, C. R., and Klimchuk, J. A.: 1999, *Astrophys. J.* **510**, 485–493.
Babcock, H. W., and Babcock, H. D.: 1955, *Astrophys. J.* **121**, 349–366.
Berger, M. A.: 1984, *Geophys. Astrophys. Fluid Dynam.* **30**, 79–104.
Berger, M. A., and Field, G. B.: 1984, *J. Fluid Mechan.* **147**, 133–148.
Berger, T. E., Loefdahl, M. G., Shine, R. S., and Title, A. M.: 1998, *Astrophys. J.* **495**, 973–983.
Bhatnagar, A.: 1996, *Ap&SS* **243**, 105–112.
Bravo, S.: 1996, *Advances in Space Research* **17**, 285–288.
Brown, D. S., Nightingale, R. W., Alexander, D., Schrijver, C. J., Metcalf, T. R., Shine, R. A., *et al.*: 2003, *Sol. Phys.* **216**, 79–108.
Brown, M. R., Canfield, R. C., and Pevtsov, A. A. (eds.): 1999, Magnetic Helicity in Space and Laboratory Plasmas.
Bruzek, A.: 1967, *Sol. Phys.* **2**, 451–461.
Burlaga, L., Fitzenreiter, R., Lepping, R., Ogilvie, K., Szabo, A., Lazarus, A., *et al.*: 1998, *J. Geophys. Res.* **103**, 277–286.

Canfield, R. C., Hudson, H. S., and McKenzie, D. E.: 1999, *Geophys. Res. Lett.* **26**, 627–630.
Canfield, R. C., Hudson, H. S., and Pevtsov, A. A.: 2000, *IEEE Transactions on Plasma Science*, December 2000 **28**, 1786–1794.
Canfield, R. C., and Pevtsov, A. A.: 1998, *ASP Conf. Ser. 140: Synoptic Solar Physics* pp. 131–143.
Canfield, R. C., Priest, E. R., and Rust, D. M.: 1974, *High Altitude Observatory, CO: Flare Related Magnetic Field Dynamics*, pp. 361–371.
Chen, J., Howard, R. A., Brueckner, G. E., Santoro, R., Krall, J., Paswaters, S. E., *et al.*: 1997, *Astrophys. J.* **490**, L191.
Chen, J., and Krall, J.: 2003, *J. Geophys. Res. (Space Phys.)* 2–1.
Chen, P. F., and Shibata, K.: 2000, *Astrophys. J.* **545**, 524–531.
Cliver, E. W., St. Cyr, O. C., Howard, R. A., and McIntosh, P. S.: 1994, *IAU Colloq. 144: Solar Coronal Struct.* pp. 83–89.
Cremades, H., and Bothmer, V.: 2004, *A&A* **422**, 307–322.
Démoulin, P., Mandrini, C. H., van Driel-Gesztelyi, L., Thompson, B. J., Plunkett, S., Kovári, Z., *et al.*: 2002, *A&A* **382**, 650–665.
Delannée, C., and Aulanier, G.: 1999, *Sol. Phys.* **190**, 107–129.
Démoulin, P., Henoux, J. C., Mandrini, C. H., and Priest, E. R.: 1997, *Sol. Phys.* **174**, 73–89.
Dere, K. P., Brueckner, G. E., Howard, R. A., Michels, D. J., and Delaboudiniere, J. P.: 1999, *Astrophys. J.* **516**, 465–474.
DeVore, C. R.: 2000, *Astrophys. J.* **539**, 944–953.
Dobrzycka, D., Cranmer, S. R., Raymond, J. C., Biesecker, D. A., and Gurman, J. B.: 2002, *Astrophys. J.* **565**, 621–629.
Engvold, O.: 1988, *Solar and Stellar Coronal Structure and Dynamics*, pp. 151–169.
Fárník, F., Hudson, H. S., Karlický, M., and Kosugi, T.: 2003, *A&A* **399**, 1159–1166.
Fan, Y.: 2001, *Astrophys. J.* **554**, L111–L114.
Feynman, J., and Martin, S. F.: 1995a, *J. Geophys. Res.* **100**, 3355–3367.
Feynman, J., and Martin, S. F.: 1995b, *J. Geophys. Res.* **100**, 3355–3367.
Filippov, B. P., Gopalswamy, N., and Lozhechkin, A. V.: 2001, *Sol. Phys.* **203**, 119–130.
Forbes, T. G.: 2000, *J. Geophys. Res.* **105**, 23153–23166.
Forbes, T. G., J. A. Linker, *et al.*: 2006, *Space Sci. Rev.* this volume, doi: 10.1007/s11214-006-9019-8.
Gaizauskas, V., Harvey, K. L., Harvey, J. W., and Zwaan, C.: 1983, *Astrophys. J.* **265**, 1056–1065.
Gaizauskas, V., Zirker, J. B., Sweetland, C., and Kovacs, A.: 1997, *Astrophys. J.* **479**, 448–457.
Gibson, S. E., Fletcher, L., Del Zanna, G., Pike, C. D., Mason, H. E., Mandrini, C. H., *et al.*: 2002, *Astrophys. J.* **574**, 1021–1038.
Gilbert, H. R., Holzer, T. E., Burkepile, J. T., and Hundhausen, A. J.: 2000, *Astrophys. J.* **537**, 503–515.
Glover, A., Harra, L. K., Matthews, S. A., and Foley, C. A.: 2003, *A&A* **400**, 759–767.
Gopalswamy, N.: 1999, in: Bastian, T. S., Gopalswamy, N., and Shibasaki, K. (eds.), *Proceedings of the Nobeyama Symposium, held in Kiyosato, Japan, Oct. 27–30, 1998*, NRO Report No. 479., pp. 141–152.
Gopalswamy, N.: 2003, *Adv. Space Res.* **31**, 869–881.
Gopalswamy, N.: 2005, *EOS Trans. AGU* **86**(50), 525.
Gopalswamy, N., Hanaoka, Y., and Hudson, H. S.: 2000, *Adv. Space Res.* **25**, 1851–1854.
Gopalswamy, N., Hanaoka, Y., Kosugi, T., Lepping, R. P., Steinberg, J. T., Plunkett, S., *et al.*: 1998, *Geophys. Res. Lett.* **25**, 2485–2488.
Gopalswamy, N., and Kaiser, M. L.: 2002, *Adv. Space Res.* **29**, 307–312.
Gopalswamy, N., Lara, A., Yashiro, S., and Howard, R. A.: 2003a, *Astrophys. J.* **598**, L63–L66.
Gopalswamy, N., Shimojo, M., Lu, W., Yashiro, S., Shibasaki, K., and Howard, R. A.: 2003b, *Astrophys. J.* **586**, 562–578.
Gopalswamy, N., Shimojo, M., Lu, W., Yashiro, S., Shibasaki, K., and Howard, R. A.: 2004a, *Adv. Space Res.* **33**, 676–680.

Gopalswamy, N., White, S. M., and Kundu, M. R.: 1991, *Astrophys. J.* **379**, 366–380.
Gopalswamy, N., Yashiro, S., Kaiser, M. L., Howard, R. A., and Bougeret, J.-L.: 2001, *J. Geophys. Res.* **106**(15), 29219–29230.
Gopalswamy, N., Yashiro, S., Kaiser, M. L., Thompson, B. J., and Plunkett, S.: 1999, in: Bastian, T. S., Gopalswamy, N., and Shibasaki, K. (eds) *Proceedings of the Nobeyama Symposium, held in Kiyosato, Japan, Oct. 27-30, 1998, NRO Report No. 479,* pp. 207–210.
Gopalswamy, N., Yashiro, S., Krucker, S., and Howard, R. A.: 2005, Dere, K., Wang, J., and Yan, Y. (eds.), *IAU Symposium*, pp. 367–373.
Gopalswamy, N., Yashiro, S., Krucker, S., Stenborg, G., and Howard, R. A.: 2004b, *Journal of Geophysical Research (Space Physics)* **109**(A18), 12105.
Gosling, J. T., Birn, J., and Hesse, M.: 1995, *Geophys. Res. Lett.* **22**, 869–872.
Green, L. M., Démoulin, P., Mandrini, C. H., and Van Driel-Gesztelyi, L.: 2003, *Sol. Phys.* **215**, 307–325.
Green, L. M., López fuentes, M. C., Mandrini, C. H., Démoulin, P., Van Driel-Gesztelyi, L., and Culhane, J. L.: 2002, *Sol. Phys.* **208**, 43–68.
Hagyard, M. J., Moore, R. L., and Emslie, A. G.: 1984, *Adv. Space Res.* **4**, 71–80.
Hanaoka, Y., Kurokawa, H., Enome, S., Nakajima, H., Shibasaki, K., Nishio, M., *et al.*: 1994, *PASJ* **46**, 205–216.
Harvey, K. L., and Harvey, J. W.: 1976, *Sol. Phys.* **47**, 233–246.
Hori, K., and Culhane, J. L.: 2002, *A&A* **382**, 666–677.
House, L. L., Illing, R. M. E., Sawyer, C., and Wagner, W. J.: 1981, *BASS* **13**, 862.
Howard, R. A., Sheeley, N. R., Michels, D. J., and Koomen, M. J.: 1985, *J. Geophys. Res.* **90**, 8173–8191.
Hudson, H. S., Acton, L. W., Harvey, K. L., and McKenzie, D. E.: 1999, *Astrophys. J.* **513**, L83–L86.
Hundhausen, A. J.: 1993, *J. Geophys. Res.* **98**, 13177–13200.
Illing, R. M. E., and Hundhausen, A. J.: 1985, *J. Geophys. Res.* **90**(9), 275–282.
Jackson, B. V., and Sheridan, K. V.: 1979, *Proc. Astron. Soc. Aust.* **3**, 383–386.
Jackson, B. V., Sheridan, K. V., Dulk, G. A., and McLean, D. J.: 1978, *Proc. Astronom. Soc. Aust.* **3**, 241.
Kaufmann, P., Giménez de Castro, C. G., Makhmutov, V. S., Raulin, J., Schwenn, R., Levato, H., *et al.*: 2003, *J. Geophys. Res. (Space Phys.)* **108**, 5–1.
Khan, J. I., and Hudson, H. S.: 2000, *Geophys. Res. Lett.* **27**, 1083–1086.
Klein, K.-L., Aurass, H., Soru-Escaut, I., and Kalman, B.: 1997, *A&A* **320**, 612–619.
Kopp, R. A., and Pneuman, G. W.: 1976, *Sol. Phys.* **50**, 85–98.
Krall, K. R., Smith, J. B., Hagyard, M. J., West, E. A., and Cummings, N. P.: 1982, *Sol. Phys.* **79**, 59–75.
Kundu, M. R., Gopalswamy, N., Saba, J. L. R., Schmelz, J. T. S., and Strong, K. T.: 1987, *Sol. Phys.* **114**, 273–288.
Kundu, M. R., and McCullough, T. P.: 1972, *Sol. Phys.* **24**, 133–141.
Kuperus, M., and van Tend, W.: 1981, *Sol. Phys.* **71**, 125–139.
Kusano, K., Maeshiro, T., Yokoyama, T., and Sakurai, T.: 2002, *Astrophys. J.* **577**, 501–512.
Lara, A., Gopalswamy, N., and Deforest, C.: 2000, *Geophys. Res. Lett.* **27**, 1435–1438.
Leamon, R. J., Canfield, R. C., Blehm, Z., and Pevtsov, A. A.: 2003, *Astrophys. J.* **596**, L255–L258.
Leamon, R. J., Canfield, R. C., and Pevtsov, A. A.: 2002, *J. Geophys. Res. (Space Phys.)*, pp. 1–1.
Leka, K. D., Canfield, R. C., McClymont, A. N., and van Driel-Gesztelyi, L.: 1996, *Astrophys. J.* **462**, 547.
Liggett, M., and Zirin, H.: 1984, *Sol. Phys.* **91**, 259–267.
Linker, J. A., Lionello, R., Mikić, Z., and Amari, T.: 2001, *J. Geophys. Res.* **106**, 25165–25176.
Linker, J. A., Mikić, Z., Lionello, R., Riley, P., Amari, T., and Odstrcil, D.: 2003a, *Phys. Plasmas* **10**, 1971–1978.

Linker, J. A., Mikić, Z., Riley, P., Lionello, R., and Odstrcil, D.: 2003b, in: Velli, M., Bruno, R., and Malara, F. (eds.), *AIP Conference Proceedings: Solar Wind Ten*, American Institute of Physics, New York, vol. 679, pp. 703–710.

Lionello, R., Mikić, Z., Linker, J. A., and Amari, T.: 2002, *Astrophys. J.* **581**, 718–725.

Litvinenko, Y. E., and Martin, S. F.: 1999, *Sol. Phys.* **190**, 45–58.

Longcope, D. W.: 2004, *Astrophys. J.* **612**, 1181–1192.

Low, B. C.: 1996, *Sol. Phys.* **167**, 217–265.

Low, B. C., and Hundhausen, J. R.: 1995, *Astrophys. J.* **443**, 818–836.

Low, B. C., and Zhang, M.: 2002, *Astrophys. J.* **564**, L53–L56.

Mackay, D. H., Gaizauskas, V., Rickard, G. J., and Priest, E. R.: 1997, *Astrophys. J.* **486**, 534–549.

MacQueen, R. M., and Fisher, R. R.: 1983, *Sol. Phys.* **89**, 89–102.

Magara, T., and Longcope, D. W.: 2001, *Astrophys. J.* **559**, L55–L59.

Marqué, C.: 2004, *Astrophys. J.* **602**, 1037–1050.

Marqué, C., Lantos, P., and Delaboudinière, J. P.: 2002, *A&A* **387**, 317–325.

Martens, P. C., and Zwaan, C.: 2001, *Astrophys. J.* **558**, 872–887.

Martin, S. F.: 1998a, *Sol. Phys.* **182**, 107–137.

Martin, S. F.: 1998b, *Sol. Phys.* **182**, 107–137.

Martin, S. F., Livi, S. H. B., and Wang, J.: 1985, *Aust. J. Phys.* **38**, 929–959.

Marubashi, K.: 1986, *Adv. Space Res.* **6**, 335–338.

Metcalf, T. R., Jiao, L., McClymont, A. N., Canfield, R. C., and Uitenbroek, H.: 1995, *Astrophys. J.* **439**, 474–481.

Metcalf, T. R., Leka, K. D., and Mickey, D. L.: 2004, *Am. Astron. Soc. Meeting Abstracts* **204**.

Mikić, Z., and Lee, M.: 2006, *Space Sci. Rev.,* this volume, doi: 10.1007/s11214-006-9012-2.

Moon, Y.-J., Chae, J., Choe, G. S., Wang, H., Park, Y. D., Yun, H. S., *et al.*: 2002, *Astrophys. J.* **574**, 1066–1073.

Moon, Y.-J., Choe, G. S., Wang, H., and Park, Y. D.: 2003, *Astrophys. J.* **588**, 1176–1182.

Munro, R. H., Gosling, J. T., Hildner, E., MacQueen, R. M., Poland, A. I., and Ross, C. L.: 1979, *Sol. Phys.* **61**, 201–215.

Nindos, A., and Zhang, H.: 2002, *Astrophys. J.* **573**, L133–L136.

Nitta, N. V., and Hudson, H. S.: 2001, *Geophys. Res. Lett.* **28**, 3801–3804.

November, L. J., and Simon, G. W.: 1988, *Astrophys. J.* **333**, 427–442.

Pevtsov, A. A., and Canfield, R. C.: 1999, *Measurement Techniques in Space Plasmas Fields*, pp. 103.

Pick, M., Forbes, T. G., Mann, G., *et al.*: 2006, *Space Sci. Rev.*, this volume doi: 10.1007/s11214-006-9021-1.

Plunkett, S. P., Michels, D. J., Howard, R. A., Brueckner, G. E., St. Cyr, O. C., Thompson, B. J., *et al.*: 2002, *Adv. Space Res.* **29**, 1473–1488.

Plunkett, S. P., Thompson, B. J., Howard, R. A., Michels, D. J., St. Cyr, O. C., Tappin, S. J., *et al.*: 1998, *Geophys. Res. Lett.* **25**, 2477–2480.

Plunkett, S. P., Vourlidas, A., Šimberová, S., Karlický, M., Kotrč, P., Heinzel, P., *et al.*: 2000, *Sol. Phys.* **194**, 371–391.

Poland, A. I., and MacQueen, R. M.: 1981, *Sol. Phys.* **71**, 361–379.

Priest, E. R.: 1984, *Solar magneto-hydrodynamics*. Geophysics and Astrophysics Monographs, Reidel, Dordrecht, 1984.

Priest, E. R., Gaizauskas, V., Hagyard, M. H., Schmahl, E. J., and Webb, D. F.: 1986, *Energetic Phenomena on the Sun*, pp. 1–1.

Roussev, I. I., Sokolov, I. V., Forbes, T. G., Gombosi, T. I., Lee, M. A., and Sakai, J. I.: 2004a, *Astrophys. J.* **605**, L73–L76.

Roussev, I. I., Sokolov, I. V., Forbes, T. G., Gombosi, T. I., Lee, M. A., and Sakai, J. I.: 2004b, *Astrophys. J.* **605**, L73–L76.

Rust, D. M., and Kumar, A.: 1994a, *Sol. Phys.* **155**, 69–97.
Rust, D. M., and Kumar, A.: 1994b, *ESA SP-373: Solar Dynamic Phenomena and Solar Wind Consequences, the Third SOHO Workshop*, pp. 39.
Rust, D. M., and Kumar, A.: 1996, *Astrophys. J.* **464**, L199–L202.
Saito, K., and Tandberg-Hanssen, E.: 1973, *Sol. Phys.* **31**, 105–121.
Sheeley, N. R., Walters, J. H., Wang, Y.-M., and Howard, R. A.: 1999, *J. Geophys. Res.* **104**, 24739–24768.
Simnett, G. M.: 1999, *The Many Faces of the Sun: A Summary of The Results from NASA's Solar Maximum Mission*, pp. 201.
Srivastava, N., Schwenn, R., and Stenborg, G.: 1999, *ESA SP-446: 8th SOHO Workshop: Plasma Dynamics and Diagnostics in the Solar Transition Region and Corona*, pp. 621.
St. Cyr, O. C., Burkepile, J. T., Hundhausen, A. J., and Lecinski, A. R.: 1999, *J. Geophys. Res.* **104**, 12493–12506.
Sterling, A. C., Hudson, H. S., Thompson, B. J., and Zarro, D. M.: 2000, *Astrophys. J.* **532**, 628–647.
Sterling, A. C., and Moore, R. L.: 2001a, *Astrophys. J.* **560**, 1045–1057.
Sterling, A. C., and Moore, R. L.: 2001b, *J. Geophys. Res.* **106**, 25227–25238.
Subramanian, P., and Dere, K. P.: 2001, *Astrophys. J.* **561**, 372–395.
Tandberg-Hanssen, E.: 1995, *The Nature of Solar Prominences*. Astrophysics and Space Science Library, vol. 199, Kluwer Academic Publishers, Dordrecht, c1995.
Tappin, S. J., and Simnett, G. M.: 1997, *ESA SP-415: Correlated Phenomena at the Sun, in the Heliosphere and in Geospace*, pp. 117–120.
Thompson, B. J., Plunkett, S. P., Gurman, J. B., Newmark, J. S., St. Cyr, O. C., and Michels, D. J.: 1998, *Geophys. Res. Lett.* **25**, 2465–2468.
Tripathi, D., Bothmer, V., and Cremades, H.: 2004, *A&A* **422**, 337–349.
Vaiana, G. S., Krieger, A. S., and Timothy, A. F.: 1973, *Sol. Phys.* **32**, 81–116.
van Ballegooijen, A. A., Priest, E. R., and Mackay, D. H.: 2000, *Astrophys. J.* **539**, 983–994.
Vourlidas, A., Buzasi, D., Howard, R. A., and Esfandiari, E.: 2002, *ESA SP-506: Solar Variability: From Core to Outer Frontiers*, pp. 91–94.
Wang, H.: 1993, *ASP Conf. Ser. 46: IAU Colloq. 141: The Magnetic and Velocity Fields of Solar Active Regions*, pp. 323–332.
Wang, Y.-M., and Sheeley, N. R.: 1999, *Astrophys. J.* **510**, L157–L160.
Wang, Y.-M., and Sheeley, N. R.: 2002, *Astrophys. J.* **575**, 542–552.
Wang, Y.-M., Sheeley, N. R., Socker, D. G., Howard, R. A., Brueckner, G. E., Michels, D. J., *et al.*: 1998, *Astrophys. J.* **508**, 899–907.
Webb, D. F., Krieger, A. S., and Rust, D. M.: 1976, *Sol. Phys.* **48**, 159–186.
Webb, D. F., Lepping, R. P., Burlaga, L. F., DeForest, C. E., Larson, D. E., Martin, S. F., *et al.*: 2000, *J. Geophys. Res.* **105**, 27251–27260.
Webb, D. F., Nolte, J. T., Solodyna, C. V., and McIntosh, P. S.: 1978, *Sol. Phys.* **58**, 389–396.
Welsch, B. T., DeVore, C. R., and Antiochos, S. K.: 2005, *Astrophys. J.* **634**, 1395–1404.
Welsch, B. T., Fisher, G. H., Abbett, W. P., and Regnier, S.: 2004, *Astrophys. J.* **610**, 1148–1156.
Wills-Davey, M. J., and Thompson, B. J.: 1999, *Sol. Phys.* **190**, 467–483.
Wolfson, R., Conover, C., and Illing, R. M. E.: 1987, *BASS* **19**, 931.
Wolfson, R., and Saran, S.: 1998, *Astrophys. J.* **499**, 496–503.
Wood, B. E., Karovska, M., Chen, J., Brueckner, G. E., Cook, J. W., and Howard, R. A.: 1999, *Astrophys. J.* **512**, 484–495.
Yang, G., and Wang, H.: 2002, *Solar-Terrestrial Magnetic Activity and Space Environment*, pp. 113–116.
Zarro, D. M., Sterling, A. C., Thompson, B. J., Hudson, H. S., and Nitta, N.: 1999, *Astrophys. J.* **520**, L139–L142.
Zwaan, C.: 1985, *Sol. Phys.* **100**, 397–414.

MULTI-WAVELENGTH OBSERVATIONS OF CMES AND ASSOCIATED PHENOMENA

Report of Working Group F

M. PICK[1,*], T. G. FORBES[2], G. MANN[3], H. V. CANE[4], J. CHEN[5], A. CIARAVELLA[6], H. CREMADES[7], R. A. HOWARD[8], H. S. HUDSON[9], A. KLASSEN[3], K. L. KLEIN[1], M. A. LEE[2], J. A. LINKER[10], D. MAIA[11], Z. MIKIC[10], J. C. RAYMOND[12], M. J. REINER[13], G. M. SIMNETT[14], N. SRIVASTAVA[15], D. TRIPATHI[7], R. VAINIO[16], A. VOURLIDAS[8], J. ZHANG[17], T. H. ZURBUCHEN[18], N. R. SHEELEY[8] and C. MARQUÉ[8]

[1]*LESIA, UMR 8109 CNRS, Observatoire de Paris, Meudon, France*
[2]*Institute for the Study of Earth, Oceans, and Space, Univ. of New Hampshire, Durham, NH, USA*
[3]*Astrophysikalisches Institut Potsdam, Potsdam, Germany*
[4]*Laboratory for High Energy Astrophysics, NASA/GSFC, Greenbelt, MD, USA; and Bruny Island Radio Spectrometer, Tasmania, Australia*
[5]*Plasma Physics Division, US Naval Research Laboratory, Washington, DC, USA*
[6]*INAF Osservatorio Astronomico di Palermo, Palermo, Italy*
[7]*Max-Planck-Institut für Sonnenforschung, Katlenburg-Lindau, Germany*
[8]*E. O. Hulburt Center for Space Research, Naval Research Laboratory, Washington, DC, USA*
[9]*Space Sciences Laboratory, University of California, Berkeley, CA, USA*
[10]*Science Applications International Corporation, San Diego, CA, USA*
[11]*CICGE, Observatório Astronómico Prof. Manuel de Barros, Faculdade de Ciências da Universidade do Porto, Vila Nova de Gaia, Portugal*
[12]*Harvard-Smithsonian Center for Astrophysics, Cambridge, MA, USA*
[13]*Center for Solar Physics and Space Weather, Catholic University of America, Washington, DC; and NASA Goddard Space Flight Center, Greenbelt, MD, USA*
[14]*School of Physics and Space Research, University of Birmingham, U.K.*
[15]*Udaipur Solar Observatory, Physical Research Laboratory, Udaipur, India*
[16]*Department of Physical Sciences, University of Helsinki, Helsinki, Finland*
[17]*Center for Earth Observing and Space Research, Institute for Computational Sciences, George Mason University, Fairfax, VA, USA*
[18]*Dept. of Atmospheric, Oceanic, and Space Sciences, Univ. of Michigan, Ann Arbor, MI, USA*
*(*Author for correspondence: E-mail: monique.pick@obspm.fr)*

(Received 2 December 2004; Accepted in final form 6 June 2006)

Abstract. This chapter reviews how our knowledge of CMEs and CME-associated phenomena has been improved, since the launch of the SOHO mission, thanks to multi-wavelength analysis. The combination of data obtained from space-based experiments and ground based instruments allows us to follow the space-time development of an event from the bottom of the corona to large distances in the interplanetary medium. Since CMEs originate in the low solar corona, understanding the physical processes that generate them is strongly dependant on coordinated multi-wavelength observations. CMEs display a large diversity in morphology and kinematic properties, but there is presently no statistical evidence that those properties may serve to group them into different classes. When a CME takes place, the coronal magnetic field undergoes restructuring. Much of the current research

Space Science Reviews (2006) 123: 341–382
DOI: 10.1007/s11214-006-9021-1

is focused on understanding how the corona sustains the stresses that allow the magnetic energy to build up and how, later on, this magnetic energy is released during eruptive flares and CMEs. Multi-wavelength observations have confirmed that reconnection plays a key role during the development of CMEs. Frequently, CMEs display a rather simple shape, exhibiting a well known three-part structure (bright leading edge, dark cavity and bright knot). These types of events have led to the proposal of the "standard model" of the development of a CME, a model which predicts the formation of current sheets. A few recent coronal observations provide some evidence for such sheets. Other more complex events correspond to multiple eruptions taking place on a time scale much shorter than the cadence of coronagraph instruments. They are often associated with large-scale dimming and coronal waves. The exact nature of these waves and the physical link between these different manifestations are not yet elucidated. We also discuss what kind of shocks are produced during a flare or a CME. Several questions remain unanswered. What is the nature of the shocks in the corona (blast-wave or piston-driven?) How they are related to Moreton waves seen in Hα? How they are related to interplanetary shocks? The last section discusses the origin of energetic electrons detected in the corona and in the interplanetary medium. "Complex type III-like events," which are detected at hectometric wavelengths, high in the corona, and are associated with CMEs, appear to originate from electrons that have been accelerated lower in the corona and not at the bow shock of CMEs. Similarly, impulsive energetic electrons observed in the interplanetary medium are not the exclusive result of electron acceleration at the bow shocks of CMEs; rather they have a coronal origin.

1. Introduction

M. PICK, T. FORBES, G. MANN

Since their discovery in 1971 (Tousey, 1973), coronal mass ejections (CMEs) have been observed and analyzed by numerous ground-based and spaceborne white-light coronagraphs. The Solar Maximum Mission (SMM) in particular, enabled detailed analyses of the basic properties of CMEs to be made for nearly a complete solar cycle. The nearly continuous solar coverage provided by the very sensitive SOHO/LASCO coronagraph, which represents a very significant advancement over all previous spaceborne coronagraphs, has greatly benefited from the comprehensive solar and heliospheric data made available by experiments on a vast array of Earth-orbiting and heliospheric spacecrafts (TRACE, ACE, GOES, Yohkoh, Ulysses, Voyager, Wind) as well as by ground-based instruments (magnetographs, Hα telescopes, radio imaging and spectral instruments, and coronagraphs). Comparison of the CME observations from SOHO instruments with complementary observations in the low corona have, for the first time, made it possible to continuously follow the detailed progression of CME phenomena from the low corona to the interplanetary medium. In this report, we have selected topics in which joint coronal observations obtained using complementary observational and analysis techniques have been crucial in shaping our current understanding of CMEs and CME-associated phenomena.

CMEs are complex, large-scale magnetic structures, expelled from the sun and made visible by their enhanced plasma density. The full dynamical evolution of a CME includes three phases: (1) an initiation phase, (2) an acceleration phase, and

(3) a propagation phase. The period of continuous acceleration is quite variable: it can last from a few minutes to several hours and the CME, which becomes fully developed during this period, can travel a distance from a fraction of a solar radius to several solar radii. Although CMEs are a complex and varied phenomena, can we identify distinct classes of CMEs, based on their dynamic behavior and/or magnetic structure? Are the disparities observed between different CMEs due to the complexities of the coronal structures?

CMEs are often associated with eruptive prominences or disappearing filaments on the solar disk. In these cases, the CMEs most often contain three distinct regions: a bright front that surrounds a dark cavity and an inner bright core. While it is sometimes assumed that the leading edge is a compressive wave front, there is little evidence for that except for a few shock wave observations. In many cases the "leading edge" may just be coronal plasma piled up in front of the CME as it moves outward. The interior structure of CMEs has often been identified as a flux rope (see Schwenn *et al.*, 2006, this volume). However modeling of the CME formation is a controversial topic: is the flux rope the cause or a consequence of the eruption? The so-called "standard model" (see Figure 6 of Hudson *et al.*, 2006), this volume) predicts the formation of a current sheet. Do recent coronal observations provide evidence for the formation of such a current sheet?

CMEs frequently have a much more complex structure. They involve multiple magnetic flux systems and neutral lines. They are often associated with flares and/or on-disk manifestations, which have been observed in EUV as large-scale dimmings and coronal waves, and in soft X-rays as large scale and trans-equatorial loops. The exact nature of the coronal waves is unknown: are these waves different from the well-known Moreton waves identified as the chromospheric trace of a blast wave? We note that while Moreton waves are systematically associated with type II radio bursts, EUV coronal waves may have no association. What is the physical link between EUV waves and CMEs? Do they represent restructuring of the corona?

One common view, which emerged in the 1980s, is that CMEs are not directly caused by flares and that flares are not necessarily the dominant energetic phenomena in the corona. They are now recognized as two different aspects of a common magnetic energy instability and release, but the physics behind the association remains poorly understood: Can the currently available observations provide an unambiguous description of their respective temporal and spatial evolution and their respective source of energy release? For example, radio imaging observations provide information on the location of regions of magnetic interaction in the corona. These regions are the sites of electron acceleration that generate various radio signatures often observed in the decimeter-meter (dm-m) wavelength domain; electron beams escaping from these regions produce type III radio bursts, which trace the open magnetic field and coronal loop-like shapes as trans-equatorial loops (J and U bursts). The high cadence of radio observations may allow us to follow the progression and development of CMEs in the corona.

The large-scale restructuring of the corona observed during the development of CMEs suggests that for a given event, different populations of energetic electrons (and possibly protons) may be accelerated by different physical mechanisms at different times and from different coronal sites. Do recent observations contradict the commonly-held view that CME bow shocks are the exclusive source generating large SEP events, observed in the interplanetary medium?

The existence of coronal shocks was recognized early from the interpretation of radio type II bursts. Two types of shocks have been considered to produce the type II bursts: Coronal shocks which generate metric type II bursts, currently attributed to blast waves originating in the flare-region; and CME-driven shocks, which produce hectometric-kilometric type II bursts during their propagation in the interplanetary medium. As coronal and interplanetary type II bursts can be produced during the same event, there is a continuing controversy on the relationship between type II bursts (coronal shocks), flares, and CMEs. At what altitude can CME-driven shocks form? Can we detect shocks by remote techniques other than radio observations such as EUV, X-ray and white light?

A summary of the questions and conclusions by this working group is presented in the following sections.

2. Coordinated Multi-Wavelength Signatures of CME Development

R. A. Howard, N. Sheeley, J. Zhang, N. Srivastava, G. Simnett, J. Chen, M. Pick, C. Marqué

To understand CME phenomena, it is essential to utilize observations in addition to those from the white-light coronagraph. While attempts have been made to develop a proxy for the coronagraph, the clear identification of when and where a CME has occurred is still best made with the coronagraph. However, the coronagraph measures only the density properties of the CMEs. The combination of all the space-based experiments currently available and the extensive network of ground based instruments present an unprecedented opportunity to characterize CME phenomena all the way from the surface of the Sun out to interplanetary space. Furthermore, the ground-based telescopes are able to acquire data at a much higher cadence than the space-based telescopes. These multi-wavelength observations can tell us: (1) What is the sequence of events below the coronagraph occulting disk? (2) What occurs first and in what time order? (3) What structures are involved in the energy release?

2.1. Diversity of CMEs

Coronal mass ejections display diversity in morphology and kinematic properties. Frequently, large and bright CMEs display a three-part structure, namely the bright frontal or the leading edge, followed by a dark cavity and a bright knot often associated with prominence material. Well-organized helical structures within CMEs have often been observed by LASCO (Dere *et al.*, 1999), CDS (Pike and Mason, 2002), and UVCS (Antonucci *et al.*, 1997; Ciaravella *et al.*, 2000). The classification of a CME in terms of its structure and morphology was first done using SOLWIND white-light observations (Howard *et al.*, 1985). These structural classes included loop, curved front, halo, spike, double spike, streamer blowouts and, as the names suggest, are based on the shape of the CMEs. They termed another class of CMEs as "complex," which were irregular in shape and did not fit in the other sub-classes. It is now becoming clear that many events are in fact multiple eruptions closely spaced in time and projected on the plane of the sky.

Another type of classification scheme based on signatures of field line disconnection was discussed by Webb and Cliver (1995). They found that about 10 % of all CMEs had disconnection features, defined as transient large-scale, concave-outward bright regions usually following the CME leading edge. Such disconnected structures were previously observed in SMM images as U-shaped features by McComas *et al.* (1991). The disconnected features of CMEs have been found to retain their shapes out to a distance of 28 solar radii in LASCO images, as shown by Simnett *et al.* (1997), while other observations show that some CMEs show little evidence of disconnection (Chen *et al.*, 1997, 2000). Although a CME often refers to a large-scale phenomenon, narrow CMEs have been reported (Gilbert *et al.*, 2001; Wang and Sheeley, 2002; Yashiro *et al.*, 2003; Dobrzycka *et al.*, 2003) as have very faint CMEs (Lyons and Simnett, 2001). At the smaller end of the scale, Sheeley *et al.* (1997) have studied tiny eruptions which they called "blobs". Because coronagraph observations yield two-dimensional (2-D) projections of intrinsically three-dimensional (3-D) structures, the true 3-D morphology of observed CMEs is not yet definitely established. Nevertheless, all existing 3-D models indicate that after the eruption, CMEs can be understood as expanding flux ropes (Chen *et al.*, 1997; Wu *et al.*, 1999; Amari *et al.*, 1999; Krall *et al.*, 2001; Tokman and Bellan, 2002).

A classification scheme based on kinematic properties was first done by MacQueen and Fisher (1983). Based on inner coronal observations (i.e. from 1.2 to $2.4 R_{\odot}$ by the ground-based MK3 coronagraph), they found that large flare-associated events exhibit high speeds (>700 km s^{-1}) and little acceleration (0.3 m s^{-2}) with height while the eruptive-filament-associated events start with low speeds of 10–20 km s^{-1} and exhibit accelerations from 0–50 m s^{-2}. Based on outer corona observations (from 2.0 to $30.0 R_{\odot}$ by the LASCO C2/C3 coronagraphs), Sheeley *et al.* (1999) also suggested the existence of two classes of events: the so-called impulsive and gradual CMEs. Impulsive CMEs show constant speed typically greater than 750 km s^{-1}, implying a short but strong acceleration in the inner corona. Gradual CMEs display a persistent and weak acceleration with their leading edges accelerating gradually to speeds in the range 400–600 km s^{-1} before leaving the LASCO field of view. Based on combined LASCO C1/C2/C3 observations, Srivastava *et al.*

(1999) presented a number of gradual or balloon-type events that start off with very low speeds (<100 km s^{-1}) below 3 solar radii and attain a maximum value of acceleration ranging between 5–25 m s^{-2}, which occurs between 3–6 $R_{\odot}$. These observations indicated that gradual-type events generally drift away in the slow solar wind. A similar classification scheme was proposed by Andrews and Howard (2001) but with different names: Type C (Constant speed) events versus Type A (Acceleration) events. Based on a large number of LASCO CME events, St. Cyr *et al.* (1999) and Moon *et al.* (2002) made a statistical study about the differences between flare-associated CMEs and eruptive-filament-associated CMEs and found a tendency for flare-associated CMEs to be faster.

Nevertheless, there is not yet convincing statistical evidence to suggest the existence of two distinct classes of events. In terms of speed, it has been consistently shown that there is a continuous distribution of speed from tens of km s^{-1} to about 2500 km s^{-1} with a single peak at about 300–400 km s^{-1} (Howard *et al.*, 1985; Hundhausen *et al.*, 1994; St. Cyr *et al.*, 2000; Yashiro *et al.*, 2003). However, the peak is in part due to the selection criteria used to define a CME in these studies, which favor the bright, large-scale events. In terms of acceleration, (Zhang *et al.*, 2004) demonstrated that the magnitude of the acceleration in the inner corona can vary by three orders of magnitude, i.e., from a few m s^{-2} to several thousand m s^{-2}. Further, the duration of the acceleration can vary by the same magnitude, i.e., from a few minutes to several hours. The existing data do not support a bimodal distribution of CME acceleration.

Despite the failure to identify objective signatures that place CMEs in separate statistical classes, we know that all CMEs are not alike. In fact, in studying different kinds of CMEs we have been able to gain insight into the physical mechanisms.

In the next several sections, we give an overview of different types of CMEs and some physical insights that have been gained by the multi-spectral approach.

2.2. Small Exploding Flare Events with Reconnection

The initiation of a CME in EUV was first detected by EIT (Dere *et al.*, 1997). In this paper a small, compact bubble is seen to form and propagate outward. It was seen in the LASCO/C1 Fe XIV channel and then subsequently in the LASCO/C2 as a small enhancement along a pre-existing streamer. Innes *et al.* (1999) used several different telescopes to investigate the relationship between a CME and an X2 flare (see Figure 1). Innes *et al.* compared SOHO/CDS EUV spectrometer data, GOES X-ray flux, Yohkoh/SXT images, SOHO/EIT Fe XII images, Meudon Hα filtergraph images, Pic du Midi Hα coronagraph images and finally the LASCO/C1 Fe XIV, C2 and C3 images. In their interpretation, a filament started to rise, before the flare, forcing anti-parallel magnetic fields to reconnect and the flaring to start. A small CME started about 20 minutes after the flare and filament eruption and in between two X-ray jets. Innes *et al.* suggested that reconnection at the site of one of

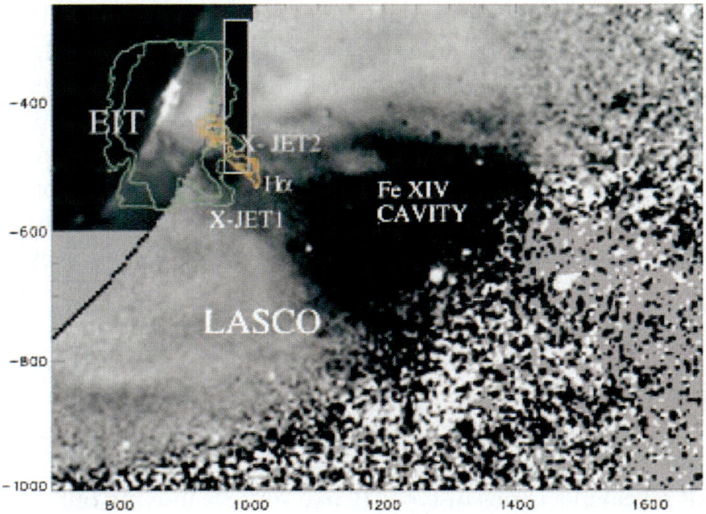

Figure 1. April 9, 1997 event. Spatial relationship between a CME and an X2 flare, as revealed by several telescopes. Shown are LASCO C1 5303 Å difference images ($>1 R_\odot$) taken before and after the flare, as well as soft X-ray contours (shown in green), Hα coronagraph contours (red) and an EIT 195 Å image ($< 1.1 R_\odot$) taken after the flare. The position of the CDS raster is outlined in white (Innes *et al.*, 1999).

the jets (at one footpoint) triggered reconnection at the other footpoint (producing the second jet).

Complimentary observations of these explosive, flare-related CMEs have been obtained with the TRACE, Yohkoh/SXT and GOES12/SXI instruments (McKenzie and Hudson, 1999; Gallagher *et al.*, 2002; Innes *et al.*, 2003; Asai *et al.*, 2004). These observations have revealed inward motions in the hot (15 MK) plasma cloud that forms in the region evacuated by the CME. In a time of about 30–40 min, the elements of the cloud move inward and cool from ~15 MK to <1 MK, becoming visible as the familiar structures we call flare loops. One by one, the individual loops in the lower-temperature emission lines disappear and are replaced by higher loops moving in from above. This accounts for the rising locus that is typically observed and often seen as radio continuum. Kerdraon *et al.* (1983) observed stationary type IV radio emission in the position of the legs of a CME, presumably in the region of post-eruptive restructuring of the corona where the flare loops were being formed. Kahler and Hundhausen (1992) suggested these type IV bursts could be associated with newly formed streamers.

Comparing SOHO/LASCO and EIT images of narrow, high speed CMEs (called jets) during the rising phase of the sunspot cycle, Wang *et al.* (1998) found that narrow (~2–4 degrees), fast (~600–1000 km s^{-1}) white-light jets in the C2 and C3 fields of view were the extensions of EUV jets seen by EIT. These jets appeared when small bipoles emerged in polar coronal holes. Wang *et al.* (1998) suggested that the jets originated when the closed field of the newly emerged flux reconnected with the open fields of the polar holes (interchange reconnection), thereby allowing material to slip outward along the open field lines. In a second study, Wang and Sheeley (2002) made a similar comparison near sunspot maximum. They found that in this epoch, when polar fields were weak and the polar holes were gone, the jets came from places where newly emerging flux penetrated coronal holes well outside the Sun's polar caps. UVCS and EIT observations of polar jets show relatively low temperatures, consistent with heating at the base of the jets (Dobrzycka *et al.*, 2002).

In a separate study, Kahler *et al.* (2001) noted that a 960 km s^{-1} jet was associated with an impulsive solar energetic particle event at the Wind spacecraft. They suggested that such jet-like CMEs might correspond to the SXT X-ray jets and type III radio bursts observed previously by Raulin *et al.* (1996), and that these type III burst electrons may provide information about the magnetic topology of CMEs that are accompanied by impulsive flares. The Nançay Radioheliograph (NRH) images for this event revealed a series of type III-like sources moving outward at the same position angle as the C2 LASCO feature (Pick *et al.*, 2003).

2.3. CMEs and Prominence Eruptions

Prominence eruptions (EPs) frequently accompany CMEs. Gilbert *et al.* (2000) and Gopalswamy *et al.* (2003) found that more than 72% of Hα erupting prominences are associated with CMEs. Gopalswamy *et al.* also noted that the latitude distributions of EPs and CMEs both peaked at the equator around sunspot minimum.

An example of a class of CME events that is not usually associated with impulsive disk phenomena was discussed by Srivastava *et al.* (2000). For this event (displayed in Figure 2), they combined Hα images from Helio-Research (courtesy of Sara Martin), 17 GHz images from the Nobeyama Radioheliograph, LASCO/C1 Fe XIV/Fe X images, and LASCO/C2 and C3 white-light images. They found that features in the images moved very gradually upward ($<$50 km s^{-1}), with the prominence activating and moving outward, for many hours ($\sim$18) before the second phase started. At that time, the CME/prominence system accelerated quickly to over 500 km s^{-1} before decelerating down to the slower solar wind speed when it left the field of view of LASCO/C3. The minimum value of the acceleration corresponds to the slowly moving structures identified by Sheeley *et al.* (1997). They further noted that there is no observational evidence for magnetic reconnection or impulsive energy release associated with this CME. Howard *et al.* (1986) showed that this class of events did not exhibit a solar-cycle dependence, indicating that the cause of such eruptions occurs at a constant rate throughout the cycle. Differential rotation and the subsequent shearing of coronal fields has been explored as possible CME initiation mechanism (e.g. Linker and Mikic, 1995; Forbes *et al.*, 2006).

This kind of event usually generates the three-part structure. In the dm-dam wavelength domain, thermal emission is detected in association with the filament

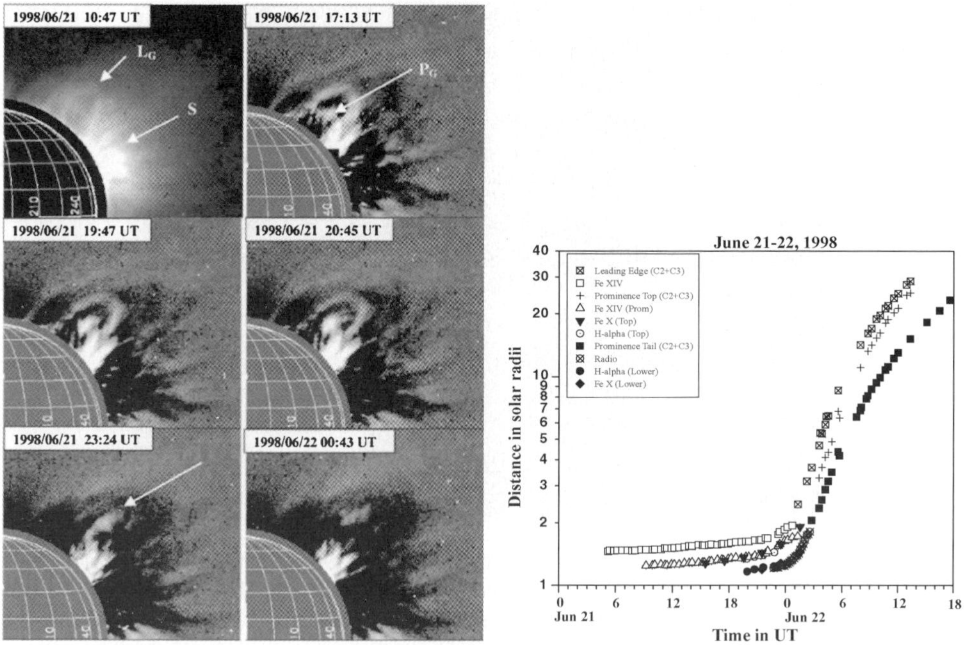

Figure 2. June 21–22, 1998 event. *Left*: Time-lapse images taken by LASCO-C1 coronagraph in Fe XIV emission line. The field of view is 1.1–3 $R_\odot$. All the images shown here have been subtracted from a reference image taken before the occurrence of the CME, except the image taken at 10:47 UT. The 10:47 UT frame is an on-line image with a nearby continuum and shows the bright streamer adjacent to the CME. *Right*: Plot of height (on a log scale) against time for different features of the CME: the leading edge, the prominence top, and prominence tail. These features were tracked using several instruments at different wavelengths (Srivastava *et al.*, 2000).

cavity; thermal emission is weak and often masked by the presence of non-thermal radio burst emission. Metric observations are, however, able in favorable cases to probe the filament cavity and make the link with the CME cavity (Marqué *et al.*, 2002). CME thermal detection in radio has a great potential interest because of the ability to determine mass and density of a CME (Sheridan *et al.*, 1978; Gopalswamy and Kundu, 1992). Despite these earlier attempts, a clear distinction between thermal and non-thermal contributions remains difficult to establish.

2.4. LARGE-SCALE CMEs

The development in the low corona of large flare/CME events is often fast and is accompanied by an opening of the magnetic field over a large region. Large scale dimming observed in EUV and soft X-rays and disappearing transequatorial loops are signatures of the opening of the magnetic field. Li *et al.* (2001) have developed a technique to show the field lines that have opened up as a result of the CME event.

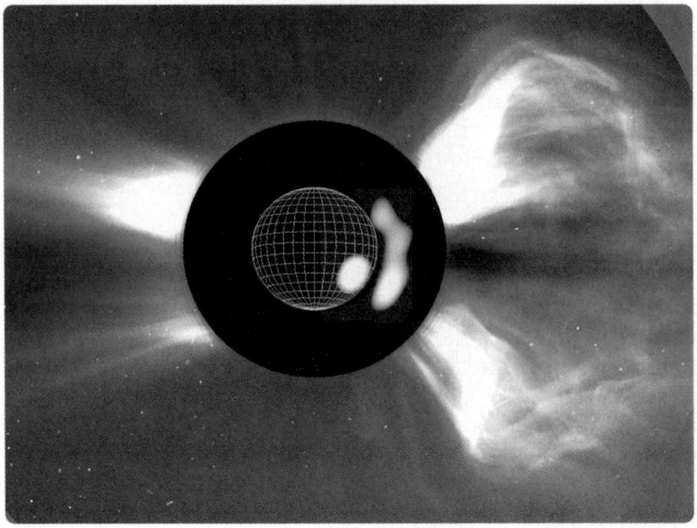

Figure 3. November 6, 1997 event. Composite image of a LASCO/C2 CME image with a radio NRH image at 164 MHz showing that there is a close correspondence between the extent in latitude of the CME seen by LASCO- C2 and the sites of radio emission (adapted from Maia *et al.*, 1999).

Coronagraphs observe limb CMEs better than halo CMEs. For the large, fast CME that occurred on the west limb on 6 November 1997 (see Figure 3), Maia *et al.* (1999) related the radio observations to the optical observations seen all the way down to $1.1 R_{\odot}$ by the LASCO/C1, which observed the CME in Fe XIV and Fe X emission. This CME expanded laterally from a relatively small angular size, in the vicinity of the flare site and reached its full size in the low corona within 5 minutes. A close association between the location of radio emission and the white-light CME structure was seen. It is interesting to compare these observations with those of an on-disk event because the two perspectives give us a more complete picture. Pojholainen *et al.* (2001) for the on-disk event on 2 May 1998 linked, both spatially and temporally, a propagating disturbance to the successive radio source components of a coronal type II burst and to an Hα Moreton wave propagating away from the flare region. Khan and Hudson (2000) suggested that, in the case of the 2 May 1998 event, the wave also causes the destabilization of the interconnecting loop structure seen in X-rays. Both these limb and on-disk events showed a similar spreading of radio sources. The full expansion of these events took place in the low corona within 5–15 minutes.

A well-studied event is the Bastille-day event, which occurred on 14 July 2000, and for which a number of papers have been published in a single volume in Solar Physics (special issue, 204/1–2, 2001). This event illustrates many points that will be discussed in the following sections. During July, the active region 9077 produced intense activity, which culminated in an X7.5 flare and a major CME

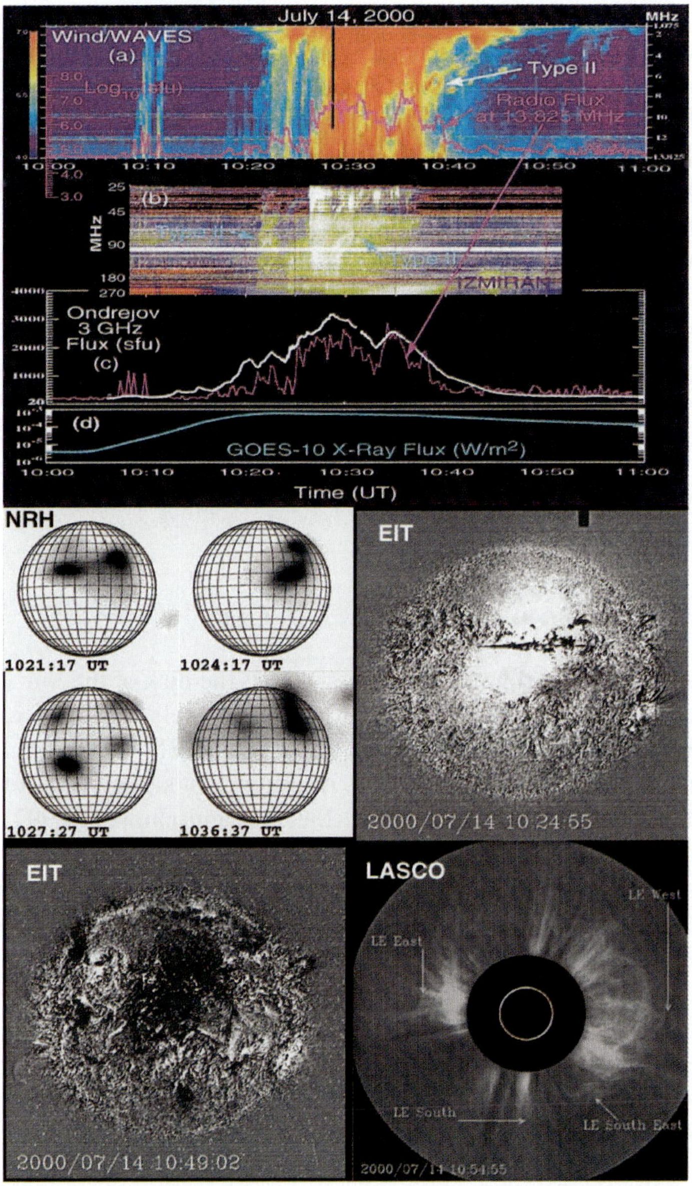

Figure 4. July 14, 2000 event. *Upper panel*:(a) Dynamic spectra of the Wind/WAVES radio emission; (b) IZMIRAN dynamic spectrum of the metric radio emissions; (c) Intensity versus time of the Ondrejov decimeter data of 3 GHz and of the flux measured by Wind at 13.825 MHz ; (d) GOES X-ray flux. *Middle and lower panels*: Images from the NRH at 164 MHz, EIT at 195 Å and LASCO-C2 illustrating the development of the CME above the solar disk and above the solar limb (references in Solar Physics, special issue, 204-1-2, 2001).

on July 14. Accompanying this activity were radio emission, energetic particles, interplanetary shocks, and interplanetary magnetic clouds. Figure 4 displays a series of observations obtained at different wavelengths. The LASCO images show that besides the loop-like CME on the west limb, there are several bright discrete features all around the Sun. Most of the solar disk was affected by the CME development. The EIT observations show a brightening of a large area of the corona followed by a dimming over roughly the same region, suggesting that the bright material had been ejected. The development of the radio event is reminiscent of that of the May 02 1998 event: In less than 15 minutes, large-scale loops spanned the entire disk, apparently in association with spectral drift type II emission. The type II emission was accompanied by very intense, complex, long duration type-III like radio emission over a very large spectral range (Figure 4; upper panel a, b, c). The close similarity in the 3 GHz and 13.82 MHz profiles of the radio emission indicates that radio emission detected by WAVES at kilometric wavelengths was produced by electrons accelerated in the coronal regions seen in the NRH images (see Section 6.3).

2.5. EIT Waves, Coronal Dimming and CMEs

EIT waves are large disturbances that travel across the disk of the Sun (Moses *et al.*, 1997; Thompson *et al.*, 1998). An unambiguous correlation between EIT waves and CMEs was found by Biesecker *et al.* (2002). The EIT-wave/shock association is considered in more detail in Section 5. An important signature of on-disk CMEs (i.e. halos) are the soft X-ray dimmings (Sterling and Hudson, 1997) observed in the Yohkoh/SXT. The dimming is ascribed to a mass deficit, $\sim 10^{14}$ g, which is consistent with the amount of material that erupts. In contrast to the EIT wave or Moreton wave, the X-ray dimming appears suddenly like transient coronal holes on the Sun's disk. Additional discussion of EIT waves can be found in Section 5.2.

3. Magnetic Reconnection

T. G. Forbes, H. S. Hudson

3.1. Theoretical Framework

The concept of magnetic reconnection arises from the idea that the topology of the magnetic field plays an important role in the forces acting on the plasma (Biskamp, 2000; Priest and Forbes, 2000). In a non-conducting fluid, or vacuum, the topology of the field has little theoretical value, and the magnetic field can be represented exactly in terms of a potential. Within the framework of MHD theory as applied to a magnetized plasma, however, dynamical effects do arise. Currents can flow within the plasma, stressing the field away from the relaxed state. Relaxation nearer to this unstressed state, if on a short time scale, is equivalent to the breakdown of the adiabatic invariants of the particle motion (e.g., (Northrop, 1966). This reflects the physical definition of a magnetic field line in terms of particle motion. The concept of reconnection inherently requires a breakdown of the ideal MHD approximation.

When a CME takes place, the coronal field undergoes a radical restructuring on a time scale short relative to the resistive time scale of the medium in what is, for the most part, a low-β plasma. Prior to the onset of the CME, the coronal field has gradually accumulated stresses related to forces acting below the photosphere. In some models, the coronal magnetic structure is presumed to be organized into topologically distinct cells divided by surfaces termed separatrices. A flare and/or CME corresponds to a transfer of magnetic flux between the cells. In the case of a CME the restructuring implies the opening of some magnetic flux, in the sense that previously closed loops become greatly extended. The main theoretical problem of flares and CMEs is to explain not only why energy is released, but also why the corona is able to sustain the stresses that allow an energy build-up (see in this volume (Mikić and Lee, 2006; Forbes *et al.*, 2006).

3.2. Relevant Observations

At present we have many solar signatures, in flares and in CME counterparts in the low corona, consistent with the general idea of magnetic reconnection. This list below could be extended in the near future, although necessarily only via indirect remote-sensing techniques. It is striking that reconnection signatures are relatively rare in the coronagraph field of view so that if reconnection plays a major role in CMEs it is likely to do so at lower altitudes. It is not yet possible to say whether magnetic reconnection plays an important role in the onset of an event. In particular we do not know if reconnection triggers a flare or simply results from a flare. Even more fundamentally, since MHD physics cannot self-consistently describe the all-important particle populations, we know that additional physics describing kinetic effects is important. At the present time very little theoretical development has occurred to incorporate such kinetic effects within the larger framework of the MHD flows.

3.2.1. *Ribbon Morphology*

The existence and motions of Hα flare ribbons strongly motivated the basic idea of the coronal reconnection models. The fact that the outer edge of the ribbon propagates by the lighting up of the existing chromospheric fibril structure is a strong indicator that the flare energy source is continually moving onto new field lines as required by reconnection. Furthermore, the existence of a one to two arcsecond wide band of strongly red shifted Hα emission along the outer edges of the ribbons (Svestka *et al.*, 1980) also supports the reconnection model. The outer edges of the ribbons are expected to be strongly heated by conduction electrons or energetic particles propagating along field lines mapping directly to the reconnection site.

The heating produces a downward flow of cool plasma in the chromosphere and an upward flow of heated plasma in the corona (Doschek *et al.*, 1983). Recent EUV spectroscopic imaging from SOHO/CDS has provided strong evidence that the corresponding upflow of heated plasma does exist at the outer edges as required (Czaykowska *et al.*, 1999).

3.2.2. *Hard X-ray Timing*

The hard X-ray footpoint sources exhibit strongly correlated fluctuations, consistent with the acceleration of electrons high in the corona. This result, originally due to Yohkoh/HXT, has now been extended by the finding that the footpoint spectra also match (Krucker and Lin, 2002). Coronal particle acceleration suggests coronal energy release, since we know that the particles (>10 keV electrons) contain a large fraction of the total energy.

3.2.3. *Cusp-Shaped Structure*

Soft X-ray images, such as those from Yohkoh/SXT, often show cusp-shaped features, especially in the late phase of a CME-related solar flare. The cusp may have an elevated temperature (Tsuneta *et al.*, 1992) and thus this morphology points to an energy source in the solar corona – *ergo*, large-scale reconnection along the lines of many models.

3.2.4. *"Above-the-Loop-Top" Source*

The discovery by Masuda *et al.* (1994) of an *above-the-loop-top* hard X-ray source suggested the reconnection geometry more directly. The interpretation considers a fast reconnection outflow jet from a reconnection point high in the corona, which then impinges on closed field lines and particle acceleration by shock waves or turbulence resulting from this.

3.2.5. *Supra-Arcade and LASCO Downflows*

Soft X-ray and EUV observations frequently show a fan of spikes extending above the loop system of a flare arcade. McKenzie and Hudson (1999) found dark features moving downward through this fan. They refer to these features as "voids" and interpret them as three-dimensional equivalents of the magnetic islands produced by the tearing mode. Voids occur during both the impulsive and gradual phases of flare evolution (Innes *et al.*, 2003; Asai *et al.*, 2004). In addition to the rapid motion of the voids, a general downward flow pattern exists in the arcade structure even as its upper edge increases in height (Sheeley *et al.*, 2004).

Downflows of two types have been directly seen in LASCO white-light images (Sheeley *et al.*, 2004) describe features similar to the supra-arcade downflows ("tadpoles") at altitudes of 2–4 $R_\odot$, much later than the eruption and exhibiting smaller inflow speeds. They identify such inflows with a continuation of the reconnection process implied by the observations in the flare arcades. These are also void regions

(Sheeley and Wang, 2002). The other (much less frequent) type of outer coronal downflow has both upflows and downflows, and more nearly resembles the classical picture of reconnection in the sense of disconnection (Sheeley and Wang, 2002; Simnett, 2004).

3.2.6. *Sigmoid and Dimming*

Sterling and Hudson (1997) found a characteristic X-ray morphology to be associated with CME formation, as detected by the formation of "transient coronal holes," a form of X-ray dimming originally identified in Skylab soft X-ray images. The sigmoid, identified with a hot component aligned with a filament channel, evolves during the event into an arcade perpendicular to it. This suggests "dipolarization" responsible for energy release following reconnection.

3.2.7. *Inflow into X-line*

The cusp configuration, if it reflects continuing magnetic reconnection, implies an essentially horizontal inflow into the X-point (in two dimensions). This has been surprisingly hard to detect, perhaps because the inflowing plasma should be extremely low-density because it is on open or opening field lines; nevertheless Yokoyama *et al.* (2001) have found a clear case, combining both soft X-ray and EUV data, of a large-scale coronal cusp with strong evidence for just such an inflow.

3.2.8. *Bidirectional Type III Radio Bursts*

Type III radio bursts have a strong association with the impulsive energy release in a solar flare, and reveal the presence of beams of non-thermal electrons. Bidirectional type III bursts, as observed by Aschwanden *et al.* (1995), show that these beams appear to diverge from the intersection of magnetic structures. This technique of inferring magnetic field line reconnection from bi-directional type III bursts is analogous to the technique of inferring coronal or heliospheric field line reconnection from bi-directional electron heat fluxes in the in-situ solar wind (Crooker *et al.*, 2002; Simnett, 2004).

4. Observational Signatures of Current Sheets

A. CIARAVELLA, H. S. HUDSON

4.1. THEORETICAL FRAMEWORK

The "standard model" for the development of an eruptive flare or of a CME involves the expansion of closed coronal magnetic fields (e.g. Anzer and Pneuman, 1982). As this bubble of bipolar magnetic field moves outwards, an inflow occurs behind it, driven by external magnetic pressure. The outward flow creates elongated field lines with appropriate orientation (opposed fields) to form a current sheet and ultimately to reconnect, and this reconnection then explains the formation of the arcade of flare loops. Coronal observations now provide some evidence for the formation of the current sheet that the standard model envisions.

4.2. Relevant Observations

4.2.1. *CME Morphology*

CMEs typically leave vertical linear features behind them; these have often been interpreted as "legs," in other words unipolar structures that represent flux tubes. Kahler and Hundhausen (1992) however pointed out that an alternative interpretation could be made that some legs were in fact current sheets. If so, this would support the basic tenet of the standard model and would explain the lack of transverse (lateral) motion expected for true legs to display as reconnection proceeded (Webb *et al.*, 2003). Early attempts towards the same end involved searches for "disconnection" events, motivated by the 2D cartoon representation of the standard model in which magnetic fields actually must disconnect from the surface of the Sun (Illing and Hundhausen, 1983). There is some evidence for concave-up features in CME images, but instead of interpreting this evidence as disconnection the more modern view is the 3D version: the concave-up features represent the bottom of a flux rope whose ends in fact remain connected to the photosphere (see Forbes *et al.*, 2006, this volume).

4.2.2. *Spectroscopy*

Because magnetic reconnection may be occurring in the current sheet and enabling energy release, the presence of a thin spike of high-temperature material beneath a retreating CME would match the expectation of the standard model. This closely describes recent spectroscopic observations by Ciaravella *et al.* (2002) and Ko *et al.* (2003), in which narrow, bright features appeared in Fe XVIII 974 Å. An illustration of the 1998 March 23 current sheet observation (Ciaravella *et al.*, 2002) is given in Figure 5, compared with an illustration of the current sheet in the Lin and Forbes (2000) model. A similar structure was observed on 2002 January 8. Figure 6 shows a composite image of EIT, MLSO-MK4[1], CDS and Fe XVIII 974 Å line intensity distribution along the UVCS slit. The high-temperature narrow feature observed by UVCS just above the post-CME loop cusp has been interpreted as the current sheet.

These current sheets refer to heights of $1.5R_{\odot}$ and exibit very similar characteristics. Both have very long duration: 20 hours in the 1998 March 23 event and about 2 days in the 2002 January 8. Their spatial width in the Fe VXIII emission is $0.1R_{\odot}$ and $0.2R_{\odot}$, respectively. Temperatures inside the current sheet are $4-6\times10^{6}$ K and densities in the range $10^{7}-10^{8}$ cm^{-3}. In both cases the analysis of the elemental

[1]The Mauna Loa Solar Observatory Mark 4 K-coronameter.

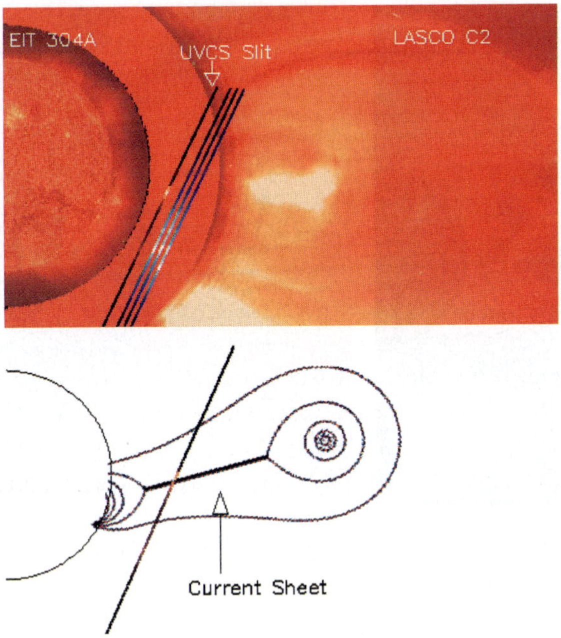

Figure 5. Composite image of EIT 304 Å (22:24 UT), LASCO C2 (12:33 UT) and line intensity distribution along the UVCS slit (16:56 UT) taken on March 23, 1998. The arrow indicates the true position of the slit and the intensity distribution shown is from the Fe XVIII 974 Å line. Three other slits are plotted for comparison and show (from left to right) the line intensity distribution of Si XII 499 Å, OVI 1032 Å, and Lyβ. The UVCS data used in this figure were taken with the slit pointed at $1.45 R_{\odot}$ and position angle $PA = 245°$. The position of the peak in the Fe XVIII line is $PA = 257°$ and height $1.55 R_{\odot}$. The different times were chosen to combine the data into a single image. At the time of the observation, the LASCO CME was much larger, but similar in shape. The flux-rope CME and the associated current sheet as predicted by the Lin and Forbes (2000) model are drawn below the composite image. Superimposed is the line intensity distribution of Fe XVIII along the slit (Ciaravella *et al.*, 2002).

abundances shows a large FIP effect[2] in which the coronal iron abundance in the 1998 March 23 current sheet, is enhanced by a factor of 7 relative to its photospheric value. The strong FIP effect in the sheets provides evidence that plasma flowing inside is from coronal origin (Ciaravella *et al.*, 2002; Ko *et al.*, 2003). Also, in the 2002 January 8 event, blobs are seen flowing along the current sheet; these were interpreted as a signature of reconnection in a non-uniform plasma by Ko *et al.* (2003).

The Lin and Forbes (2000) model provides a description of the current sheet geometry, as shown in Figure 5. Reconnection along the current sheet could help the flux rope to escape outwards. The model does not predict the physical parameters of the plasma in the sheet but the general properties such as length, position and duration of the sheet, agree with the UVCS observations. The MHD simulation of

[2]Enhancement of elements with high first ionization potential.

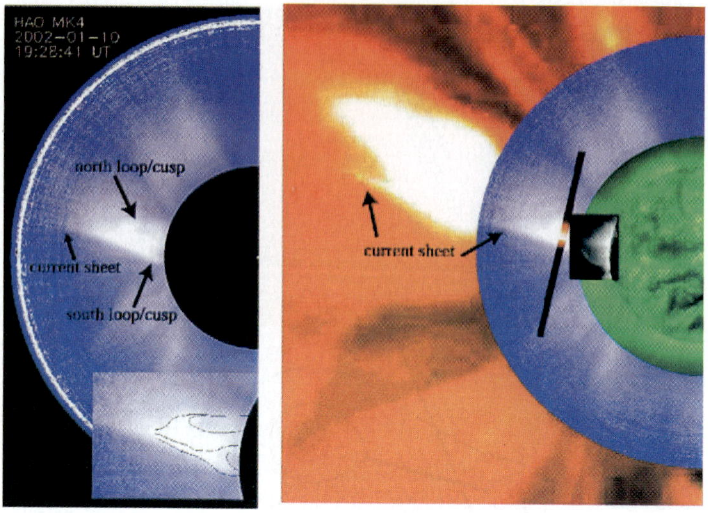

Figure 6. Left: MLSO MK4 white-light image at 19:28 UT, January 10, showing that the two loop/cusp systems still exist 2 days after the CME. The inset shows our perception of what the two loop/cusp systems might look like. *Right*: Composite image from EIT 195 Å (19:36 UT), CDS Fe XVI (small rectangle at the limb), MK4 (19:28 UT), and C2 (20:26 UT) images of January 10. Superposed on the image is the projection of the UVCS slit on the plane of the sky (the width is not to scale) showing the spatial distribution of the [Fe XVIII] 974 emission. It is clear that the high-temperature emission is along the direction of the current sheet (Ko *et al.*, 2003).

Linker *et al.* (2003) show the self-consistent formation of a similar current sheet (see Figure 3 of Forbes *et al.*, 2006, this volume).

4.2.3. *X-ray Evidence*

X-ray imaging observations of solar flares have provided evidence of a second high energy source located above the lower source. This upper source is at the right altitude to be located at the upper tip of the current sheet predicted by the standard model. Furthermore, the temperature distribution within this source is reversed to that of the lower source as would be consistent with the inverted geometry of the field at the upper tip of the current sheet (Sui and Holman, 2003). In other words, the gradient is opposite to that found at the loop top by Masuda *et al.* (1994), and together, the two sources are consistent with a hot current sheet (at small emission measure) lying between them.

Both the UV and X-ray observations add the element of plasma temperature to the streamer geometry. At the boundary between oppositely-directed magnetic fields one must have a current layer; these new data show that such a current system in CME-related structures actively heats the plasma to temperatures normally found only in solar flares. This clearly points to magnetic reconnection in the standard model as a mechanism to enable the heating to happen. The theoretical problem (e.g., Webb *et al.*, 2003) is that current theories do not predict "how hot or dense the current sheet should be;" thus at present we have no quantitative means to link these observations to the rate of reconnection and thus to flare effects. More sophisticated models, planned for the future, may make such comparisons possible (see Mikić and Lee, 2006, this volume).

5. Coronal Shocks and Waves

M. Reiner, J. C. Raymond, M. Pick, D. Maia, H. S. Hudson,
A. Klassen, K. L. Klein, G. Mann

5.1. Introduction

Coronal shock observations are important because they provide information on the characteristics and dynamics of energetic phenomena on the Sun, such as solar flares and CMEs (see in this volume Hudson *et al.*, 2006; Schwenn *et al.*, 2006). CME shocks are also involved in accelerating Solar Energetic Particles (SEPs) that are observed in interplanetary space (see in this volume (Cane and Lario, 2006; Klecker *et al.*, 2006; Mikić and Lee, 2006; Forbes *et al.*, 2006); see also (Reames, 1999). The interplanetary extension of coronal shocks, that are driven by CMEs, can significantly affect the space weather environment when they encounter Earth. Coronal shocks can presently only be observed remotely. The primary observational signature of these shocks are the radio emissions they generate as they propagate through the solar corona. These radio signatures, called metric type II bursts, have been observed from ground-based radio observatories since the early 1950s. More recently, a few coronal shocks have been observed and studied in UV, X-ray and white-light images (Vourlidas *et al.*, 2003; Raymond *et al.*, 2000).

When these shocks propagate into interplanetary space, they produce low-frequency kilometric type II radio emissions (see Forsyth *et al.*, 2006, this volume). Kilometric type II emissions are unambiguously identified with shocks that are driven by CMEs and can sometimes be observed in situ (Bale *et al.*, 1999). In-situ shock observations provide a better understanding of the nature of the radio emission mechanism, as well as the interplanetary plasma conditions and shock geometries that favor the generation of the type II radio emissions (see Forbes *et al.*, 2006, this volume). While type II radio bursts can be generated by either blast-wave or driven shocks, it is expected that only driven shocks can propagate into interplanetary space because blast-wave shocks quickly weaken with distance and are unlikely to escape the lower corona. Hence in the solar corona the problem of shock association is more complex than in interplanetary space. There has been a long-standing controversy as to whether coronal or metric type II bursts originate from CME-driven (piston-driven) shocks and/or from blast-wave shocks (Cliver *et al.*, 1999).

Radio observations of solar type II bursts were first made by ground-based radio spectrographs, operating at metric wavelengths (Wild, 1950). These early observations therefore provided only spectral information on these coronal radio sources. Later, radioheliographs were constructed whose images provided direct information on the spatial locations and heights of the coronal radio sources. Simultaneous imaging by coronagraphs and radioheliographs permitted a comparison of the type II sources with coronal structures seen in white light, albeit in somewhat different spatial regions.

The rest of this section concentrates on recent observations of coronal shocks detected at radio, X-ray, UV and optical wavelengths and their relationships to other solar features. Recent observations of coronal shocks at optical and UV wavelengths have provided unique diagnostics for the dynamics of CMEs and for the physical parameters and processes in coronal shocks.

5.2. Shocks in the Corona: Metric Type II Bursts and Coronal Waves

An example of a metric type II radio burst is illustrated in Figure 7. This figure shows a dynamic spectrum, which is a plot of the radio intensity as a function of frequency and time. The metric type II burst appears as relatively narrow banded spectral features that drift slowly to lower frequencies with time. The spectral characteristics of metric type II bursts can often be very complex (see Nelson and Melrose, 1985

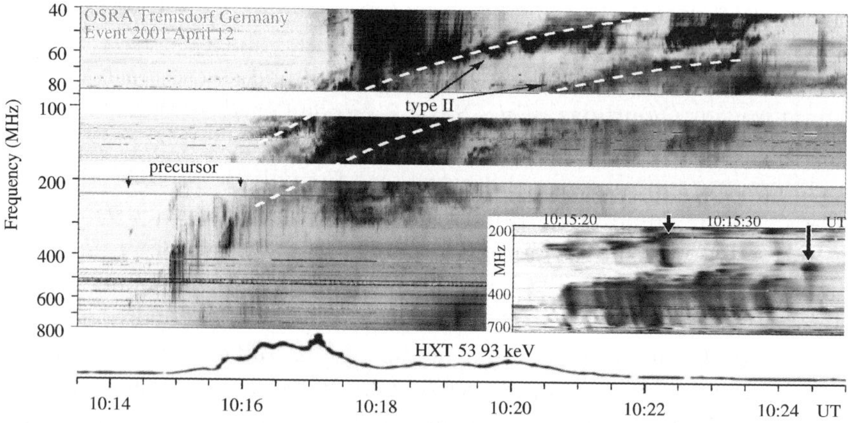

Figure 7. Dynamic radio spectrum and hard X–ray emission (53–93 keV) of the event on 12 April 2001. The event starts with a harmonic type II precursor, followed by a type II burst with multi-lane structure. The "backbone" of the fundamental and harmonic type II emissions is indicated by the white dashed curves. The precursor is cotemporal with the hard X–ray emission (lower panel) and consists mainly of sequences of fast reverse slope (RS) drift bursts, details of which are evident on the expanded time scale spectrum shown in the insert on the right (two of the RS bursts are indicated by the arrows) (adapted from Klassen *et al.*, 2003).

and Mann, 1995 for reviews). In addition to a common fundamental-harmonic band structure, the metric type II bursts also typically exhibit band splittings, multiple lanes and sometimes a so-called herringbone structure.This structure appears as a series of rapidly drifting, short-duration bursts that can drift to higher or lower frequencies. The frequency drift corresponds to the "backbone" of the herringbone structure. While such type II spectral features provide evidence of a coronal shock wave, they do not directly yield unambiguous information on the nature and origin of the shock.

In 1960, solar Hα observations revealed the existence of flare-associated waves, known as Moreton waves, that appeared as arc-shaped fronts propagating away from the flare site at speeds of the order of 1000 km s^{-1} (Moreton, 1960). The origin of these waves was attributed to a shock propagating through the corona, while sweeping over the chromosphere (Uchida, 1960). Noting that type II radio bursts were closely associated in time with Moreton waves, Uchida (1974) unified the interpretation of type II bursts and Moreton waves by demonstrating that both could be simultaneously produced by a propagating "weak" fast-mode shock. The observed slow frequency drift of the type II emission was interpreted as radiation generated by the plasma emission mechanism as the shock propagated through the decreasing coronal density. In Figure 7, the two drifting bands (outlined by the white dashed curves), with a frequency ratio close to 2, correspond to type II radiation generated in the corona at the fundamental and harmonic of the plasma frequency.

It was originally believed that the MHD shock that generated these metric type II bursts were blast-wave shocks produced by an associated solar flare. When CMEs were first discovered (Tousey, 1973), it was then suggested that these metric type II bursts might in fact be generated by the piston-driven shocks associated with CMEs (Gosling *et al.*, 1976). To compare the location or height of the radio source with the white-light CME, it was necessary to use a coronal density model (e.g. Saito *et al.*, 1970) to convert an observed radio frequency to its corresponding coronal height. It was also necessary to extrapolate the observations from the spatial region of the metric type II burst (in the low corona) to the spatial region of the coronagraph observations (in the high corona). To match the coronal shock dynamics implied by the frequency drift rate of the type II radio emissions to the CME dynamics derived from the coronagraph observations, it was found necessary to conclude that the metric type II bursts were generated in enhanced density regions in the corona (Robinson and Stewart, 1985). One possible way out of this dilemma was to assume that metric type II bursts were produced by distinct coronal blast-wave shocks that were not directly related to the CME and its associated shock (Wagner and MacQueen, 1983). However, Steinolfson (1985) suggested an alternative explanation, namely that the type II radio sources were generated along the flanks of the CME-driven shock. (For a possible direct optical detection of a shock along the flank of a CME, see Vourlidas *et al.* (2003) and Section 5.5).

Spectral analyses depend critically on the density model. With the development of radioheliographs, radio images were obtained which could be directly

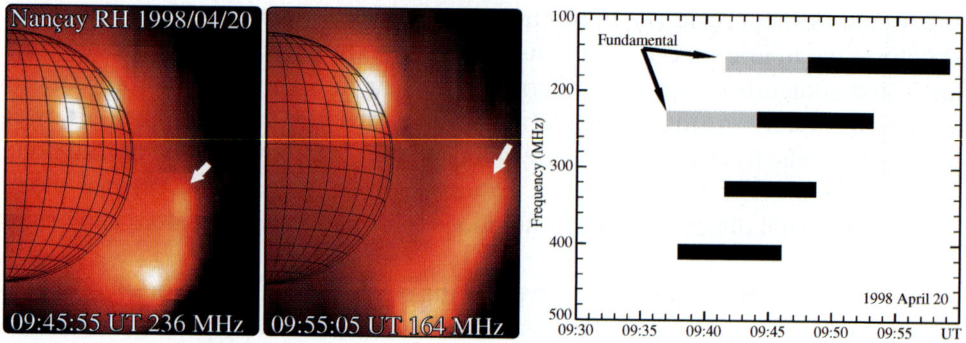

Figure 8. April 20, 1998 event. (a) NRH images showing the outward progression of a weak type II-like radio source (indicated by the arrows) in the southwest quadrant at two different frequencies. (The source in the northwest quadrant is a stationary noise storm). (b) A radio spectrum constructed from consecutive images of the type II-like radio sources (adapted from Maia *et al.*, 2000).

compared to the white-light images for the limb events and no assumption about a coronal density model was necessary. In general, comparisons between radio and coronagraph image data from the Solar Maximum Mission (SMM) and more recently from SOHO/LASCO have indicated that the type II burst sources were often located well below the CME leading edge (Gergely *et al.*, 1983; Pick, 1999), who listed most of the relevant analysis performed with Clark-lake, Culgoora and Nançay radioheliographs.

However, Maia *et al.* (2000) observed a few events in the Nançay Radioheliograph (NRH) images where type II-like signatures exhibited the properties expected for a CME-driven shock. They found sources that move radially outward and closely match both the position and velocity of the CME leading edge. One of the best cases is shown in Figure 8. The CME on 1998 April 20 was fast (>1400 km s^{-1}), and as such one would expect a shock associated with this event to develop in the corona at relatively low altitudes. Figure 8 (left) shows images of an extended radio feature at two NRH frequencies at about the time that the CME, observed later by LASCO, was propagating through this region of the corona. Its leading edge, measured at 10:07 UT and extrapolated to the time of the NRH image at 164 MHz, coincided with the location of the radio feature. Observed at four NRH frequencies, the extended radio sources moved radially outwards with an initial projected velocity of about 1000 km s^{-1}. A spectrum constructed from the observations at these frequencies is shown in Figure 8 (right). The regions shown in gray at 164 and 236 MHz correspond to weaker emissions. A large increase in intensity was later observed, at about the time where the black regions in the figure first appear. This spectrum of the outward moving radio feature is clearly type II-like, with evidence for radio emissions at both the fundamental and harmonic (compare with Figure 7).

Multi-wavelength coronal studies, based on imaging observations in EUV, X-ray, Hα and radio, have significantly improved our understanding of the origin and

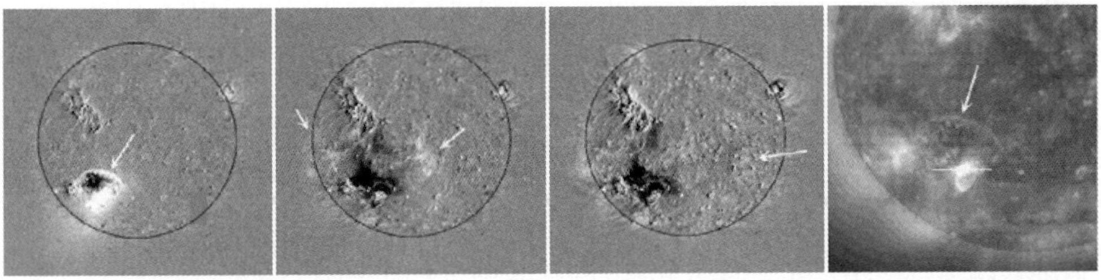

Figure 9. Example of an EIT wave from September 24 1997. The first three panels show successive images at 02:49, 03:03 and 03:23 UT with a pre-event image digitally differenced from them. Arrows indicate the EIT- wave front(s). The last panel shows a subfield of the first panel (undifferenced), showing an example of a sharp brightening (from Biesecker *et al.*, 2002).

nature of the shocks producing type II radio emissions. There is mounting evidence to suggest that many, if not most, metric type II bursts are produced by coronal shocks that are probably not the pure blast waves originally envisioned by Uchida (1974).

An important factor in this new understanding was the discovery of the large-scale coronal EUV (EIT) and SXT waves by the SOHO/EIT and Yohkoh/SXT telescopes, respectively. The EIT waves (Thompson *et al.*, 1998) typically appear as diffuse brightenings propagating away from the site of active regions; their speeds, which are difficult to accurately determine due to the poor temporal cadence of the observations, appear to be a few hundreds of km s^{-1}. An example of an EIT wave is shown in Figure 9. The exact nature of these waves is not yet fully understood. EIT waves have a strong association with CMEs (Biesecker *et al.*, 2002) and often with EUV dimmings, which are usually restricted to the region traced by the transit of Moreton waves. Biesecker *et al.* (2002) established that EIT waves are not generally well correlated with type II bursts, except a small subclass ($\sim$7%) that have sharp bright fronts. They proposed that these sharp wave fronts may be the coronal counterpart of Moreton waves.

This suggestion was confirmed for a few events for which EIT and Moreton waves and type II-like bursts appeared in close spatial and temporal coincidence (Pohjolainen *et al.*, 2001). Moreover, this and subsequent studies established that, for this class of flare/CME events, the observed latitudinal expansion of the CME was directly related to the propagation of the Moreton wave. This suggests that the coronal shock triggers destabilization and magnetic interaction of coronal structures as it propagates through the corona. Examining this correlation more carefully, Warmuth *et al.* (2004) found that all EIT waves with sharp wave fronts decelerated with the rate of deceleration decreasing with increasing time and distance. They pointed out that the characteristics of these coronal waves, their rapid deceleration, the broadening of the intensity profile and the decrease of amplitude with increasing distance and time, are characteristics of blast waves, rather than piston driven shocks. Additional evidence was provided by Khan and Aurass (2002), Narukage *et al.* (2002) and Hudson *et al.* (2003) who showed that the coronal waves detected by the Yohkoh/SXT telescope also have the expected relationship to Moreton waves and type II bursts. These analyses of the wave propagation, especially of directional changes, led them to conclude that the disturbances were blast waves related to a flare and they argued that the type II burst was generated by the same blast wave.

However, not all observations of type II emission are consistent with pure blast waves. Klein *et al.* (1999) observed a type II source to appear above an expanding soft X-ray loop in SXT images, while a neighbouring loop did not change. This difference between the loops suggested that the proximity of the type II source and the expanding loop was not coincidental, but that the expanding loop outlined a peculiar magnetic structure which was the driver of the type II emitting shock wave. A second argument against a pure blast wave picture is the statistical result by Pearson *et al.* (1989) who found no significant indication that the type II producing flares were particularly impulsive. These findings suggest that type II shocks are initially not pure blast waves, although they may become blast waves during the course of the event. The frequent observation of type II radio precursors also suggests that the type II shocks are initially piston driven (Klassen *et al.*, 1999, 2003). Indeed, such precursor spectral signatures (illustrated in Figure 7) occur during the impulsive flare phase and consist of sequences of fast-drift bursts or pulsations whose overall frequency envelope drifts at a normalized drift rate. This rate is similar to that of the subsequent metric type II burst, and it typically displays a similar harmonic structure. The fast-drift bursts that constitute the precursor trace the propagating disturbance before it excites type II emissions. This implies the existence of electron beams that propagate towards the expanding loops and the hard X–ray (HXR) sources. Klassen *et al.* (2003) found that the precursor of the event shown in Figure 7 was associated with an expanding loop system, with speeds of expansion that varied from 200 to 700 km s^{-1}. They suggested that the type II precursor is a signature of a moving reconnection process that occurred above the expanding soft X–ray loops, which drive the precursor and generate the subsequent type II burst.

5.3. The Link between the Corona and the Interplanetary Medium: Decametric-Hectometric Type II Bursts

The above discussions demonstrate the difficulty of determining conclusively whether a given coronal shock, detected solely by its ground-based radio signatures, is a piston-type shock being driven by the CME (or an associated small-scale disturbance) or is instead an undriven blast-wave shock associated with a flare. Another approach that may help to resolve the blast wave versus CME-driven shock controversy is to establish the relationship of the coronal type II bursts to their low-frequency counterparts that are observed by radio receivers on interplanetary spacecraft (see Forsyth *et al.* 2006, this volume). It is well established that these counterparts are directly driven by CMEs (Cane *et al.*, 1987).

An important observational advance was provided by the radio receivers of the WAVES experiment on the Wind spacecraft, which for the first time observed solar radio emissions at high-frequency and high-time resolution in the decametric-to-hectometric (D-H) wavelength band from 1 to 14 MHz. Leaving only a relatively small frequency gap between the spaceborne and ground-based radio observations, the Wind/WAVES observations offer a unique opportunity to establish the possible causal relationship between the coronal (metric) type II bursts and the interplanetary (kilometric) type II emissions. However, since imaging at these low frequencies is not possible, only spectral comparisons can, at present, be made.

When Wind was launched in 1994, during solar minimum, no solar type II radio emissions were observed in the D-H band, although during that time many metric type II bursts were observed by ground-based observatories. This led Gopalswamy *et al.* (1998) to conclude that the metric type II bursts were produced exclusively by blast-wave shocks, associated with flares, that did not propagate into the high corona and interplanetary medium. However, by 1997, as the maximum of solar cycle 23 approached, distinct slow frequency-drifting type II emissions began to appear in the D-H wavelength range and they became more abundant as solar maximum was reached (Kaiser *et al.*, 1998; Gopalswamy *et al.*, 2000). The spectral characteristics of these D-H type II bursts were quite diverse and were usually observed in association with major flares associated with fast CMEs.

It was found that these D-H type II bursts often naturally extended to the low-frequency kilometric type II emissions, known to be generated by CMEs (Reiner and Kaiser, 1999b; Reiner *et al.*, 2000a, 2003). However, it was more difficult to establish unequivocally their causal relationship to the metric type II bursts, partly because of the frequency gap (14–20 MHz) to the ground-based observations (Reiner *et al.*, 2000a; Gopalswamy *et al.*, 2000; Reiner *et al.*, 2003). The best way to establish this causal relationship is to require that the dynamics implied by the observed frequency drift rates of the metric and decametric type II bursts should agree, i.e., a metric type II burst is related to a decametric type II burst if and only if both the frequency drift rate *and* the projected origin time in these two frequency regimes correspond. Since the frequency drift rate of the type II burst depends on the frequency, this of necessity can only be done within the context of a coronal density model. However, if it is found that their frequency drift rates are a continuation of each other over a significant time period and frequency range, then one can be reasonably confident that the metric and decametric type II emissions were produced by the same coronal shock.

An example is illustrated in Figure 10. The D-H type II emissions and a metric type II burst, which were observed over a particularly wide frequency range, are shown in the (space-based) WAVES and (ground-based) Culgoora dynamic spectra, respectively. Also shown in Figure 10 are the height-time measurements for the

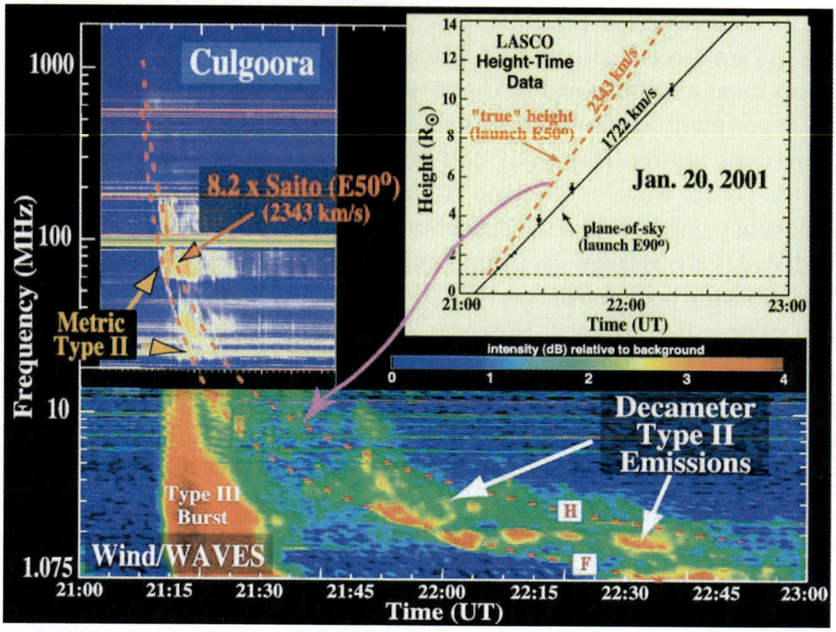

Figure 10. Dynamic spectra showing the decametric and metric frequency-drifting type II radio emissions associated with the 20 January 2001 CME event. The measured CME height-time data were used to obtain a frequency-time track, which was then simultaneously fit to the frequency drift of the metric and decametric type II radio emissions. It was found that to get a good fit it was necessary to assume a CME launch angle of E50° with an enhanced Saito coronal density model, suggesting that the radio emissions originated in high density streamers (adapted from Reiner *et al.*, 2003a).

corresponding LASCO CME. By requiring consistency between the frequency drift rate of the type II radio emissions and the height-time data from the CME, Reiner *et al.* (2003) were able to fix the scale of the coronal density model (Saito *et al.*, 1970) and to establish, within the context of this model, that the frequency drift of the D-H emissions was a continuation of the frequency drift of the metric type II emissions, as indicated by the red dashed curves in Figure 10. Thus, at least for this event, the metric and D-H type II emissions were both generated by the same CME-driven shock, which must have formed very low in the corona, in agreement with the results of Maia *et al.* (2000). Dulk *et al.* (1999) using a similar approach reached similar conclusions. By contrast, there are other cases where, using the same technique, it was found that the metric and D-H type II emissions were not dynamically related (Reiner *et al.*, 2000a).

5.4. Coronal Shock Interactions

Shocks in the corona can also interact with coronal structures. The propagating CME-driven shocks are often observed to deflect coronal streamers and to interact with coronal material from a preceeding CME (see Schwenn *et al.* 2006, this volume). These shock interactions can also have radio consequences (Gopalswamy *et al.*, 2001).

Mancuso and Abbo (2004) presented a rare observation of a type II burst associated with an X20 flare on April 2, 2001 accompanied by a very fast (>2000 km s^{-1}) CME. The type II emission appeared at frequencies near 100 MHz and subsequently drifted toward both higher and lower frequencies. Mancuso and Abbo (2004) modeled this event as the result of the CME shock impact on a streamer at some distance above the solar surface, and, as the CME *encountered* the streamer, shocks moved both up and down along the streamer axis. The SOHO/LASCO and SOHO/UVCS observations supports this model. This observation demonstrated that type II emission does not necessarily occur at the outermost part of a CME, but can arise from the CME flanks if conditions are right. Emission from a streamer suggests that high density or perhaps relatively high plasma β may be important in determining which parts of the shock produce radio emission. The analysis of Reiner *et al.* (2003) also suggest that the type II emissions may be generated in portions of the shock that pass through the coronal streamers.

5.5. Optical Observations

Direct optical observations of shocks associated with CMEs are very difficult and at best can only be achieved in rare situations, but they can be made at heliocentric distances of a few solar radii where shocks are expected to form. A major difficulty is discriminating between plasma compressed by a shock and plasma compressed by an expanding magnetic structure.

The best direct white-light observation of a CME/shock is that of Vourlidas *et al.* (2003) (for details see Schwenn *et al.* 2006, this volume). A 800–1000 km s^{-1} jet-like CME (about 20° opening angle) on 1999 April 2 showed an extremely sharp edge, as expected for a shock, and its shape was that expected for a bow shock driven by a jet. While there was no reported type II radio emission to confirm the shock interpretation, this CME produced a clear deflection of a streamer when the "shock" reached it and Vourlidas *et al.* (2003) provided a simulation to support their interpretation of the sharp white-light feature as a fast-mode MHD shock.

5.6. Ultraviolet Observations

There have also been about 10 detections of CME-driven shocks in the ultraviolet with SOHO/UVCS (Ciaravella *et al.*, 2006). Raymond *et al.* (2000) observed an event on 1998 June 11. Its speed was 1200 km s^{-1} based on LASCO images and the timing of the event seen at $1.75 R_{\odot}$ by UVCS. Furthermore, a type II burst was observed at Nançay during the time that the event was seen by UVCS, and the density obtained from the type II frequency agreed with the pre-CME density obtained from the ratio of O VI emission lines. Both the type II frequency width and the observed brightening in O VI and Si XII lines indicated a modest compression. The O VI line profile obtained by UVCS implied an oxygen kinetic temperature much higher than the proton temperature, as has been observed in heliospheric shocks (Berdichevsky *et al.*, 1997) and supernova remnant shocks (Raymond *et al.*, 1995; Laming *et al.*, 1996). Raymond *et al.* (2000) noted that Wind/WAVES observed type II emission simultaneous with that observed at Nançay, but 6 times lower in frequency, implying a density nearly 40 times smaller. The UVCS density, and therefore the Nançay type II emission, pertain to the streamer that lay directly above the CME site. The Wind/WAVES emission probably originated in the lower density region between the streamers to the southeast of the CME axis. Another possible interpretation is that Wind/WAVES emissions corresponded to a shock that originated earlier. It is also interesting that the shock speed estimated from the Nançay frequency drift rate and models of the average coronal density was only about half the speed obtained from LASCO, even though the density agreed with that obtained by UVCS. This was attributed to the difference between the actual streamer density distribution and the average coronal models, but perhaps non-radial motion of the section of the shock that produced type II emission also affected the velocity estimate. However, the type II drift rate in this complex event is uncertain.

Mancuso *et al.* (2002) observed another SOHO/UVCS event on 2000 March 3 that produced type II emission observed at Hiraiso, Culgoora and Bruny Island simultaneous with the appearance of the shock in the UVCS spectrograph slit at $1.5R_{\odot}$. The UVCS data is shown in Figure 11. The densities inferred from the type II frequency and UVCS line ratios obtained in an earlier synoptic scan were in

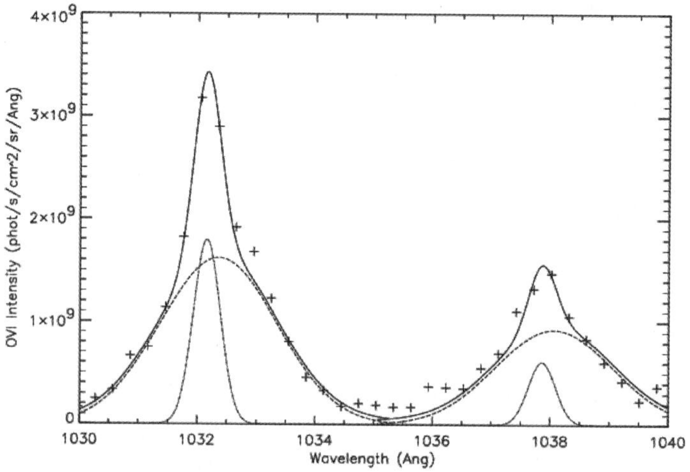

Figure 11. O VI 1032, 1037 Å doublet detected on March 3, 2000 during the passage of the shock front. The $1/e$ width of the shocked material indicates a speed of 400 km/s as compared with 70 km/s of the unshocked pre-CME corona (adapted from Mancuso *et al.*, 2002).

agreement at 1.2 to 1.5×10^7 cm^{-3}. The shock speed obtained from the frequency drift and the pre-CME density structure obtained from UVCS was 1100 $km\,s^{-1}$, in reasonable agreement with the 920 $km\,s^{-1}$ speed observed by LASCO in the plane of the sky. The Bruny Island radio data suggests that the type II drift rate has been overestimated by as much as 25% in this complex event, which would still be in reasonable agreement with the LASCO speed. As for the 1998 June 11 event, the compression was a modest factor of 1.8 as estimated from type II frequency splitting and UVCS observed strong broadening of the O VI emission lines. The O VI line widths agreed well with the expected velocity width for the observed shock speed and compression ratio in the model of Lee and Wu (2000). While the radio and UV observations are remarkably consistent, the lack of an interplanetary shock signature from Wind/WAVES leaves open the possibility of a blast wave origin for the type II emission.

An important application of UV data to the study of shocks is the derivation of pre-CME density structures for use in shock speed estimates. They can also be used to derive magnetic field limits from the requirement that for a piston-type shock to appear the CME speed must exceed the fast mode speed. This yielded lower limits to the plasma β at 1.9 $R_\odot$ of 0.01 to 0.47 for 37 radio bursts observed in 1999. Similarly, during the 2002 April 21 X1.5 class event, the lack of type II emission until after the CME passed the UVCS slit at $1.63 R_\odot$ gave a lower limit to the magnetic field at this height of 1 G (Raymond *et al.*, 2003).

6. Radio Diagnostics of Coronal Electron Acceleration and In-Situ Electrons

M. Reiner, D. Maia, H. Cane, L. K. Klein

6.1. Introduction

Solar radio emissions are diagnostics of the acceleration of coronal electrons from suprathermal to relativistic energies. These remote electromagnetic signatures of coronal electrons, observed at high cadence in dynamic spectra and in images, provide information on the precise timing of solar acceleration processes. They also provide information on the locations and physical characteristics of the electron acceleration sites and their relationships to coronal structures. The number and diversity of these radio sources, extending from millimeter to decameter wavelengths, suggest a wide variety of coronal acceleration processes related to solar energetic events observed in white-light, EUV, and X-rays. Some of these accelerated electrons are trapped in closed coronal magnetic structures, while others freely escape along open magnetic field structures into interplanetary space, where they can be observed in situ. These remote radio observations can facilitate tracing the in-situ signatures of solar energetic particles back to their solar origin.

Recent multiwavelength observations indicate that, for a given solar event, different populations of energetic protons and electrons may be accelerated by different physical mechanisms at different times and from different coronal sites. Until recently, the prevalent view was that large SEP events, observed in interplanetary space, were the exclusive result of particle acceleration at the bow shock of CMEs. However, from the association of both CMEs and radio emissions observed at low coronal altitudes with the complex radio emissions observed in the 1–14 MHz range, there is now ample evidence to suggest that coronal processes can also contribute to the SEP production (see also Klecker *et al.*, 2006, this volume). We focus here mainly on the solar origin of near-relativistic impulsive electron events. For these events, the release time at the sun can be accurately determined and directly compared to the various radio signatures. The inferred solar injection times of these near-relativistic in-situ electrons are often found to be significantly delayed with respect to the radio signatures generated by the low-energy electrons in the corona.

6.2. Diverse Radio Diagnostics of Electron Acceleration in the Corona

We discuss here those radio signatures that are necessary for understanding the causal link between energetic electrons in the corona, detected by their remotely observed radio signatures, and those detected in situ in the interplanetary medium.

6.2.1. *Persistent Sources of Suprathermal Electrons*

Interplanetary type III radio storms detected at kilometric wavelengths are closely related with noise storms at metric wavelengths, which are a common form of activity observed in the absence of flares (Bougeret *et al.*, 1984; Kayser *et al.*, 1987). These storm emissions are related to the dynamical evolution of the coronal magnetic field (Bentley *et al.*, 2000), and at onset, are located in regions of coronal restructuring observed in white light (Kerdraon *et al.*, 1983) or in soft X-rays (e.g. (Lantos *et al.*, 1981). Type III storms can persist for days and likely contribute to the population of electrons extending to 2 keV in the interplanetary medium.

6.2.2. *Radio Emission from Electron Beams*

Solar type III radio bursts are observed over a very wide range of frequencies ($\geq$1 GHz – 10 kHz) and are emitted by $\leq$20 keV electrons accelerated at the sun, at various altitudes (sometimes as high as $1\,R_\odot$). These electron beams have access to open field lines that allow them to escape into interplanetary space, where they can be detected in situ as impulsive electron events (Lin, 1998). Metric type III bursts generally consist of groups of individual bursts (typically tens of bursts over a period of one or more minutes) that overlap and merge at decreasing frequencies so that by $\sim$1 MHz they generally appear as a single burst with a typical duration of ~5 minutes. Metric type III bursts are generally not well correlated with hard X-ray (HXR) emissions (Kane, 1981), although when both emissions are associated, the temporal correlation is excellent, indicating a common source of electron acceleration (Raoult *et al.*, 1985).

6.2.3. *Type IV Bursts*

Type IV bursts generally consist of a broadband continuum that can extend from several GHz to ~10 MHz, and are most often associated with large flare/CME events (see Schwenn *et al.* 2006, this volume). Type IV continua are often associated with gradual HXR bursts, and are generally followed by a post-gradual emissions (stationary type IV bursts) that may last for tens of minutes to hours after the end of the HXR and microwave emissions. Such long lasting radio emission is a signature of the acceleration of suprathermal electrons in the middle corona, continuing for hours after the flare (Trottet, 1986). Moving type IV sources are due to suprathermal to MeV electrons injected into expanding magnetic arches, behind the CME leading edges. Some other narrower type IV moving sources are often associated with eruptive prominences.

6.3. Complex Type III-Like Bursts at D-H Wavelengths

While many type III radio bursts have simple, well-defined intensity-time profiles at hectometric wavelengths, a significant number of type III events are very intense, complex and of long-duration at decametric-to-kilometric wavelengths. First identified and studied in the 1980's, it was noted that they had a good temporal correspondence with the metric type II bursts, leading (Cane *et al.*, 1981) to suggest that the electrons that produced these complex type III bursts were accelerated by coronal shocks (and they were therefore called shock associated (SA) events). On the other hand, some authors noted that the durations of the complex type III bursts were similar to those of associated continuum radio and HXR emissions, and argued in favor of a low corona electron acceleration process (Kundu and Stone, 1984; Klein and Trottet, 1994). Nevertheless, it was commonly accepted, from supporting statistical studies (Kahler *et al.*, 1989), that the complex type III events were generated by electrons accelerated at coronal shocks.

Cane *et al.* (2002) noted that some complex type III events had clear metric type III emissions at frequencies higher than those of the corresponding type II burst and Reiner *et al.* (2000b) demonstrated that for some complex type III-like events there was a good temporal correspondance between the duration and intensity variations of the time profiles of the hectometric and decimeter/microwave emissions at 1–3 GHz, as illustrated in Figure 12. This correspondence suggests that both emissions were generated by different electron populations accelerated in the lower corona over extended time periods. Reiner and Kaiser (1999a) reported that the radiation characteristics of the complex type III-like emissions observed at high time and frequency resolution in the D-H range (1–14 MHz) (see Figure 12) were quite

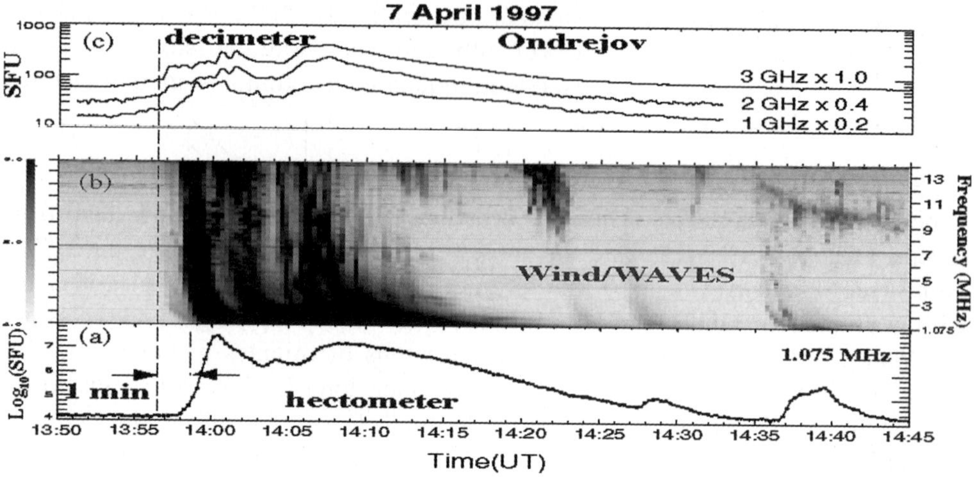

Figure 12. Comparison of the complex type III burst observed on 1997 April 7 by Wind/WAVES in the frequency range from 1 to 14 MHz, with the radio emissions observed by the Ondřejov observatory from 1 to 3 GHz. Note the similarity in the shape and duration of the intensity-time profiles in these widely separated frequency regimes (adapted from Reiner *et al.*, 2000b).

unusual and suggested that some of these complex radiation characteristics might be related to the electron beams propagating through the highly disturbed corona behind the associated CME. Reiner *et al.* (2001) observed that these complex type III-like emissions were usually associated with major flare/CME events, while (Cane *et al.*, 2002) showed that >20 MeV SEP events were associated with a class of long-duration type III bursts (called type III-*l*), which include the complex type III-like bursts discussed here. Nevertheless, although many of these intense complex type III bursts are most likely generated from electrons accelerated low in the corona, there may still be some that result from electrons accelerated by coronal shocks (e.g. Bougeret *et al.* 1998).

6.4. Coronal Radio Diagnostics of "Delayed" In-Situ Electrons

A different approach to inferring the origin of solar electrons is based on the analysis of in-situ electrons observed by spacecraft. Systematic studies on the relative timing of coronal electron release and of associated phenomena (Krucker *et al.*, 1999; Haggerty and Roelof, 2002) found that the inferred solar release times of electrons could be significantly delayed relative to the onset of the type III burst that result from the escape of suprathermal electrons into interplanetary space. Such delays, ranging from 0 to more than 40 minutes, are quite inconsistent with the expectation that the electron beams detected in situ should be the generators of the type III radiation (Lin *et al.*, 1973).

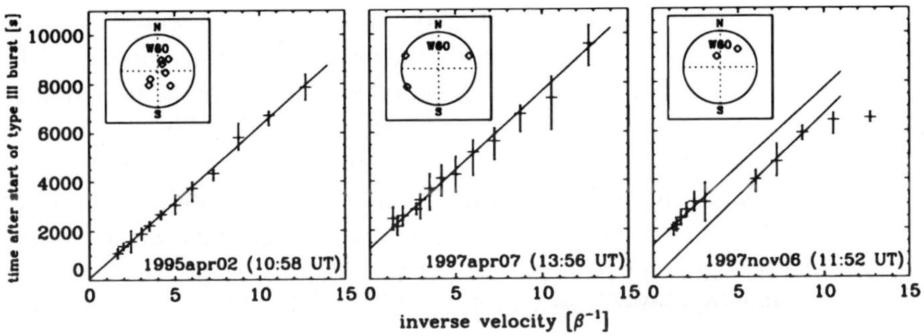

Figure 13. Illustrative examples of the three different classes of in-situ impulsive electron events that have very different velocity dispersion characteristics. In all cases, the light propagation time of 8.3 minutes was added to permit direct comparison of the injection times with the times of the observed remote radio emission at 1 AU (adapted from Krucker *et al.*, 1999).

The Krucker *et al.* (1999) study, which analyzed suprathermal to near-relativistic impulsive electrons, found that impulsive electron events could be classified into three categories depending on the characteristics of the observed velocity dispersion in the beam made evident on $1/\beta$ plots such as illustrated in Figure 13. First, the onsets at all energies lie on the same straight line with an inferred release time that coincides with the onset of the type III radio emissions. Second, the onsets at different energies all lie along the same straight line but the inferred injection time is significantly delayed relative to the onset time of the type III radiation. Third, the low-energy electron velocity dispersion extrapolates to a time that coincides with the onset time of the type III radio emissions, while the velocity dispersion of the high-energy (near-relativistic) electrons extrapolates to a solar injection time delayed by up to 40 minutes from the type III emissions. Explanations advanced for this phenomenon can be divided into two major categories: (1) the “delays” reflect distinct coronal acceleration processes (multiple populations), (2) the “delays” result from propagation effects in the interplanetary medium (single population).

6.4.1. *Multiple Populations*

Krucker *et al.* (1999) found good agreement between the nominal Parker spiral length, based on the measured solar wind speed, and the path length deduced from the measured velocity dispersion in the electron onset times at different energies (the so called $1/\beta$ method – see Figure 13). This led them to conclude that the delays observed in the times of the first-arriving electrons were not due to propagation effects but rather reflected differences in the timing of the electron acceleration at the Sun. They thus proposed that the delayed in-situ electrons corresponded to a population of electrons that were released at the Sun (without producing a type III burst) later than the suprathermal electrons (which were not detected in situ) that generated the type III radiation. Based on the correlation they found with the occurrence of EIT waves (see Schwenn *et al.* 2006, this volume), Krucker *et al.* (1999) suggested that EIT waves were somehow at the origin of the phenomenon. Klassen *et al.* (2002), on the other hand, found a temporal association of the near-relativistic electron events with coronal shocks (type II bursts) that had associated CMEs.

Simnett *et al.* (2002) also favored a multiple population scenario but regarded any low-corona association as a secondary effect. Based on trends observed between the CME speeds and the corresponding electron fluxes, spectral hardness and delays, they suggested that most of the near-relativistic electrons were accelerated by the CME-driven shock (possibly from a seed population) and were released at a distance from 2 to $3R_{\odot}$.

From in-situ observations near 1 AU, it is known that shocks can accelerate electrons to keV energies and possibly tens of keV, but there is no information on the efficiency of shock acceleration closer to the Sun. Furthermore, arguments on efficiency of electron acceleration to near-relativistic energies by coronal shocks are not conclusive: Mann and Luehr (1994) claim that it should be possible under special conditions, while (Klein *et al.*, 2003b) argued that estimates on the numbers and energy content of type II shock-accelerated electrons between a few keV and several tens of keV remain inconclusive in terms of proving or disproving electron acceleration to high-energies by coronal shocks.

The complexity of coronal processes may also play an important role in the "delays". Detailed studies for all of the Haggerty and Roelof (2002) events for which there were imaging observations from the Nançay Radioheliograph (NRH) were conducted by Pick *et al.* (2003) and Maia and Pick (2004). They identified two classes of electron events: radio-simple (events associated only with type III bursts) and radio-complex (events associated with broadband continua and/or coronal type II bursts). The radio-simple events were either not associated with CMEs or associated with slow CMEs, while the radio-complex events were associated with fast CMEs (velocity in general exceeding 800 km s^{-1}). The inferred release times for the in-situ impulsive events associated with radio-simple emissions fell within the time of the metric type III group. An example of a radio complex event, with complex type III emission at decametric frequencies and multiple moving sources at metric frequencies is shown in Figure 14. For the radio-complex events, the inferred release times for the near-relativistic electrons did not generally occur within the type III burst group, but, more significantly, were always close to the start or times of abrupt modifications during the development of the radio event, providing evidence for new sources of coronal acceleration at heights below $1R_{\odot}$. The simplest interpretation of these analyses is that the delayed acceleration of the near-relativistic electrons results from coronal restructuring at variable heights behind the leading-edge of the CME. For some events a new site of electron acceleration may have been triggered by a shock, consistent with the association with EIT waves found by Krucker *et al.* (1999). Similar conclusions were independently reached from preliminary results using NRH and Wind/3DP observations (Klein *et al.*, 2003a).

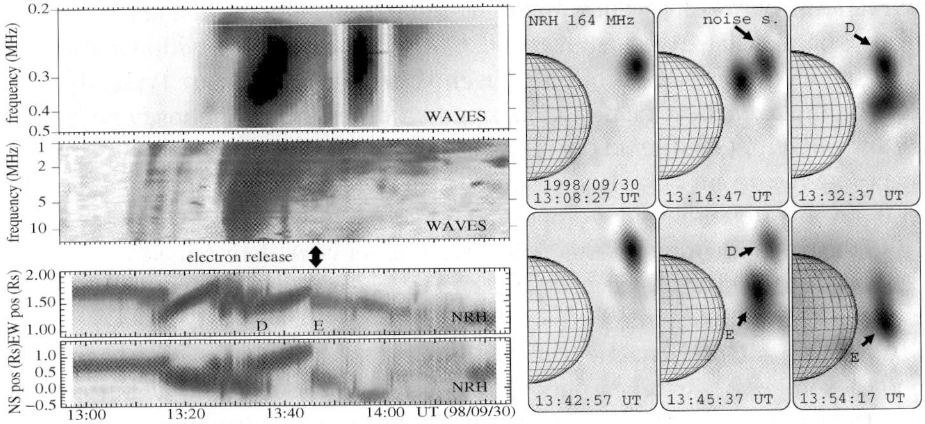

Figure 14. 1998 September 30. (left) Comparison of the emission observed by the NRH (bottom) and WAVES (top). The vertical white bands seen in in WAVES spectrum are due to the saturation. The one-dimensional plot shows a series of radio sources. D marks a moving continuum, followed by a stationary labeled E. The arrow indicates the inferred release time at the sun for the electrons with energies above 100 keV. Note that this time coincides with the sudden disappearance of D, the onset of E, and with the low frequency type III burst detected by WAVES. (right) Panel of images showing the positions of the sources seen by the NRH (adapted from Maia and Pick, 2004).

6.4.2. *Single Population*

Cane (2003) made a careful analysis of the interplanetary type III bursts, associated with the Haggerty and Roelof (2002) electron events, whose spectral shape, near the local plasma frequency at the spacecraft, is indicative that they are emitted close to the spacecraft. Cane (2003) found that the "radio drift times" (the time interval from the onset of the type III to the time when the emission extrapolates to the plasma line) appear to correlate with the electron "delays" and that the electron delays tend to be longer for higher local plasma density. The author argues that the simplest explanation for these correlations is that the energetic in-situ electrons, and the ones at the origin of the type III burst, are of the same population, and that the observed delays arise from underestimating the particle propagation times. These analyses disagree with the $1/\beta$ analyses which assume a scatter-free propagation path in the solar wind. Reconciliation of the disagreement between the two techniques is a topic for future investigation.

7. Summary

T. Forbes, G. Mann, M. Pick

Since CMEs originate in the low solar corona, understanding the physical processes that generate them is strongly dependent on coordinated multi-wavelength observations. Observations in white light, radio, optical lines, UV, and X-rays have already established that most, but not all, CMEs are accelerated within a solar radius of the surface of the Sun. It has also been shown that CMEs are typically associated with the formation of X-ray loops and chromospheric ribbons, which may, or may not, be bright enough to constitute a classical flare. However, many questions concerning the origin of CMEs remain to be answered. For example, the processes that build-up stresses in the coronal magnetic field, as well as the processes that lead to the destabilization of this field have not yet been determined.

The multi wavelength observations of the last decade have confirmed that reconnection plays a key role in the relaxation of the magnetic field disrupted by a CME. During a CME, magnetic field lines mapping from the ejected plasma to the photosphere are stretched outward to form a current sheet in which field lines reconnect to form closed, magnetic loops. During the reconnection process, power extracted from the magnetic field drives flows and heats the plasma. Much of the released energy is channeled down to the solar surface and leads to the heating and evaporation of the chromospheric plasma. Although the occurrence of reconnection in the aftermath of CMEs has now been firmly established, whether reconnection also plays a significant role before, or during, the onset of a CME remains unknown. Observations of small to moderately sized flares suggest that reconnection can trigger the eruption of fast CMEs. Much of current research is focused on looking for reconnection signatures just before onset. We anticipate that as more information is gathered about the flows and heating produced by reconnection in the corona, we will eventually learn how rapid reconnection occurs and what the dominant physical processes are.

Opening of transequatorial loops typically occurs during the development of large-scale CMEs. The opening of the loops leaves a region dimmed in EUV and multiple loop systems are often involved. CMEs reach their full angular extension in the low corona, through successive magnetic field interactions that produce energetic electrons. The associated emitting radio sources spread through the solar disk in a few minutes, much faster than the cadence of the coronagraph instruments. The EUV and Soft X-ray dimmings observed on the disk open the possibility of estimating the mass in these ejecta and of determining the source regions relative to the surface field. The EUV dimmings that are observed result from the removal of the hot coronal material by the CME, so they trace out the footpoints of the magnetic field lines that are opened by the eruption of the field.

What kind of shocks form during a flare or a CME also remains an unanswered question. That shocks do form is without doubt, but how many there are during a given coronal event and whether they are of the blast-wave or driven (i.e. piston-driven) type is still unknown. How these are related to the driven shocks seen in interplanetary space is also not clear. At the moment it appears that type II radio bursts can be generated by different sources. Radio observations suggest that CME-driven shocks can be formed low in the corona. As more data have been obtained the picture seems to have become even more confused in some respects. For example, the relationship between CME-generated waves seen in UV by EIT to the Moreton waves seen in Hα by ground based observatories is uncertain. A high observing cadence may be necessary to resolve this. The fact that shocks have been rarely detected in white light is not yet fully understood.

It is well known that shock waves are able to accelerate particles to high energies. There are, however, a few doubts that large SEP events observed in the interplanetary medium are the exclusive result of particle acceleration at the bow shocks of CMEs. In this chapter, we restricted our discussion to the current understanding of the origin of energetic electrons. It is well accepted that the appearance of type II radio bursts shows that electrons must be accelerated by coronal shock waves. But, the role that shocks play in the corona in the production of energetic electrons is not clear. This is true for impulsive electron events observed in the interplanetary medium and also for energetic electrons which strike the solar surface. Most of the complex type III events detected at hectometric wavelengths in the high corona, which are also associated with CMEs, appear to be generated from electrons accelerated low in the corona. For these events, as well as for impulsive quasi-relativistic electron events, these electrons might also be generated by the reconnection process in the corona directly, by the reconnection electric field, or indirectly by the turbulence in the reconnection outflows. How these electrons escape into the interplanetary medium is an open question.

During the last decade the availability of high quality data from numerous spacecraft such as Yohkoh, ULYSSES, SOHO, TRACE, Wind, ACE, and RHESSI has deeply changed our picture of the processes occurring on the Sun. Nevertheless, many problems remain unsolved and will require much effort in the future in order to be resolved. This is especially true for CME initiation, the relationship between the primary energy release and shock waves, and the sources of energetic electrons and ions. The launch of Solar-B, STEREO (Solar Terrestrial Relations Observatory), and SDO (Solar Dynamic Observatory) in the near future should lead to a significant advance in resolving these outstanding issues.

Acknowledgements

We wish to thank Prof. Frank McDonald for useful comments. T.G. Forbes' contribution to this work was supported by grants ATM-0327512, ATM-0422764, and ATM-0518218 from the US National Science Foundation; NASA grant NNH05-AA131, and the US Dept. of Defense MURI program on Space Weather.

References

Amari, T., Luciani, J. F., Mikic, Z., and Linker, J.: 1999, *Astrophys. J.* **518**, L57.
Andrews, M.D., and Howard, R. A.: 2001, *Space Sci. Rev.* **95**, 147.

Antonucci, E., Kohl, J. L., Noci, G., Tondello, G., Huber, M. C. E., Gardner, L. D., *et al.*: 1997, *Astrophys. J.* **490**, L183.

Anzer, U., and Pneuman, G. W.: 1982, *Sol. Phys.* **79**, 129.

Asai, A., Yokoyama, T., Shimojo, M., and Shibata, K.: 2004, *ApJ* **605**, L77.

Aschwanden, M. J., Benz, A. O., Dennis, B. R., and Schwartz, R. A.: 1995, *ApJ* **455**, 347.

Bale, S. D., Reiner, M. J., Bougeret, J.-L., Kaiser, M. L., Krucker, S., Larson, D. E., *et al.*: 1999, *Geophys. Res. Lett.* **26**, 1573.

Bentley, R. D., Klein, K.-L., van Driel-Gesztelyi, L., Démoulin, P., Trottet, G., Tassetto, P., *et al.*: 2000, *Sol. Phys.* **193**, 227.

Berdichevsky, D., Geiss, J., Gloeckler, G., and Mall, U.: 1997, *J. Geophys. Res.* **102**, 2623.

Biesecker, D. A., Myers, D. C., Thompson, B. J., Hammer, D. M., and Vourlidas, A.: 2002, *Astrophys. J.* **569**, 1009.

Biskamp, D.: 2000, *Magnetic reconnection in plasmas*, Cambridge, UK: Cambridge University Press.

Bougeret, J.-L., Fainberg, J., and Stone, R. G.: 1984, *A&A* **136**, 255.

Bougeret, J.-L., Zarka, P., Caroubalos, C., Karlický, M., Leblanc, Y., Maroulis, D., *et al.*: 1998, *Geophys. Res. Lett.* **25**, 4103.

Cane, H. V.: 2003, *Astrophys. J.* **598**, 1403.

Cane, H. V., Erickson, W. C., and Prestage, N. P.: 2002, *J. Geophys. Res. (Space Phys.)* 14.

Cane, H. V., and Lario, D.: 2006, *Space Sci. Revs.*, this volume, 10.1007/s11214-006-9011-3.

Cane, H. V., Sheeley, N. R., and Howard, R. A.: 1987, *J. Geophys. Res.* **92**, 9869.

Cane, H. V., Stone, R. G., Fainberg, J., Steinberg, J. L., Hoang, S., and Stewart, R. T.: 1981, *Geophys. Res. Lett.* **8**, 1285.

Chen, J., Howard, R. A., Brueckner, G. E., Santoro, R., Krall, J., Paswaters, S. E., *et al.*: 1997, *Astrophys. J.* **490**, L191.

Chen, J., Santoro, R. A., Krall, J., Howard, R. A., Duffin, R., Moses, J. D., *et al.*: 2000, *Astrophys. J.* **533**, 481.

Ciaravella, A., Raymond, J. C., and Kahler, S. W.: 2006, *Astrophys. J.* submitted.

Ciaravella, A., Raymond, J. C., Li, J., Reiser, P., Gardner, L. D., Ko, Y.-K., *et al.*: 2002, *Astrophys. J.* **575**, 1116.

Ciaravella, A., Raymond, J. C., Thompson, B. J., van Ballegooijen, A., Strachan, L., Li, J., *et al.*: 2000, *Astrophys. J.* **529**, 575.

Cliver, E. W., Webb, D. F., and Howard, R. A.: 1999, *Sol. Phys.* **187**, 89.

Crooker, N. U., Gosling, J. T., and Kahler, S. W.: 2002, *J. Geophys. Res. (Space Phys.)* **107**(A2), 3.

Czaykowska, A., de Pontieu, B., Alexander, D., and Rank, G.: 1999, *ApJ* **521**, L75.

Dere, K. P., Brueckner, G. E., Howard, R. A., Koomen, M. J., Korendyke, C. M., Kreplin, R. W., *et al.*: 1997, *Sol. Phys.* **175**, 601.

Dere, K. P., Brueckner, G. E., Howard, R. A., Michels, D. J., and Delaboudiniere, J. P.: 1999, *Astrophys. J.* **516**, 465.

Dobrzycka, D., Cranmer, S. R., Raymond, J. C., Biesecker, D. A., and Gurman, J. B.: 2002, *Astrophys. J.* **565**, 621.

Dobrzycka, D., Raymond, J. C., Biesecker, D. A., Li, J., and Ciaravella, A.: 2003, *Astrophys. J.* **588**, 586.

Doschek, G. A., Cheng, C. C., Oran, E. S., Boris, J. P., and Mariska, J. T.: 1983, *Astrophys. J.* **265**, 1103.

Dulk, G. A., Leblanc, Y., and Bougeret, J.: 1999, *Geophys. Res. Lett.* **26**, 2331.

Forbes, T. G., *et al.*: 2006, *Space Sci. Revs.*, this volume, 10.1007/s11214-006-9019-8.

Forsyth, R. J., *et al.*: 2006, *Space Sci. Revs.*, this volume, 10.1007/s11214-006-9022-0.

Gallagher, P. T., Dennis, B. R., Krucker, S., Schwartz, R. A., and Tolbert, A. K.: 2002, *Sol. Phys.* **210**, 341.

Gergely, T. E., Kundu, M. R., and Hildner, E.: 1983, *Astrophys. J.* **268**, 403.

Gilbert, H. R., Holzer, T. E., Burkepile, J. T., and Hundhausen, A. J.: 2000, *Astrophys. J.* **537**, 503.
Gilbert, H. R., Serex, E. C., Holzer, T. E., MacQueen, R. M., and McIntosh, P. S.: 2001, *Astrophys. J.* **550**, 1093.
Gopalswamy, N., Kaiser, M. L., Lepping, R. P., Kahler, S. W., Ogilvie, K., Berdichevsky, D., *et al.*: 1998, *J. Geophys. Res.* **103**, 307.
Gopalswamy, N., Kaiser, M. L., Thompson, B. J., Burlaga, L. F., Szabo, A., Vourlidas, A., *et al.*: 2000, *Geophys. Res. Lett.* **27**, 1427.
Gopalswamy, N., and Kundu, M. R.: 1992, *ApJ* **390**, L37.
Gopalswamy, N., Shimojo, M., Lu, W., Yashiro, S., Shibasaki, K., and Howard, R. A.: 2003, *Astrophys. J.* **586**, 562.
Gopalswamy, N., Yashiro, S., Kaiser, M. L., Howard, R. A., and Bougeret, J.-L.: 2001, *ApJ* **548**, L91.
Gosling, J. T., Hildner, E., MacQueen, R. M., Munro, R. H., Poland,A. I., and Ross, C. L.: 1976, *Sol. Phys.* **48**, 389.
Haggerty, D. K., and Roelof, E. C.: 2002, *Astrophys. J.* **579**, 841.
Howard, R. A., Sheeley, N. R., Michels, D. J., and Koomen, M. J.: 1985, *J. Geophys. Res.* **90**, 8173.
Howard, R. A., Sheeley, N. R., Michels, D. J., and Koomen, M. J.: 1986, in: *ASSL Vol. 123: The Sun and the Heliosphere in Three Dimensions*. p. 107.
Hudson, H, Bougeret, J.-L., and Burkepile, J.: 2006, *Space Sci. Revs.*, this volume, 10.1007/s11214-006-9009-x.
Hudson, H. S., Khan, J. I., Lemen, J. R., Nitta, N. V., and Uchida, Y.: 2003, *Sol. Phys.* **212**, 121.
Hundhausen, A. J., Burkepile, J. T., and St. Cyr, O. C.: 1994, *J. Geophys. Res.* **99**, 6543.
Illing, R. M. E., and Hundhausen, A. J.: 1983, *J. Geophys. Res.* **88**, 10210.
Innes, D. E., Inhester, B., Srivastava, N., Brekke, P., Harrison, R. A., Matthews, S. A., *et al.*: 1999, *Sol. Phys.* **186**, 337.
Innes, D. E., McKenzie, D. E., and Wang, T.: 2003, *Sol. Phys.* **217**, 247.
Kahler, S. W., Cliver, E. W., and Cane, H. V.: 1989, *Sol. Phys.* **120**, 393.
Kahler, S. W., and Hundhausen, A. J.: 1992, *J. Geophys. Res.* **97**, 1619.
Kahler, S. W., Reames, D. V., and Sheeley, N. R.: 2001, *Astrophys. J.* **562**, 558.
Kaiser, M. L., Reiner, M. J., Gopalswamy, N., Howard, R. A., St. Cyr, O. C., Thompson, B. J., *et al.*: 1998, *Geophys. Res. Lett.* **25**, 2501.
Kane, S. R.: 1981, *Astrophys. J.* **247**, 1113.
Kayser, S. E., Bougeret, J., Fainberg, J., and Stone, R. G.: 1987, *Sol. Phys.* **109**, 107.
Kerdraon, A., Pick, M., Trottet, G., Sawyer, C., Illing, R., Wagner, W., *et al.*: 1983, *ApJ* **265**, L19.
Khan, J. I., and Aurass, H.: 2002, *A&A* **383**, 1018.
Khan, J. I., and Hudson, H. S.: 2000, *Geophys. Res. Lett.* **27**, 1083.
Klassen, A., Aurass, H., Klein, K.-L., Hofmann, A., and Mann, G.: 1999, *A&A* **343**, 287.
Klassen, A., Bothmer, V., Mann, G., Reiner, M. J., Krucker, S., Vourlidas, A., *et al.*: 2002, *A&A* **385**, 1078.
Klassen, A., Pohjolainen, S., and Klein, K.-L.: 2003, *Sol. Phys.* **218**, 197.
Klecker, B., *et al.*: 2006, *Space Sci. Revs.*, this volume, 10.1007/s11214-006-9018-9.
Klein, K., Khan, J. I., Vilmer, N., Delouis, J., and Aurass, H.: 1999, *A&A* **346**, L53.
Klein, K.-L., Krucker, S., and Trottet, G.: 2003a, *Adv. Space Res.* **32**, 2521.
Klein, K.-L., Schwartz, R. A., McTiernan, J. M., Trottet, G., Klassen, A., and Lecacheux, A.: 2003b, *A&A* **409**, 317.
Klein, K.-L., and Trottet, G.: 1994, In: *AIP Conf. Proc. 294: High-Energy Solar Phenomena – a New Era of Spacecraft Measurements*. pp. 187.
Ko, Y., Raymond, J. C., Lin, J., Lawrence, G., Li, J., and Fludra, A.: 2003, *Astrophys. J.* **594**, 1068.
Krall, J., Chen, J., Duffin, R. T., Howard, R. A., and Thompson, B. J.: 2001, *Astrophys. J.* **562**, 1045.
Krucker, S., Larson, D. E., Lin, R. P., and Thompson, B. J.: 1999, *Astrophys. J.* **519**, 864.
Krucker, S., and Lin, R. P.: 2002, *Sol. Phys.* **210**, 229.

Kundu, M. R., and Stone, R. G.: 1984, *Adv. Space Res.* **4**, 261.
Laming, J. M., Raymond, J. C., McLaughlin, B. M., and Blair, W. P.: 1996, *Astrophys. J.* **472**, 267.
Lantos, P., Kerdraon, A., Rapley, G. G., and Bentley, R. D.: 1981, *A&A* **101**, 33.
Lee, L. C., and Wu, B. H.: 2000, *Astrophys. J.* **535**, 1014.
Li, Y., Luhmann, J. G., Mulligan, T., Hoeksema, J. T., Arge, C. N., Plunkett, S. P., *et al.*: 2001, *J. Geophys. Res.* 25103.
Lin, J., and Forbes, T. G.: 2000, *J. Geophys. Res.* **105**, 2375.
Lin, R. P.: 1998, *Space Sci. Rev.* **86**, 61.
Lin, R. P., Evans, L. G., and Fainberg, J.: 1973, *Astrophys.* **14**, 191.
Linker, J. A., and Mikic, Z.: 1995, in: *Solar Wind Conference*. pp. 60.
Linker, J. A., Mikić, Z., Lionello, R., Riley, P., Amari, T., and Odstrcil, D.: 2003, *Phys. Plasmas* **10**, 1971.
Lyons, M. A., and Simnett, G. M.: 2001, *Solar Phys.* **200**, 203.
MacQueen, R. M., and Fisher, R. R.: 1983, *Solar Phys.* **89**, 89.
Maia, D., Pick, M., Vourlidas, A., and Howard, R.: 2000, *ApJ* **528**, L49.
Maia, D., Vourlidas, A., Pick, M., Howard, R., Schwenn, R., and Magalhães, A.: 1999, *J. Geophys. Res.* **104**, 12507.
Maia, D. J. F., and Pick, M.: 2004, *Astrophys. J.* **609**, 1082.
Mancuso, S., and Abbo, L.: 2004, *A&A* **415**, L17.
Mancuso, S., Raymond, J. C., Kohl, J., Ko, Y.-K., Uzzo, M., and Wu, R.: 2002, *A&A* **383**, 267.
Mann, G.: 1995, *Lecture Notes in Physics*. Springer Verlag, Berlin, vol. 444, pp. 183.
Mann, G., and Luehr, H.: 1994, *ApJS* **90**, 577.
Marqué, C., Lantos, P., and Delaboudinière, J. P.: 2002, *A&A* **387**, 317.
Masuda, S., Kosugi, T., Hara, H., Tsuneta, S., and Ogawara, Y.: 1994, *Nature* **371**, 495.
McComas, D. J., Phillips, J. L., Hundhausen, A. J., and Burkepile, J. T.: 1991, *Geophys. Res. Lett.* **18**, 73.
McKenzie, D. E., and Hudson, H. S.: 1999, *ApJ* **519**, L93.
Mikić, Z., and Lee, M. A.: 2006, *Space Sci. Revs.*, this volume, 10.1007/s11214-006-9012-2.
Moon, Y. J., Choe, G. S., Wang, H., Park, Y. D., Gopalswamy, N., Yang, G., *et al.*: 2002, *Astrophys. J.* **581**, 694.
Moreton, G. E.: 1960, *AJ* **65**, 494.
Moses, D., Clette, F., Delaboudiniere, J.-P., Artzner, G. E., Bougnet, M., Brunaud, J., *et al.*: 1997, *Sol. Phys.* **175**, 571.
Narukage, N., Hudson, H. S., Morimoto, T., Akiyama, S., Kitai, R., Kurokawa, H., *et al.*: 2002, *ApJ* **572**, L109.
Nelson, G. J., and Melrose, D. B.: 1985, *Type II bursts*, Solar Radiophysics: Studies of Emission from the Sun at Metre Wavelengths, pp. 333.
Northrop, T. G.: 1966, in: *ASSL Vol. 5: Radiation Trapped in the Earth's Magnetic Field*. p. 26.
Pearson, D. H., Nelson, R., Kojoian, G., and Seal, J.: 1989, *Astrophys. J.* **336**, 1050.
Pick, M.: 1999, in: Bastian, T. S., Gopalswamy, N., and Shibasaki, K. (eds.), *Proceedings of the Nobeyama Symposium, held in Kiyosato, Japan, Oct. 27-30, 1998* NRO Report No. 479, p. 187.
Pick, M., Maia, D., Wang, S. J., Lecacheux, A., and Hawkins, S. E.: 2003, *Adv. Space Res.* **32**, 2527.
Pike, C. D., and Mason, H. E.: 2002, *Sol. Phys.* **206**, 359.
Pohjolainen, S., Maia, D., Pick, M., Vilmer, N., Khan, J. I., Otruba, W., *et al.*: 2001, *Astrophys. J.* **556**, 421.
Priest, E., and Forbes, T. (eds.): 2000, 'Magnetic reconnection: MHD theory and applications'.
Raoult, A., Pick, M., Dennis, B. R., and Kane, S. R.: 1985, *Astrophys. J.* **299**, 1027.
Raulin, J. P., Kundu, M. R., Hudson, H. S., Nitta, N., and Raoult, A.: 1996, *A&A* **306**, 299.
Raymond, J. C., Blair, W. P., and Long, K. S.: 1995, *ApJ* **454**, L31.

Raymond, J. C., Ciaravella, A., Dobrzycka, D., Strachan, L., Ko, Y.-K., Uzzo, M., *et al.*: 2003, *Astrophys. J.* **597**, 1106.
Raymond, J. C., Thompson, B. J., St. Cyr, O. C., Gopalswamy, N., Kahler, S., Kaiser, M., *et al.*: 2000, *Geophys. Res. Lett.* **27**, 1439.
Reames, D. V.: 1999, *Space Sci. Rev.* **90**, 413.
Reiner, M. J., and Kaiser, M. L.: 1999a, *Geophys. Res. Lett.* **26**, 397.
Reiner, M. J., and Kaiser, M. L.: 1999b, *J. Geophys. Res.* **104**, 16979.
Reiner, M. J., Kaiser, M. L., and Bougeret, J.-L.: 2001, *J. Geophys. Res.* **106**, 29989.
Reiner, M. J., Kaiser, M. L., Plunkett, S. P., Prestage, N. P., and Manning, R.: 2000a, *ApJ* **529**, L53.
Reiner, M. J., Karlický, M., Jiřička, K., Aurass, H., Mann, G., and Kaiser, M. L.: 2000b, *Astrophys. J.* **530**, 1049.
Reiner, M. J., Vourlidas, A., Cyr, O. C. S., Burkepile, J. T., Howard, R. A., Kaiser, M. L., *et al.*: 2003, *Astrophys. J.* **590**, 533.
Robinson, R. D., and Stewart, R. T.: 1985, *Sol. Phys.* **97**, 145.
Saito, K., Makita, M., Nishi, K., and Hata, S.: 1970, *Ann Tokyo Astronomical Observatory* **12**, 51.
Schwenn, R., *et al.*: 2006, *Space Sci. Revs.*, this volume, 10.1007/s11214-006-9016-y.
Sheeley, N. R., J., Wang, Y.-M., Hawley, S. H., Brueckner, G. E., Dere, K. P., Howard, R. A., *et al.*: 1997, *Astrophys. J.* **484**, 472.
Sheeley, N. R., Walters, J. H., Wang, Y.-M., and Howard, R. A.: 1999, *J. Geophys. Res.* **104**, 24739.
Sheeley, N. R., and Wang, Y.-M.: 2002, *Astrophys. J.* **579**, 874.
Sheeley, N. R., Warren, H., and Wang, Y.-M.: 2004, *Astrophys. J.* **616**, 1224.
Sheridan, K. V., Jackson, B. V., McLear, D. J., and Dulk, G. A.: 1978, *Proc. Astron. Soc. Aust.* **3**, 249.
Simnett, G. M.: 2004, *A&A* **416**, 759.
Simnett, G. M., Roelof, E. C., and Haggerty, D. K.: 2002, *Astrophys. J.* **579**, 854.
Simnett, G. M., Tappin, S. J., Plunkett, S. P., Bedford, D. K., Eyles, C. J., Cyr, O. C. S., *et al.*: 1997, *Solar Phys.* **175**, 685.
Srivastava, N., Schwenn, R., Inhester, B., Martin, S. F., and Hanaoka,Y.: 2000, *Astrophys. J.* **534**, 468.
Srivastava, N., Schwenn, R., Inhester, B., Stenborg, G., and Podlipnik, B.: 1999, *Space Sci. Rev.* **87**, 303.
St. Cyr, O. C., Burkepile, J. T., Hundhausen, A. J., and Lecinski, A. R.: 1999, *J. Geophys. Res.* **104**, 12493.
St. Cyr, O. C., Plunkett, S. P., Michels, D. J., Paswaters, S. E., Koomen, M. J., Simnett, G. M., *et al.*: 2000, *J. Geophys. Res.* 18169.
Steinolfson, R. S.: 1985, *Wash. DC Am Geophys. Union Geophys. Monogr. Ser.* **35**, 1.
Sterling, A. C., and Hudson, H. S.: 1997, *ApJ* **491**, L55.
Sui, L., and Holman, G. D.: 2003, *ApJ* **596**, L251.
Svestka, Z., Martin, S. F., and Kopp, R. A.: 1980, in: Dryer, M., and Tandberg-Hanssen, E. (eds.), *IAU Symp. 91: Solar and Interplanetary Dynamics*. pp. 217.
Thompson, B. J., Plunkett, S. P., Gurman, J. B., Newmark, J. S., St. Cyr, O. C., and Michels, D. J.: 1998, *Geophys. Res. Lett.* **25**, 2465.
Tokman, M., and Bellan, P. M.: 2002, *Astrophys. J.* **567**, 1202.
Tousey, R.: 1973, in: Rycroft, M. J., and Runcorn, S. K. (eds.), *Space Research XIII, Proceedings of the 15th Plenary Meeting of COSPAR*, Madrid, Spain, 10–24 May 1972, Akademie-Verlag, Berlin, pp. 713.
Trottet, G.: 1986, *Sol. Phys.* **104**, 145.
Tsuneta, S., Hara, H., Shimizu, T., Acton, L. W., Strong, K. T., Hudson, H. S., *et al.*: 1992, *PASJ* **44**, L63.
Uchida, Y.: 1960, *PASJ* **12**, 376.
Uchida, Y.: 1974, *Sol. Phys.* **39**, 431.

Vourlidas, A., Wu, S. T., Wang, A. H., Subramanian, P., and Howard, R. A.: 2003, *Astrophys. J.* **598**, 1392.

Wagner, W. J., and MacQueen, R. M.: 1983, *A&A* **120**, 136.

Wang, Y.-M., and Sheeley, N. R.: 2002, *Astrophys. J.* **575**, 542.

Wang, Y.-M., Sheeley, N. R., Walters, J. H., Brueckner, G. E., Howard, R. A., Michels, D. J., *et al.*: 1998, *ApJ* **498**, L165.

Warmuth, A., Vršnak, B., Magdalenić, J., Hanslmeier, A., and Otruba, W. : 2004, *A&A* **418**, 1101.

Webb, D. F., Burkepile, J., Forbes, T. G., and Riley, P.: 2003, *J. Geophys. Res. (Space Phys.)* 6.

Webb, D. F., and Cliver, E. W.: 1995, *J. Geophys. Res.* **100**, 5853.

Wild, J. P.: 1950, *Aust. J.Phys.* **3**, 541.

Wu, S. T., Guo, W. P., Michels, D. J., and Burlaga, L. F.: 1999, *J. Geophys. Res.* **104**, 14789.

Yashiro, S., Gopalswamy, N., Michalek, G., and Howard, R. A.: 2003, *Adv. Space Res.* **32**, 2631.

Yokoyama, T., Akita, K., Morimoto, T., Inoue, K., and Newmark, J.: 2001, *ApJ* **546**, L69.

Zhang, J., Dere, K. P., Howard, R. A., and Vourlidas, A.: 2004, *ApJ* **604**, 420.

ICMES IN THE INNER HELIOSPHERE: ORIGIN, EVOLUTION AND PROPAGATION EFFECTS

Report of Working Group G

R. J. FORSYTH[1,*], V. BOTHMER[2], C. CID[3], N. U. CROOKER[4], T. S. HORBURY[1], K. KECSKEMETY[5], B. KLECKER[6], J. A. LINKER[7], D. ODSTRCIL[8], M. J. REINER[9], I. G. RICHARDSON[10], J. RODRIGUEZ-PACHECO[3], J. M. SCHMIDT[11] and R. F. WIMMER-SCHWEINGRUBER[12]

[1]*The Blackett Laboratory, Imperial College London, London, UK*
[2]*Institute for Astrophysics, University of Göttingen, Göttingen, Germany*
[3]*Alcalá Space Research Group, Dpto. Física, Univ. de Alcalá, Madrid, Spain*
[4]*Center for Space Physics, Boston University, Boston, Massachusetts, USA*
[5]*KFKI Research Institute for Particle and Nuclear Physics, Budapest, Hungary*
[6]*MPI für Extraterrestrische Physik, Garching, Germany*
[7]*SAIC, San Diego, CA, USA*
[8]*NOAA Space Environment Center, Boulder, CO, USA*
[9]*The Catholic University of America and NASA/GSFC, Greenbelt, MD, USA*
[10]*NASA Goddard Space Flight Center, Greenbelt, MD, USA*
[11]*International University Bremen, Bremen, Germany*
[12]*Institut für Experimentelle und Angewandte Physik, Extraterrestrische Physik, Christian-Albrechts-Universität zu Kiel, Kiel, Germany*
*(*Author for correspondence: E-mail: r.forsyth@imperial.ac.uk)*

(Received 7 February 2006; Accepted in final form 24 May 2006)

Abstract. This report assesses the current status of research relating the origin at the Sun, the evolution through the inner heliosphere and the effects on the inner heliosphere of the interplanetary counterparts of coronal mass ejections (ICMEs). The signatures of ICMEs measured by *in-situ* spacecraft are determined both by the physical processes associated with their origin in the low corona, as observed by space-borne coronagraphs, and by the physical processes occurring as the ICMEs propagate out through the inner heliosphere, interacting with the ambient solar wind. The solar and *in-situ* observations are discussed as are efforts to model the evolution of ICMEs from the Sun out to 1 AU.

Keywords: coronal mass ejections, magnetic clouds, solar wind, interplanetary shocks

1. Introduction

A fundamental problem in understanding the physics of CMEs is our limited knowledge of their physical properties. Remote sensing observations, such as those from coronagraphs, do not provide quantitative values of coronal plasma and magnetic field parameters of CMEs, while by the time that *in-situ* observations are made out in the heliosphere, ICMEs have already experienced substantial evolution and interaction with the ambient solar wind (e.g., Klein and Burlaga, 1982; Bothmer and Schwenn, 1994, 1998; Crooker and Horbury, 2006, this volume). Furthermore,

DOI: 10.1007/s11214-006-9022-0

attempts to make direct associations between *in-situ* parameters and coronagraph images suffer from the fact that those ICMEs that are intercepted by near-Earth spacecraft usually originate as Earth-directed front-side halos for which the CME structure and speed of propagation are most difficult to determine with current coronagraphs located near Earth. Another complication is that a spacecraft takes measurements along a trajectory through the ICME. In simple cases, such as a magnetic cloud, the large-scale ICME structure can be inferred from the observations assuming a suitable model, although it has to be borne in mind that such observations characterise only the middle segment near the apex of the much larger flux ropes (e.g., Figure 2 of Zurbuchen and Richardson, 2006, this volume). In other cases, in particular where the spacecraft only skims the ICME, it may be unclear how to interpret the observations in terms of ICME structure.

In this chapter, we focus on relating ICMEs to their CME origins, in particular, those origins that are assumed to involve flux rope formation (e.g., Mikic and Lee, 2006, this volume), and discuss ICME evolution with heliocentric distance, including the effects on the ambient solar wind. Important questions addressed include:

- How do the features of ICMEs observed *in-situ* reflect their solar origins?
- How does the structure of ICMEs change as they propagate through the inner heliosphere?
- How are ICMEs decelerated and/or accelerated on their journey out to 1AU?
- How do ICME shocks form and evolve?
- How do ICMEs interact with other solar wind streams, either high speed flows from coronal holes or other ICMEs, and how do compound streams form?
- What are the solar cycle variations of the above phenomena?
- Can we model ICME evolution in the inner heliosphere?

The data available for addressing these questions consists of solar observations, primarily from space-borne coronagraphs, and of *in-situ* data from various spacecraft. Sections 2 and 3 discuss ICME parameters that reflect their solar origins and how ICME parameters evolve with distance from the Sun. Section 4 addresses ICME interactions with the structured solar wind, and Section 5 describes some solar cycle variations. Section 6 discusses attempts to model CME and ICME evolution through the inner heliosphere, presenting a number of case studies. Finally, Section 7 addresses Type II radio emissions and the interplanetary scintillation technique, which provide information that bridges the gap between solar and *in-situ* observations.

2. Relating ICMEs to CMEs

V. Bothmer, N. U. Crooker, and J. Rodriguez-Pacheco

Research that attempts to relate features in ICMEs to features in CMEs focuses on the simplest forms of each, the ICME with flux-rope structure (e.g., Zurbuchen and

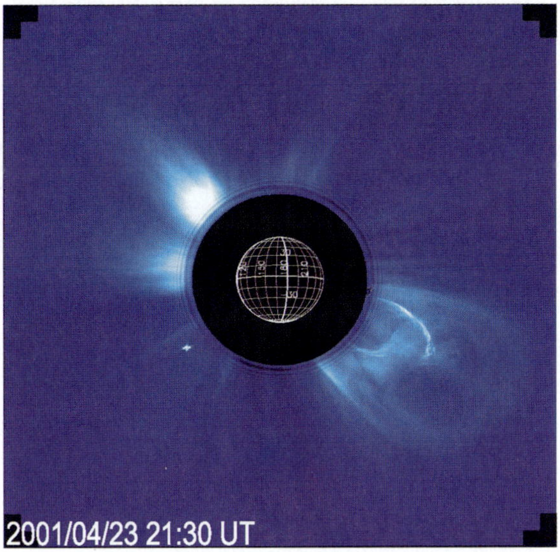

Figure 1. CME with flux-rope structure (Cremades and Bothmer, 2004).

Richardson, 2006, this volume) and the three-part CME (e.g., Hudson *et al.*, 2006, this volume). This section discusses how certain aspects of these forms appear to survive the kinematic and dynamic distortions that CMEs undergo as they expand out into the spherical geometry of the heliosphere and become ICMEs.

CMEs with the typical three-part structure are made up of a leading outward moving bright front followed by a dark cavity and finally a bright core of filament plasma at its trailing edge (Hundhausen, 1988). Figure 1 from Cremades and Bothmer (2004) shows an example. While to date it is not clear how the different CME parts evolve in the heliosphere and what signatures they give in the *in-situ* data, it is generally assumed that the bright front corresponds to the sheath of compressed solar wind while the dark cavity comprises the flux rope structure observed in magnetic clouds and reflects their low gas pressure balanced by high magnetic pressure. The dark cavity in Figure 1 is a particularly good example of this expectation since its circular shape resembles the cross-section of a cylindrical flux rope. What becomes of the cool, dense filament plasma, which can cover large areas in coronagraph images, is an open question, since evidence (such as exceptionally low ion charge states) appears extremely rarely in in situ data (see Wimmer-Schweingruber *et al.*, 2006, this volume, Zurbuchen and Richardson, 2006, this volume, and Crooker and Horbury, 2006, this volume).

The flux rope structure of magnetic clouds also shows what is probably the clearest imprint of the solar origin of ICMEs (see, also, Crooker and Horbury, 2006, this volume and Wimmer-Schweingruber *et al.* (this volume, cf. their Figure 6)). Figure 2 reviews how the magnetic field structure at the sites of solar prominences

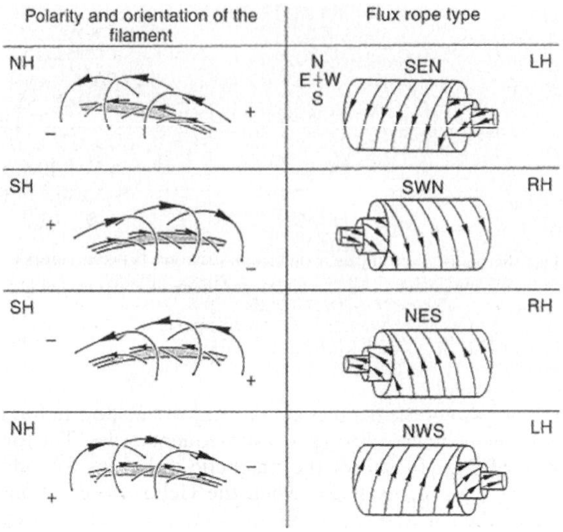

Figure 2. Magnetic field structure of solar prominences and magnetic clouds (Bothmer and Schwenn, 1994).

relates to the structure of the associated clouds according to Bothmer and Schwenn (1994). For prominences in the southern solar hemisphere, the axial field should be right-handed, independent of the solar cycle, whereas the direction of the arcade of loops overlying the prominence that later may be identified as the (I)CME's circular field lines reflects the dipolar component of the solar field, which reverses during the solar cycle. Bothmer and Schwenn (1994, 1998) and Bothmer and Rust (1997) found a good correlation between the magnetic structure at the site of disappearing filaments and the subsequent magnetic clouds (ICMEs) observed by the Helios spacecraft.

To give an example of how the pattern in Figure 2 can be tested against data, we use the well-studied event of January 1997. Figure 3 shows the *in-situ* cloud data. The rotation of the magnetic field vector varies from south to west to north. Based on this right-handed signature, one would expect the cloud to originate from the southern hemisphere. Figure 4, adapted from Bothmer (2003), shows the source region of the (I)CME. It indeed lies in the southern hemisphere, and the field polarity pattern shown in the magnetogram is consistent with a right-handed structure and a southward leading field. The cloud configuration thus agrees with the scenario proposed in Figure 2, and the source region of the CME is consistent with the results of Dere *et al.* (1999) and Cremades and Bothmer (2004), who have shown that CMEs arise from bipolar regions, either active or decaying ones, or from parts of it.

Less clear is the degree to which the tilt of the cloud axis reflects the orientation of the configuration at the source. The Yohkoh soft X-ray image on the right of Figure 4 shows hot coronal loops in the source region with enhanced magnetic flux, which are generally interpreted as evidence of reconnection in the wake of a CME, although in

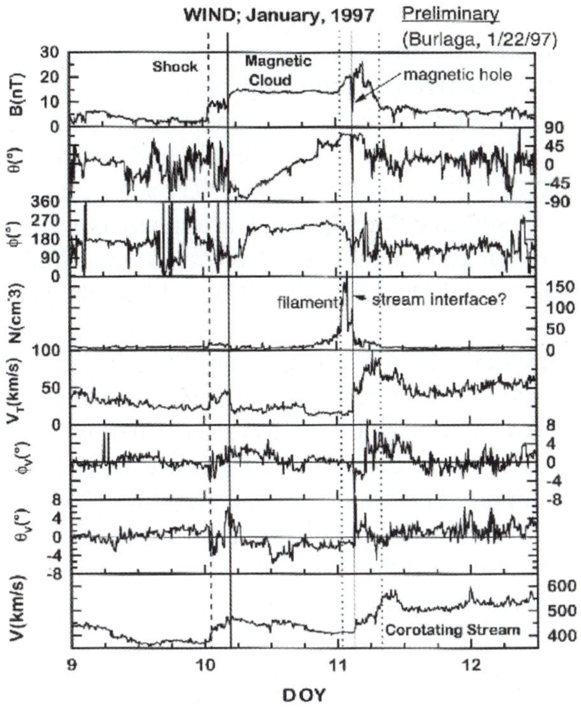

Figure 3. The magnetic cloud (ICME) in January 1997 (adapted from Burlaga *et al.*, 1998).

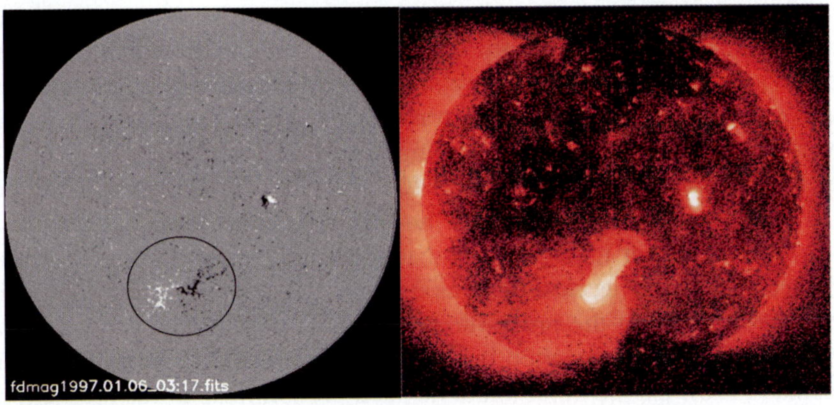

Figure 4. SOHO/MDI magnetogram (left) and Yohkoh soft X-ray images (right) showing the source region (circle) of the January 6, 1997 CME. White colors in the magnetogram indicate positive magnetic polarity (field lines pointing away from the sun).

this case the intensity of the signature was marginal (Webb *et al.*, 1998). The arcade of loops arches over the neutral line between the positive and negative polarity regions in the magnetogram, and this structure presumably aligns with the flux rope that becomes the magnetic cloud. It appears to be highly inclined with respect to the ecliptic plane, but Zhao and Hoeksema (1997) found an inclination of only 27 ° for the portion of the neutral line from which the associated filament erupted. From their statistically-derived relationship based upon 14 published values of filament and associated cloud-axis inclinations, they predicted a 20 ° cloud-axis tilt for the January 1997 event. This tilt is slightly higher than the flux-rope model results of Burlaga *et al.* (1998) and Bothmer (2003), whose values range from 3 ° to 15 °, depending on the selected boundaries. The agreement is thus reasonably good, although cases with poor agreement have also been found (Webb *et al.*, 2000). The 14 cases used by Zhao and Hoeksema (1997) yield a relatively high correlation coefficient of 0.76.

The flattening of cloud-axis tilts compared to filament tilts apparent both in the January 1997 case and in the formulation of Zhao and Hoeksema (1997) may be the result of a global deflection of CMEs toward the heliomagnetic equator by the fast solar wind emanating from coronal holes. Clear evidence for such a systematic deflection has been reported by Cremades and Bothmer (2004). At the time of the January 1997 event, polar coronal holes were present, and the CME was interacting with a fast solar wind stream. It thus created a shock wave which had not existed at the time of its lift-off (as inferred from flaring and radio wave signatures) (Burlaga *et al.*, 1998). This interaction may naturally explain the flattening of the axis as observed in the heliosphere.

As reviewed by Crooker and Horbury (2006, this volume), magnetic clouds reportedly carry not only the imprint of filament axes at low solar altitudes but the imprint of the coronal streamer belt at high altitudes, as well. This tendency most likely reflects the influence of the dipolar component of the solar magnetic field, at least during the quieter phases of the solar cycle. The dipolar component dominates the helmet arcade constituting the streamer belt and is also apparent in the smaller arcades overlying the filaments, illustrated in Figure 2. During solar maximum, however, the influence of the dipolar component is apparent only far from the Sun, where the heliospheric current sheet (HCS) marking the heliomagnetic equator remains coherent on the global scale and turns over as the solar field reverses its polarity. Because of its weakness near the Sun, the overturning streamer belt may have no control over cloud axis orientation at solar maximum. Some control might be deduced from the predominance of high cloud-axis inclinations at and beyond solar maximum reported by Mulligan *et al.* (1998), following Zhao and Hoeksema (1996). On the other hand, no such predominance was found in a more recent study by Huttunen *et al.* (2005), in which high cloud-axis inclinations appear to be distributed with roughly equal probability throughout the solar cycle. The reason for these conflicting results may be due to CMEs with origins outside the helmet streamer belt, in closed-field regions surrounded by open regions of the same polarity, a configuration more common during solar maximum (Zhao and Webb, 2003; Liu and Hayashi, 2006).

Figure 5 from Rodriguez-Pacheco *et al.* (2005) supports the view that the streamer belt has little influence over magnetic cloud axes at solar maximum. On a

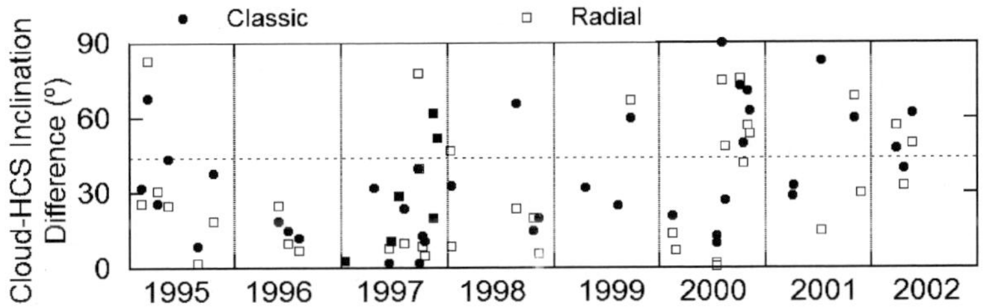

Figure 5. Differences between inclinations of magnetic cloud axes and the HCS predicted for the time of cloud encounter from classic and radial source surface maps from the Wilcox Observatory (Rodriguez-Pacheco *et al.*, 2005).

case-by-case basis during the rise in activity from solar minimum (late 1996) to maximum (late 2000), Figure 5 shows that the difference between the cloud-axis inclination calculated from a force-free cylindrical flux rope model by R. P. Lepping (`http://lepmfi.gsfc.nasa.gov/mfi/mag_cloud_pub1p.html`) and the inclination of the HCS predicted from the corresponding source surface map (both classic and radial) increases substantially. The predicted HCS inclinations rise, as expected, but the cloud axis inclinations show little solar cycle variation, being predominantly low throughout the period. Although the lack of clouds with higher axis inclinations at solar maximum probably reflects a selection bias, and an unbiased data set might well show the tendency for higher elevations found by Mulligan *et al.* (1998) and Huttunen *et al.* (2005), these results clearly demonstrate that the streamer belt orientation does not govern axis inclinations of many clouds at solar maximum, consistent with the weak influence of the dipole component of the solar magnetic field at that time.

In a related effort, Blanco *et al.* (2003) reported the results of a preliminary comparison between cloud axis orientations and *in-situ*, rather than predicted, HCS orientations. They found that only about half of their 17 cases had elevation angle differences of less than 45°, consistent with only weak streamer belt control of magnetic cloud axis orientation.

3. Evolution of ICME Parameters

I. G. Richardson, C. Cid, N. U. Crooker, T. S. Horbury, B. Klecker, J. Rodriguez-Pacheco and R. F. Wimmer-Schweingruber

3.1. Average Properties of ICMEs

The most comprehensive observations of ICMEs in the inner heliosphere currently available are those made by the Helios 1 and 2 spacecraft at 0.3 – 1 AU. Bothmer and Schwenn (1998) examined 46 Helios magnetic clouds, and their results are summarized in Table I. They concluded that the mean density within these clouds

TABLE I

Dependence of ICME parameters on heliocentric distance R (AU)

	BS98[a]	This work	L05[b]	W05[c]
R (AU)	0.3–1.0	0.3–1.0	0.3–5.4	0.3–5.4
S (AU)	$0.24R^{0.78}$	$0.31R^{0.53}$	$0.25R^{0.92}$	$0.19R^{0.61}$
n (cm^{-3})	$6.47R^{-2.4}$	$7.03R^{-2.18}$	$6.16R^{-2.32}$	$6.7R^{-2.4}$
V (km/s)		$483R^{0.07}$	$458R^{0.002}$	$456R^{-0.003}$
T (10^3 K)		$44.3R^{-0.83}$	$35.4R^{-0.32}$	$29.2R^{-0.74}$
Mean B (nT)	$17.7R^{-1.73}$[d]	$10.3R^{-1.31}$	$7.35R^{-1.40}$	$8.3R^{-1.52}$
V_{ex}		$39.7R^{-0.16}$	$57.5R^{-0.12}$	$0.12V_{ICME}R^{-0.39}$
γ_p			1.14 ± 0.03	

[a]Bothmer and Schwenn (1998) magnetic clouds.
[b]Liu *et al.* (2005).
[c]Wang *et al.* (2005).
[d]Axial magnetic field (M. Leitner, personal communication, 2004).

decreased with heliocentric distance slightly faster than the $n = 6.1R^{-2.1}$ cm^{-3} variation found by Helios generally in the solar wind (Schwenn, 1990). Based on an analysis that included magnetic clouds observed by Pioneer 10 and Voyagers 1/2 beyond 1 AU, the radial size ($S = \int V_{sw}dt$ during cloud passage) was found to increase as $S = (0.24 \pm 0.01) \times R^{(0.78\pm0.1)}$ AU. Bothmer and Schwenn (1998) noted that the expected $R^{-2.56}$ mass density dependence agrees reasonably well with that observed, assuming clouds have a cylindrical cross-section, the flux-tube length is proportional to R, and mass is conserved within the flux tube. The axial magnetic fields inferred from force-free fits to these magnetic clouds declined as $B = 17.7 \times R^{-1.73}$ nT (M. Leitner, personal communication, 2004).

Figure 6 shows the radial dependence of the mean values of several parameters within an expanded sample of 103 Helios ICMEs during 1975–1980, including some non-cloud events not considered by Bothmer and Schwenn (1998) (e.g., Cane *et al.*, 1997). The results are also summarized in Table I together with those of the recent papers of Liu *et al.* (2005) and Wang *et al.* (2005), which include both Helios and Pioneer Venus Orbiter (PVO) observations within the inner heliosphere and Ulysses observations out to 5.4 AU. Their results refer to "ICME-like" structures identified primarily using the abnormally low proton temperature signature.

All these studies show that ICMEs expand with heliocentric distance, though the dependence varies from $\sim R^{0.5}$ to $R^{0.9}$. The average size at 1 AU is ~ 0.25 AU. The radial dependences in ICME plasma density are in good agreement ($\sim R^{-2.3}$), and again the density declines only slightly faster than in the general solar wind at Helios (as also found by González-Esparza *et al.* (1998) in ICMEs beyond 1 AU). The mean ICME speed has no significant radial variation ($R^{-0.003}$ to $R^{0.07}$), suggesting that, on average, there is little acceleration or deceleration between 0.3 AU and

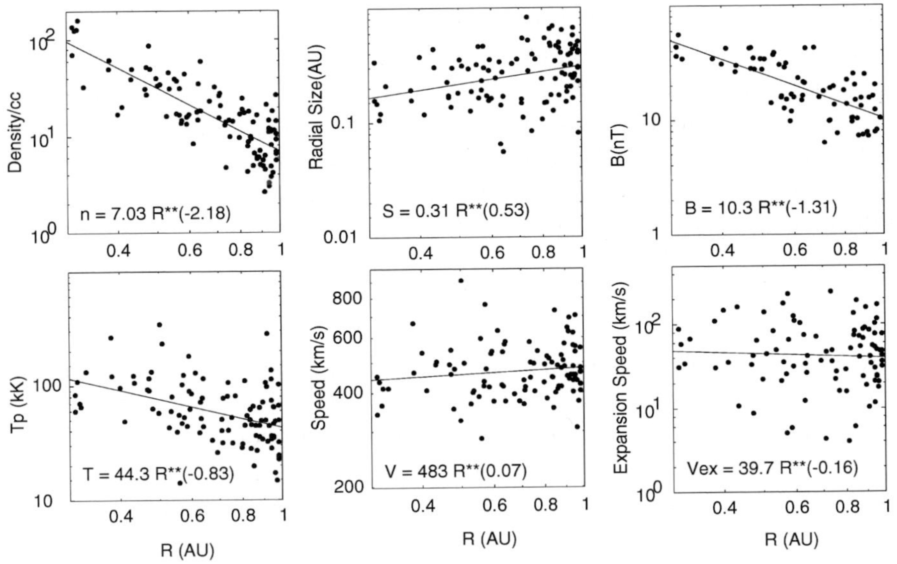

Figure 6. Variation of event-averaged ICME parameters with heliocentric distance observed by the Helios 1/2 spacecraft in 1975–1980, plotted in a log-log format.

the outer heliosphere. Mean ICME speeds are also similar to the average solar wind speed measured during the Helios missions (481 km/s; Schwenn, 1990). The proton temperature declines with distance ($R^{-0.3}$ to $R^{-0.8}$). This is comparable to, but slightly slower than the $\sim R^{-0.9}$ dependence found generally in Helios 1 solar wind data by Totten *et al.* (1995). Wang *et al.* (2005) note that this result is contrary to the expectation that adiabatic cooling due to ICME expansion will lead to faster cooling of ICME plasma. The mean magnetic field intensity declines as $\sim R^{-1.3}$ to $R^{-1.5}$. This rate of decrease is slower than for both the Parker-spiral magnetic field – relative to the Parker field, the mean field for the ICMEs in Figure 6, for example, increases as $R^{(0.44\pm0.16)}$ – and the axial fields inferred for the Bothmer and Schwenn (1998) magnetic clouds by M. Leitner (private communication, 2004). ICME expansion speeds (estimated as half the difference in solar wind speeds inside the ICME leading and trailing edges) have no significant radial dependence. Average values ($\sim$ 40–60 km/s) are around half the Alfvén speed in the ICME ($V_A \sim 85R^{-0.2}$ km/s for average values of the field and density for the ICMEs in Figure 6) as previously noted by Klein and Burlaga (1982). Liu *et al.* (2005) estimate the proton polytropic index (γ_p) to be 1.14. Totten *et al.* (1995) note that the radial dependencies in the solar wind density ($n \propto R^{-\beta}$) and proton temperature ($T \propto R^{-\delta}$) are related to the proton polytropic index by $\gamma = 1 + \delta/\beta$, obtaining $\gamma \sim 1.46$ for the Helios solar wind observations. This is comparable to $\gamma \sim 1.38$ obtained in the same way for the ICMEs in Figure 6.

Radial variations may also be examined in individual ICMEs observed by multiple spacecraft located at similar heliolongitudes but different radial distances.

Unfortunately, such events are rare. One example was observed by Helios 2 (0.4 AU, W5°) on April 23–24, 1979, and by IMP 8 (1 AU) on April 25–26. The estimated radial dependencies ($B \sim R^{-1.07}$; $T \sim R^{-0.2}$; $n \sim R^{-1.43}$; $V \sim R^{0.07}$; and $S \sim R^{0.31}$) overall are slightly weaker, though generally not inconsistent with, those inferred from the statistical results. Russell *et al.* (2003) discuss the properties of two magnetic clouds observed both at ACE (at 1 AU) and NEAR (at ~1.7 AU). They find that the flux rope radii vary as $R^{0.96}$ or $R^{0.84}$, consistent with the results in Table I.

3.2. ICME Speeds

Figure 6 and Table I suggest that average ICME speeds at 0.3–1 AU show little radial variation. Nevertheless, ICMEs evidently do accelerate or decelerate after leaving the Sun, since transit speeds through the inner heliosphere typically differ from, though are correlated with, measured *in-situ* speeds or the speeds of the associated CMEs at the Sun (e.g., Schwenn, 1986; Cliver *et al.*, 1990; Lindsay *et al.*, 1999). For example, Lindsay *et al.* (1999) inferred that *in-situ* speeds at ~ 0.7–1 AU are related to the CME speed ($V_{\rm CME}$) by $V = 360 + 0.25 V_{\rm CME}$ and tend to converge to the speed of the ambient solar wind. Gopalswamy *et al.* (2000) summarized the relationship between *in-situ* and CME speeds in terms of a constant ICME acceleration (a in m/s^2) during transit to 1 AU, given by $a = 1.41 - 0.0035 V_{\rm CME}$. This implies that CMEs with speeds > (<) 405 km/s decelerate (accelerate) en route to 1 AU. (See Section 3.1 in Forbes *et al.* (2006, this volume) for a theoretical discussion of ICME speeds.)

Figure 7 shows observed travel times to 1 AU (defined by the arrival of the ICME-associated interplanetary shock) for 75 events examined by Schwenn *et al.* (2005). The solid line shows a fit to the transit times given by $T_{tr} = 203 - 20.77 \ln(V_{\rm CME})$. For comparison, the dashed line assumes propagation at a constant speed $V_{\rm CME}$. Again, the tendency for fast CMEs to decelerate, and slow CMEs to accelerate, is evident. The transit times of individual shocks, however, do show considerable scatter about the fitted line. Schwenn *et al.* (2005) discuss several factors that influence transit times. For example, CME expansion speeds measured against the plane of the sky do not necessarily correspond to speeds along the Sun-Earth line, though they are evidently correlated to some extent. In particular, Dal Lago *et al.* (2003) and Schwenn *et al.* (2005) estimate that the radial expansion speeds of CMEs are typically ~88% of the lateral expansion speeds. Similarly, Gopalswamy *et al.* (2001a) concluded that the plane of the sky halo CME speed "seems a reasonable representation of the CME initial speed." Travel times also depend on whether the arrival time of the shock (Schwenn *et al.*, 2005), ICME material (Gopalswamy *et al.*, 2000, 2001), or first ICME-related disturbance (Cane and Richardson, 2003a) is considered. In the latter case, Cane and Richardson (2003a) estimate that mean 1 AU transit speeds range from $V_T \sim 0.4 V_{\rm CME}$ up to $V_T \sim 400 + 0.8 V_{\rm CME}$, implying transit times of ~1.1 to 2.9 days for a 1500 km/s CME. For comparison, the constant speed assumption gives 1.16 days and the Gopalswamy *et al.* (2000) model

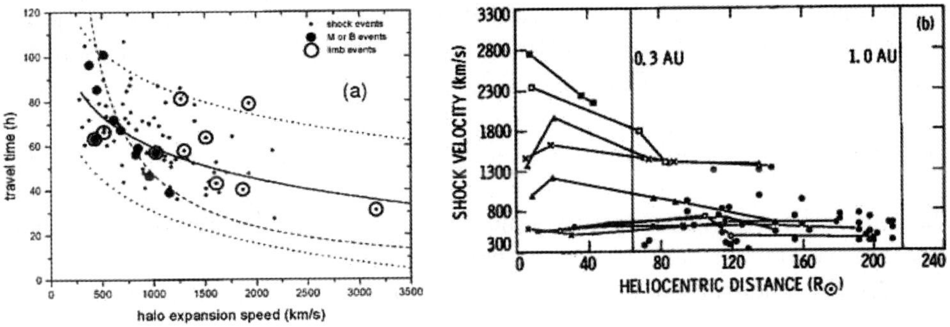

Figure 7. (a) Observed ICME-driven shock transit times vs. LASCO CME expansion speeds (Schwenn *et al.*, 2005) compared with a fit to mean transit times (solid line; dotted lines indicate 2 standard deviations from fit) and assuming constant speed (dashed line). Large solid dots indicate ICME transit times for events without upstream shocks that are not included in the fit. (b) Shock speeds vs. heliocentric distance measured in situ at Helios 1/2 (•) and by radio Doppler scintillation observations (lines; heavy lines = associated with strong flares) (Woo, 1988).

~ 1.4 days for the time of ICME arrival. Interestingly, the Schwenn *et al.* (2005) formula gives a longer (2.1 day) mean shock transit time. Apparently this is because it is better constrained by observations of high-speed events than the Gopalswamy *et al.* (2000) model, which was only based on CMEs with speeds below 1100 km/s and converges to the constant speed assumption at high speeds. For ICMEs associated with geomagnetic storms, Zhang *et al.* (2003) estimate $T_{tr} = 96 - V_{\rm CME}/21$ hours, though this formula evidently cannot be applied to exceptionally fast CMEs, since $T_{tr} \rightarrow 0$ as $V_{\rm CME} \rightarrow 2016$ km/s. In summary, it is arguable whether methods based on halo CME speeds can give reasonably reliable forecasts of ICME or shock arrival times for space weather purposes, given the large scatter in transit speeds for similar CME speeds (e.g., Cane and Richardson, 2003b; Schwenn *et al.*, 2005).

3.3. Shocks

Several types of observations provide information on the propagation of ICME-driven shocks in the inner heliosphere. The tracking of shocks via type II radio emissions is discussed in Section 5. Shock speeds have been measured in situ by the Helios spacecraft at 0.3 to 1 AU (e.g., Sheeley *et al.*, 1985). Within the unexplored region between the Sun and the orbits of the Helios spacecraft, shock speeds may be inferred from coronagraph CME observations and by Doppler scintillation measurements of spacecraft radio signals along sight lines passing close to the Sun. Combining the latter measurements with Helios observations (e.g., Woo, 1988; Woo and Schwenn, 1991) suggests that shocks generally decelerate before reaching Helios, as shown in Figure 7(b). In particular, shocks associated with strong flares (heavy curves) show high initial speeds and rapid deceleration near the Sun. Using CME and Helios observations, Cane *et al.* (1986) concluded that relatively slower shocks result from CMEs associated with filament eruptions accompanied by weak X-ray and microwave bursts. Typically, slower shocks near the Sun show less variation in speed with heliocentric distance.

3.4. Waves, Turbulence and Discontinuities Within ICMEs

The solar wind is pervaded by waves and turbulence on a wide range of scales (see, for example, Marsch (1991) for a comprehensive review). An active turbulent cascade transfers energy from large-scale waves (time scales of hours in the spacecraft frame) which originate in the corona and dissipates it at kinetic scales by heating the plasma. The waves and turbulence can be detected as broadband fluctuations in the magnetic field and plasma velocity and density. The turbulent cascade is less developed in high-speed wind from coronal holes than in slow wind, and power levels tend to be higher in fast wind. These differences probably result from the different conditions in the corona where fast and slow wind streams originate. There are also large numbers of sharp changes in magnetic field direction (discontinuities), many of which are tangential (Knetter *et al.*, 2004), having no magnetic field threading through the plane of the structure.

It has long been known that the amplitude of fluctuations within ICMEs tends to be considerably lower than in the ambient solar wind (Zurbuchen and Richardson, 2006, this volume). However, there have been few detailed studies of the nature of fluctuations within ICMEs. Ruzmaikin *et al.* (1997) compared the spectral index of fluctuations in fast solar wind, slow wind and within ICMEs. They found that fluctuations in ICMEs were statistically similar to those in slow wind, but not to those in fast wind. They argued that the nature of the fluctuations depends on conditions in the corona where the solar wind originates and therefore that solar wind in slow streams and ICMEs probably originates in similar, most likely closed magnetic field regions in the corona.

Leamon *et al.* (1998) considered the "geometry" of the fluctuations – that is, whether turbulent energy was largely in wave vectors parallel or perpendicular to the magnetic field direction – within and around a magnetic cloud, in both the inertial range of the turbulent cascade and the dissipation scales. While the results were complicated and rather difficult to interpret, they, like those of Ruzmaikin *et al.* (1997), are consistent with turbulence in ICMEs being more like the dynamically old fluctuations in slow wind than fluctuations in fast wind.

In contrast to the low level of magnetic field and plasma variability within magnetic clouds, the sheath of solar wind plasma upstream of fast ICMEs tends to contain large-amplitude fluctuations as a result of being compressed and shocked by the ICME. These enhanced fluctuations may produce large negative values of the GSM Z component of the magnetic field, which, when combined with the increased density in the compressed sheath, can make these regions highly geoeffective (Crooker, 2000; see also Figure 1 of Zurbuchen and Richardson, 2006, this volume).

Vasquez *et al.* (2001) performed a detailed study of discontinuities within an ICME which contained multiple ejecta and showed that there were many tangential discontinuities within the boundaries between the ejecta. Such discontinuities are interesting not only in terms of the large-scale structure of ICMEs – since they form topological boundaries between different plasma regions – but also because energetic particle diffusion coefficients are much lower in their presence. Discontinuities are also important in the upstream sheath region: Intriligator *et al.* (2001) argued that discontinuities dramatically reduced particle diffusion across corotating interaction regions. A similar effect should be present within ICME sheaths, perhaps helping to explain the effectiveness of the sheath region in cosmic ray Forbush decreases (e.g., Burlaga, 1991).

4. ICME Dynamics and Interactions

I. G. RICHARDSON, N. U. CROOKER, D. ODSTRCIL AND J. M. SCHMIDT

4.1. ICME-STREAM AND ICME-ICME INTERACTIONS

The ambient solar wind through which ICMEs propagate is divided into intervals of slow ($\lesssim$400 km/s) solar wind and faster flows emerging from coronal holes which produce a quasi-stationary pattern that corotates with the Sun. Regions of compressed plasma (corotating interaction regions, CIRs) form at the leading edges of high-speed streams as they collide with the preceding slower solar wind (for a comprehensive review see the ISSI volume by Balogh *et al.*, 1999). One effect of high-speed streams on ICMEs is that, since ICME speeds tend to converge to the ambient solar wind speed, ICME travel speeds may be higher when propagating through high-speed streams. However, ICMEs are only infrequently embedded in high-speed flows, presumably because ICME sources rarely if ever lie in the weak field regions underlying coronal holes. For example, only $\sim$8% of the ICMEs identified by Cane and Richardson (2003a) at 1 AU arrived during passage of a high-speed stream. A few examples have been observed by Ulysses within high-speed flows above polar coronal holes (e.g., Gosling *et al.*, 1998) although these may have expanded into the fast wind from adjacent high-field sources rather than directly from coronal hole sources (e.g., Hammond *et al.*, 1995). More frequently, ICMEs are found near high-speed stream leading edges or in slow, interstream solar wind. These typically overlie the solar active regions and meandering solar neutral line from which CMEs arise (e.g., Crooker and Cliver, 1994). A particularly interesting situation occurs when the ICME forms the slower-speed plasma immediately preceding the fast stream. Plasma within the trailing edge of the ICME may be compressed in this situation, and, if the embedded magnetic field is directed southward, enhancement of the southward field by compression may lead to stronger geomagnetic effects than would have occurred in the absence of the interaction with the

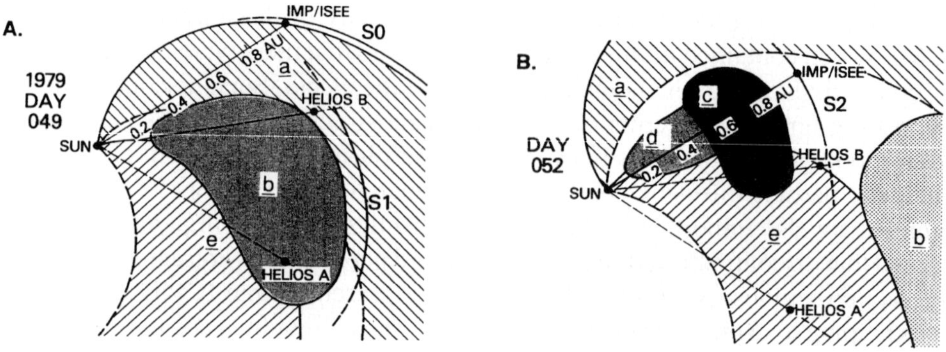

Figure 8. Sketch of an evolving complex stream in February, 1979 (Behannon *et al.*, 1991).

fast stream (e.g., Zhao, 1992; Cane and Richardson, 1997; Fenrich and Luhmann, 1998; Crooker, 2000).

Interactions between more than one ICME may occur, in particular at times of elevated solar activity levels. Figure 2(c) of Zurbuchen and Richardson (2006, this volume) illustrates two aspects of such interactions. First, the ICME shown apparently consists of two components, which suggests that the ICME may have been formed by the interaction of two individual ICMEs. Second, the associated shock is traveling through plasma associated with a preceding ICME, as indicated, for example by depressed proton temperatures and enhanced Fe charge states, rather than ambient solar wind. For further discussion of ICME-ICME interactions and boundaries within ICMEs, see (Wimmer-Schweingruber *et al.*, 2006, this volume).

Burlaga (1975) pointed out that many high-speed flows are contiguous but composed of multiple components, either streams or ICMEs, applying the term "compound streams" to such regions. Figure 8 shows a sketch of a complex, evolving compound stream inferred from observations by the Helios and near-Earth spacecraft and interplanetary scintillation data (Behannon *et al.*, 1991). Structures "a" and "e" are corotating high-speed streams, while "b" to "d" are transient structures.

4.2. Interplanetary Reconnection

Field lines on the surface of a CME with the structure of a magnetic flux tube propagating within the radial magnetic field near the Sun can reconnect with the external magnetic field if oppositely-directed magnetic field lines are squeezed together due to the flux tube motion (e.g., McComas *et al.*, 1994; Moldwin *et al.*, 1995; Rogers *et al.*, 2000). Figure 9 (Schmidt, 2000; Schmidt and Cargill, 2003) shows a flux tube with field lines rotating in an anticlockwise sense, projected onto the plane in which the x-axis is in the ecliptic plane and the z-axis is parallel to the solar rotation axis. An outward-directed, distorted, radial solar magnetic field is assumed in the northern hemisphere and an inward-directed field in the southern hemisphere, separated by a current sheet in the ecliptic plane. Due to the sense

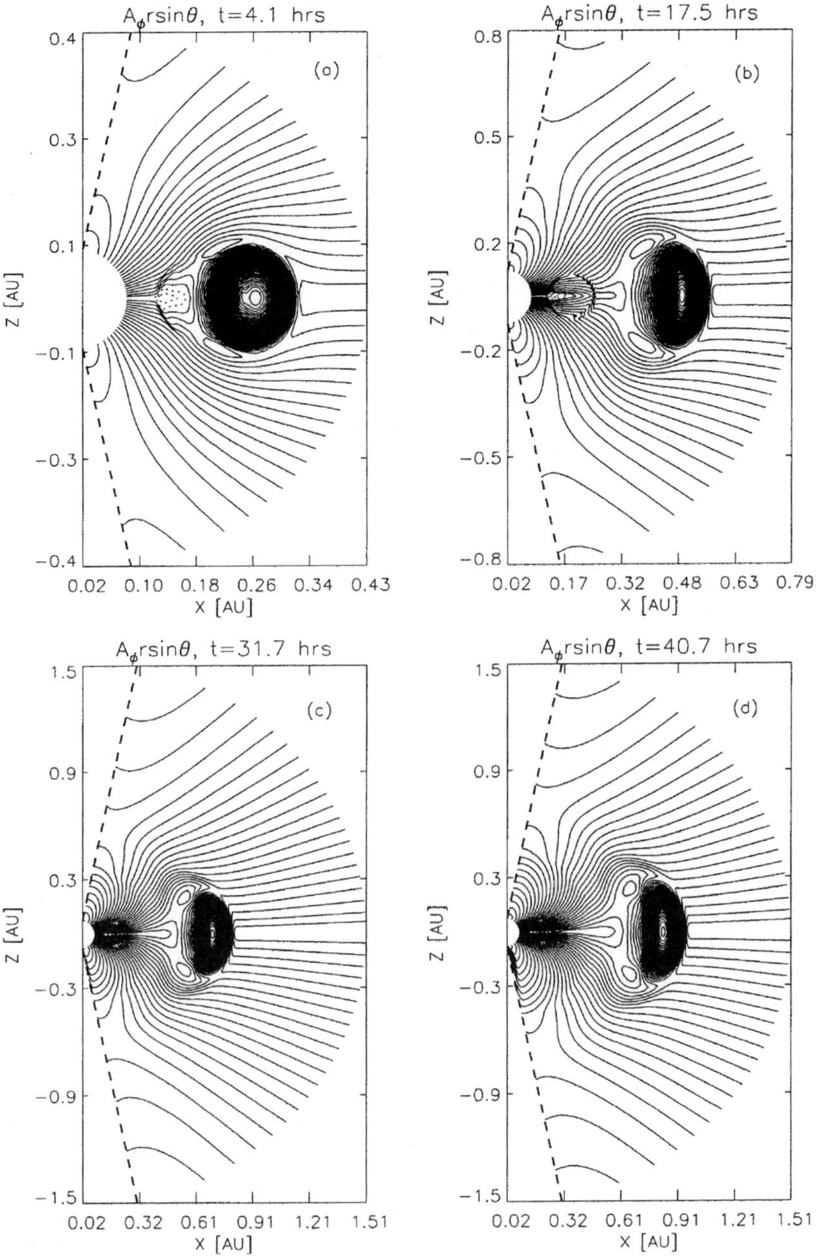

Figure 9. A magnetic flux tube propagating initially at 1.5 times the solar wind speed in a current sheet. Reconnection regions develop that are symmetric with respect to the ecliptic plane (Schmidt, 2000; Schmidt and Cargill, 2003).

of rotation of the flux tube magnetic field, reconnection occurs at the northern and southern leading edges of the tube symmetrically about the ecliptic plane. When the flux-tube reaches about 1 AU (after 40.7 hours), reconnection has reduced its size by almost one third. An equivalent fraction of the internal material, when released from the flux tube, contributes to local heating and momentum transfer to the solar wind plasma. Reconnection becomes more effective as the flux tube speed and field strength increase relative to their values in the solar wind. When the magnetic field of the flux tube is as weak as the external field, the tube completely loses its magnetic form by the time it reaches Earth's orbit.

4.3. Modeling ICME Propagation in a Structured Medium

Numerical simulations provide an important tool for understanding ICME interactions with the ambient medium. Two and a half-D MHD simulations have shown significant distortions of the shock front when propagating along the heliospheric current sheet (Odstrcil *et al.*, 1996; Hu, 1998). Simulations with a 2-D hydrodynamic model have demonstrated that the parts of a single ICME straddling both high- and low-speed flows would evolve radically differently in the two regions (Riley *et al.*, 1997). The 3-D interactions between transient and corotating structures produce a rich set of dynamic phenomena. A single interplanetary disturbance can have radically different appearances at various locations. The appearance also depends critically upon the CME launch location with respect to the streamer belt flow (Odstrcil and Pizzo, 1999a,b).

An example of a model CME flux tube propagating through a structured solar wind is shown in Figure 10 (Schmidt and Cargill, 2001). The flux tube is initially in slow, low-latitude solar wind with speeds of about 300 km/s. It is then given an initial negative meridional velocity that drives it into higher-latitude, high-speed (600 km/s) solar wind (the division between the low- and high-speed wind is at 45° latitude). The first two panels of Figure 10 show that the penetration of the

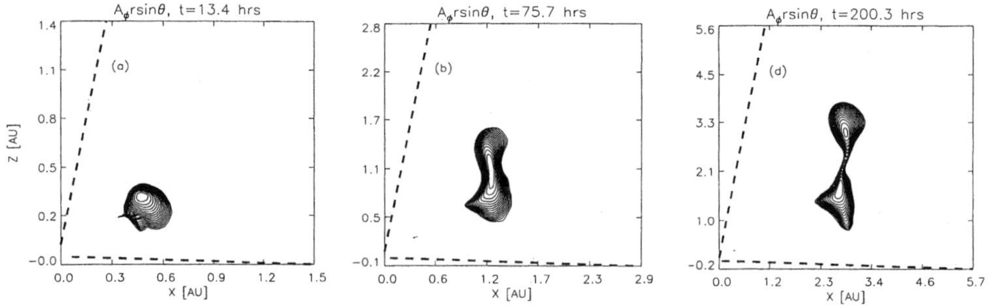

Figure 10. The evolution of a magnetic flux tube initially located in slow, low-latitude, solar wind that is given an initial meridional velocity (−300 km/s) that takes it into the region of high-latitude, high-speed wind (Schmidt and Cargill, 2001). The tube is projected onto the plane defined by the in-ecliptic x-axis and rotation axis of the Sun z.

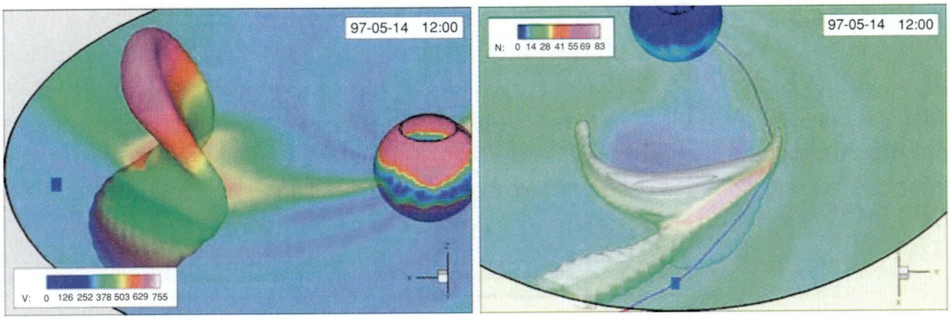

Figure 11. Numerical simulation of ICME associated with the 12 May 1997 solar event (adapted from Odstrcil *et al.*, 2004).

flux tube into the high-speed wind is very slow. This is because an additional $\mathbf{j} \times \mathbf{B}$ force acts on the tube when it enters the high-speed wind, opposing its motion. As soon as the tube crosses the boundary, however, a portion begins to pull away in the high-speed wind, although it never completely separates owing to magnetic tension forces along a corridor of linking field lines. In addition, the flux tube stretches meridionally because pressure gradients are smaller in the meridional than in the radial direction.

Odstrcil *et al.* (2004) found that even relatively small-scale structures in the background solar wind may play an important role in the interplanetary evolution of transient disturbances. Figure 11 shows a numerical simulation of the ICME associated with the 12 May 1997 solar event injected into a structured solar wind. In the left-hand panel the colour scale indicates flow speed, the translucent plane represents the equatorial plane, and the injected ICME is indicated by the iso-surface at 6 cm^{-3}. The position of Earth is shown by the blue box. In the right-hand panel, the color scale indicates plasma density. The density structure defined by a white iso-surface at 30 cm^{-3} outlines both the compressed ICME structure and a corotating interaction region (CIR) threaded by a magnetic field line (blue). The results show that: (a) injected material undergoes substantial latitudinal distortion caused by the large-scale, bi-modal velocity structure of the background solar wind; and (b) an interplanetary shock formed in the slow streamer belt is modified when it merges with the CIR caused by fast flow from an equatorward extension of the southern coronal hole. These effects can be observed by *in-situ* and remote white-light observations (see Section 6, Figures 13 and 14).

5. Solar Cycle Variations

I. G. RICHARDSON

The occurrence rate of ICMEs essentially follows the ~11-year solar activity cycle (e.g., Lindsay *et al.*, 1994; Cane and Richardson, 2003a). This was evident

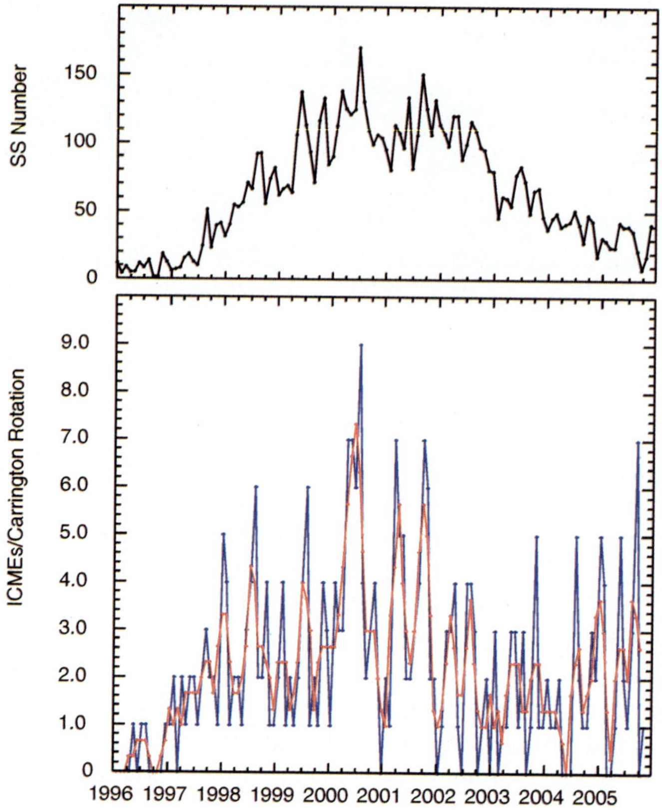

Figure 12. Monthly sunspot number in 1996–2005 (upper panel) together with the ICME rate/Carrington rotation (plus 3-rotation running means) at 1 AU, updated from Cane and Richardson (2003a).

indirectly even before the space era, for example, from the rate of "sporadic" geomagnetic storms, now known to be predominantly associated with ICMEs and their related shocks (e.g., Gosling *et al.*, 1991; Cliver, 1995; Richardson *et al.*, 2001 and references therein). Several studies have tracked the occurrence rate of various ICME signatures throughout the solar cycle. For example, Gosling *et al.* (1992) noted that bi-directional suprathermal electron strahls were observed in $\sim$1% of the near-Earth solar wind at solar minimum, rising to $\sim$15% around solar maximum, when $\sim$4 events/month were identified (Gosling, 1990). Figure 12 shows the ICME rate/solar rotation (plus 3-rotation running averages) in the near-Earth solar wind estimated by Cane and Richardson (2003a) since 1996 and updated/revised to the end of 2005. The ICMEs are identified principally from examination of the solar wind plasma and magnetic field observations. Note that the rate increased by an order of magnitude from $\sim$0.3 ICMEs/rotation in 1996 to $\sim$3/rotation in 1998–2002, though with several brief intervals with higher rates during this time that are associated with periods of exceptionally high solar activity. Interestingly, these variations are quasi-periodic with a dominant period of ∼166 days (Cane and Richardson, 2003a; Richardson and Cane, 2005), similar to the "154-day" periodicity intermittently present in various solar and interplanetary phenomena during cycle 23 and earlier solar cycles (e.g., Cane *et al.*, 1998; Dalla *et al.*, 2001, and references therein). The sunspot number is also shown, illustrating that there is no simple correlation between solar activity levels and the ICME rate, such as that between solar activity and the CME rate reported by Webb and Howard (1994) (see, also, Schwenn *et al.*, 2006, this volume). In particular, there was a temporary decline in the ICME rate in 1999, yet solar activity continued to increase. Furthermore, the ICME rate has remained relatively high in 2004–2005 (with occasional brief intervals of higher rates due to bursts of enhanced solar activity) despite the substantial decline in sunspot number (Richardson and Cane, 2005).

Given that an ICME takes ∼1 day to pass a spacecraft, it is clear from the occurrence rates in Figure 12 that even at solar maximum, ICMEs generally do not dominate the near-Earth solar wind, except possibly during brief periods of exceptionally high solar activity. For example, on average, the Cane and Richardson (2003a) ICMEs were observed for ∼1% of the time during 1996, increasing to ∼16% in 2000–2001 (cf. Gosling *et al.*, 1992). Typically at solar maximum, the near-ecliptic solar wind at ∼1 AU is approximately equally divided between (a) ICMEs and the associated post-shock flows; (b) fast streams from coronal holes, and (c) slow, interstream solar wind (Richardson *et al.*, 2002). Furthermore, although ICMEs tend to have stronger than average magnetic field strengths, ICMEs typically do not dominate mean interplanetary magnetic field strengths at solar maximum. In particular, fields carried by ICMEs do not appear to be responsible for the increase in mean IMF strength as activity levels increase because this increase is present in the ambient solar wind, outside of ICMEs (Richardson *et al.*, 2002). Smith and Phillips (1997) estimate that removing ICMEs would decrease the average IMF by only ∼8%. On the other hand, Crooker *et al.* (2004) suggest that the "legs" of ICMEs whose leading edges have passed far out into the heliosphere may contribute to the increase in mean IMF strength and may be difficult to distinguish from ambient solar wind.

During brief intervals of unusually high solar activity, it is possible that the ICMEs interact with themselves and other solar wind structures, such as CIRs, to form so-called "global merged interaction regions" (GMIR)s. These shell-like structures with intense magnetic fields that encircle the Sun and extend to fairly high latitudes have been suggested as the cause of step-like decreases in the long-term (11-year) modulation of galactic cosmic rays (Burlaga, 1995, and references therein). See Gazis *et al.* (2006, this volume) for further discussion of GMIRs. In addition, Cliver and Ling (2001) and Cliver *et al.* (2003) have argued that ICMEs, in particular the strong magnetic fields in the "tail" of the field distribution that are predominantly associated with ICMEs, drive long-term modulation. However, the role of ICMEs in long-term modulation (cf. Gazis *et al.*, 2006, this volume) is still a topic of debate – see, e.g., Wibberenz *et al.* (2002) for an alternative viewpoint.

6. Comparison of Observations and Models

D. ODSTRCIL, C. CID, J. A. LINKER AND J. M. SCHMIDT

Efforts to date have been devoted mainly toward increasing the sophistication of numerical models and to improving understanding of ambient and transient disturbances. Relatively few papers have been published on the numerical simulation of observed events. This is due mostly to the inherent difficulties of such a task and by the lack of reliable observational data to initialize the numerical models. Nevertheless, such activities are vital for supporting the analysis of various *in-situ* and remote observations and for the development of space weather forecasting capabilities. This section describes recent achievements in numerical modeling of specific heliospheric events.

6.1. MODELING A CME FROM SUN TO EARTH

The simplest models are empirical, with the aim being to predict the arrival of CMEs at Earth (Gopalswamy *et al.*, 2001a; Schwenn *et al.*, 2005). More sophisticated numerical MHD models have the potential to predict the solar wind density, mean temperature, and components of the flow velocity and magnetic field. A successful match with observations requires not only an adequate physical model and numerical resolution but also reliable observations to drive the computations. Note that both ambient solar wind and transient disturbances have to be well-replicated in the modeling process, since their 3-D interactions can significantly modify their structure en route to Earth.

Significant progress has been made in simulating the ambient solar wind (Linker *et al.*, 1999; Usmanov *et al.*, 2000; Riley *et al.*, 2001; Arge *et al.*, 2002; Hayashi *et al.*, 2003; Roussev *et al.*, 2003). The simulation of 3-D transient disturbances is less mature, and various models have been used to emulate the CME launch and propagation (Detman *et al.*, 1991; Odstrcil and Pizzo, 1999a; Groth *et al.*, 2000; Vandas *et al.*, 2002; Odstrcil *et al.*, 2004; Manchester *et al.*, 2004a; Roussev *et al.*, 2004). This section reviews three case studies.

12 MAY 1997 EVENT

The 12 May 1997 halo-CME event has been chosen for detailed studies by the scientific community (`http://solarmuri.ssl.berkeley.edu`, `http://www.bu.edu/cism`, `http://www.shineorg.org`). Solar and heliospheric background conditions at that time were relatively simple, thereby facilitating analysis and modeling. However, because the photospheric vector magnetograms for that period are unfortunately of low quality, self-consistent "data-driven" simulation of the solar eruption has proven to be unusually challenging (Z. Mikic, private communication). As a consequence, "data-inspired" simulations have been

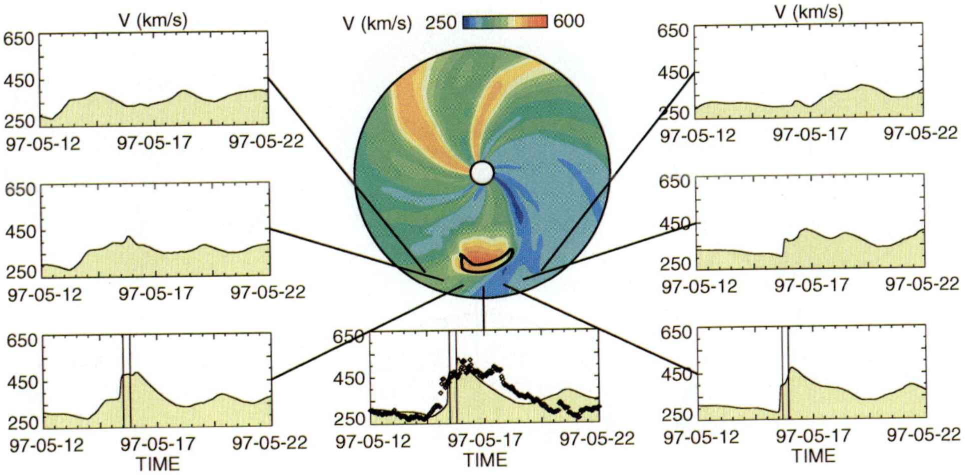

Figure 13. Simulated multi-point *in-situ* observations of a transient disturbance at different positions in the equatorial plane at 1 AU (Odstrcil *et al.*, 2005). The central top image shows the solar wind radial velocity in the equatorial plane on 15 May 1997 at 00:00 UT. The seven plots show the temporal evolution of the solar wind radial velocity at the different observing positions as indicated by solid lines. The vertical lines indicate the extent of the simulated ICME material, and black dots show Wind spacecraft observations at Earth.

realized (Odstrcil *et al.*, 2004, 2005) to analyze this event, to provide global context, and to set a benchmark for further modeling. In these 3-D MHD simulations, the background solar wind was determined from the SAIC (Riley *et al.*, 2001) or WSA (Arge *et al.*, 2002) coronal model, and the transient disturbance was determined from the cone model (Zhao *et al.*, 2002).

Numerical results show that the 12 May 1997 ICME interacts with the leading edge of a fast stream (see Figure 11 in Section 4.3). This results in a substantial latitudinal distortion of the injected material, a strong density compression within the heliospheric streamer belt, merging of an interplanetary shock with the CIR, and modification of the magnetic connectivity. Figure 13 shows the temporal evolution of the solar wind velocity at various positions at 1 AU. Comparison with Wind observations (plot in the middle of the bottom row) shows that it is becoming feasible to reproduce the parameters of the ambient solar wind and to estimate the arrival of interplanetary shocks and coronal ejecta. The shock stand-off distance from the driving ejecta and the shock front inclination are difficult to match because even relatively small-scale solar wind structures can significantly affect the appearance of transient disturbances. Additional work is necessary to specify smaller-scale structures, more accurate locations, and the temporal evolution of the streamer boundaries in the corona (Odstrcil *et al.*, 2005). Figure 14 shows synthetic images of the white light scattered by the solar wind density structures. Such images, which show a large latitudinal distortion of the ICME with localized bright spots at the

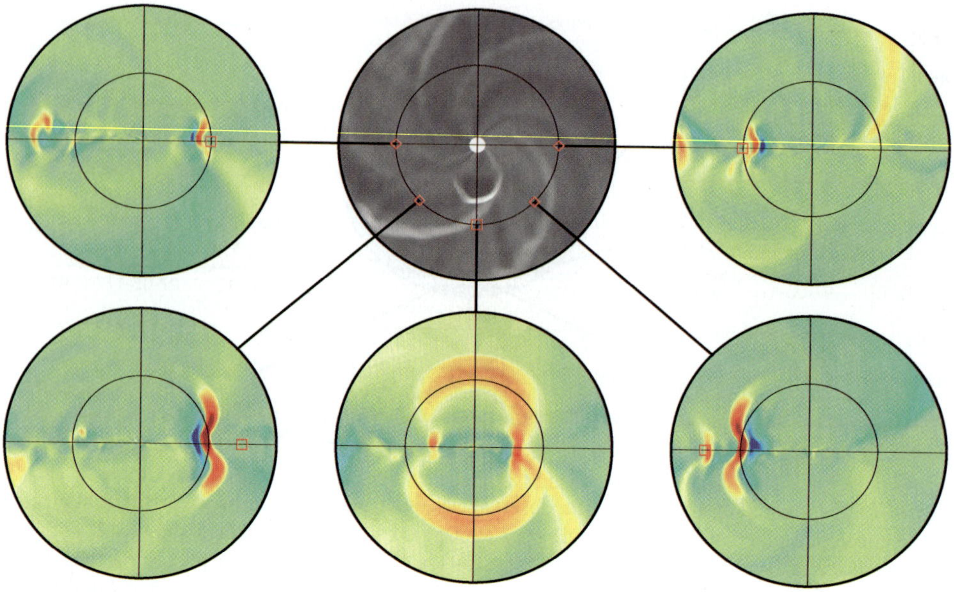

Figure 14. Simulated multi-perspective remote white-light observations of a transient disturbance from various positions in the equatorial plane at 1 AU. The central top image shows the distribution of the solar wind density scaled by $(R_A/r)^2$ in the equatorial plane on 14 May 1997 at 12:00 UT. The remaining five images show synthetic difference images of the total brightness (generated at 12:00 and 06:00 UT on May 14) as viewed from the respective observing positions indicated by solid lines. The Earth position is shown by a red square.

slow streamer, may be compared with SMEI and upcoming STEREO observations from missions with global imaging capabilities.

1– 2 May 1998 Events

The 1–2 May 1998 CME events have also been chosen for detailed studies by the scientific community (`http://solarmuri.ssl.berkeley.edu`, `http://csem.engin.umich.edu`, `http://www.shineorg.org`). These events are more complex than the 12 May 1997 event; however, the availability of high-quality vector magnetograms favors the initialization of simulations ab-initio. The University of Michigan team has recently simulated these events by the BATS-R-US code with two different initiation models. Both models start by deriving an ambient state of the global corona and solar wind from synoptic magnetograms observed by the Wilcox Solar Observatory.

In the model of the 1 May 1998 CME (Manchester *et al.*, 2004b), a Gibson-Low magnetic flux rope (Gibson and Low, 1998) is placed in the helmet streamer of the pre-event active region. Initially the flux rope is in a state of force imbalance and

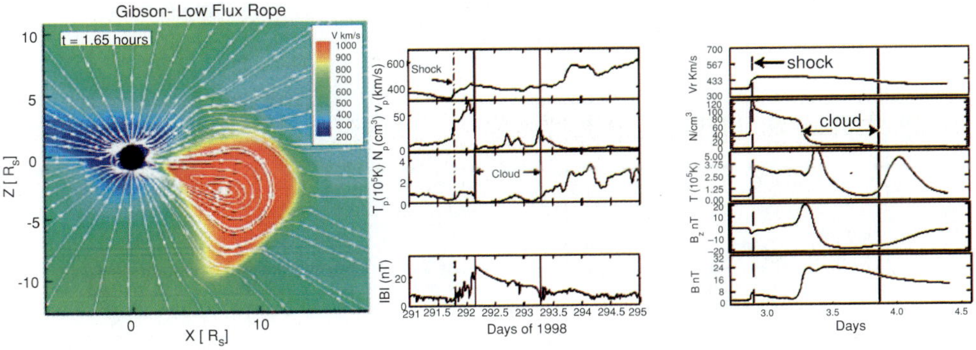

Figure 15. Simulated CME in the meridional plane (*left*) and temporal evolution of plasma parameters at Earth (*right*). The CME is shown 1.65 hours after initiation; the color scale represents the flow velocity, and white lines show the projected magnetic field lines. Observed (*left*) and simulated (*right*) temporal profiles show, from top to bottom, the radial velocity, proton number density, temperature, magnetic field southward component B_z and magnitude (adapted from Manchester *et al.*, 2004b).

expands at a rate approximating the observed speed. The bow shock ahead of the flux rope as well as the current sheet behind it are well-resolved. Figure 15 shows the magnetic field disturbed by the ICME and a comparison of plasma parameters at Earth with observations.

In the model of the 2 May 1998 CME (Roussev *et al.*, 2004), the solar eruption is initiated by slowly evolving the boundary condition for the horizontal magnetic field at the Sun until a critical point is reached where the configuration loses equilibrium. At this point the field erupts, and a flux rope is ejected with a maximum speed in excess of 1000 km/s. A shock forms in front of the flux rope, and it reached a fast-mode Mach number in excess of 4 at $5R_S$. Diffusive-shock-acceleration theory predicts a distribution of solar energetic protons with a cut-off energy of about 10 GeV (Roussev *et al.*, 2004; Forbes *et al.*, 2006, this volume).

6.2. Multiple Events

On 10 June 2000 between 16:30 and 19:30 UT, the Wind/WAVES instrument detected an extremely narrow-band radio type-II burst which was flanked by intense radio type-III bursts (Gopalswamy *et al.*, 2001b). That event was associated with the collision of a slow, dense CME with a fast, less dense CME approaching from behind, as can be seen in the LASCO coronagraph images (inverted, see caption) in Figure 16.

Although not all CME collision events are so intense that they can give rise to a radio signal burst (e.g., Richardson *et al.*, 2003), it has been argued that there is a strong correlation with such events (e.g., Gopalswamy *et al.*, 2001b, 2002). The top panels of Figure 16 show results from a simulation of the observed collision event. The simulation box, indicated by the thick dashed lines, comprises the field of view of the C2 and C3 LASCO coronagraphs, and the geometrical dimensions

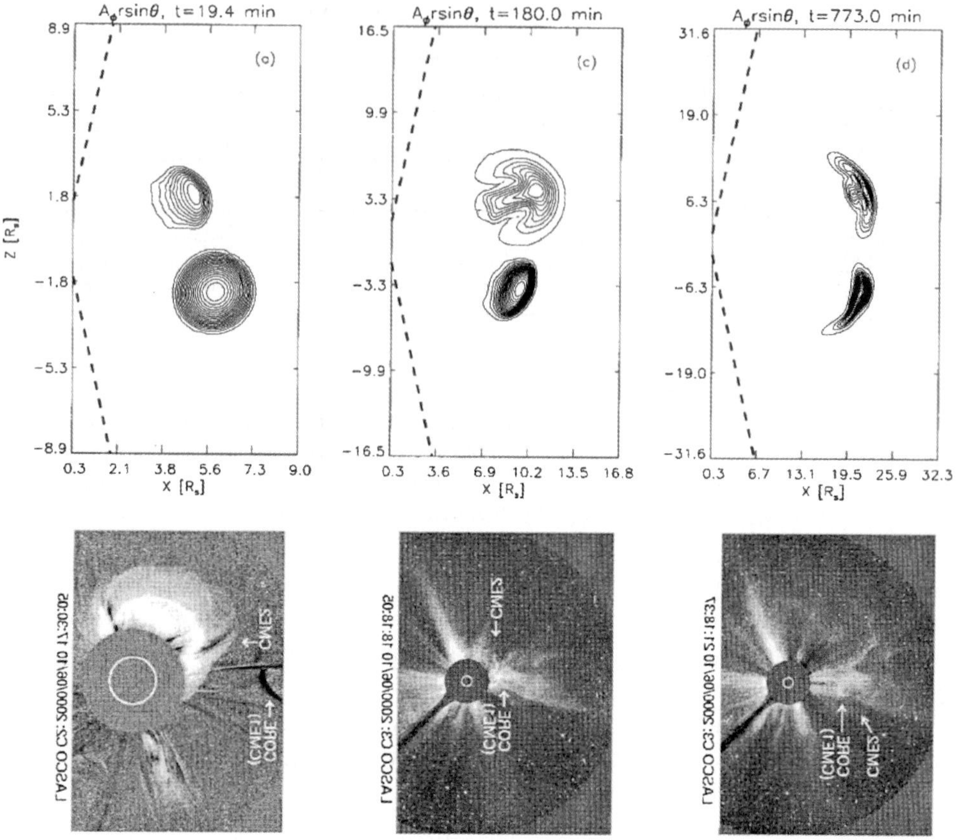

Figure 16. Contour plots (top panels) of the vector potential at three times during the evolution of a shock interaction between two colliding CMEs which are separated by a meridional angle of 40°. The plots model the 10 June 2000 events in the LASCO images (bottom panels), which have been inverted to match. They show that the fast CME2 catches up with the slow CME1 (first panel), that CME1 is deflected and flattened by the shock driven by CME2 (second panel), and that CME2 and CME1 then propagate outward together (third panel) (adapted from Schmidt and Cargill, 2004; Gopalswamy *et al.*, 2001b).

and velocities of the CMEs are taken from the coronagraph observations. We see that the fast, less dense (upper) CME overtakes the slow, dense (lower) CME and that the slow CME is deflected into the southern hemisphere due to an interaction with the forward shock created by the fast CME. The shock hits the slow CME along its northern edge. This impact flattens the slow CME at that location, and the shock, when it penetrates into the denser material of the slow CME, steepens significantly due to the reduced Alfvèn speed there. We find that the steepened shock persists for a long time. This circumstance favors strong acceleration of particles, provided there is an adequate population of seed particles in the ambient plasma.

7. Remote Sensing of ICMEs

7.1. Remote Radio Sensing and Tracking of ICMEs

M. J. Reiner

Type II radio emissions are remote signatures of coronal and interplanetary shocks (see Pick *et al.*, 2006, this volume) that are generated at the plasma frequency and its harmonic, implying that the observed frequency is directly related to the plasma density in the source region ($f_p(\mathrm{kHz}) = 9\sqrt{n_p(\mathrm{cm}^{-3})}$). It was noted that coronal (metric) type II radio bursts, generated by coronal shocks, were often associated with the occurrence of sudden commencement geomagnetic storms on Earth. This led to speculation that these coronal shocks likely extended into interplanetary space, where, due to the falloff of the plasma density with distance, they might generate radio emissions at low frequencies, below that of Earth's ionospheric cutoff at about 20 MHz. The first clear detection of a low-frequency type II burst, at kilometric wavelengths, was made by the spaceborne radio receivers on the IMP-6 spacecraft (Malitson *et al.*, 1973).

Subsequently, with better radio instrumentation on the ISEE-3 spacecraft, Cane *et al.* (1982) and Lengyel-Frey *et al.* (1989) studied the observational characteristics of these kilometric type II emissions. The example shown in Figure 17 is characterized by broad diffuse emissions that drift in frequency from ~1 MHz at 07:00 UT to 200 kHz by 10:00 UT. The smooth diffuse nature of this radiation suggests that it was generated at the harmonic of the plasma frequency (Lengyel-Frey *et al.*, 1985).

These type II radio emissions are generated by electrons accelerated at shocks. The observation of the fundamental/harmonic structure of these emissions confirmed their generation by the plasma emission mechanism (Lengyel-Frey *et al.*, 1985). While the precise physical origin of the coronal shocks that generate the metric type II bursts is not yet firmly established (see Pick *et al.*, 2006, this volume), the kilometric type II bursts observed in the interplanetary medium are unambiguously associated with CME-driven shocks (Cane *et al.*, 1987). *In-situ* observations have established that the type II radiation is generated in the upstream region of the CME-driven shock (Hoang *et al.*, 1992) by processes similar to those that generate radio emissions in Earth's electron foreshock (Bale *et al.*, 1999; Knock *et al.*, 2001).

Kilometric type II radio emissions often extend as low as 20 or 30 kHz, corresponding to plasma densities from 5 to 10 cm^{-3}, which are typical of the densities measured at 1 AU. This suggests that the radio-producing CME-driven shocks extend well into interplanetary space and that the frequency drift rate of the associated type II emissions can therefore be used to track these CME/shocks through the interplanetary medium, beyond the limit of the white-light coronagraph observations (see Schwenn *et al.*, 2006, this volume). To facilitate this interplanetary tracking, Reiner *et al.* (1997, 1998) displayed the radio dynamic spectrum as the inverse of

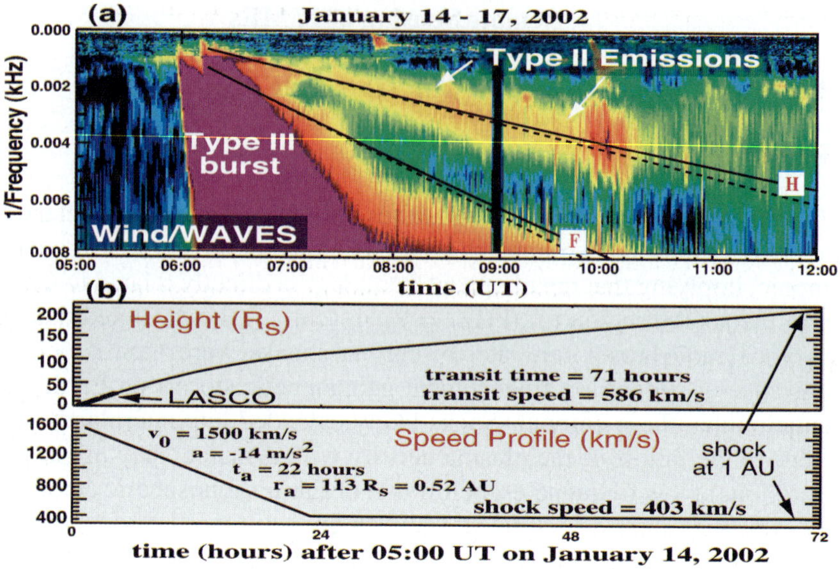

Figure 17. (a) Dynamic radio spectrum (plot of intensity as a function of inverse frequency and time) showing frequency drifting type II radio emissions for an event on January 14, 2002. On this plot a straight (dashed) line corresponds to radio emissions generated by a shock propagating at a constant speed. The solid curves correspond to type II radio emissions generated by a shock that is decelerating at a constant rate of 14 m/s^2. (b) Speed profile and height-time dependence derived from the *in-situ* and radio data for a shock propagating through the interplanetary medium in 71 hours and arriving at 1 AU at a speed of 400 km/s, as described in the text (adapted from Reiner *et al.*, 2003).

the frequency versus time, as in Figure 17. Since the interplanetary density falls off as $1/R^2$ (R is the heliocentric distance), the inverse of the frequency is proportional to R. Thus, in this representation a CME/shock propagating at a constant speed will produce frequency drifting type II emissions that lie along a straight line, originating from the solar liftoff time. Deviations from a straight line may indicate acceleration or deceleration. In Figure 17, the deviation of the frequency drift from the straight dashed line indicates that the associated shock was decelerating through the outer corona.

Deriving the complete speed profile of a CME/shock from the type II radio data is difficult due to the complexities of the type II emissions, to its often sporadic and fragmented nature and to the unknown scale of the interplanetary density falloff in the source region. Reiner *et al.* (2001, 2003) proposed a simple model based on the typical behavior of CME/shocks deduced from the radio and white-light observations. The LASCO CME measurements indicate that when CMEs decelerate they tend to do so at an approximately constant rate. On the other hand, the radio observations indicate that far out in the interplanetary medium the CME-driven shocks propagate at an approximately constant speed (Reiner *et al.*, 1999). Thus, Reiner *et al.* (2001) assumed that, in general, a CME shock in the outer corona will first decelerate at a constant rate, then propagate at a constant speed to 1 AU (see Figure 7(b) and Woo, 1988). In this simple model, a family of speed profiles corresponding to different CME initial speeds can be deduced from the observed transit time of the CME/shock and from the speed of the shock at 1 AU, directly measured from the *in-situ* plasma parameters. Reiner *et al.* (2001) then select out of this family of speed profiles the one that gives the best fit to the frequency drift of the observed radio data. This technique was used to obtain the speed profile shown in Figure 17(b) for the type II event illustrated in Figure 17(a). For this particular type II event it was found that the best fit to the frequency drift of the type II (solid curve in Figure 17(a)) corresponded to the speed profile with an initial CME speed of 1500 km/s (which was essentially the same as the CME plane-of-sky speed). The unique solution suggested by the simultaneous fit to all these data then corresponded to the CME decelerating at a constant rate of 14 m/s^2 for 22 hours (corresponding to a propagation distance of 0.5 AU), then propagating at a constant speed of 400 km/s to 1 AU. The corresponding height-time curve is also shown in Figure 17(b).

Until now, the remote type II radio and the IPS observations described below have provided the only means of tracking the interplanetary transport of CMEs and their shocks. With the recent launch of the 'all sky' camera on SMEI (Eyles *et al.*, 2003), we have for the first time some white-light observations with which to compare these low-frequency and IPS radio observations (see, e.g., Reiner *et al.*, 2005).

7.2. IPS Observations of ICMEs

R. J. Forsyth

As discussed in the previous sections, coronagraphs provide a multitude of observational information on CMEs as they begin their journey out from the Sun, and a wealth of data is also available on ICMEs from 0.3 AU outwards from *in-situ* spacecraft observations. However, if we are trying to follow and understand the evolution of ICMEs as they propagate out from the Sun through the inner heliosphere, there is a key gap in observational information between $\sim 30 R_s$ and 0.3 AU. There has been growing use in recent years of the Interplanetary Scintillation (IPS) technique to provide remote sensing of ICMEs as they traverse this inner region.

IPS employs multiple antennas to measure the scintillation of distant astronomical radio sources, such as quasars (e.g., Hewish *et al.*, 1964), enabling information to be derived about the interplanetary medium through which the radio signals are passing. Two commonly used techniques are applied to study ICMEs. The first uses multiple radio sources to make sky maps of the level of turbulent fluctuations in the solar wind. The turbulence level in the sheath region ahead of ICMEs is found to be higher than the ambient, allowing these regions to be tracked as they propagate through the maps (e.g., Gapper *et al.*, 1982). The second technique applies a cross-correlation analysis to scintillation data of the same source from multiple observing stations to derive the speed of the solar wind in the region through which the signals pass (Armstrong and Coles, 1972). This can be used to remote sense the transient high speed solar wind streams created by ICMEs. The radio frequencies at which the observations are made determines the distance range from the Sun which can be usefully explored. For example, the Nagoya University system (e.g., Tokumara *et al.*, 2005) at 327 MHz samples the distance range from $\sim$0.2–0.9 AU, while the European EISCAT (ionospheric radar) system operating at $\sim$930 MHz can probe as close to the Sun as ~ 18–$30 R_s$ (e.g., Breen *et al.*, 2002). However, the EISCAT system tends to be used for IPS in relatively short campaigns when not in use for ionospheric studies. Thus while it has been successfully used in comparing the speeds of fast and slow solar wind streams close to the Sun with *in-situ* observations (e.g., Breen *et al.*, 2002), it has not yet been seriously used to study an ICME propagating through the inner regions of the heliosphere. The advantage of IPS in being able to sample a wide range of distances in the gap between solar and *in-situ* observations is counterbalanced by difficulties in analysis and interpretation due to the measurement being an integration of the scintillation occurring all the way along the line of sight of the antenna.

Recent IPS investigations of ICMEs have focussed on case studies of individual events, attempting to make the link with one or the other or both of solar observations and *in-situ* observations of the events. Two studies have investigated the IPS signature of the 14 July 2000 CME associated with an X-class solar flare, the so-called "Bastille Day" event, using the Ooty radio telescope in India and the Nagoya system in Japan, both operating at 327 MHz (Manoharan *et al.*, 2001; Tokumara *et al.*, 2003). By tracking the zones of enhanced turbulence across sky maps, Manoharan *et al.* (2001) were able to deduce that the speed of the ICME declined slowly out to $\sim 100 R_s$ and then much more rapidly beyond this distance, suggesting increased interaction with the ambient solar wind. Tokumara *et al.* (2003) were able to model the IPS observations to deduce that the interplanetary disturbance had a toroidal shape. Tokumara *et al.* (2005) identified 10 interplanetary disturbances in Nagoya IPS data during a 19 day period of intense solar activity in October-November 2003. They were able to identify an interplanetary disturbance associated with all shock events observed *in-situ* at 1 AU during this period as well as possible links to solar flare events, although not on a one-to-one basis. For two particular events it was possible to establish the full chain of events through from solar origin to *in-situ* observation. A similar end-to-end to study was made of an April 2000 event by Jadav *et al.* (2005).

8. Conclusions

In this chapter we have reported on our present-day knowledge of ICMEs in the inner heliosphere, relating observations of their origin in the solar corona to *in-situ* spacecraft observations and discussing the evolution that takes place in between.

- Attempts to relate the coronal and *in-situ* observations (Section 2) to date have focussed on the relationship between the three-part CME and the interplanetary magnetic flux rope of the ICME. While it is still not fully clear how the different parts of the CME evolve and what signatures they lead to in the *in-situ* data, certain aspects, such as the handedness and orientation of the coronal flux rope, do appear to survive the interactions and distortions that affect the CMEs as they expand out into the heliosphere.

- Studies of the evolution of ICME parameters (Section 3) confirm that ICMEs expand as they travel out through the heliosphere and can both decelerate and accelerate compared to the speeds of the corresponding CMEs observed by coronagraphs. However, calculations of the travel time from the corona to 1 AU based on these speeds give results with a considerable scatter.
- Some of the more complex *in-situ* observations can be understood by the interaction of ICMEs with both the stream structure of the solar wind and with other ICMEs of differing speeds (Section 4). Numerical simulations have proven a valuable aid to understanding these interactions.
- Solar cycle effects are apparent in ICME occurrence rates (Section 5) and to some extent in magnetic cloud axis orientations (Section 2).
- Progress has been made in the numerical modeling and comparison to observations of selected example ICMEs (Section 6), the primary difficulty being the availability of reliable solar data to initialise the models.
- Type II radio emissions and the interplanetary scintillation technique (Section 7) provide a means of tracking ICMEs and their associated shock waves through the region of space between the corona and 1 AU.

Despite the material presented in this chapter, there still remain many open questions on the detailed correspondence between the features and phenomena revealed in the solar and in situ observations. Some of these will be explored in the future by missions which approach closer to the Sun. These will certainly shed new light on the evolution of (I)CMEs. In the nearer future, simultaneous *in-situ* and coronagraph observations in high temporal and spatial resolution will allow detailed comparisons of the *in-situ* and remote sensing observations. The cadence of the instruments will also allow better study of the evolution of CMEs in the low corona. Observations in this region are also important for the reliable interpretation of *in-situ* characteristics subsequently measured in the interplanetary medium. Opportunities will also be available to measure the same ICME *in-situ* at different points in the inner heliosphere. Differences observed should help us to understand the evolution that takes place due the interaction of ICMEs with the structure already existing in the ambient solar wind.

Acknowledgements

N. U. Crooker acknowledges support from NASA under grant no. NNG05GD97G. J. Rodriguez-Pacheco acknowledges the support of the Spanish Ministerio de Educación y Ciencia under grant no. ESP2002-04379-C02-02.

References

Arge, C. N., Odstrcil, D., Pizzo, V. J., and Mayer, L.: 2002, in Velli, M., Bruno, R., and Malara, F. (eds.), *Solar Wind Ten*, Vol. 679 of *AIP Conference Proceedings*, Melville, NY, pp. 190–193.

Armstrong, J., and Coles, W. A.: 1972, *J. Geophys. Res.* **77**, 4602–4610.

Bale, S. D., Reiner, M. J., Bougeret, J.-L., Kaiser, M. L., Krucker, S., Larson, D. E., *et al.*: 1999, *Geophys. Res. Lett.* **26**, 1573–1576.

Balogh, A., Gosling, J. T., Jokipii, J. R., Kallenabch, R., and Kunow, H.: 1999, *Space Sciences Series of ISSI* Vol. 7, Kluwer, Dordrecht.

Behannon, K. W., Burlaga, L. F., and Hewish, A.: 1991, *J. Geophys. Res.* **96**, 21,213–21, 225.

Blanco, J. J., Cid, C., Rodriguez-Pacheco, J., Hidalgo, M., Nieves-Chinchilla, T., and Sequeriros, J.: 2003, *Geophys. Res. Abs.* **5**, 00354.

Bothmer, V.: 2003, *ESA Special Pub.* **535**, 419–428.

Bothmer, V., and Schwenn, R.: 1994, *Space Sci. Rev.* **70**, 215–220.

Bothmer, V., and Rust, D. M.: 1997, in Crooker, N. U., Joselyn, J. A., and Feynman, J. (eds.), *Coronal Mass Ejections*, Vol. 99 of *AGU Geophys. Mon.*, Washington, DC, pp. 139–146.

Bothmer, V., and Schwenn, R.: 1998, *Ann. Geophysicae* **16**, 1–24.

Breen, A. R., Riley, P., Lazarus, A. J., Canals, A., Fallows, R. A., and Linker, J., *et al.*: 2002, *Ann. Geophysicae* **20**, 1291–1309.

Burlaga, L. F.: 1975, *Space Sci. Rev.* **17**, 327–352.

Burlaga, L. F.: 1991, in Schwenn, R., and Marsch, E. (eds.), *Physics of the Inner Heliosphere 2*, Springer-Verlag, Berlin, Heidelberg, New York, pp. 1–22.

Burlaga, L. F.: 1995, *Interplanetary Magnetodynamics*, Oxford University Press, New York and Oxford.

Burlaga, L. F., Fitzenreiter, R., Lepping, R., Ogilvie, K., Szabo, A., Lazarus, A., *et al.*: 1998, *J. Geophys. Res.* **103**, 277–285.

Cane, H. V., and Richardson, I. G.: 1997, *J. Geophys. Res.* **102**, 17,445–17,449.

Cane, H. V., and Richardson, I. G.: 2003a, *J. Geophys. Res.* **108**, 1156, doi:10.1029/2002JA009817.

Cane, H. V., and Richardson, I. G.: 2003b, *Geophys. Res. Lett.* **30**, 2233, doi:10.1029/2003GL017685.

Cane, H. V., Stone, R. G., Fainberg, J., Steinberg, J. L., and Hoang, S.: 1982, *Solar Physics* **78**, 187–198.

Cane, H. V., Kahler, S. W., and Sheeley, N. R., Jr.: 1986, *J. Geophys. Res.* **91**, 13,321–13, 329.

Cane, H. V., Sheeley, N. R., Jr., and Howard, R. A.: 1987, *J. Geophys. Res.* **92**, 9869–9874.

Cane, H. V., Wibberenz, G., and Richardson, I. G.: 1997, *J. Geophys. Res.* **102**, 7075–7086.

Cane, H. V., Richardson, I. G., and von Rosenvinge, T. T.: 1998, *Geophys. Res. Lett.* **25**, 4437–4440.

Cliver, E. W.: 1995, *Eos Trans. AGU* **76(8)**, 75.

Cliver, E. W., and Ling, A. G.: 2001, *Astrophys. J. Lett.* **556**, 432–437.

Cliver, E. W., Feynman, J., and Garrett, H. B.: 1990, *J. Geophys. Res.* **95**, 17,103–17,112.

Cliver, E. W., Ling, A. G., and Richardson, I. G.: 2003, *Astrophys. J.* **592**, 574–579.

Cremades, H., and Bothmer, V.: 2004, *Astron. Astrophys.* **422**, 307–322.

Crooker, N. U.: 2000, *J. Atmos. Sol. Terr. Phys.* **62**, 1071–1085.

Crooker, N. U., and Cliver, E. W.: 1994, *J. Geophys. Res.* **99**, 23,383–23,390.

Crooker, N. U., and Horbury, T. S.: 2006, *Space Sci. Rev.* this volume, 10.1007/s11214-006-9014-0.

Crooker, N. U., Forsyth, R., Rees, A., Gosling, J. T., and Kahler, S. W.: 2004, *J. Geophys. Res.* **109**, A06110, doi:10.1029/2004JA010426.

Dalla, S., Balogh, A., Heber, B., and Lopate, C.: 2001, *J. Geophys. Res.* **106**, 5721–5730.

Dal Lago, A., Schwenn, R., and Gonzalez, W. D.: 2003, *Adv. Space Res.* **32**, 2637–2640.

Dere, K. P., Brueckner, G. E., Howard, R. A., and Michels, D. J.: 1999, *Astrophys. J.* **516**, 465–474.

Detman, T. R., Dryer, M., Yeh, T., Han, S. M., Wu, S. T., and McComas, D. J.: 1991, *J. Geophys. Res.* **96**, 9531–9540.

Eyles, C. J., Simnett, G. M., *et al.*: 2003, *Solar Phys.* **217**, 319–347.

Fenrich, F. R., and Luhmann, J. G.: 1998, *Geophys. Res. Lett.* **25**, 2999–3002.

Forbes, T. G., Linker, J. A., *et al.*: 2006, *Space Sci. Rev.*, this volume, 10.1007/s11214-006-9019-8.

Gapper, G. R., Hewish, A., Purvis, A., and Duffet-Smith, P. J.: 1982, *Nature* **296**, 633–636.

Gazis, P. R., Balogh, A., *et al.*: 2006, *Space Sci. Rev.*, this volume, 10.1007/s11214-006-9023-z.

Gibson, S., and Low, B. C.: 1998, *Astrophys. J.* **493**, 460–473.

González-Esparza, J. A., Neugebauer, M., Smith, E. J., and Phillips, J. L.: 1998, *J. Geophys. Res.* **103**, 4767–4773.

Gopalswamy, N., Lara, A., Lepping, R. P., Kaiser, M. L., Berdichevsky, D., and St. Cyr, O. C.: 2000, *Geophys. Res. Lett.* **27**, 145–148.

Gopalswamy, N., Lara, A., Yashiro, S., Kaiser, M. L., and Howard, R. A.: 2001a, *J. Geophys. Res.* **106**, 29,207–29,217.

Gopalswamy, N., Yashiro, S., Kaiser, M. L., Howard, R. A., and Bougeret, J.-L.: 2001b, *Astrophys. J.* **548**, L91–L94.

Gopalswamy, N., Yashiro, S., Michalek, G., Kaiser, M. L., Howard, R. A., Reames, D. V., *et al.*: 2002, *Astrophys. J.* **572**, L103–L107.

Gosling, J. T.: 1990, in Russell, C. T., Priest, E. R., and Lee, L. C. (eds.), *Physics of Magnetic Flux Ropes*, Vol. 58 of *AGU Geophys. Mon.*, Washington, DC, pp. 343–364.

Gosling, J. T., McComas, D. J., Phillips, J. L., and Bame, S. J.: 1991, *J. Geophys. Res.* **96**, 7831–7839.

Gosling, J. T., McComas, D. J., Phillips, J. L., and Bame, S. J.: 1992, *J. Geophys. Res.* **97**, 6531–6535.

Gosling, J. T., Riley, P., McComas, D. J., and Pizzo, V. J.: 1998, *J. Geophys. Res.* **103**, 1941–1954.

Groth, C. P. T., DeZeeuw, D. L., Gombosi, T. I., and Powell, K. G.: 2000, *J. Geophys. Res.* **105**, 25,053–25,078.

Hammond, C. M., Crawford, G. K., Gosling, J. T., Kojima, H., Phillips, J. L., Matsumoto, H., *et al.*: 1995, *Geophys. Res. Lett.* **22**, 1169–1172.

Hayashi, K., Kojima, M., Tokumaru, M., and Fujiki, K.: 2003, *J. Geophys. Res.* **108**, doi:10.1029/2002JA009567.

Hewish, A., Scott, P. F., and Willis, D.: 1964, *Nature* **203**, 1214–1217.

Hoang, S., Pantellini, F., Harvey, C. C., Lacombe, C., Mangeney, A., Meyer-Vernet, N., *et al.*: 1992, in Marsch, E., and Schwenn, R. (eds.), *Solar Wind Seven*, Pergamon Press, New York, pp. 465–468.

Hu, Y. Q.: 1998, *J. Geophys. Res.* **103**, 14,631–14,641.

Hudson, H. S., Bougeret, J.-L., and Burkepile, J.: 2006, *Space Sci. Rev.*, this volume, doi: 10.1007/s11214-006-9009-x.

Hundhausen, A. J.: 1988, in Pizzo, V. J., Holzer, T. E., and Sime, D. G. (eds.), *Proceedings of the Sixth International Solar Wind Conference, Rep. NCAR/TN-306*, Natl. Cent. for Atmos. Res., Boulder, pp. 181–214.

Huttunen, E., Schwenn, R., Bothmer, V., and Koskinen, H. E. J.: 2005, *Ann. Geophysicae* **23**, 625–641.

Jadav, R. M., Iyer, K. N., Joshi, H. P., and Vats, H. O.: 2005, *Planet. Space Sci.* **53**, 671–679.

Intriligator, D. S., Jokipii, J. R., Horbury, T., Intriligator, J. M., Forsyth, R., Kunow, H., *et al.*: 2001, *J. Geophys. Res.* **106**, 10,625–10,634.

Klein, L. W., and Burlaga, L. F.: 1982, *J. Geophys. Res.* **87**, 613–624.

Knetter, T., Neubauer, F. M., Horbury, T., and Balogh, A.: 2004, *J. Geophys. Res.* **109**, 10.1029/2003JA010099.

Knock, S. A., Cairns, I. H., Robinson, P. A., and Kuncic, Z.: 2001, *J. Geophys. Res.* **106**, 25,041–25,051.

Leamon, R. J., Smith, C. W., and Ness, N. F.: 1998, *Geophys. Res. Lett.* **25**, 2505–2508.

Lengyel-Frey, D., Stone, R. G., and Bougeret, J.-L.: 1985, *Astron. Astrophys.* **151**, 215–221.

Lengyel-Frey, D., and Stone, R. G.: 1989, *J. Geophys. Res.* **94**, 159–167.
Lindsay, G. M., Russell, C. T., Luhmann, J. G., and Gazis, P.: 1994, *J. Geophys. Res.* **99**, 11–17.
Lindsay, G. M., Luhmann, J. G., Russell, C. T., and Gosling, J. T.: 1999, *J. Geophys. Res.* **104**, 12,515–12,523.
Linker, J. A., Mikic, Z., Biesecker, D. A., Forsyth, R. J., Gibson, S. E., Lazarus, A. J., *et al.*: 1999, *J. Geophys. Res.* **104**, 9809–9830.
Liu, Y., and Hayashi, K.: 2006, *Astrophys. J.* **640**, 1135–1141.
Liu, Y., Richardson, J. D., and Belcher, J. W.: 2005, *Planet. Space Sci.* **53**, 3–17.
Malitson, H. H., Fainberg, J., and Stone, R. G.: 1973, *Astrophys. Lett.* **14**, 111–114.
Manchester IV, W. B., Gombosi, T. I., Roussev, I., DeZeeuw, D. L., Sokolov, I. V., Powell, K. G., *et al.*: 2004a, *J. Geophys. Res.* **109**, doi:10.1029/2002JA009672.
Manchester IV, W. B., Gombosi, T. I., Roussev, I., Ridley, A., DeZeeuw, D. L., Sokolov, I. V., *et al.*: 2004b, *J. Geophys. Res.* **109**, doi:10.1029/2003JA010150.
Manoharan, P. K., Tokumaru, M., Pick, M., Subramanian, P., Ipavich, F. M., Schenk, K., *et al.*: 2001, *Astrophys. J.* **559**, 1180–1189.
Marsch, E.: 1991, in Schwenn, R., and Marsch, E. (eds.), *Physics of the Inner Heliosphere 1*, Springer-Verlag, Berlin, Heidelberg, New York, pp. 159–241.
McComas, D. J., Gosling, J. T., Hammond, C. M., Moldwin, M. B., Philips, J. L., and Forsyth, R. J.: 1994, *Geophys. Res. Lett.* **21**, 1751–1754.
Mikic, Z., and Lee, M. A.: 2006, *Space Sci. Rev.*, this volume, doi: 10.1007/s11214-006-9012-2.
Moldwin, M. B., Phillips, J. L., Gosling, J. T., Scime, E. E., McComas, D. J., Balogh, A., *et al.*: 1995, *J. Geophys. Res.* **100**, 19,903–19,910.
Mulligan, T., Russell, C. T., and Luhmann, J. G.: 1998, *Geophys. Res. Lett.* **25**, 2959–2962.
Odstrcil, D., Dryer, M., and Smith, Z.: 1996, *J. Geophys. Res.* **101**, 19,973–19,986.
Odstrcil, D., and Pizzo, V. J.: 1999a, *J. Geophys. Res.* **104**, 483–492.
Odstrcil, D., and Pizzo, V. J.: 1999b, *J. Geophys. Res.* **104**, 493–503.
Odstrcil, D., Riley, P., and Zhao, X. P.: 2004, *J. Geophys. Res.* **109**, A02116, doi:10.1029/2003JA010135.
Odstrcil, D., Pizzo, V. J., and Arge, C. N.: 2005, *J. Geophys. Res.*, **110**, A02106, doi:10.1029/2004JA010745.
Pick, M., Forbes, T. G., *et al.*: 2006, *Space Sci. Rev.*, this volume, doi: 10.1007/s11214-006-9021-1.
Reiner, M. J., Kaiser, M. L., Fainberg, J., Bougeret, J.-L., and Stone, R. G.: 1997, *Proceedings of the 31st ESLAB Symposium, ESA SP-415*, 183–188.
Reiner, M. J., Kaiser, M.L., Fainberg, J., and Stone, R. G.: 1998, *J. Geophys. Res.* **103**, 29,651–29,664.
Reiner, M. J., Kaiser, M. L., Fainberg, J., and Stone, R. G.: 1999, in Habbal, S. R., Esser, R., Hollweg, J. V., and Isenberg, P. A. (eds.) *Solar Wind Nine*, Vol. 471 of *AIP Conference Proceedings* Woodbury, NY, pp. 653–656.
Reiner, M. J., Kaiser, M. L., Karlický, M., Jiřička, K., and Bougeret, J.-L.: 2001, *Solar Phys.* **204**, 123–139.
Reiner, M. J., Kaiser, M. L., and Bougeret, J.-L.: 2003, in Velli, M., Bruno, R., and Malara, F. (eds.), *Solar Wind Ten*, Vol. 679 of *AIP Conference Proceedings*, Melville, NY, pp. 152–155.
Reiner, M. J., Jackson, B. V., Webb, D. F., Mizuno, D. R., Kaiser, M. L., and Bougeret, J.-L.: 2005, *J. Geophys. Res.* **110**, doi:10.1029/ 2004JA010943.
Richardson, I. G., and Cane, H. V.: 2005, *Geophys. Res. Lett.* **32**, L02104, doi:10.1029/2004GL021691.
Richardson, I. G., Cliver, E. W., and Cane, H. V.: 2001, *Geophys. Res. Lett.* **28**, 2569–2572.
Richardson, I. G., Cane, H. V., and Cliver, E. W.: 2002, *J. Geophys. Res.* **107**, 10.1029/2001JA000504.
Richardson, I., Lawrence, G. R., Haggerty, D. K., Kucera, T. A., and Szabo, A.: 2003, *Geophys. Res. Lett.*, **30**, 8014, doi:10.1029/2002GL016424.

Riley, P., Gosling, J. T., and Pizzo, V. J.: 1997, *J. Geophys. Res.* **102**, 14,677–14,686.
Riley, P., Linker, J. A., and Mikic, Z.: 2001, *J. Geophys. Res.* **106**, 15,889–15,901.
Rodriguez-Pacheco, J., Blanco, J. J., and Crooker, N. U.: 2005, unpublished manuscript.
Rogers, B. N., Drake, J. F., and Shay, M. A.: 2000, *Geophys. Res. Lett.* **27**, 3157–3160.
Roussev, I. I., Gombosi, T. I., Sokolov, I. V., Velli, M., Manchester IV, W., DeZeeuw, D. L., *et al.*: 2003, *Astrophys. J.* **595**, L57–L61.
Roussev, I. I., Sokolov, I. V., Forbes, T. G., Gombosi, T. I., Lee, M. A., and Sakai, J. I.: 2004, *Astrophys. J.* **605**, L73–L76.
Russell, C. T., Mulligan, T., and Anderson, B. J.: 2003, in Velli, M., Bruno, R., and Malara, F. (eds.), *Solar Wind Ten*, Vol. 679 of *AIP Conference Proceedings*, Melville, NY, pp. 121–124.
Ruzmaikin, A., Feynman, J., and Smith, E. J.: 1997, *J. Geophys. Res.* **102**, 19,653–19,759.
Schmidt, J. M.: 2000, *Sol. Phys.* **197**, 135–184.
Schmidt, J. M., and Cargill, P. J.: 2001, *J. Geophys. Res.* **106**, 8283–8289.
Schmidt, J. M., and Cargill, P. J.: 2003, *J. Geophys. Res.* **108**, 1023, doi:10.1029/2002JA009325.
Schmidt, J. M., and Cargill, P. J.: 2004, *Annales Geophys.* **22**, 2245–2254.
Schwenn R.: 1986, *Space Sci. Rev.* **44**, 139–168.
Schwenn, R.: 1990, in Schwenn, R., and Marsch, E. (eds.), *Physics of the Inner Heliosphere 1*, Springer-Verlag, Berlin, Heidelberg, New York, pp. 99–181.
Schwenn, R., Dal Lago, A., Huttunen, E., and Gonzales, W. D.: 2005, *Ann. Geophysicae* **23**, 1033–1059.
Schwenn, R., Raymond, J. C., *et al.*: 2006, *Space Sci. Rev.*, this volume, doi: 10.1007/s11214-006-9016-y.
Sheeley, N. R., Jr., Howard, R. A., Koomen, M. J., Michels, D. J., Schwenn, R., Mühlhäser, K. H., *et al.*: *J. Geophys. Res.* **90**, 163–175.
Smith, C. W., and Phillips, J. L.: 1997, *J. Geophys. Res.* **102**, 249–261.
Tokumaru, M., Kojima, M., Fujiki, K., Yamashita, M., and Yokobe, A.: 2003, *J. Geophys. Res.* **108**, 1220, doi:10.1029/2002JA009574.
Tokumaru, M., Kojima, M., Fujiki, K., Yamashita, M., and Baba, D.: 2005, *J. Geophys. Res.* **110**, A01109, doi:10.1029/2004JA010656.
Totten, T. L., Freeman, J. W., and Arya, S.: 1995, *J. Geophys. Res.* **100**, 13–17.
Usmanov, A. V., Goldstein, M. L., Besser, B. P., and Fritzer, J. M.: 2000, *J. Geophys. Res.* **105**, 12,675–12,696.
Vandas, M., Odstrcil, D., and Watari, S.: 2002, *J. Geophys. Res.* **107**, doi:10,1029/2001JA005068.
Vasquez, B. J., Farrugia, C. J., Markovskii, S. A., Hollweg, J. V., Richardson, I. G., Ogilvie, K. W., *et al.*: 2001, *J. Geophys. Res.* **106**, 29,283–29,298.
Wang, C., Du, D., and Richardson, J. D.: 2005, *J. Geophys. Res.* **110**, doi:10.1029/2005JA011198.
Webb, D. F., and Howard, R. A.: 1994, *J. Geophys. Res.* **99**, 4201–4220.
Webb, D. F., Cliver, E. W., Gopalswamy, N., Hudson, H. S., and St. Cyr, O. C.: 1998, *Geophys. Res. Lett.* **25**, 2469–2472.
Webb, D. F., Lepping, R. P., Burlaga, L. F., DeForest, C. E., Larson, D. E., Martin, S. F., *et al.*: 2000, *J. Geophys. Res.* **105**, 27,251–27,259.
Wibberenz, G., Richardson, I. G., and Cane, H. V.: 2002, *J. Geophys. Res.* **107**, doi: 10.1029/2002JA009461.
Wimmer-Schweingruber, R. F., Crooker, N. U., *et al.*: 2006, *Space Sci. Rev.*, this volume, doi:10.1007/s11214-006-9017-x.
Woo, R.: 1988, *J. Geophys. Res.* **93**, 3919–3926.
Woo, R., and Schwenn, R.: 1991, *J. Geophys. Res.* **96**, 21,227–21,244.
Zhao, X. P.: 1992, *J. Geophys. Res.* **97**, 15,051–15,055.

Zhao, X. P., and Hoeksema, J. T.: 1996, *J. Geophys. Res.* **101**, 4825–4834.
Zhao, X. P., and Hoeksema, J. T.: 1997, *Geophys. Res. Lett.* **24**, 2965–2968.
Zhao, X. P., and Webb, D. F.: 2003, *J. Geophys. Res.* **108**, 1234, doi:10.1029/2002JA009606.
Zhao, X. P., Plunkett, S. P., and Liu, W.: 2002, *J. Geophys. Res.* **107**, doi:10,1029/2001JA009143.
Zhang, J., Dere, K. P., Howard, R. A., and Bothmer, V.: 2003, *Astrophys. J.* **582**, 520–533.
Zurbuchen, T. H., and Richardson, I. G.: 2006, *Space Sci. Rev.*, this volume, doi: 10.1007/s11214-006-9010-4.

ICMES AT HIGH LATITUDES AND IN THE OUTER HELIOSPHERE

Report of Working Group H

P. R. GAZIS[1,*], A. BALOGH[2], S. DALLA[3], R. DECKER[4], B. HEBER[5], T. HORBURY[2], A. KILCHENMANN[6], J. KOTA[7], H. KUCHAREK[8], H. KUNOW[5], D. LARIO[9], M. S. POTGIETER[9], J. D. RICHARDSON[10], P. RILEY[11], L. RODRIGUEZ[12], G. SISCOE[13] and R. VON STEIGER[6]

[1]*SJSU Foundation, NASA Ames Research Center, Moffett Field, CA, USA*
[2]*Imperial College, London, England*
[3]*University of Manchester, Manchester, England*
[4]*Johns Hopkins University/Applied Physics Lab, Laurel, MD, USA*
[5]*Christian-Albrechts-Universität zu Kiel, Kiel, Germany*
[6]*International Space Science Institute, Bern, Switzerland*
[7]*University of Arizona, Tucson, AZ, USA*
[8]*University of New Hampshire, Durham, NH, USA*
[9]*Potchefstroom University, Potchefstroom, South Africa*
[10]*Massachusetts Institute of Technology, Center for Space Research, Cambridge, MA, USA*
[11]*Science Applications International Corporation, San Diego, CA, USA*
[12]*Max-Planck-Institut für Sonnensystemforschung, Katlenburg-Lindau, Germany*
[13]*Boston University, Boston, MA, USA*
*(*Author for correspondence, E-mail: pgazis@mail.arc.nasa.gov)*

(Received 31 January 2006; Accepted in final form 19 May 2006)

Abstract. Interplanetary coronal mass ejections (ICMEs) propagate into the outer heliosphere, where they can have a significant effect on the structure, evolution, and morphology of the solar wind, particularly during times of high solar activity. They are known to play an important role in cosmic ray modulation and the acceleration of energetic particles. ICMEs are also believed to be associated with the large global transient events that swept through the heliosphere during the declining phases of solar cycles 21 and 22. But until recently, little was known about the actual behavior of ICMEs at large heliographic latitudes and large distances from the Sun. Over the past decade, the Ulysses spacecraft has provided in situ observations of ICMEs at moderate heliographic distances over a broad range of heliographic latitudes. More recently, observations of alpha particle enhancements, proton temperature depressions, and magnetic clouds at the Voyager and Pioneer spacecraft have begun to provide comparable information regarding the behavior of ICMEs at extremely large heliocentric distances. At the same time, advances in modeling have provided new insights into the dynamics and evolution of ICMEs and their effects on cosmic rays and energetic particles.

Keywords: ICMEs, outer heliosphere, coronal mass ejections

1. Introduction

The structure and evolution of interplanetary coronal mass ejections (ICMEs) in the inner heliosphere in the vicinity of the solar equatorial plane has been the subject of many observations over the past four decades (see Crooker and Horbury, 2006,

Space Science Reviews (2006) 123: 417–451
DOI: 10.1007/s11214-006-9023-z

this volume; von Steiger and Richardson, 2006, this volume). Much less is known about the behavior of ICMEs at high heliographic latitudes and large distances from the Sun. The evolution of ICMEs has been the subject of numerous models (Burlaga *et al.*, 1986; Whang and Burlaga, 1986; Gosling *et al.*, 1995c; Odstrcil and Pizzo, 1999; Wang *et al.*, 2001, and many others). As ICMEs propagate into the outer heliosphere, they can merge with co-rotating interaction regions (CIRs) or other ICMEs to produce merged interaction regions (MIRs, Burlaga *et al.*, 1986; Whang and Burlaga, 1986) or global merged interaction regions (GMIRs, Burlaga *et al.*, 1984, 1993; Whang, 1991; Burlaga, 1995; Zank, 1999). MIRs and GMIRs are important large scale structures in the outer heliosphere. They are known to be associated with energetic particle variations (Burlaga, 1995; Decker *et al.*, 1995; Decker and Krimigis, 2003; Neugebauer and Goldstein, 2003; Cane and Lario, 2006, this volume) and the modulation of galactic cosmic rays (GCRs, see Burlaga *et al.*, 1986; McDonald *et al.*, 1994; and many others).

Until recently, there were few direct observations of ICMEs in the outer heliosphere and at high heliographic latitudes, and most comparisons of models with observations were restricted to limited periods associated with spacecraft alignments. Magnetic clouds (Burlaga *et al.*, 1982; Burlaga and Behannon, 1982) have provided one possible signature of ICMEs in the outer heliosphere, but observations from the inner heliosphere (Richardson and Cane, 1993) suggest that magnetic clouds might only be detected in a minority of events and may not be a suitable signature with which to conduct large-scale surveys. This has made it difficult to address questions regarding the evolution of ICMEs, their effects on the dynamics and evolution of solar wind streams and the large-scale structure and morphology of the heliosphere, and how these effects might vary over course of a solar cycle. In particular, in the absence of direct observations of ICMEs at large heliocentric distances, the relationship between ICMEs, MIRs, and GMIRs has been to some extent a matter of speculation.

Recently Paularena *et al.* (2001) and Wang and Richardson (2004) demonstrated that it is possible to identify ejecta associated with ICMEs in the outer heliosphere using alpha particle and temperature measurements. Observations from the inner heliosphere (Richardson and Cane, 1993) suggest these signatures may be sufficient to identify a majority of ICMEs. The past decade has also seen significant advances in modeling the dynamics and evolution of ICMEs at high heliographic latitudes (Riley, 1999; Linker *et al.*, 2002; Cargill and Schmidt, 2002) and large heliocentric distances (Zank and Mueller, 2003; Wang and Richardson, 2004) and the effects of ICMEs on energetic particles and cosmic rays.

This chapter will discuss the following recent developments in the study of ICMEs: (1) Ulysses observations of ICMEs and their effects on energtic particles and cosmic rays at high heliographic latitudes, (2) observations of ICMEs in the outer heliosphere from the Voyager and Pioneer spacecraft, (3) the relationship between ICMEs and large global transient events associated with Forbush decreases, and (4) the interpretation of in situ observations using modelling.

2. Observations at Moderate Heliocentric Distances and all Latitudes

2.1. Latitudinal Survey

R. von Steiger

For a survey of the distribution of ICMEs as a function of heliographic latitude we have the Ulysses mission. It has completed two full revolutions around the Sun on a high-inclination orbit, the first one around solar minimum and the second one around solar maximum. The observations of ICMEs, as taken from the list kept at Los Alamos (`http://swoops.lanl.gov/cme_list.html`), are plotted in Figure 3 in the next subsection. Obviously the distribution of ICMEs is very different along the two orbits.

On the solar minimum orbit only three ICMEs occurred at high latitudes, and these were all of the new class of overexpanding ICMEs defined by Gosling *et al.* (1995a), embedded in the dominating polar high-speed streams. It is not meaningful to construct a statistics of latitudinal distribution from only three events. At solar minimum, it is safe to say, almost all ICMEs occur at low latitudes, where also the streamer belt is confined.

The situation is radically different at solar maximum (cf. right panel of Figure 3). Superficially it looks as if ICMEs can be seen at all latitudes along the Ulysses maximum orbit, in line with naïve expectations from the fact that CMEs occur essentially uniformly at all position angles at the Sun (Gopalswamy *et al.*, 2006, this volume). However, we recall from McComas *et al.* (1998) that, surprisingly, the ICME rate appeared to drop as Ulysses climbed to high southern latitudes even though solar activity was still on the rise (see also Figure 6 of von Steiger and Richardson, 2006, this volume). This impression is confirmed in Figure 1, where we have plotted the monthly ICME rate as a function of latitude. The result clearly indicates an ICME rate that is three times higher at low latitudes than it is at the highest latitudes reached by Ulysses.

Of course the above result has not been obtained simultaneously at all latitudes and is therefore potentially influenced by changes in the solar activity level, at least during the slow latitude scan (solid curve in Figure 1. However, solar activity was on the rise during the entire time of this scan from low to high latitudes, such that the latitude effect might even be larger than reported here.

A possibility for simultaneously comparing low-latitude and high-latitude ICME rates is given by comparing data from Ulysses and from an in-ecliptic spacecraft such as ACE. This has been done by Lepri *et al.* (2001), as reproduced in Figure 2. The two rates agree when Ulysses was at low to mid latitudes ($\leq 30^\circ$) but progressively deviate from each other as Ulysses reached high latitudes, as expected from Figure 1. However, when we scale the Ulysses ICME rate by dividing it by the cosine of its heliographic latitude the two rates can be brought back into agreement, at least very roughly so.

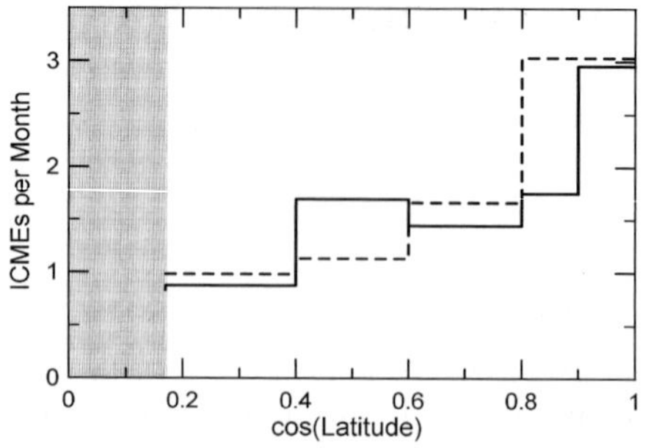

Figure 1. Number of ICMEs per month observed at Ulysses as a function of the cosine of latitude. The solid curve is from the southbound slow latitude scan (DOY 350-1997 to 329-2000) and the dashed curve is from the fast latitude scan (DOY 329-2000 to 284-2001). The subsequent northern slow latitude scan was disregarded since there the latitude distribution is masked by the declining solar activity. Both latitude scans clearly indicate that the ICME rate is latitude dependent by as much as a factor of three between the lowest and the highest latitudes. (Latitudes with a cosine < 0.17 are not reached by Ulysses.)

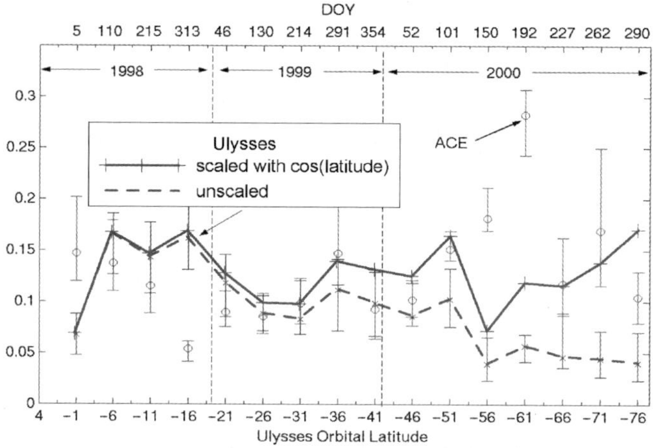

Figure 2. ICME rates as estimated from the average iron charge state measured with Ulysses-SWICS ($\times$, dashed line) and with ACE-SWICS ($\circ$). When the former is divided by the cosine of the Ulysses latitude ($+$, solid line) the two rates are brought back roughly into agreement. Figure adapted from Lepri *et al.* (2001).

It thus appears that at solar maximum activity CMEs occur at all latitudes uniformly, but when propagating into the interplanetary medium those from high latitudes are deflected to lower latitudes, leading to the observed disparity of ICME rates at low versus high latitudes by as much as a factor of three. This implies that, as is the case for fast streams at solar minimum, there is a superradial expansion of the high-latitude solar wind (or at least the high-latitude ICMEs) also at solar maximum.

2.2. Cosmic Ray Latitudinal Variations

B. Heber

Short term decreases in the galactic cosmic ray (GCR) flux were first observed by Forbush (1937). Simpson (1954) showed that these Forbush decreases had their cause in the interplanetary medium. Two classes of events, "recurrent" and "non-recurrent", have been found. While the first class of event is associated with CIRs (see for example Heber *et al.*, 1999; Richardson, 2004, and references therein) the second class is associated with ICMEs (Cane, 2000). The latter are discussed by Cane (2000) and by Klecker *et al.* (2006, this volume). Here we focus on the influence of ICMEs on GCR intensities for latitudes above 40° by using measurements of >250 MeV protons by the Kiel Electron Telescope (KET) onboard the Ulysses spacecraft (Simpson *et al.*, 1992). The solar minimum and maximum out-of-ecliptic paths are displayed in Figure 3. Ulysses heliographic latitude is shown as a function of radial distance. Gosling and Reisenfeld (`http://swoops.lanl.gov/cme_list.html`) identified 41 out-of-ecliptic ICMEs. These are indicated by '⋆'s in the figure. While it is well known that ICMEs are more numerous during solar maximum (38 cases) than at solar minimum (3 cases) we learned from Ulysses that they can occur at all latitudes (even though their distribution is not uniform). This has enabled us to study GCR intensity variations at high heliographic latitudes.

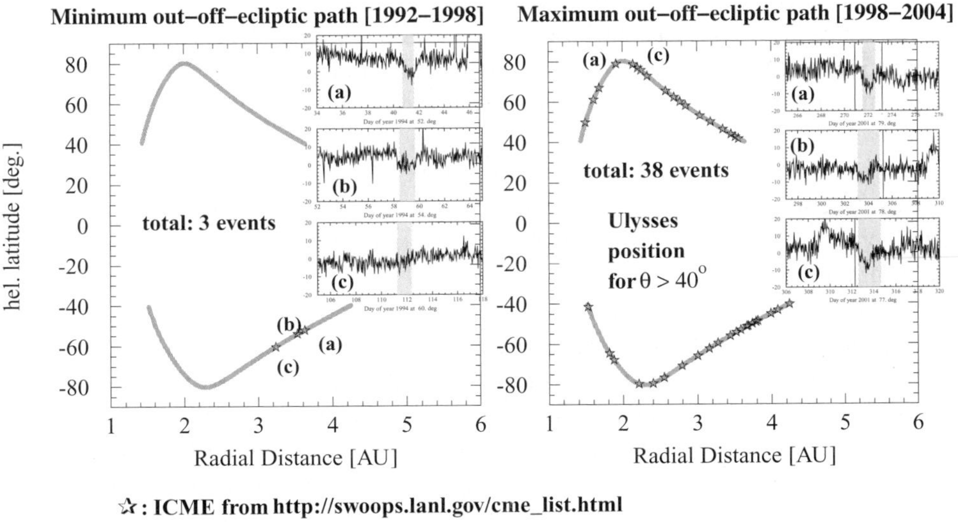

Figure 3. Ulysses minimum and maximum out-of-ecliptic trajectory: Heliographic latitude as function of radial distance. The ⋆ display time periods when an ICME has been observed. The two insets show the hourly averaged galactic cosmic ray variation during six selected events.

The insets in Figure 3 show the count rate variation during six selected ICME events. The three GCR intensity decreases on the right correspond to the events (1), (2) and (3) in panel (d) of Figure 5 (Cane and Lario, 2006, this volume). The three intensity decreasess on the left hand side are the three events observed during the solar minimum out-of-ecliptic orbit (Bothmer *et al.*, 1997). Two of these events show an intensity decrease of about 6%.

Around solar maximum Ulysses identified 38 ICMEs. Following Cane (2000) the corresponding intensity decreases show different characteristics. The "classical" Forbush decreases consist of two step-decreases, as displayed in panel (d) of Figure 3 in Cane and Lario (2006, this volume). The first step occurs in the turbulent magnetic field region behind a shock wave generated by a fast ICME while the second step is associated with the closed magnetic field line geometry within the ICME plasma. But single step intensity decreases correlated with either the shock or the ICME crossing are also observed. (The list of shocks were taken from Gosling and Forsyth, http://www.sp.ph.ic.ac.uk/Ulysses/shocklist.txt. Note that this shock list starts from 1996.)

Because the KET instrument can only determine the hourly averaged GCR intensities within an accuracy of ~3%, the smallest detectable decreases therefore are limited to approximately ~4%. This may be why only one classical two-step Forbush decrease has been found. During 16 (~40%) and 8 (~20%) of the events, an intensity decrease was caused by the ICME-plasma or the shock wave, respectively. However, during the remaining 16 (~40%) events, the galactic cosmic rays show no systematic variation during the ICME. For these 16 events a single detector counter with relative sensitivity $\Delta C/C \approx 0.25\%$ has been used to verify these "null" result. Because this counter is sensitive to protons >30 MeV and electrons >1 MeV, it cannot be used when the intensity decrease is accompanied by an energetic particle event (3 events). In another 5 cases the channel indicates a small decrease. This suggests that between 8 and 11 ICMEs are not accompanied by cosmic ray intensity decreases.

In Figure 4 the observed intensity decreases are shown on the left and right hand side as function of Ulysses radial distance and heliographic latitude. The open and filled symbols in the right hand panel correspond to observations in the southern and northern hemisphere, respectively.

Due to the statistical limitations the detection limit is 4%. Figure 4 suggests that (1) the amplitudes of the cosmic ray decreases varied strongly from event to event and that there were no obvious correlations with either (2) radial distance or (3) heliographic latitude at moderate heliocentric distances (between 1–5 AU), though Cane *et al.* (1994) have reported that there might be a tendency for the amplitude of these events to decrease with increasing radial distance in the inner heliosphere.

Bothmer *et al.* (1997) reported cosmic ray intensity decreases occurred in assso-ciation with 3 ICMEs which were observed in the fast solar wind from the southern polar coronal hole. In 2002 an additional 5 ICMEs were identified in the fast solar wind stream coming from the northern polar coronal hole. For all 8 events an

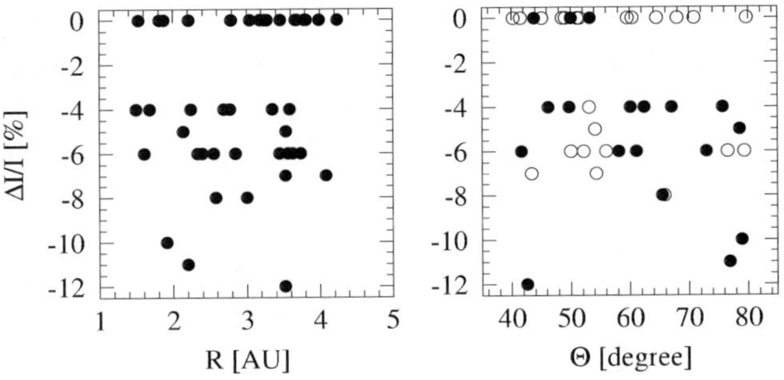

Figure 4. Amplitude of the GCR intensity decreases as a function of radial distance and heliographic latitude. The open and filled symbols in the right hand panel correspond to observations in the southern and northern hemisphere, respectively.

intensity decrease has been detected. Wibberenz *et al.* (1998) suggest that for these ICMEs the over-expansion of the high latitude ejecta probably results in efficient adiabatic cooling with a significant intensity decrease. It is interesting to note, that the number of ICMEs associated with a significant intensity decreases was larger in 2001/2002 when the spacecraft was in the northern hemisphere than in 2000/2001 when the spacecraft was in the southern hemisphere. Whether this is a spatial effect or caused by a larger number of ICME within the solar cycle can not be answered from the Ulysses observations alone.

2.3. Energetic Particle Response to ICMEs at High Heliographic Latitudes

D. Lario

In situ observations of both energetic particles and ICMEs at high heliographic latitudes only became possible after the Ulysses spacecraft began its polar orbit around the Sun. Energetic particle signatures associated with the passage of ICMEs at high heliographic latitudes are diverse (Bothmer *et al.*, 1995; Malandraki *et al.*, 2001; Malandraki *et al.*, 2003; Lario *et al.*, 2004). Clear differences have been observed between those ICMEs propagating within high-speed streams and those ICMEs propagating within slow solar wind streams. For those ICMEs propagating at high heliolatitudes and within slow solar wind streams, energetic particle signatures range from intensity depressions observed throughout the passage of the ICME (Malandraki *et al.*, 2003) to energetic particle enhancements observed within the ICMEs and due to the injection of solar energetic particles (SEPs) by unrelated solar events (Armstrong *et al.*, 1994; Malandraki *et al.*, 2001). By contrast, energetic particle signatures observed during the passage of ICMEs at high heliographic latitudes and when Ulysses was immersed in polar coronal hole solar wind flows

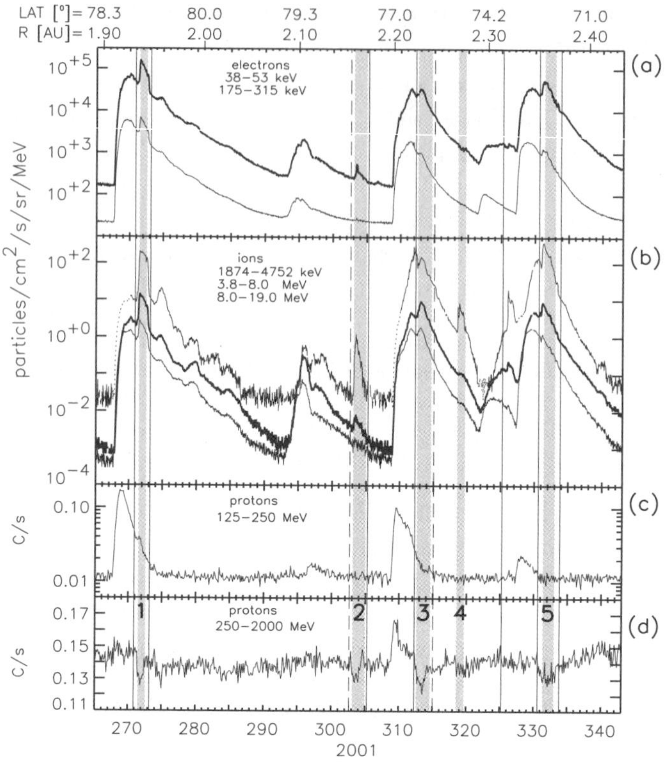

Figure 5. From top to bottom. (a) 1-hour spin-average near-relativistic electron intensities as measured by the DE system of the HI-SCALE instrument on board Ulysses. (b) 1-hour spin-average ion intensities as measured by the LEMS120 telescope of the HI-SCALE instrument (top trace) and the COSPIN/LET system on board Ulysses. (c–d) 3-hour high-energy proton count rates as measured by the COSPIN/KET in the energy range 125–250 MeV (panel c) and 250–2000 MeV (panel d). Solid and dashed vertical lines indicate the passage of shocks and wave disturbances measured using solar wind and magnetic field data. Gray vertical bars indicate the passage of ICMEs.

showed low-energy particle intensity enhancements (Bothmer *et al.*, 1995; Lario *et al.*, 2004).

Figure 5 shows energetic particle data throughout the Ulysses second northern polar passage (September–November 2001). During this time interval, Ulysses remained immersed in polar coronal hole solar wind flow ($\gtrsim$700 km/s) and observed the passage of five ICMEs (Reisenfeld *et al.*, 2003a). Low-energy ($\lesssim$8 MeV) ion and ($\lesssim$50 keV) electron intensity enhancements were observed at the entry of Ulysses into these five ICMEs. By contrast, high energy (250–2000 MeV) protons, mostly of galactic origin, showed clear depressions during the passage of these ICMEs. Lario *et al.* (2004) interpreted the low-energy particle intensity enhancements observed at the entry of Ulysses into these five high-heliolatitude ICMEs as due to (1) the lack of an intense shock-accelerated population propagating outside the ICMEs, (2) the efficient confinement of low-energy ions within the ICMEs, and (3) the effects that local magnetic field structures have on the particle transport within and around the ICMEs. Whereas at 1 AU and in the ecliptic plane, low-energy ion intensities usually peak at the arrival of fast shocks and decrease at the entry of the spacecraft into the ICMEs (Richardson, 1997), for these five ICMEs at high heliographic latitude, as well as for the solar minimum over-expanding ICME observed by Ulysses at about 54°S and at 3.5 AU (Bothmer *et al.*, 1995), the highest intensities were observed during the passage of the ICMEs. At high heliographic latitudes and within solar wind streams, the shocks driven by the ICMEs (if any) or by the over-expansion of the ICMEs are not efficient accelerators of energetic particles. Depending on the magnetic field configuration of the ICMEs, energetic particles may remain confined within the ICMEs, and therefore intra-ICME particle intensities decrease with a longer time-scale than those particles propagating outside the ICMEs, resulting in the intensity enhancements observed during the passage of these ICMEs over the spacecraft (details can be found in Lario *et al.*, 2004).

2.4. The Solar Origin of ICMEs Observed at Ulysses

L. Rodriguez

To find the region on the solar disk from which an ICME emanated is a complex task. If the spacecraft detecting the ICME has an orbit like Ulysses, then this becomes even more difficult.

As a first step, it is necessary to consider the relative position of Ulysses and the near-Earth spacecraft. For white light coronagraph studies, Ulysses-Earth quadratures, i.e. when the Ulysses-Sun-Earth angle is near to 90 degrees, are best suited. This happens about twice per year, and presents a chance for observing limb events. For chromospheric and low coronal studies, smaller Ulysses-Sun-Earth angles can be used, CMEs which originate near disk center can be observed and a clear view on the eruption processes can be attained. In order to identify a possible candidate event on the Sun, a simple ballistic travel time approach is normally applied, by using the speed of the ICME and the heliocentric distance of Ulysses.

A severe problem is the ambiguity which arises when correlating ICMEs observed at Ulysses with events observed at the Sun. The Ulysses ICME list (http://swoops.lanl.gov/cme_list.html) created by the SWOOPS team consists of ca. 150 events, while the SOHO list comprises thousands of them. This results in one ICME at Ulysses having several candidates near the Sun, especially during periods of high solar activity. The use of additional in-situ observations (when available) of the same ICME by a near-Earth spacecraft could help to restrict the number of solar candidates.

Several authors have identified the solar counterparts of Ulysses' ICMEs, with diverse objectives such as energetic particle behavior (e.g., Bothmer *et al.*, 1996; Simnett, 2003), flux rope modeling (e.g., Watari *et al.*, 2002), tracking disturbances by interplanetary scintillation (e.g., Janardhan *et al.*, 1997), expansion of solar wind electrons (e.g., Skoug *et al.*, 2000). Studies with Yohkoh data provided support to the idea that flux ropes in interplanetary space are the result of reconnection processes within the rising CME (Gosling *et al.*, 1995a). On the other hand, Weiss *et al.* (1996) and Lemen *et al.* (1996) found no clear correlation between X-ray and interplanetary signatures. The acceleration and evolution of CMEs as they propagate into interplanetary space were reviewed by Funsten *et al.* (1999), Gosling *et al.* (1995a) and Reisenfeld *et al.* (2003a,b) who used high latitude ICMEs to provide hints on the great latitudinal expansion that ICMEs may undergo. Probably the study involving most separate point measurements is the one by Richardson *et al.* (2002), who observed a ICME from the Sun (Yohkoh) passing through the Earth (Wind), sampled later by Ulysses and finally arriving at Voyager 2. Table I provides information on several recent correlative studies of Ulysses ICMEs and their solar identification.

It is worth noticing that the CMEs in Table I which could be identified as presenting a three-part structure at the Sun (H. Cremades, private communication) also had a magnetic cloud (MC) structure at Ulysses (defined by the corresponding author, or by Rodriguez *et al.*, 2004). The inverse relation was not found, partially due to the lack of coronagraph images before SOHO and because of the presence of many halo CMEs, for then it is not possible to infer clearly any internal structure.

These kinds of correlated observations are the ones probably best suited to answer many of the questions raised by Schwenn (1996) in his "catalogue of open questions"; nevertheless most of them remain still unanswered.

3. Radial Evolution of ICMEs in the Outer Heliosphere

3.1. Radial Survey

J. D. Richardson

The radial evolution of ICMEs was discussed by von Steiger and Richardson (2006, this volume). To summarize, ICMEs expand with radial distance to 10–15 AU, then maintain a constant average width of about 2 AU. ICMEs expand on average by a factor of 6–8 outside 1 AU. The speed gradient across the ICMEs decreases with distance consistent with the halt of the ICME expansion. In this section we discuss the evolution of the plasma within the ICMEs compared to that in the ambient solar wind.

Lists of ICMEs from Helios 1 and 2, WIND, ACE, Ulysses and Voyager 2 were compiled using similar identification schemes. Liu *et al.* (1985) used the criteria of low-temperatures (Richardson and Cane, 1993) and high helium abundances (Neugebauer and Goldstein, 2003) to identify ICMES in data from the first five spacecraft. Wang and Richardson (2004) used the temperature criterion

TABLE I
Sources of interplanetary CMEs

Date (Ulysses)	Date (Sun)	Solar observation	Reference
1992-03-14	1992-02-26	Yohkoh/SXT	Lemen *et al.*, 1996
1992-05-09	1992-04-17	Yohkoh/SXT	Lemen *et al.*, 1996
1992-11-09	1992-10-30	Ooty Radio Tel.	Janardhan *et al.*, 1997
		Yohkoh/SXT	Lemen *et al.*, 1996
1992-11-14$^{(MC)}$	1992-10-31/11-02	Ooty Radio Tel.	Janardhan *et al.*, 1997
1992-12-15$^{(MC)}$	1992-11-29	Yohkoh/SXT	Bothmer *et al.*, 1996
			Weiss *et al.*, 1996
1993-06-09$^{(MC)}$	1993-05-31	Yohkoh/SXT	Bothmer *et al.*, 1996; Gosling *et al.*, 1994a, 1995a, 1998; Weiss *et al.*, 1996
1993-07-20	1993-07-09	Yohkoh/SXT	Lemen *et al.*, 1996
1994-02-09$^{(MC)}$	1994-02-01	Yohkoh/SXT	Bothmer *et al.*, 1995, 1996; Weiss *et al.*, 1996; Lemen *et al.*, 1996
1994-02-27	1994-02-20	Yohkoh/SXT	Bothmer *et al.*, 1995, 1996; Gosling *et al.*, 1994b, 1995c, 1998; Lemen *et al.*, 1996; Weiss *et al.*, 1996
1994-04-20	1994-04-14	Yohkoh/SXT	Alexander *et al.*, 1996; Bothmer *et al.*, 1995, 1996; Gosling *et al.*, 1994b; Hudson *et al.*, 1996; Weiss *et al.*, 1996; Lemen *et al.*, 1996
1996-10-14$^{(MC)}$	1996-10-05$^{(3p)}$	SOHO/EIT/LASCO	Funsten *et al.*, 1999; Gosling *et al.*, 1998; Hudson *et al.*, 1996; Watari *et al.*, 2002
1996-12-10$^{(MC)}$	1996-11-28$^{(3p)}$	SOHO/EIT/LASCO	Funsten *et al.*, 1999
1997-01-08$^{(MC)}$	1996-12-21/25	SOHO/EIT/LASCO	Funsten *et al.*, 1999
	1996-12-21$^{(3p)}$	SOHO/LASCO	Watari *et al.*, 2002
1997-11-13$^{(MC)}$	1997-10-23	SOHO/LASCO	Watari *et al.*, 2002
1998-03-21$^{(MC)}$	1998-02-28	SOHO/EIT/LASCO	Skoug *et al.*, 2000
1998-10-10$^{(MC)}$	1998-09-23	Yohkoh/SXT	Richardson *et al.*, 2002
		GOES/Soft X-ray	
2001-05-10	2001-05-07	SOHO/LASCO	Simnett, 2003
2001-10-29$^{(MC)}$	2001-10-24	SOHO/EIT/LASCO	Reisenfeld *et al.*, 2003b
2001-11-08$^{(MC)}$	2001-11-04	Sacramento Peak	Reisenfeld *et al.*, 2003a
		SOHO/LASCO	
2001-11-26	2001-11-22	SOHO/LASCO	Reisenfeld *et al.*, 2003a,b

(MC) ICMEs identified by the corresponding authors or by Rodriguez *et al.*, 2004 as MC (3p) CMEs which show a 3-part structure on LASCO (H. Cremades, private communication).

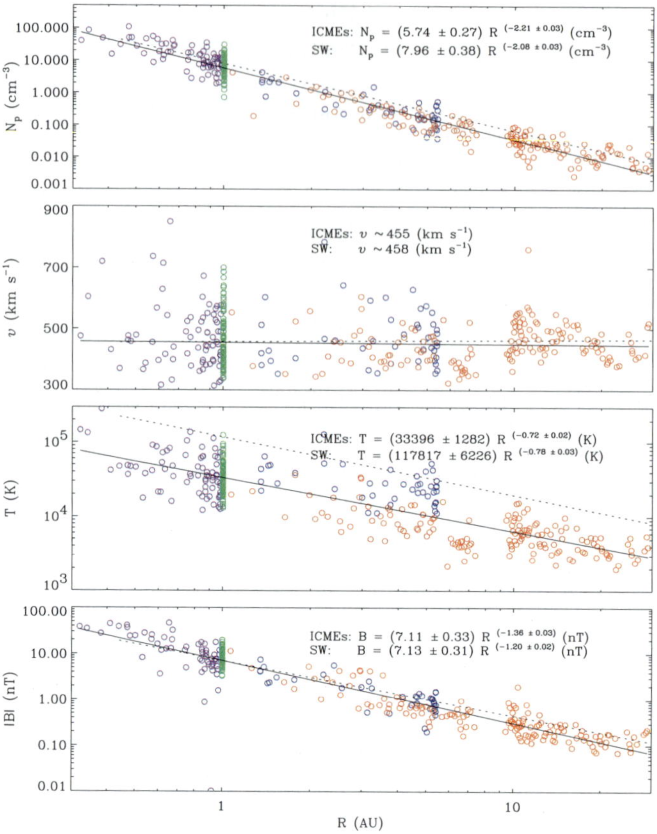

Figure 6. Radial profile of solar wind and IMF parameters observed in ICMEs at differenbt spacecraft (open symbols and solid line) and in the undisturbed solar wind (dashed line) at Voyager 2. From the top, the panels show solar wind density, speed, proton temperature, and $|B|$.

supplemented by the other plasma and magnetic field data to identify ICMEs at Voyager 2 out to 30 AU. The expansion of ICMEs leads to the expectation that the density, magnetic field strength, and temperature should decrease faster in ICMEs than in the ambient, non-ICME solar wind.

Figure 6 shows radial profiles of solar wind parameters inside ICMEs at different spacecraft and in the non-ICME solar wind at Voyager 2. The open symbols show the average values for each plasma parameter within each ICME; these ICME parameters are fit to a power law shown by the solid line. The ICME times are removed from the Voyager 2 solar wind data, and the remaining Voyager 2 solar wind data are also fit with a power law shown by the dashed line. These fits are summarized in Table II.

The density in the ambient solar wind varies with solar cycle and heliolatitude but on average decreases as R^{-2} to at least 70 AU (Richardson *et al.*, 2004) The ICME density in Figure 6 decrease as $R^{-2.21}$, faster than the ambient solar wind.

TABLE II
Radial variation of SW and IMF parameters

	ICMEs	Undisturbed solar wind
N_p (cm^{-3})	5.74 ± 0.27	7.96 ± 0.38
V (km/s)	455	458
T (K)	$33398 \pm 1282\ R^{-0.71 \pm 0.02}$	$117817 \pm 6226\ R^{-0.78 \pm 0.03}$
B (nT)	7.11 ± 0.33	7.13 ± 0.31

The constant in the fit is also less in ICMEs than in the ambient solar wind; since the densities within ICMEs at 1 AU are roughly equal to those in the surrounding plasma (Crooker *et al.*, 2000), the lower value is likely also the result of the ICME expansion beyond 1 AU.

The speeds inside and outside ICMEs are nearly identical and do not change significantly with distance out to 30 AU. The variation of ICME speed decreases with distance as the ICMEs are entrained in the surrounding solar wind flow.

The temperatures in ICMEs are much lower than in the solar wind, a result forced by use of the low-temperature criterion for ICME selection. The temperature of the ambient solar wind decreases as $R^{-0.72}$, consistent with values in the literature (Gazis *et al.*, 1994; Richardson *et al.*, 1995). The temperatures inside ICMEs decrease less quickly than in the solar wind, regardless of the expected adiabatic cooling that should accompany the ICME expansion. The ICME plasma must be heated preferentially (on a percentage basis) compared to the solar wind plasma. The mechanism for this heating is not understood, although the electron heat flux and heating via damping of magnetic fluctuations are being investigated.

The magnetic field magnitude also varies with solar cycle but on average decreases roughly as Parker predicts, as R^{-1} at large distances and faster near the Sun. The magnetic field strength within ICMEs decreases more rapidly than in the solar wind, consistent with expansion of the ICMEs. Thus, at least qualitatively, all the parameters except the temperature change as expected for an expanding structure.

The ICMEs should expand in the perpendicular as well as the radial direction; this hypothesis has not yet been tested. Multiple spacecraft would be needed for such a study.

Liu *et al.* (1985) investigated the thermodynamic structure of ICMEs from 0.3 to 5.3 AU. They showed that the value of gamma, the index used under the assumption of a polytropic equation of state, is roughly constant with distance inside ICMEs. The value for γ_p is about 1, less than in the solar wind where it is 1.5 (Totten *et al.*, 1995) so that the expansion of the ICMEs behaves as an isothermal process. The significance, if any, of this quasi-isothermal temperature profile remains to be determined, but it supports the suggestion that the ICME plasma must be heated preferentially compared to the solar wind plasma.

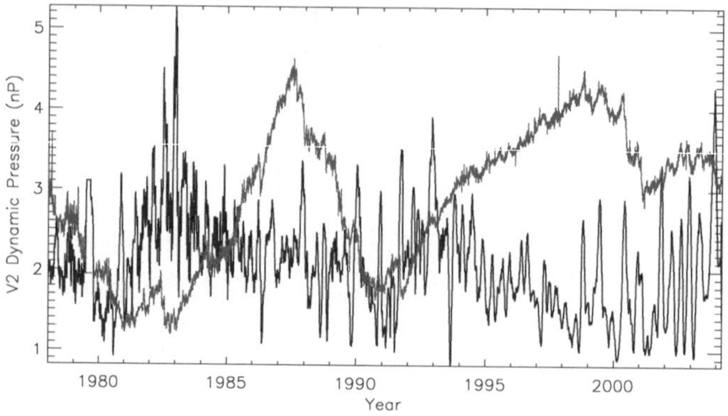

Figure 7. Solar wind dynamic pressure and cosmic ray counting rate for the entire Voyager 2 mission.

3.2. Outer Heliosphere Modeling

J. D. Richardson

The introductory chapter showed an example of a pair of ICMEs which resulted in the formation of a MIR. Figure 7 shows the solar wind dynamic pressure and cosmic ray counting rate for the entire Voyager 2 mission. MIRs are regions of high magnetic field intensity and are often formed when ICMEs compress plasma ahead of them. The enhanced magnetic field impedes diffusion of the energetic particles, resulting in step-like decreases in the cosmic ray counting rates. The figure shows that only a few large MIRs were present near the solar maximum in 1982. Four major MIRs were observed near solar maximum after 1990. But beginning in 1999, the most recent maximum was dominated by MIRs. We believe most of the MIRs during the recent maximum were driven by large ICMEs and that MIRs continue to merge as they move outward.

Richardson *et al.* (2004) have simulated this evolution using a 1-D MHD model of the solar wind propagation which included the effects of pickup ions (Wang *et al.*, 2000, 2001). They used ACE plasma and magnetic field data as input and propagated the solar wind out to Voyager 2. The model predicts the correlated structure which is observed, but not the timing. They cite this lack of agreement as evidence that transient ICMEs drive the MIRs, so that the timing of the MIRs differs with heliolongitude and depends on the ICME history at each longitude.

3.3. Energetic Particle Response to ICMEs at Large Heliocentric Distances

D. Lario, R. B. Decker

To study the radial evolution of the energetic particle responses to the passage of ICMEs, it is essential to analyze those cases when the same ICME has been observed

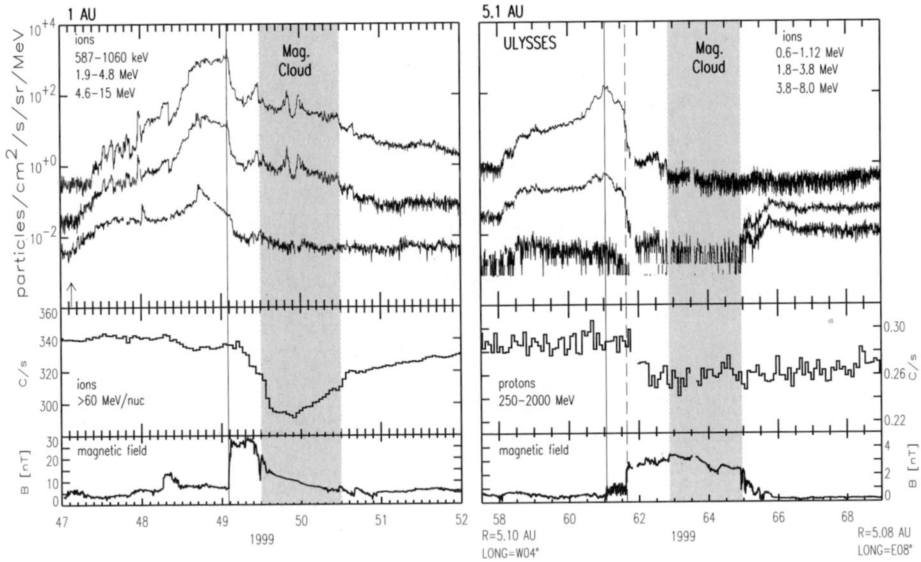

Figure 8. Ion intensities and magnetic field magnitude as measured by the ACE and IMP-8 spacecraft (left panel) and the Ulysses spacecraft (right panel) during an event in February 1999 when the same ICME was observed first at 1 AU and later at 5.1 AU. Solid vertical lines indicate the passage of shocks and gray vertical bars the passage of ICMEs.

at different heliocentric distances. Figure 8 shows one of the few cases where the same ICME was first observed at 1 AU by the ACE spacecraft and later at 22°S and 5.1 AU by the Ulysses spacecraft (Lario *et al.*, 2001). Both low-energy (<8 MeV) ion and cosmic ray intensities decreased at the entry of both spacecraft into the high magnetic field region associated with the passage of the ICME. Low-energy ion intensities peaked around the passage of the preceding shocks (as a result of local shock-acceleration) but decreased during the passage of the ICME. Bidirectional ~1 MeV ion flows were observed within the ICME at 1 AU whereas at 5.1 AU ion flows were mainly isotropic (see details in Lario *et al.*, 2001).

At larger heliocentric distances, it is difficult to observe single and isolated ICMEs. The individual solar energetic particle (SEP) events observed at 1 AU also lose their identity at large heliocentric distances because of both the superposition of multiple energetic particle events and the continuous acceleration of particles by large-scale interplanetary disturbances driven by the ICMEs. In addition, the coalescence of multiple ICMEs originating during periods of intense levels of solar activity results in the formation of MIRs (Burlaga, 1995). Figure 9 shows data from the Voyager 2 spacecraft at 34.6 AU during 10 days of May 1991 centered around the arrival of an interplanetary shock observed at ~0230 UT on day 146 (Decker *et al.*, 1995). Both before and roughly a day after the shock passage there was no response above background for ion channels with energies >500 keV. However, early (~0400 UT) on day 147 these channels increased rapidly over only a few

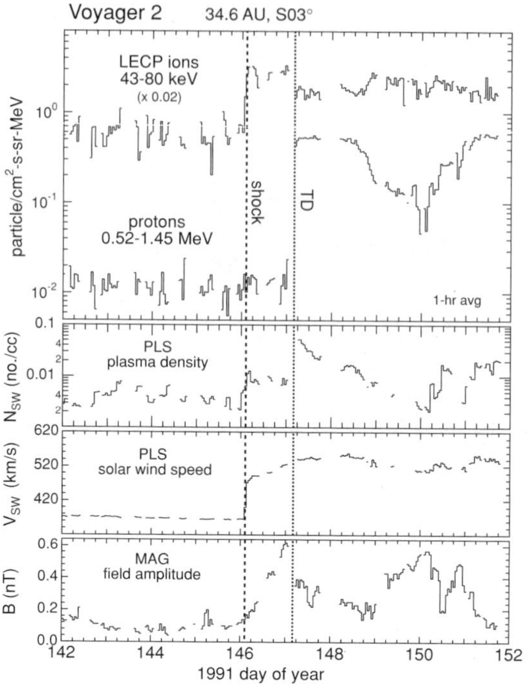

Figure 9. Voyager 2 data during the May 1991 event showing from top to bottom: intensities of 43–80 keV and 0.52–1.45 MeV protons as measured by the LECP instrument; solar wind density and speed as measured by the Plasma Science experiment; and magnetic field magnitude as measured by the magnetometer on board V2. The shock passage (dashed line) is at ~0230 UT on day 146. Dotted line at ~0400 UT on day 147 indicates the passage of a tangential discontinuity (TD) that marks the boundary of the driver gas. Details can be found in Decker *et al.* (1995).

hours. This abrupt increase of the energetic ion intensity (and plasma density) has been interpreted as a result of the entry of the spacecraft into the magnetically confined structure that drove the interplanetary shock (i.e., the ejecta; Decker *et al.*, 1995). At lower energies ($\lesssim$200 keV) ion intensities peaked in the shell confined between the shock and the plasma driver. These low-energy ions showed abrupt intensity changes across the TD early on day 147. A possible interpretation of this event is that (1) the shock was only able to locally accelerate ions at low energies ($\lesssim$200 keV) but not at higher energies, and (2) the energetic particles observed after the TD were effectively confined within the ejecta. This event did not have a counterpart at larger heliocentric distances (45 AU) and northern latitudes (31°N) where Voyager 1 was located. This event is thus a clear example of a transient interplanetary disturbance that remained confined to a limited latitudinal range.

Figure 10 shows the arrival at Voyager 2 of two strong interplanetary shocks in September 1991 and October 2001 (Decker *et al.*, 1995; Decker and Krimigis, 2003). Ion intensity increases associated with the shock in September 1991 are clearly dependent on the energy considered. Low-energy channels ($\lesssim$80 keV) show

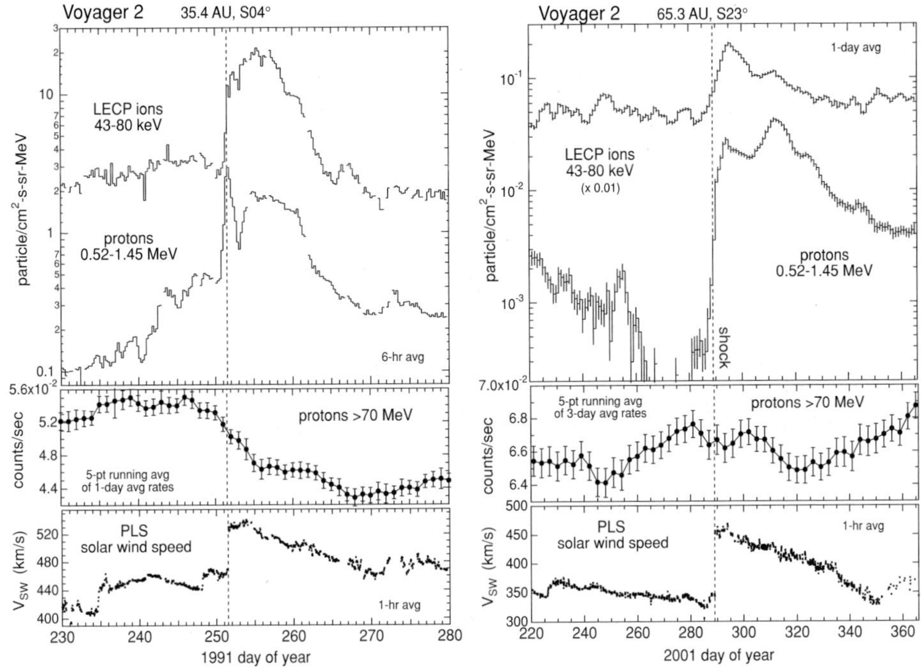

Figure 10. Voyager 2 data showing from top to bottom: intensities of 43–80 keV and 0.52–1.45 MeV protons; count rates of >70 MeV protons; and solar wind speed during the passage of two GMIRs in September 1991 (left panels) and October 2001 (right panels). Dashed vertical lines indicate the passage of interplanetary shocks.

a step-like increase after the shock passage, whereas the higher energies have localized peaks centered close to the shock passage followed by a notch and then a later increase about 2 days after the shock passage. It is tempting to interpret this new increase of particle intensities as the entry of Voyager 2 into the driver gas or ejecta where low-energy ions remained confined (similar to the event shown in Figure 9). Ion intensity increases associated with the shock in October 2001 show a considerable level of energy-dependent spatial structure (Decker and Krimigis, 2003). The bulk part of the particles were observed in the downstream region of the shock. Both events shown in Figure 10 were also observed by Voyager 1, well separated from Voyager 2 in radius, latitude and longitude, and therefore were associated with the passage of GMIRs. Cosmic ray depressions were observed in both events shown in Figure 10, although the event in October 2001 occurred during the recovery phase of solar cycle 23 and the effects of the GMIR were not as pronounced as during September 1991.

To summarize, ICMEs at large heliocentric distances are often associated with MIRs or GMIRs. Energetic ion intensities display significant structure that is markedly different from event to event. Low-energy ion intensities tend to peak at or after the passage of the shocks (if at all). In contrast to observations in the inner heliosphere where low-energy ion intensities decrease during the passage of ICMEs (Figure 8), in the outer heliosphere high ion intensities may be observed during the passage of the gas forming the MIRs or GMIRs (Figures 9 and 10). The characteristics of energetic particle enhancements observed during the passage of these structures depends on the balance between particle acceleration at the shocks and particle confinement, energy loss, and transport within these structures (see discussion in Lario *et al.*, 2004).

4. Large Transient Events in the Outer Heliosphere

As observed by several spacecraft during 1982 and 1991, the heliosphere was swept by a series of large global transient events. The 1982 events (Lockwood and Weber, 1984; VanAllen, 1987) occurred during the declining phase of solar cycle 21 and the 1991 events (Webber and Lockwood, 1993; VanAllen, 1993; McDonald *et al.*, 1994) occurred during the declining phase of solar cycle 22. A similar but weaker series of events was observed in 2000/2001 during the declining phase of solar cycle 23 (Wang *et al.*, 2001; Burlaga *et al.*, 2003a,b).

These events were characterized by increases in solar wind speed, density, temperature, and IMF magnitude over timescales of 10–100 days (McDonald *et al.*, 1994; Richardson *et al.*, 2004) that were associated with energetic particle enhancements, cosmic ray modulation, and the large Forbush decreases of 1982 and 1991 (Webber and Lockwood, 1993; McDonald *et al.*, 1994). It has also been suggested (McNutt, 1988; Gurnett *et al.*, 1993). that they are associated with the 2–3 kHz heliospheric radio emission.

Burlaga *et al.* (1984, 1993) suggested that ICMEs and CIRs could merge to form expanding quasi-spherical shells which they termed GMIRs. These GMIRs would be associated with regions of intense magnetic field that would act as diffusive barriers to produce step decreases in cosmic ray intensity as proposed by Perko and Fisk (1983).

In practice, the term GMIR has not always been used consistently by different observers, and it has been applied to a variety of different large transient events. It remains to be determined whether all of these are similar. The discussion in this section shall be restricted to the limited number of large global transient events that were associated with large ($>10\%$) Forbush decreases.

The formation of GMIRs has been the subject of numerous models and is discussed in detail by Burlaga (1995). These simulations are necessarily limited in scope. In particular, it remains to be determined what distinguished the large global transient events from smaller events and why these events were restricted to limited time periods during the descending phase of the solar cycle. But in view of their extended duration and large angular extent, it is reasonable to assume that the large global transient events that involve succession of streams and ICMEs that merge to produce a quasi-spherical shell or shells. This is represented schematically in Figure 11.

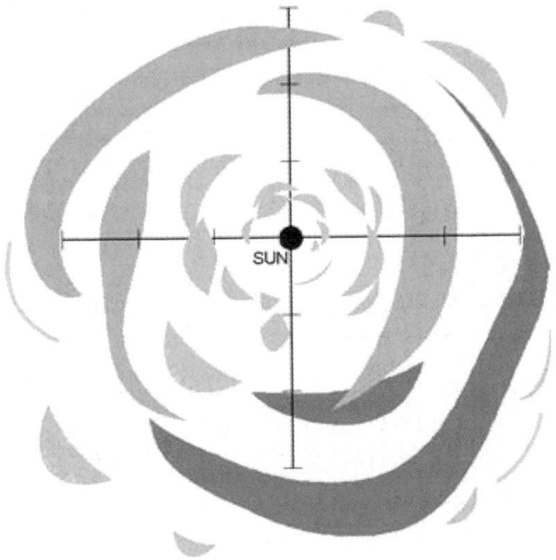

Figure 11. Schematic representation of the combination of multiple ICMEs and CIRs to form a quasi-spherical shell.

4.1. The Structure of Large Transient Events

P. R. Gazis

Throughout 1991, a succession of large global transient events were observed by multiple spacecraft throughout the heliosphere, from the heliocentric distance of 0.72 AU to greater than 53 AU. These events involved two large Forbush decreases that were observed around day 84 and day 162 at Earth. They were described in detail by McDonald *et al.* (1994).

Figure 12 shows a comparison of solar wind speed observed at PVO, IMP, Ulysses, Voyager 2, and Pioneer 10 along with the cosmic ray count rate observed by the Climax neutron monitor for 1991 and early 1992. No attempt has been made to delay data to account for solar wind travel time. Each spacecraft observed a succession of peaks in solar wind speed over a 200-day time period. The onset of this activity at different spacecraft (around day 60 at PVO, day 70 at IMP and Ulysses, day 170 at Voyager 2, and day 240 at Pioneer 10) was consistent with the expansion of a quasi-spherical shell with a speed of $\simeq$600 km/s. The epochs of this activity along with the heliocentric distances, heliographic latitudes, and travel times for the different spacecraft are summarized in Table III.

The 1991 events were observed over a broad range of heliographic longitudes. They were not uniform with longitude. In particular, they were somewhat more pronounced in the direction of Pioneer 10 (McDonald *et al.*, 1994). The detailed structure of these events evolved with heliocentric distance. At heliocentric distances less than 5 AU, the events took the form of a succession of peaks in solar

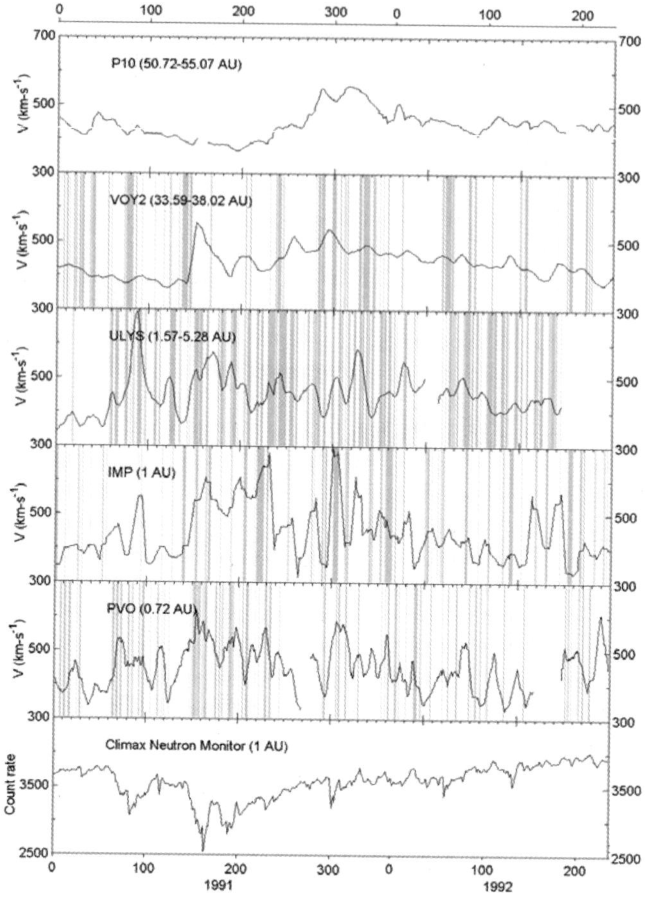

Figure 12. Solar wind speeds and cosmic ray intensities for 1991 and 1992. >From the top, panels show 10-day running averages of solar wind speeds from Pioneer 10, Voyager 2, Ulysses, IMP, PVO, and daily-averaged cosmic ray intensities from the Climax neutron monitor.

wind speed with durations on the order of 10 days or less. At larger distances from the sun, these peaks began to merge to form two much broader structures with durations on the order of 50 days.

These events were associated with a succession of ICMEs. The location of possible ICMEs observed at PVO, IMP, Ulysses, and Voyager 2, as determined by depressions in proton temperature, is shown by the gray bars in the figure. The pattern of these events was complex, and comparisons between observations at different spacecraft are complicated by uncertainties in the measurement of temperatures, but several broad trends can be observed: (1) the pattern of events observed at different spacecraft was different, and the differences cannot be reconciled by a simple shift in the time axis to account for solar wind propagation; (2) ICMEs appeared to be more frequent in the vicinity of peaks in solar wind speed, particularly at

TABLE III
Locations and transit times for the events of 1991

Spacecraft	Epoch	R (AU)	Heliolongitude	Transit time from IMP
PVO	60–230	0.72	65°–320°	−9d to–1d
IMP	70–230	1.00	170°–325°	n/a
Ulysses	70–240	2.34–4.11	124°–148°	1d to 21d
Voyager 2	170–300	34.82–35.77	282°–282°	101d to 95d
Pioneer 10	240–340	52.46–53.19	73°– 73°	158d to 154d

PVO and IMP; and (3) the frequency of ICMEs appeared to be lower at Voyager 2, which suggests that ICMEs may merge and/or decay as they are convected to larger heliocentric distances.

Figure 13 shows time series of solar wind, IMF, and energetic particle measurements from Voyager 2 between days 50 and 300 of 1991. (Note that this encompasses the 10-day interval shown in Figure 9). Voyager 2 observed two Forbush decreases during this time period, in the vicinity of days 147 and 251. These events were characterized by increases in solar wind speed, density, and temperature that endured for more than 40 days. This is substantially longer than a solar rotation period, and consistent with the suggestion that the large global transient events were formed by the merging of a succesion of events of shorter duration.

Both of these events were accompanied with increases in the magnitude of the IMF. Presumably these increases were associated with the diffusive barrier that was responsible for Forbush decreases. It is noteworthy that no Forbush decreases were associated with the increases in solar wind speed, density and temperature that began near days 180, and 280 that were not accompanied by increases in the IMF magnitude. Cosmic ray modulation issues are discussed in greater detail in the next section.

The large global transient events were accompanied by sizable increases in the flux of low-energy particles. These are discussed in detail by McDonald *et al.* (1994). and elsewhere in this chapter. It remains to be determined whether these increases in flux were associated with solar energetic particle events that persisted to large heliocentric distances or were due to local acceleration. McDonald *et al.* (1994) concluded that these events were consistent with local acceleration.

The shaded regions in the bottom panel indicate times of regions of low proton temperature that may be associated with ICMEs (Wang and Richardson, 2004). Three points should be noted. First, each large global transient event was associated with a cluster of possible ICMEs. This is consistent with the picture in (Figure 11). Second, the possible ICMEs tended to occur before the large global transient events themselves. This is unlikely to be due to selection effects associated with the detection scheme, for the large global transient events are characterized by high solar wind speeds and temperatures, where unusually low temperatures should

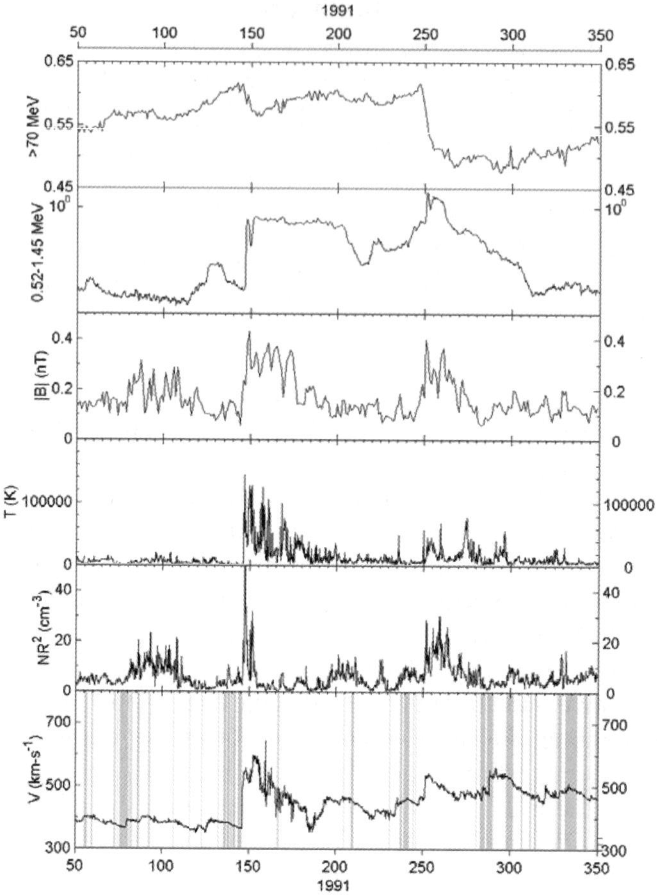

Figure 13. Solar wind and energetic particle data from Voyager 2 for the large global tranbsient events of 1991. From the top, panels show intensity of >70 MeV cosmic rays, 0.52–1.45 protons, IMF magnitude, and solar wind temperature, density, and speed.

be easier to observe. Finally, ICMEs did not always seem to be associated with large transient events. In particular, the clusters of possible ICMEs that occurred after days 60, 270, and 360 did not seem to be followed by GMIRs.

4.2. Modulation Issues Associated with Propagation of ICMEs Through the Heliosphere

M. S. Potgieter

Long-term modulation of galactic cosmic rays in the heliosphere shows an 11-year cycle, anti-correlated with solar activity. A 22-year cycle is also evident, with some periods of maximum intensity following a plateau-like and some peaked-like profiles. In an attempt to model these features (Potgieter and le Roux, 1992) illustrated that a time-dependent numerical model (based on Parker's transport equation) with gradient, curvature and current sheet drifts (Jokipii *et al.*, 1977) could reproduce these observations for solar minimum periods. They assumed the waviness of the heliospheric current sheet (HCS) as the only time dependent parameter. Later, le Roux and Potgieter (1995) showed that their model could not reproduce observations for increased solar activity with changes in the HCS as the only time dependent parameter, particularly true when large and prominent step decreases occurred (McDonald *et al.*, 1981). To simulate intensities during moderate to high solar activity requires propagating diffusion barriers, first introduced by Perko and Fisk (1983). The extreme form of these diffusion barriers are GMIRs as introduced by Burlaga *et al.* (1993; see reference therein). It was illustrated by le Roux and Potgieter (1995) that it was possible to simulate a complete 11-year modulation cycle including the large steps by a combination of drifts and GMIRs in a comprehensive time-dependent model. The period during which the GMIRs affect long-term modulation depends on their rate of occurrence, the radius of the heliosphere, the speed with which they propagate, their spatial extent and amplitude, and the background diffusion coefficients they encounter. Large-scale drifts, on the other hand, seem to dominate the minimum modulation periods so that during an 11-year cycle a transition must occur from a period dominated by drifts to a period dominated by propagating structures (e.g., Potgieter and Ferreira, 2001).

The GMIR simulations were done for radial distances >20 AU allowing enough time for merging of large structures (transients) to take place. Cane *et al.* (1999) argued that the step decreases observed at Earth could not be caused by GMIRs occurring later and beyond 10 AU. Instead, they suggested that time-dependent global changes in the heliospheric magnetic field alone might be responsible for long-term modulation. Modeling done by le Roux and Fichtner (1999) also showed that a series of GMIRs only could not reproduce the observed level of modulation. This could only be achieved by adding some global, long-term variation in e.g., the diffusion coefficients and/or in the HCS. The basic concept of Cane *et al.* (1999) was used by Potgieter and Ferreira (2001) by changing all the diffusion coefficients in a time-dependent model to reflect the time-dependent changes in the measured HMF magnitude at Earth. These changes were convected outwards at the solar wind speed to form effective propagating diffusive barriers, changing their frequency with the solar cycle. This approach could only simulate the 11-year cycle at neutron monitor energies. For rigidities < 5 GV it resulted in far less modulation than what was observed so that they developed a compound approach where the diffusion coefficients scale inversely to the power of the magnetic field magnitude, dependent on the energies used Ferreira and Potgieter (2004). They also found that some merging between neighboring propagating diffusion barriers was necessary for the model to simulate the large steps e.g., in 1981–1983 and 1991.

It has not yet been shown what the physical relation is between the large-scale modulation transients (barriers), supposedly the causes of long-term modulation, and ICMEs, e.g., how many ICMEs should occur for a GMIR to develop if ICMEs are indeed the primary cause of GMIRs.

4.3. Modulation Issues Associated with Passage of an ICME Into the Heliosheath

J. Kota

Large ICMEs form GMIRs in the outer heliosphere. Such propagating disturbances have been observed in the distant heliosphere directly and indirectly through their effects on anomalous and galactic cosmic rays. Cosmic rays, owing to their mobility, can furnish information on ICMEs well after they passed the farthermost spacecraft.

McDonald *et al.* (2000) studied the response of galactic and anomalous cosmic rays to the passage of a large GMIR in the 1991 March/June period. These events were very intense, the June event caused one of the largest Forbush decreases observed at Earth. The effects of the GMIR were well documented to 50 AU. The GMIR caused simultaneous steplike decreases in both the ACR and GCR fluxes as the disturbance passed the Voyager 1 spacecraft, which was at 46 AU that time. This decrease was followed by a slow recovery as the GMIR propagated further out. When comparing the response of GCR and ACR, McDonald *et al.* (2000) found a remarkable difference between the temporal variation of GCR and ACR: the recovery of GCR took considerably longer than the recovery of ACR. This observation indicates that the disturbance propagating into the heliosheath does not affect anomalous cosmic rays but still can impede the flux of galactic cosmic rays. An important implication is that a large part of cosmic-ray modulation takes place in the heliosheath. The inferred transit-time the GMIR took to reach the termination shock, made possible to estimate the location of the termination shock at 88.5 $\pm$ 7 AU.

Le Roux and Fichtner (1999) modeled the passage of GMIRs through the termination shock into the heliosheath and consequent effects on cosmic rays in a spherically symmetric model. In another work, Jokipii and Kóta (2001) applied a two-dimensional, time-dependent numerical code to model the effect of a barrier propagating through the termination shock into the heliosheath on ACR and GCR. These simulation results were broadly consistent with the observations of McDonald *et al.* (2000) and supported their suggestion that the propagation of the disturbance beyond the shock is the explanation of the different recovery of ACR and GCR, respectively.

5. Interpretation of In Situ Observations Using Modeling

5.1. Modeling of ICMEs from the Critical Point to 5 AU

P. Riley

Models of the propagation and evolution of ICMEs provide an important insight into their dynamics and are a valuable tool in the interpretation of interplanetary in situ observations. In addition, they represent a virtual laboratory for exploring conditions and regions of space that are not conveniently or currently accessible by spacecraft.

In this section we summarize recent advances in modeling the properties and evolution of ICMEs out to moderate distances in the solar wind. We describe the current state of research and we suggest what topics will likely be important for models to address in the future. We focus on the physics described by the models and not specifically on the models themselves. Given the need for brevity, references will be selective and illustrative, rather than comprehensive. Other reviews that complement the present one have been given by Linker *et al.* (2002), Cargill and Schmidt (2002), and Riley (1999). While we emphasize fluid and MHD modeling in this report, we note that other modeling approaches have been used with success. The extension of force-free flux rope fitting (Lepping *et al.*, 1990) to include the effects of expansion (Osherovich *et al.*, 1993; Marubashi, 1997) and multiple spacecraft. Mulligan *et al.* (2001), for example, have provided improved descriptions of this important subset of ICMEs (see Klecker *et al.*, 2006, this volume, for more details). Hybrid codes have also been used to model the interaction of fast ICMEs with the ambient solar wind allowing limited ion-kinetic effects to be explored (Riley *et al.*, 1998).

Since the basic mechanism(s) by which CMEs erupt at the Sun (Forbes, 2000, Klimchuk, 2001) is (are) not well known, it is therefore not surprising that models developed to investigate the initiation and evolution of CMEs near the Sun and the evolution of ICMEs in the solar wind tend to be idealized. In fact, to make problems tractable, significant approximations must be made. For example, consider the placement of the inner radial boundary. For many years, this was chosen to be beyond the outermost critical point (e.g., Hundhausen and Gentry, 1968; Dryer *et al.*, 1989; Riley *et al.*, 1997; Odstrcil and Pizzo, 1999; Cargill and Schmidt, 2002; Vandas and Watari, 2002). Modeling the propagation and evolution of ICMEs beyond this point is a much simpler task than including the initiation process and evolution through the lower corona. Given accurate boundary conditions at $\simeq 20 - 30R_S$, the physics of the medium is simpler and better understood, and the magnetofluid equations used to describe the system are easier to solve. Further, the minimum time step required to advance the solutions are also typically much larger than would be required if the lower corona were included. Unfortunately, it is difficult to measure the plasma and magnetic field properties in this region, leading to the specification of ad hoc boundary conditions. Moreover, such an approach completely avoids the question of CME initiation.

An often-used approximation is to neglect the magnetic field (e.g., Hundhausen and Gentry, 1968). Thus strictly speaking the simulations are valid only for high-β ICMEs and the characteristic speeds at which pressure disturbances propagate in the simulation is less than in the real solar wind. Obviously such studies cannot address questions related to the magnetic structure of the ICME. Nevertheless, they have proven to be extremely useful in illuminating the fundamental aspects of the processes by which both transient and corotating disturbances evolve in the solar wind (see, for example, reviews by Hundhausen, 1985, Gosling, 1996).

A major drawback of initiating ICMEs at an arbitrary boundary outside the outermost critical point is one of self-consistency. Virtually any kind of perturbation can be inserted. With this freedom comes the ability to "tweak" the parameters so that a good match is found between simulation results and observations. On one hand this can be an instructive exercise, allowing one to narrow down the initial configuration of the pulse; however, particularly when non-reversible processes such as at shocks are present, there is no guarantee that the correct one has been found. Moreover, when coupled with other questionable assumptions, such as neglecting the magnetic field and/or reducing the system to cylindrical or spherical symmetry, the initial configuration may be significantly different in reality. It is likely, for example, that magnetic pressure is responsible for driving the expansion of the so-called "over-expanding" ICMEs observed by Ulysses at high heliographic latitudes (Gosling *et al.*, 1994b, 1998). Thus the one-dimensional, gas-dynamic simulations that used enhanced density to mimic the initial high pressure were probably not accurate initial configurations, even though the dynamic evolution of the ejecta, and the development of associated disturbances are undoubtedly qualitatively correct.

Nevertheless 1-D gasdynamic simulations continue to be useful tools in probing the large-scale dynamics associated with ICME evolution. For example, they have been applied to the evolution of ICMEs at large heliocentric distances (Riley and Gosling, 1998), the acceleration of ICMEs near the Sun (Gosling and Riley, 1996), and the relationship between density and temperature within ICMEs and its implications for the polytropic index of the plasma (Riley *et al.*, 2001).

5.2. Modeling ICMEs from the Solar Surface to Earth

As we have noted, modeling the solar environment below the critical points is more complicated because information can now travel in both directions. Nevertheless several groups are modeling the Sun's extended Corona from $1 R_S$ to 1 AU, and beyond. Wu *et al.* (1999), for example, generated a CME from the eruption of a helmet streamer using an ad hoc increase in the azimuthal component of the magnetic field. The University of Michigan group (e.g., Groth *et al.*, 2000; Manchester *et al.*, 2004; Roussev *et al.*, 2004) have developed a finite-volume, AMR scheme to study the evolution of ICMEs from the Sun to Earth. The CME is "initiated" in one of several ways. Groth *et al.* (2000) applied a localized density enhancement at the solar surface, essentially mimicking a pressure pulse. In contrast, Manchester *et al.* (2004) superimposed the magnetic and density solutions of the 3-D Gibson and Low (1998) flux rope within the coronal streamer belt; the CME is then being driven by the resulting force imbalance. Roussev *et al.* (2004) used data from the Wilcox Solar Observatory to mimic the May 2, 1998 CME. The team at SAIC have focused more on the underlying mechanisms that are believed to lead to CME eruption. Linker and Mikić (1997), for example, initiated an eruption through differential rotation and followed its evolution out to 1 AU. More recently, in collaboration with SEC/NOAA, they used the process of flux cancellation (see Klecker *et al.*, 2006, this volume) to simulate the eruption and evolution of an ICME all the way from the solar surface to 5 AU (Odstrcil *et al.*, 2002; Riley *et al.*, 2003). In spite of the idealized nature of the eruption process and ambient solar wind, the solutions are remarkably rich and complex. The results were used by Riley *et al.* (2003) to interpret the plasma and magnetic field signatures of an ICME observed by both ACE and Ulysses, which were aligned in longitude, but separated significantly in radial distance and latitude. These simulations also suggested that a jetted outflow, driven by post-eruptive reconnection underneath the flux rope occurs and may remain intact out to 1 AU and beyond (Riley *et al.*, 2002).

To illustrate the general features of these global MHD models, in Figure 14 we summarize the evolution of a flux rope as it propagates through the inner heliosphere. Notice how the ejecta becomes progressively more distorted with increasing heliocentric distance. By $\sim$5 AU is has been squeezed so much at low latitudes that it has evolved into two lobes, connected by a thin band of compressed field. Models incorporating a more realistic solar wind (e.g. Odstrcil and Pizzo, 1999) show even more pronounced effects. An interesting, but relatively misunderstood phenomenon is the "pancakingŠŠ" of the ejecta as it moves away from the Sun. This has typically been interpreted as the result of the fast ejecta ploughing into slower ambient solar wind and becoming compressed. While this effect undoubtedly makes a contribution, the distortion is dominated by a much simpler kinematic process related to the spherical expansion of the solar wind (Riley and Crooker, 2004).

5.3. Modeling ICMEs at High Heliographic Latitudes

One of the fundamental discoveries of the Ulysses mission was a new class of ICMEs in the solar wind (Gosling *et al.*, 1994b). While at latitudes above 35°S, during its initial poleward excursion, Ulysses became immersed in quiescent, tenuous, high-speed solar wind and observed CME profiles that were fundamentally different from those at low latitudes. They appeared to have begun life as high-pressure pulses that coasted out with the fast ambient solar wind, driving forward and reverse shocks ahead and behind them, respectively. As with their lower-latitude counterparts,

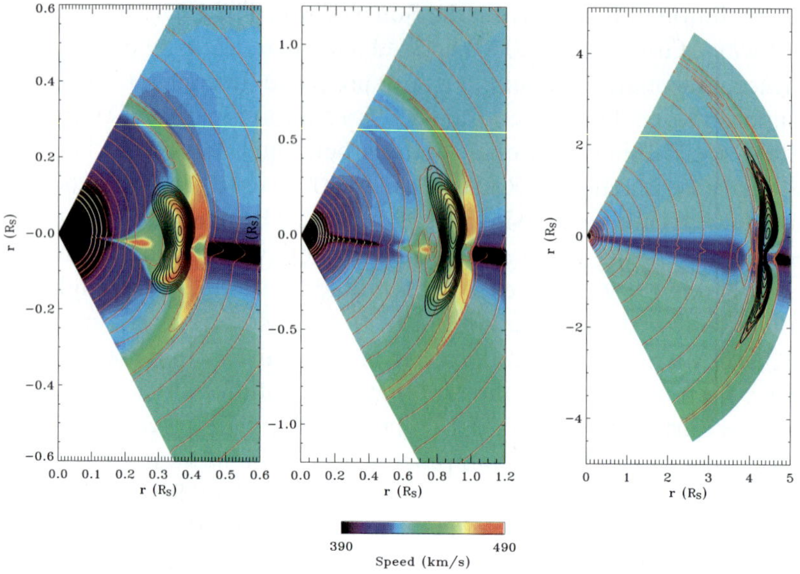

Figure 14. Evolution of a flux rope propagating through the inner heliosphere. The panels extend ±60° in latitude and from left to right, extend in heliocentric distance from the Sun to 0.6 AU, 1.2 AU, and 5 AU. The contours denote radial velocity (color); density (red lines); and magnetic field (black lines). Adapted from Riley *et al.* (2003).

some contained flux ropes while others did not. It is likely that most – if not all – of these events were high-latitude extensions of larger-scale structures. In fact at least 3 events were observed at different latitudes by two spacecraft (Hammond *et al.*, 1995; Gosling *et al.*, 1995b; Riley *et al.*, 2002). Thus the cartoons presented by Gosling *et al.* (1994b) and the simulation results by Cargill *et al.* (2000) suggesting isolated "bubbles" are almost certainly oversimplifications of structures that are considerably more complex in reality. Two- and 3-D simulations (Riley *et al.*, 1997; Odstrcil and Pizzo, 1999) have highlighted the role of the ambient solar wind in interacting with, and deforming the ejecta as it moves away from the Sun.

5.4. Modeling ICMEs in the Outer Heliosphere

J. D. Richardson

Zank and Mueller (2003) have modeled the effect of a large GMIR on the structure of the heliosphere. Figure 15 shows a simulation of the disruption of the heliosheath due to a GMIR propagating across the termination shock. The collision of the GMIR shock complex with the termination shock sets up a spatially enormous asymmetric ringing of the termination shock as it oscillates back and forth while gradually settling back into its original location. The maximum extent of the termination

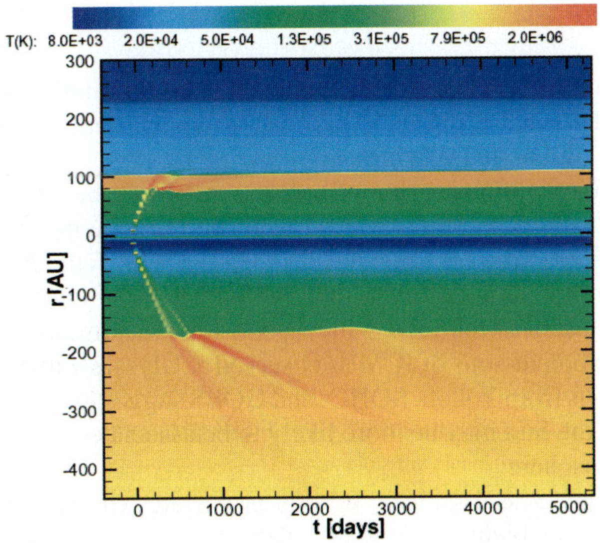

Figure 15. Space-timeplot of the plasma temperature along the stagnation axis (upwind and downwind directions) illustrating the response of the termination shock and heliopause to a shock of global extent.

shock excursions in the upwind direction is about 3 AU when it is driven outward and 5 AU when it overshoots on its recovery inward. The oscillation lasts for ~670 days. The oscillation is complicated by the overall structure of the GMIR prior to collision, which results in a long interaction time with the termination shock.

A space-time plot along the stagnation axis is shown in Figure 15. In the supersonic solar wind, the forward shock in the upwind and downwind directions propagates at an approximately constant speed, and the weaker reverse shock trails. The forward shock transmitted through the termination shock slows dramatically and the termination shock is driven outward. As is seen at the edge of the inner heliosheath, the termination shock then recovers to move inward, and shortly thereafter encounters the reduced ram pressure of the reverse shock. This causes the termination shock to accelerate inward and heat the solar wind even more. As a result, a region of strongly heated subsonic solar wind is produced and convects slowly away from the termination shock. In the upwind direction, the GMIR propagates very slowly into the shocked LISM. A reverse shock follows the leading forward shock into the LISM, and in principle, both may be responsible for the radio emission observed occasionally by the Voyager spacecraft. This scenario is repeated to a greater or lesser extent for every solar wind structure colliding with the shock. Since the recovery time from the GMIR collision is much longer (~670 days) than the observed time between MIRs (about 90 days), the heliosheath should be in a continuously disturbed state.

6. Conclusions

ICMEs have now been observed directly at high heliographic latitudes and large distances from the Sun. Results of these observations are summarized by von Steiger and Richardson (2006, this volume).

At high latitudes ICMEs are limited to a very small number of the overexpanding kind embedded in the polar high-speed streams during solar minimum. Conversely, at solar maximum ICMEs are observed at all heliographic latitudes. Still, the ICME rate is peaked at low latitudes even then, which has been interpreted as a sign of superradial expansion of the solar wind also at solar maximum.

Preliminary comparisons of ICMEs observed at Ulysses with the corresponding CME observations from Yokoh, SOHO, and GOES suggest that CMEs with a three-part structure at the Sun may be more likely to be associated with magnetic clouds in the outer heliosphere.

At large heliocentric distances, ICMEs propagate with speeds comparable to the speed of the ambient solar wind. The proton density and the magnitude of the IMF within ICMEs decrease slightly faster than in the surrounding solar wind, which suggests that ICMEs continue to expand as they are convected into the outer heliosphere. Temperatures inside ICMEs decrease more slowly than in the solar wind. This suggests that the ICMEs are preferentially heated compared to the surrounding solar wind plasma. The nature of the process or processes by which this heating might occur is an important question for the future.

Like their counterparts near the solar equator, ICMEs at high latitudes are associated with energetic particle variations. But in contrast to in-ecliptic observations at 1 AU, where low-energy particle intensities usually decrease during the passage of an ICME, at high heliographic latitudes and in high-speed solar wind streams, Ulysses observed low-energy particle intensity enhancements. At larger heliocentric distances, the Voyager and Pioneer spacecraft continued to observe low energy particle enhancements. A variety of causes have been suggested for these intensity enhancements, ranging from confinement of SEPs to local acceleration at shocks. The detailed spatial structure of these events should provide insight into relevant physical processes.

The majority of ICMEs observed by Ulysses during the high-inclination phase of its mission were associated with cosmic ray decreases. But a significant number of events did not appear to involve any variation in cosmic ray intensity. There was no obvious correlation between the characteristics of cosmic ray decreases observed at Ulysses and latitudes at which the ICMEs occurred, but Cane *et al.* (1994) have suggested that the amplitude of these events may be correlated with heliocentric distance. Cosmic ray decreases have also been observed at Pioneer and Voyager in associated with ICMEs farther from the sun. As more of these observations become available, it should become possible to characterize the radial and temporal variations of these events.

The largest variations in the intensity of energetic particles and cosmic rays were associated with the large global transient events of 1982 and 1991. Similar but smaller variations were associated with the Bastille Day and September events of 2000 (Bieber *et al.*, 2001; Burlaga *et al.*, 2003b; Wang *et al.*, 2001). While the relationship between these large global transient events and ICMEs seems firmly established, the details of this relationship remain to be determined. In particular, it is not known why these events are associated with some clusters of ICMEs and not with others, nor is it known why the largest of these events were restricted to particular solar maxima. The rate of ICMEs observed in the outer heliosphere did not appear to be significantly higher during solar maximum than it was during other portions of the solar cycle (Wang and Richardson, 2004). This suggests that the formation of large global transient events may be associated with some change in the character of CIRs and ICMEs rather than a simple change in frequency.

Models of the evolution of ICMEs have increased in realism, and can now address many of the physical processes associated the evolution of flux-rope ICMEs beyond the critical point. Predicting the path of future research is clearly speculation, but one challenge will undoubtedly involve the ability to self-consistently model ICMEs with a range of properties. Another question involves the difference between slow and fast ICMEs and whether they are generated by the same mechanism, or whether two (or more) mechanisms are responsible. Currently, self-consistent models can only produce flux-rope ICMEs. Additional work will be required to explore the underlying differences between these and ICMEs that do not contain a flux rope. In particular, is it an observational selection effect or are there intrinsically different mechanisms for producing each type?

In summary, observations from high heliographic latitudes and large distances from the Sun have finally begun to provide a picture of the structure, evolution, and distribution of ICMEs in these regions of the heliosphere. At the same time, models have advanced to the point where it may be possible to interpret this picture. We may finally be in a position to address outstanding questions related to the dynamics and evolution of ICMEs in the outer heliosphere, the physical processes associated with energetic particle enhancements, and the nature and relative importance of the diffusion boundary and other processes associated with cosmic ray modulation.

References

Alexander, D., Harvey, K. L., Hudson, H. S., Hoeksema, J. T., and Zhao, X.: 1996, in: D. Winterhalter, J. T. Gosling, S. R. Habal, W. S. Kurth, and M. Neugebauer (eds.), *Solar Wind 8*, pp. 80–83.

Armstrong, T. P., Haggerty, D., Lanzerotti, L. J., Maclennan, C. G., Roelof, E. C., Pick, M., *et al.*: 1994, *Geophys. Res. Lett.* **21**, 1747–1750.

Bieber, J. W., Dröge, W., Evenson, P., Pyle, R., Ruffolo, D., Pinsook, U., *et al.*: 2001, *Proceedings of the 27th International Cosmic Ray Conference, Hamburg 2001*, p. 3434.

Bothmer, V., Heber, B., Kunow, H., Müller-Mellin, R., Wibberenz, G., Gosling, J. T., *et al.*: 1997, *Proceedings of the 25th International Cosmic Ray Conference, Durban, South Africa, Vol. 1*, Potchefstroom, pp. 333–336.

Bothmer, V., Marsden, R. G., Sanderson, T. R., Trattner, K. J., Wenzel, K.-P., Balogh, A., *et al.*: 1995, *Geophys. Res. Lett.* **22**, 3369–3372.

Bothmer, V., Marsden, R. G., Sanderson, T. R., Trattner, K. J., Wenzel, K.-P., Balogh, A., *et al.*: 1996, in: D. Winterhalter, J. Gosling, S. R. Habal, W. S. Kurth, and M. Neugebauer (eds.), *Solar Wind 8*. AIP Conference Proceedings, 382, pp. 445–448.

Burlaga, L. F.: 1995, *Interplanetary Magnetohydrodynamics*, Oxford University Press, New York, Oxford.

Burlaga, L. F. and Behannon, K. W.: 1982, *Solar Phys.* **181**, 81.

Burlaga, L. F., Klein, L., Sheeley, N. R., Michels, D. J., Howard, R. A., Koomen, M. J., *et al.*: 1982, *Geophys. Res. Lett.* **9**, 1317.

Burlaga, L. F., Lepping, R. P., Behannon, K. W., and Klein, L. W.: 1984, *J. Geophys. Res.* **89**, 10,659–10,668.

Burlaga, L. F., McDonald, F. B., and Ness, N. F.: 1993, *J. Geophys. Res.* **98**, 1.

Burlaga, L. F., McDonald, F. B., and Habal, S. R.: 1986, *J. Geophys. Res.* **91**, 13,331.

Burlaga, L. F., Ness, N. F., McDonald, F. B., Richardson, J. D., and Wang, C.: 2003a, *Astrophys. J.* **582**, 540–549.

Burlaga, L. F., Wang, C., Richardson, J. D., and Ness, N. F.: 2003b, *Astrophys. J.* **585**, 1158–1168.

Cane, H. V.: 2000, *Space Sci. Rev.* **93**, 55–77.

Cane, H. V. and Lario, D.: 2006, *Space Sci. Rev.*, (this volume), doi: 10.1007/s11214-006-9011-3.

Cane, H. V., Richardson, I. G., and von Rosenvinge, T. T.: 1994, *J. Geophys. Res.* **99**, 21,429–21,441.

Cane, H. V., Wibberenz, G., Richardson, I. G., and von Rosenvinge, T. T.: 1999, *Geophys. Res. Lett.* **26**, 565.

Cargill, P. J., Schmidt, J., Spicer, D. S., and Zalesak, S. T.: 2000, *J. Geophys. Res.* **105**(A4), 7509.

Cargill, P. J. and Schmidt, J. M.: 2002, *Ann. Geophys.* **20**, 879.

Crooker, N. U., Shodhanand, S., Gosling, J. T., Simmerer, J., Lepping, R. P., Steinberg, J. T., *et al.*: 2000, *J. Geophys. Res.* **27**, 3769–3772.

Crooker, N. U. and Horbury, T. S.: 2006, *Space Sci. Rev.*, (this volume), doi: 10.1007/s11214-006-9014-0.

Decker, R. B. and Krimigis, S. M.: 2003, *Adv. Space Res.* **32**, 597–602.

Decker, R. B., Krimigis, S. M., McNutt, R. L., Hamilton, D. C., and Collier, M. R.: 1995, *Space Sci. Rev.* **72**, 347–352.

Dryer, M., Detman, T. R., Wu, S. T., and Han, S. M.: 1989, *Adv. Space Res.* **9**, 75–80.

Ferreira, S. E. S. and Potgieter, M. S.: 2004, *Astrophys. J.* **603**, 744–752.

Forbes, T. G.: 2000, *J. Geophys. Res.* **105**(A10), 23,153.

Forbush, S. E.: 1937, *Phys. Rev.* **51**, 1108–1109.

Funsten, H. O., Gosling, J. T., Riley, P., Cyr, O. C. S., Forsyth, R. J., Howard, R. A., *et al.*: 1999, *J. Geophys. Res.* **104**, 679–689.

Gazis, P. R., Barnes, A., Mihalov, J. D., and Lazarus, A. J.: 1994, *J. Geophys. Res.* **99**, 6561–6573.

Gibson, S. E. and Low, B. C.: 1998, *Astrophys. J.* **493**, 460.

Gopalswamy, N., Mikić, Z., *et al.*: 2006, *Space Sci. Rev.*, (this volume), doi: 10.1007/s11214-006-9020-2.

Gosling, J. T.: 1996, *Ann. Rev. Astron. Astrophys.* **34**, 35.

Gosling, J. T., Bame, S. J., McComas, D. J., Phillips, J. L., Balogh, A., and Strong, K. T.: 1995a, *Space Sci. Rev.* **72**, 133.

Gosling, J. T., Bame, S. J., McComas, D. J., Phillips, J. L., Scime, E. E., Pizzo, V. J., *et al.*: 1994a, *Geophys. Res. Lett.* **21**(3), 237.

Gosling, J. T., Birn, J., and Hesse, M.: 1995b, *Geophys. Res. Lett.* **22**(8), 869.

Gosling, J. T., McComas, D. J., Phillips, J. L., Pizzo, V. J., Goldstein, B. E., Forsyth, R. J., *et al.*: 1995c, *Geophys. Res. Lett.* **22**, 1753.

Gosling, J. T., McComas, D. J., Phillips, J. L., Weiss, L., Pizzo, V. J., Goldstein, B. E., *et al.*: 1994b, *Geophys. Res. Lett.* **21**(21), 2271.

Gosling, J. T. and Riley, P.: 1996, *Geophys. Res. Lett.* **23**(21), 2867.

Gosling, J. T., Riley, P., McComas, D. J., and Pizzo, V. J.: 1998, *J. Geophys. Res.* **103**, 1941.

Groth, C. P. T., de Zeeuw, D. L., Gombosi, T. I., and Powell, K. G.: 2000, *J. Geophys. Res.* **105**(A11), 25,053.

Gurnett, D. A., Kurth, W. S., Allendorf, S. C., and Poynter, R. L.: 1993, *Science* **262**, 199.

Hammond, C. M., Crawford, G. K., Gosling, J. T., Kojima, H., Phillips, J. L., Matsumoto, H., *et al.*: 1995, *Geophys. Res. Lett.* **22**, 1169–1172.

Heber, B., Sanderson, T. R., and Zhang, M.: 1999, *Adv. Space Res.* **23**(3), 567–579.

Hudson, H. S., Alexander, L. W. A. D., Harvey, K., Kurokawa, H., Kahler, S., and Lemen, J. R.: 1996, in: D. Winterhalter, J. T. Gosling, S. R. Habal, W. S. Kurth, and M. Neugebauer (eds.), *Solar Wind 8*. AIP Conference Proceedings, 382, p. 58.

Hundhausen, A. J.: 1985, *Collisionless Shocks in the Heliosphere: A Tutorial Review*, American Geophysical Union, Washington, DC, pp. 37–58.

Hundhausen, A. J. and Gentry, R. A.: 1968, *Astron. J.* **73**, 63.

Janardhan, P., Balasubramanian, V., and Ananthakrishnan, S.: 1997, *Proceedings of the 31st ESLAB Symposium*. ESA SP-415.

Jokipii, J. R. and Kóta, J.: 2001, *Proceedings of the 27th International Cosmic Ray Conference, Hamburg 2001*, p. 3345.

Jokipii, J. R., Levy, E. H., and Hubbard, W. B.: 1977, *Astrophys. J.* **213**, 861.

Klecker, H., Kunow, H., *et al.*: 2006, *Space Sci. Rev.*, (this volume), doi: 10.1007/s11214-006-9018-9.

Klimchuk, J. A.: 2001, *Space Weather Geophys. Monogr.*, vol. 125. Washington, DC, p. 143.

Lario, D., Decker, R. B., Roelof, E. C., Reisenfeld, D. B., and Sanderson, T. R.: 2004, *J. Geophys. Res.* **109**, A01107, (this volume), doi:10.1029/2003JA010071.

Lario, D., Haggerty, D. K., Roelof, E. C., Tappin, S. J., Forsyth, R. J., and Gosling, J. T.: 2001, *Space Sci. Rev.* **97**, 277–280.

le Roux, J. A. and Fichtner, H.: 1999, *J. Geophys. Res.* **104**, 4709.

le Roux, J. A. and Potgieter, M. S.: 1995, *Astrophys. J.* **442**, 847.

Lemen, J. R., Acton, L. W., Alexander, D., Galvin, A. B., Harvey, K. L., Hoeksema, J. T., *et al.*: 1996, in: D. Winterhalter, J. T. Gosling, S. R. Habal, W. S. Kurth, and M. Neugebauer (eds.), *Solar Wind 8*. AIP Conference Proceedings, 382, p. 92.

Lepping, R. P., Jones, J. A., and Burlaga, L. F.: 1990, *J. Geophys. Res.* **95**, 11,957.

Lepri, S. T., Fisk, T. H. Z. L. A., Richardson, I. G., Cane, H. V., and Gloeckler, G.: 2001, *J. Geophys. Res.* **106**, 29,231–29,238.

Linker, J. A. and Mikić, Z.: 1997, in: N. Crooker, J. Joselyn, and J. Feynmann (eds.), *Coronal Mass Ejections*, vol. 99. Washington, D.C., p. 269.

Linker, J. A., Mikić, Z., Riley, P., Lionello, R., and Odstrcil, D.: 2002, in M. Velli, R. Bruno, and F. Malara (eds.), *Solar Wind 10*, vol. 10. p. 703.

Liu, Y., Richardson, J. D., and Belcher, J. W.: 1985, *Plan. Sp. Sci.* p. in press.

Lockwood, J. A. and Weber, W. R.: 1984, *J. Geophys. Res.* **89**, 17.

Malandraki, O. E., Sarris, E. T., Lanzerotti, L. J., Maclennan, C. G., Pick, M., and Tsiropoula, G.: 2001, *Space Sci. Rev.* **97**, 263–266.

Malandraki, O. E., Sarris, E. T., and Tsiropoula, G.: 2003, *Ann. Geophys.* **21**, 1249–1256.

Manchester, W. B., Gombosi, T. I., Roussev, I., Ridley, A., de Zeeuw, D. L., Sokolov, I. V., *et al.*: 2004, *J. Geophys. Res.* **109**(A18), (this volume), doi: 10.1029/2003JA010150.

Marubashi, K.: 1997, in: N. Crooker, J. Joselyn, and J. Feynman (eds.), *Coronal Mass Ejections*, pp. 147–156.

McComas, D. J., Bame, S. J., Barraclough, B. L., Feldman, W. C., Funsten, H. O., Gosling, J. T., *et al.*: 1998, *Geophys. Res. Lett.* **25**, 1–4.
McDonald, F. B., Barnes, A., Burlaga, L. F., Gazis, P. R., Mihalov, J. D., and Selesnick, R. S.: 1994, *J. Geophys. Res.* **99**, 14,705–14,715.
McDonald, F. B., Heikkila, B., Lal, N., and Stone, E. C.: 2000, *J. Geophys. Res.* **105**, 1–8.
McDonald, F. B., Lal, N., Trainor, J. H., Hollebeke, M. A. I. V., and Webber, W. R.: 1981, *Astrophys. J.* **L71**, 294.
McNutt, R. L.: 1988, *Geophys. Res. Lett.* **15**, 1307.
Mulligan, T., Russell, C. T., Anderson, B. J., and Acuna, M. H.: 2001, *Geophys. Res. Lett.* **29**(23), 4417.
Neugebauer, M. and Goldstein, R.: 2003, in: Crooker, N. and Joselyn, J. A. (eds.), *Coronal Mass Ejections*. p. 245.
Odstrcil, D., Linker, J. A., Lionello, R., Mikic, Z., Riley, P., Pizzo, V. J., *et al.*: 2002, *J. Geophys. Res.* **107**, 1493.
Odstrcil, D. and Pizzo, V. J.: 1999, *J. Geophys. Res.* **104**, 28,225–28,239.
Osherovich, V. A., Farrugia, C. J., and Burlaga, L. F.: 1993, *Adv. Space Res.* **13**, 57–62.
Perko, J. S. and Fisk, L. A.: 1983, *J. Geophys. Res.* **88**, 9033.
Paularena, K. I., von Steiger, R., and Heber, B.: 2001, *Geophys. Res. Lett.* **28**, 2755–2758.
Potgieter, M. S. and Ferreira, S. E. S.: 2001, *Adv. Space Res.* **27**(3), 481.
Potgieter, M. S. and le Roux, J. A.: 1992, *Astrophys. J.* **386**, 336.
Reisenfeld, D. B., Gosling, J. T., Forsyth, R. J., Riley, P., and Cyr, O. C. S.: 2003a, *Geophys. Res. Lett.* **30**, 8031, (this volume), doi:10.1029/2003GL017155.
Reisenfeld, D. B., Gosling, J. T., Steinberg, J. T., Riley, P., Forsyth, R. J., and Cyr, O. C. S.: 2003b, in: M. Velli, R. Bruno, and F. Malara (eds.), *Solar Wind 10*. AIP Conference Proceedings, 679, pp. 210–213.
Richardson, I. G.: 1997, in: N. Crooker, J. A. Joselyn, and J. Feynman (eds.), *Coronal Mass Ejections*. AGU Geophysical Monograph, 99, pp. 189–198.
Richardson, I. G.: 2004, *Space Sci. Rev.* p. 1.
Richardson, I. G. and Cane, H. V.: 1993, *J. Geophys. Res.* **98**, 15,295–15,304.
Richardson, J. D., Paularena, K. I., Lazarus, A. J., and Belcher, J. W.: 1995, *Geophys. Res. Lett.* **22**, 325–328.
Richardson, J. D., Paularena, K. I., Wang, C., and Burlaga, L. F.: 2002, *J. Geophys. Res.* **107**(A4), (this volume), doi: 10.1029/2001JA000175.
Richardson, J. D., Wang, C., and Burlaga, L. F.: 2004, *Adv. Space Res.* **34**, 150–156.
Riley, P.: 1999, in: S. R. Habal, R. Esser, J. V. Hollweg, and P. A. Isenberg (eds.), *Solar Wind 9*, pp. 131–136.
Riley, P. and Crooker, N. U.: 2004, *Astrophys. J.* **600**, 1035–1042.
Riley, P. and Gosling, J. T.: 1998, *Geophys. Res. Lett.* **25**, 1529–1532.
Riley, P., Gosling, J. T., and Pizzo, V. J.: 1997, *J. Geophys. Res.* **102**, 14,677–14,686.
Riley, P., Gosling, J. T., and Pizzo, V. J.: 2001, *J. Geophys. Res.* **106**, 8291–8300.
Riley, P., Linker, J. A., Mikic, Z., Odstrcil, D., Pizzo, V. J., and Webb, D. F.: 2002, *Astrophys. J.* **578**, 972–978.
Riley, P., Linker, J. A., Mikic, Z., Odstrcil, D., Zurbuchen, T. H., Lario, D., *et al.*: 2003, *J. Geophys. Res.* **106**, 8291–8300.
Riley, P., Omidi, N., and Gosling, J. T.: 1998, *EOS Trans.* **79**, F697.
Rodriguez, L., Woch, J., Krupp, N., Fränz, M., von Steiger, R. J. F. R., Reisenfeld, D. B., *et al.*: 2004, *J. Geophys. Res.* **109**, A01108, (this volume), doi: 10.1029/2003JA010156.
Roussev, I. I., Sokolov, I. V., Forbes, T. G., Gombosi, T. I., Lee, M. A., and Sakai, J. I.: 2004, *Astrophys. J.* **605**, L73–L76.

Schwenn, R.: 1996, in: D. Winterhalter, J. T. Gosling, S. R. Habal, W. S. Kurth, and M. Neugebauer (eds.), *Solar Wind 8*. AIP Conference Proceedings, 382, pp. 426–429.

Simnett, G. M.: 2003, *Solar Phys.* **213**, 387.

Simpson, J. A.: 1954, *Phys. Rev.* **94**, 426–440.

Simpson, J. A., Anglin, J. D., Barlogh, A., Bercovitch, M., Bouman, J. M., Budzinski, E. E., *et al.*: 1992, *Astronomy & Astrophysics Supplements* **92**(2), 365–399.

Skoug, R. M., Feldman, W. C., Gosling, J. T., McComas, D. J., Reienfeld, D. B., Smith, C. W., *et al.*: 2000, *J. Geophys. Res.* **105**(A12), 27,269–27,275.

Totten, T. L., Freeman, J. W., and Arya, S.: 1995, *J. Geophys. Res.* **100**, 13.

VanAllen, J. A.: 1987, *Solar Phys.* **112**, 165–179.

VanAllen, J. A.: 1993, *Geophys. Res. Lett.* **20**, 2797–2800.

Vandas, M. and Watari, D. O. S.: 2002, *J. Geophys. Res.* **107**, 1236.

von Steiger, R. and Richardson, J. D.: 2006, *Space Sci. Rev.*, (this volume), doi: 10.1007/s11214-006-9015-z.

Wang, C. and Richardson, J. D.: 2004, *J. Geophys. Res.* p. in press.

Wang, C., Richardson, J. D., and Burlaga, L.: 2001, *Solar Phys.* **204**, 413–423.

Wang, C., Richardson, J. D., and Gosling, J. T.: 2000, *J. Geophys. Res.* **105**, 2337–2344.

Watari, S., Watanabe, T., and Marubashi, K.: 2002, *Adv. Space Res.* **29**(3), 451–455.

Webber, W. R. and Lockwood, J. A.: 1993, *J. Geophys. Res.* **98**, 7821–7825.

Weiss, L. A., Gosling, J. T., McAllister, A., Phillips, J. L., Hundhausen, A. J., Burkepile, J. T., *et al.*: 1996, *Astron. Astrophys.* **316**, 384.

Whang, Y. C.: 1991, *Space Sci. Rev.* **57**, 339–388.

Whang, Y. C. and Burlaga, L. F.: 1986, *J. Geophys. Res.* **91**, 13,341–13,348.

Wibberenz, G., le Roux, J. A., Potgieter, M. S., and Bieber, J. W.: 1998, *Space Sci. Rev.* **83**, 309–348.

Wu, S. T., Guo, W. P., Michels, D. J., and Burlaga, L. F.: 1999, *J. Geophys. Res.* **104**, 14,789–14,802.

Zank, G. P.: 1999, *Space Sci. Rev.* **89**, 413–688.

Zank, G. P. and Mueller, H. R.: 2003, *J. Geophys. Res.* **108**, 10.1029/2002JA009689.

CME DISTURBANCE FORECASTING

G. SISCOE[1,*] and R. SCHWENN[2]
[1]*Center for Space Physics, Boston University, Boston, MA, USA*
[2]*Max-Planck-Institut für Sonnensystemforschung, Katlenburg-Lindau, Germany*
*(*Author for correspondence: E-mail: siscoe@bu.edu)*

(Received 27 June 2006; Accepted in final form 10 July 2006)

Abstract. CME disturbances at Earth arise from the sheath that arrives in front of the ICME and from the ICME itself. The geoeffective environment is qualitatively different in the sheath than within the ICME. Consequently two types of forecast procedures using solar observations of phenomena associated with the release of the CME as input parameters have been developed to treat the two types of environment. This chapter surveys efforts that have resulted in implementable (at least in principle) forecast algorithms for sheath and ICME disturbances and discusses uncertainties associated with both.

Keywords: space weather, storm forecasting

1. Background

CMEs are the hurricanes of space weather – the storms with the greatest potential to inflict damage (e.g., Echer *et al.*, 2006). As with all storms, whether in the atmosphere or in space, the forecaster is interested in when it will start and how intense it will be.

For recent reviews of aspects of space weather that pertain to the solar and heliospheric environments, the reader might consult Joselyn (1995), Crooker (2000) and Schwenn (2006). The symptoms of space weather (including CME disturbances) as manifested through its effects on technological systems and human activities have been well described, for example, by Oldenwald (2001), Freeman (2001), and Carlowicz and Lopez (2002). This chapter applies results of CME research described in other chapters of this volume to discuss amelioration of space-weather symptoms as far as is currently possible through predicting the beginning and intensity of CME disturbances.

CME storms manifest separate magnetic and energetic particle phases. Both affect terrestrial systems, while the latter also affect spacecraft and human activities beyond the magnetosphere. The discussion here will be restricted to magnetic disturbances. These have longer lead times and, so, have greater potential for amelioration through forecast algorithms.

Space Science Reviews (2006) 123: 453–470
DOI: 10.1007/s11214-006-9024-y

2. Arrival Times of CME Disturbances

Forecasting the onset of CME disturbances from solar or coronal signatures could give one-to-four day advance warnings. The forecaster is concerned with the arrival of both the ICME shock and the ICME itself, since shocks can arrive with no following ICME ejecta (an off-center impact) and vice versa (a 'slow' ICME, subsonic with respect to the solar wind flow). Models to forecast the arrival time of ICME disturbances divide into empirical and physics-based.

2.1. Empirical Models of CME-Disturbance Arrival Time

Empirical models consist mostly of algebraic algorithms obtained by fitting curves to scatter plots of measured disturbance arrival times versus some measure of speed of a halo CME (see discussion in Section 3.2 of Forsyth *et al.*, 2006, this volume). Schwenn *et al.* (2001, 2005) define the CME "expansion speed" (V_{exp}) as the speed at which the CME expands in a direction perpendicular to its direction of propagation. As Figure 1A shows, this definition has the advantage that, unlike the apparent CME speed in the plane of the sky (V_{PS}), V_{exp} is independent of the direction of motion of the CME relative to the viewer's line of sight. Regarding the relation between V_{exp} and the actual radial speed of the front of the CME moving

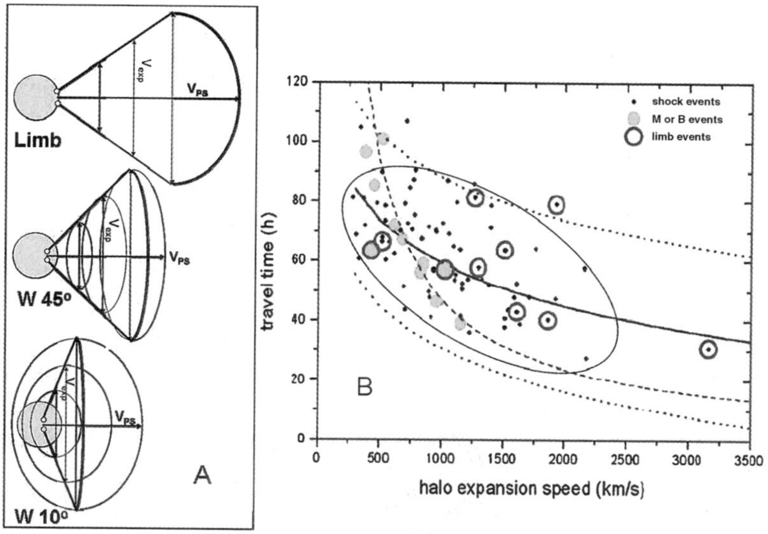

Figure 1. A. Sketch to illustrate the definition of expansion velocity V_{exp} and plane-of-sky velocity V_{PS} (after Schwenn *et al.*, 2005). B. Scatter plot of shock travel times and associated halo expansion speeds. Solid curve gives optimal fit to the functional form $y = a + b\ln(x)$. Dotted lines show two-standard deviation from the optimal fit. Dashed line gives travel time based on constant velocity. (Modified from Schwenn *et al.*, 2005).

away from the Sun (V_{rad}), Dal Lago *et al.* (2003) found $V_{rad} = 0.88 V_{exp}$ in an analysis of 57 CME-shock associations within 30° of the limb where V_{rad} could be accurately measured.

Figure 1B shows a scatter plot of 75 CME disturbance travel times versus V_{exp}. The dashed line gives the travel time based on the assumption of constant V_{rad} and the Dal Lago *et al.* relation between V_{exp} and V_{rad}. (The ellipse enclosing most of the points is a shape-and-position template for use with Figure 2.) When $V_{exp} >$ 800 km/s, most shocks arrive late relative to the constant-speed curve, implying deceleration. Schwenn *et al.* used the kinematics of viscous deceleration in a static medium to arrive at the following expression for the transit time (T_{SH}) (though this is not the actual solution of the problem),

$$T_{SH}(\mathrm{h}) = 203 - 20.77 \ln(V_{exp}(\mathrm{km/s})) \tag{1}$$

in which the constants optimally fit the data (solid curve in Figure 1B) (Schwenn *et al.*, 2005). This equation for T_{SH} represents an algorithm that in principle could give one-to-four day operational predictions of the arrival time of ICME shocks. The standard deviation of the scatter around the prediction is 14 h. The dotted lines in Figure 1b mark two standard deviations within which 95% of the points lie.

Gopalswamy *et al.* (2000, 2001) have developed an algorithm for predicting the arrival time of an ICME itself (not of its shock, if it has one) from observations of the ICME's halo-CME phase. The algorithm is based on the kinematics of constant acceleration (or deceleration) between the corona and some distance within 1 AU followed by motion at constant speed. As its initial velocity the algorithm uses the maximum plane-of-sky speed of a halo CME (V_{PS} in Figure 1A). Thus, the algorithm is

$$T_{CME,1} = \frac{-V_{PS} + \sqrt{V_{PS}^2 + 2aD}}{a}; \quad T_{CME,2} = \frac{1\ \mathrm{AU} - D}{\sqrt{V_{PS}^2 + 2aD}} \tag{2}$$

where a and D are the acceleration and the acceleration-cessation distance, and $T_{CME,1}$ and $T_{CME,2}$ are the travel times from the Sun to D and from D to 1 AU, respectively. Obviously, the total travel time is the sum of $T_{CME,1}$ and $T_{CME,2}$. The value $D = 0.76$ AU seems to give best overall results. The acceleration, a, depends on V_{PS}, since slow CMEs must accelerate up to solar wind speed and fast CMEs decelerate. Gopalswamy *et al.* (2001) have determined the dependence using concomitant CME observations and in-situ data from spacecraft at quadrature with which to identify the first signature of an arriving ICME (not its shock). They find

$$a(\mathrm{m/s^2}) = 2.193 - 0.0054\, V_{PS}(\mathrm{km/s}) \tag{3}$$

Figure 2 compares predictions of the Gopalswamy *et al.* algorithm with 47 halo CME events for which ICME signatures could be identified in Wind or ACE data (Gopalswamy *et al.*, 2001). The 18 h deviation lines contain 88% of the points. The

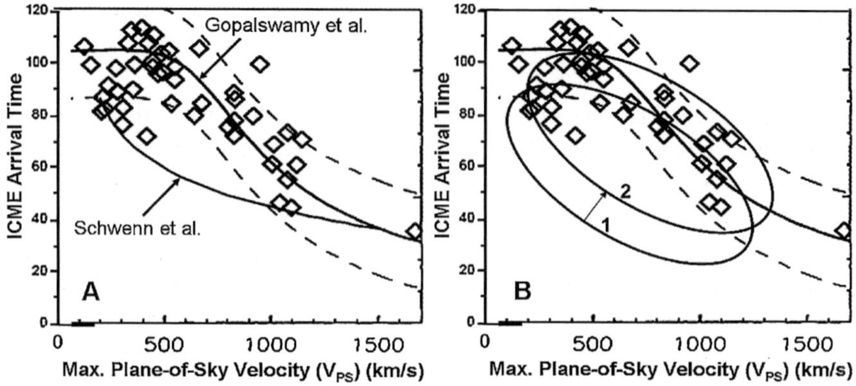

Figure 2. A. Observed ICME arrival times compared with prediction of Equations (2) and (3) (Gopalswamy *et al.* curve). Dashed lines show 18 h deviations from predicted times. The Schwenn *et al.* curve is mapped onto this figure from Figure 1 under the assumption $V_{PS} = V_{exp}/2$. B. Ellipse 1 corresponds to that in Figure 1 under the same mapping assumption. Ellipse 2 is Ellipse 1 shifted vertically to account for an average 12-h lag between shock and ICME arrivals and horizontally to enclose the maximum number of points (100 km/s) (adapted from Gopalswamy *et al.*, 2001).

flat part of the curve for $V_{PS} < 500$ km/s suggests that initially slow- and medium-speed CMEs are swept into the solar wind and are merely advected out to 1 AU and, thus, all have a typical 100 h solar wind arrival time.

Figure 2a shows the Schwenn *et al.* curve mapped from Figure 1 assuming that $V_{PS} = V_{exp}/2$, which is appropriate to a strictly circular, Sun-centered halo CME. The Schwenn *et al.* and Gopalswamy *et al.* prediction curves differ considerably for $V_{PS} < 1000$ km/s, but two corrections are to be expected since one curve refers to ICME shocks and the other to ICMEs themselves. First, ICMEs follow their shocks by typically between 6 and 12 hours (with big variations, of course) (Russell and Mulligan, 2002), which means that the Schwenn *et al.* curve should be shifted up to longer times to compare with ICME arrival times, and such a shift brings the curves closer. Second, in general a halo CME is not a circle so that in general $V_{PS} > V_{exp}/2$, and thus the Schwenn *et al.* curve should be shifted to the right to higher speeds to compare with the Gopalswamy V_{PS}-based algorithm. This also brings the curves closer.

Figure 2b carries out the mentioned shifts on the ellipse of Figure 1 (labeled 1). Clearly it poorly overlaps the data points before shifting. Shifting it 12 h up and (a not-unreasonable) 100 km/s to the right gives ellipse 2, which encloses a maximum number of points. The result is somewhat reassuring considering that the two data sets fitted in Figures 1 and 2 are completely different. Moreover, even without shifting, the two curves predict about the same arrival times for high initial speeds ($V_{PS} > 1000$ km/s), when, owing to the high speed, one expects the ICME to arrive shortly after its shock.

A major part of the differences between the two prediction curves results from the formal structure of the fitting algorithms – the Schwenn *et al.* curve has a built-in steep rise for small initial speeds and the Gopalswamy *et al.* curve has a built-in plateau for small initial speeds. Both cases are based on an analog to a simple analytic kinematic model. But even if one were to use more comprehensive analytical dynamical models as discussed in Forbes *et al.* (2006, this volume), little improvement would result. The message of the Schwenn and Gopalswamy fitting efforts is that state-of-the-art empirically-based algorithms predict arrival times of ICME disturbances with uncertainties (at the 90% level) of about plus-or-minus one day.

As was concluded earlier from Helios/Solwind studies, Schwenn *et al.* (2005) maintain that empirical algorithms (specifically, halo CME-based algorithms) are inherently incapable of reducing the stated uncertainty because it arises from variations between the Sun and Earth in the interplanetary medium into which an ICME propagates. By numerically running shock waves through different solar wind conditions, Heinemann (2002) found that that the uncertainty such differences impose on the predicted shock transit time is about plus-or-minus 25% of the predicted transit time (i.e., fast shocks have a smaller absolute arrival-time uncertainty than slow ones). It appears that the Schwenn *et al.* algorithm achieves close to the Heinemann lower limit on forecast uncertainty. Therefore, any hope to reduce the uncertainty further must lie in algorithms that can adjust an ICME's propagation speed in response to predicted variations in the upstream conditions. Such algorithms require physics-based numerical models.

2.2. Physics-Based Models of CME-Disturbance Arrival Time

Three physics-based models are currently being used to predict the arrival times of ICME shocks. They amount to different parameterizations of the physics of a shock wave propagating from a localized region near the Sun into a pre-existing solar wind. In one, the Shock Time of Arrival model (STOA) (Dryer, 1974), a shock wave is assumed to be driven at constant speed (equal to the coronal-density-dependent speed inferred from the event-associated metric type II radio frequency drift rate) for a time set by the duration of the event-associated soft X-ray emission (measured by the GOES satellite) into a Parker-solution solar wind with a speed at 1 AU equal to that measured at L1 at the time of the event. After the driving phase, the shock speed decreases with the $R^{-1/2}$ fall-off (where R is distance from the Sun) appropriate to a blast wave (Parker, 1963). This plus an assumed shock shape determine when the shock will arrive at Earth. The energy that the shock initially acquires during its prescribed launch (speed plus duration) together with the assumed shock shape and the assumed solar wind conditions determine how fast the shock weakens and, so, its strength at Earth.

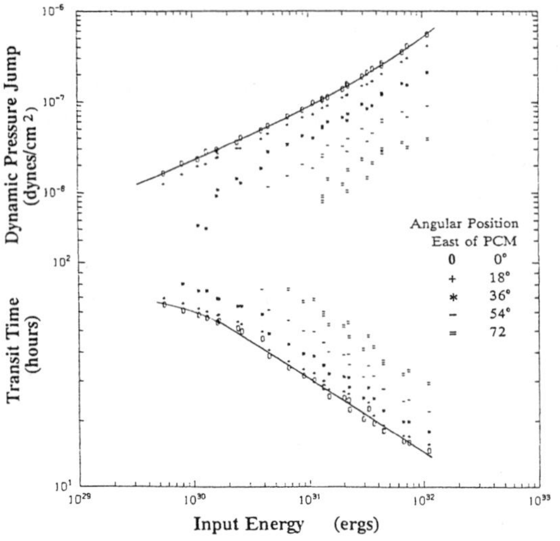

Figure 3. Results of MHD simulations of shocks propagating into a prescribed solar wind (used to parameterize the ISPM) showing how the time of arrival and shock strength at Earth depend on the initiating energy and solar longitude (from Smith and Dryer, 1990).

The second operational physics-based model, the Interplanetary Shock Propagation Model (ISPM) (Smith and Dryer, 1990), consists of analytical fits to results from a numerical MHD shock code (shown in Figure 3) using the same input parameters as STOA. Initial shock energy and location are the only input variables, not solar wind speed (unlike STOA).

The third operational physics-based model, the Hakamada-Akasofu-Fry version 2 model (HAFv2) (Fry *et al.*, 2001), uses the same data inputs as the others to specify the initial shock parameters but differs from them in using the NOAA-produced source-surface velocities near the Sun to generate an inhomogeneous solar wind. At the time of the event, over the event site, source surface velocities are replaced for a certain duration by shock-derived values. A stream-penetration-preventing kinematic algorithm is used to propagate solar wind parcels out from the source surface. The stream-penetration-preventing feature of the algorithm causes fast-moving parcels to bunch up, simulating a shock surface propagating into the inhomogeneous wind. An empirically-calibrated, gradient-threshold criterion is used to identify the shock.

The three physics-based prediction algorithms allow an assessment of the improvement physics-based makes over empirical models and the improvement that incorporating solar wind inhomogeneities makes. Fry *et al.* (2003) applied the STOA, ISPM, and HAFv2 models to 173 events to compile statistics on their performance. All three models show positive skill at the 20% level relative to predictions based on the average shock transit time for the 173 events. The root-mean-square error in predicted shock arrival times was 12.2 h, 11.2 h, and 11.6 h in the order STOA,

ISPM, HAFv2. Thus, perhaps not surprisingly, there is a 2 to 3 hour improvement compared to the 14 h RMS error for the Schwenn *et al.* algorithm. More interesting is the lack of a significant improvement between the HAFv2 model, which incorporates a constantly updated representation of the inhomogeneous solar wind into which the shock propagates, and the ISPM (with no solar wind adjustment) and STOA (with only a one-time, single speed adjustment) models. Either the kinematical treatment of stream interactions and shock identification that HAFv2 uses (which is the major difference between the models) does not adequately simulate the real situation or the error in arrival times resides in an aspect of the modeling that the three models share. This aspect may be the blast-wave formulation, which has been abandoned by many in the modeling community in favor of CME-driven shocks (see below).

Of greater concern to the forecaster than the difference between 12-h and 14-h errors in predicted arrival times are false alarms and false all-clears. Here model performance shows room for significant improvement. In all three cases, about 50% of predicted shocks do not arrive within one day of the predicted time, and after about 25% of predicted no-shocks, shocks arrive anyway.

False alarms statistics are more favorable for predictions based on halo CME signatures as used in the empirical algorithms. From a catalog of 328 entries documenting either CMEs or ICME signatures (including shocks), Schwenn *et al.* (2005) found that 85% of front-side halo CMEs were followed by an ICME disturbance at Earth. The remaining 15% of ICMEs evidently missed the Earth, which perhaps represents an irreducible false alarm rate for predictions based on halo CMEs alone.

Significant reductions in the error of predicted arrival times and of false alarms will probably not happen until full-up numerical codes become operational that self-consistently integrate the equations of motion of the entire Sun-to-Earth medium – the corona, the solar wind, the CME and the ICME. As discussed in Section 3.2 of Forbes *et al.* (2006, this volume), such codes are being constructed and tested but are still in the development stage. Readers interested in what the future offers in this area can consult reports on two ambitious space-weather-code-development projects: the Center for Integrated Space Weather Modeling (CISM) (Hughes and Hudson, 2004) and the Space Weather Modeling Framework (SWMF) (Tóth *et al.*, 2005).

3. Intensities of CME Disturbances

3.1. Geoeffective Solar Wind Parameters

Solar wind parameters most effective in causing magnetospheric storms are speed, southward-pointing magnetic field, and dynamic pressure (ρV^2) (e.g., Srivastava and Venkatakrishnan, 2004). The speed and magnetic field work in combination to generate the geoeffective component of the interplanetary electric field

($IEF = -\mathbf{V} \times \mathbf{B}$). Thus, more concisely, the geoeffective parameters are the geoeffective component of the *IEF* and ram pressure; but, as a practical matter, the speed, magnetic field, and density that make up the *IEF* and the ram pressure are separate forecast operations. (In the following, whenever the interplanetary magnetic field (IMF) is mentioned in the context of geoeffectiveness, the radial-from-the-Sun component is regarded to be zero, since it has little effect on storm intensity, and it complicates the discussion to retain it.)

The strength of the voltage across the polar cap, Φ_{PC}, (which drives ionospheric currents that produce high-latitude magnetic disturbances) is a convenient parameter to illustrate the separate roles that the *IEF* and the dynamic pressure, P_D, have in causing geomagnetic effects. This is because there is an analytic expression relating the three variables, which has the form (Siscoe *et al.*, 2002)

$$\Phi_{PC} = \frac{C_1 P_D^{1/3} IEF g(\theta)}{P_D^{1/2} + C_2 IEF g(\theta)} \tag{4}$$

where C_1 and C_2 are constants determined by theory and $g(\theta)$ is the (highly difficult-to-predict, see below) 'coupling-strength function', which depends on the angle, θ, between the IMF and the geomagnetic dipole. $g(\theta)$ ranges from unity when $\theta = 0°$ to zero when $\theta = 180°$. Figure 4 shows a plot of contours of constant Φ_{PC} in the $P_D - IEF$ plane assuming maximum coupling ($g(\theta) = 1$). For small *IEF*, Φ_{PC} is nearly independent of P_D; whereas for large *IEF*, the dependence on P_D becomes substantial while the dependence on *IEF* weakens considerably (so called transpolar potential saturation).

In a CME-induced magnetospheric storm, the disturbance arrives in two stages. First comes the ICME sheath, the wave of disturbance that precedes an ICME if it is

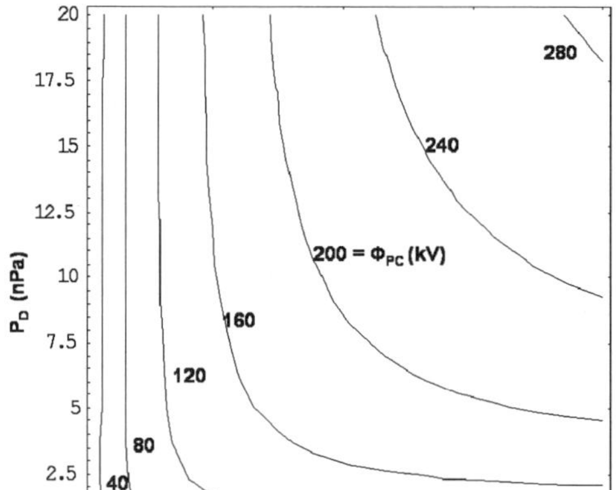

Figure 4. Contours of constant Φ_{PC} in the $P_D - IEF$ plane illustrating the control of the solar wind parameters *IEF* and P_D on a geomagnetic disturbance parameter (from Siscoe *et al.*, 2002).

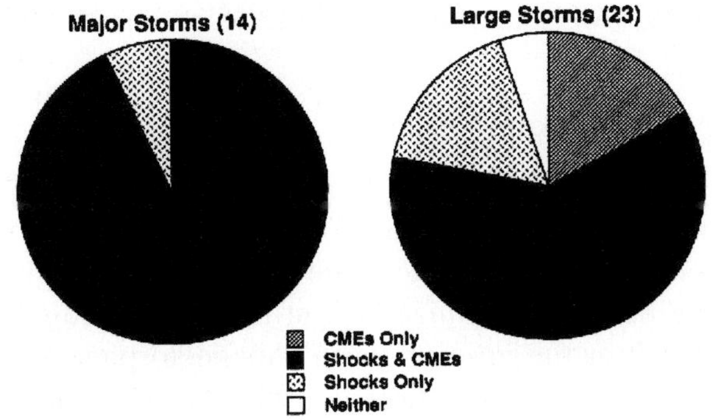

Figure 5. Pie charts showing relative occurrence frequency of 'Major Storms' (those with storm index *Kp* greater than 8-) and 'Large Storms' (*Kp* between 7- and 8-) that are caused by ICMEs without shocks, shocks without ICMEs, both together, and neither (from Gosling *et al.*, 1991).

plowing into the solar wind ahead of it, then the ICME itself, unless it is a glancing passage. As Figure 5 from Gosling *et al.* (1991) shows both the ICME sheath ('Shocks Only') and ICMEs by themselves ('CMEs only') can be geoeffective. But the one-two punch of an ICME-sheath followed by an ICME produces the most intense storms.

To illustrate what tools are currently available to predict storm intensity, it suffices to consider the ideal case in which solar indicators (halo CME, type II radio burst, X-ray duration and intensity, and the location of the associated flare or disappearing solar filament) tell the forecaster to expect a direct hit by a fast ICME and its shock. Then the forecaster considers the speed, magnetic field, and density in the ICME sheath and, separately, in the ICME body.

3.2. Geomagnetic Disturbances Induced by ICME Sheaths

Regarding the ICME sheath, Equation (5) gives an empirically-derived algorithm relating the maximum ICME-related solar wind speed, V_{Max}, at 1 AU to the shock transit time, T_{SH} (after Cliver *et al.*, 1990):

$$V_{\mathrm{Max}}(\mathrm{km/s}) = \frac{32,292}{T_{\mathrm{SH}}(\mathrm{h}) - 40} \tag{5}$$

Presumably this maximum speed is reached at the leading edge of the ICME and, thus, represents also the maximum speed in the ICME. The relevant point for forecasting is that T_{SH} can be predicted from the Schwenn *et al.* algorithm (Equation (1)). Thus, one critical, intensity-determining parameter has an implementable forecast algorithm, albeit with an uncertainty that compounds the uncertainties of Equations (1) (standard deviation of 14 h) and (5) (correlation coefficient of 0.72).

A second of the critical, intensity-determining parameters – the maximum magnetic field strength B_{Max} – also has an implementable algorithm (Owens *et al.*, 2005) that relates B_{Max} to V_{Max} of Equation (5) and, therefore, also to V_{exp} of Equation (1), as follows:

$$B_{\text{Max}}(\text{nT}) = 0.047\ V_{\text{Max}}(\text{km/s}) + 0.06 \qquad (6)$$

The uncertainty in applying Equation (6) to predict B_{Max} from V_{exp} now comprises the uncertainties in Equations (1), (5), and (6) (correlation coefficient of 0.83 (0.90 for the Owens *et al.* equation for the average sheath field strength)).

One expects the field to maximize also at the leading edge of the ICME (as for the magnetosheath at the nose of Earth's magnetosphere). Thus multiplying Equations (5) and (6) gives a prediction algorithm for the maximum value of the *IEF* in the CME sheath (but not necessarily a good prediction of the maximum geoeffective component of the *IEF* – see below). The uncertainty in this case is a fifth-order concatenation of composite uncertainties.

The third critical, intensity-determining parameter is density, n. Since post-shock flow is approximately incompressible, density could, in principle, be computed from the shock jump conditions and a prediction of the pre-shock solar wind conditions from solar data such as given by the Wang-Sheeley-Arge model (WSA) (Arge and Pizzo, 2000), where the ICME leading speed, V_{Max} would be used for the shock speed. In its present form, however, the WSA model predicts speed and IMF polarity but neither density nor magnetic field strength. Thus, one must use climatological values for these. Consequently, in a prediction algorithm for ICME-sheath density constructed from empirical formulas, the uncertainty would appear to be as great as those for V_{Max} and B_{Max}. Only in the case of a predicted fast ICME with a high-Mach-number shock would the uncertainty be reduced, for then the post-shock density is (to a good approximation) four times the pre-shock density, and the uncertainty is restricted to the uncertainty in determining the pre-shock density value. On the other hand, fast shocks are of greatest interest to the forecaster, so the situation is not as bad as it seems at first.

Regarding empirically-based algorithms for predicting the magnitudes of the intensity-determining parameters B, V, and n in an ICME sheath from near-Sun observations, one must conclude that there is a need to develop formulas that directly relate the desired quantities (B, V, and n or, better, the products BV and nV^2) to the observed solar quantities (as has been done for the shock arrival time, e.g., Equation (1)) to avoid the growth of uncertainties that results from concatenating forecast algorithms.

The physics-based forecast codes, STOA and ISPM, predict shock strength at Earth based on their estimates of initial energy in the disturbance. From the shock strength thus predicted, some estimate of ICME-sheath parameters can be computed from shock jump conditions, but again the pre-shock values needed for input to the computation must be provided by some extra-algorithmic procedure. And so, as in the empirical algorithms, there is a concatenation of uncertainties. The HAFv2 code does predict ICME-sheath parameters including B, V, and n. At present this is the most comprehensive code available for operational forecasts of geoeffective

ICME-sheath parameters; but, as of now, an assessment of its forecast skill has not been published.

The greatest uncertainty in forecasting the intensity of the disturbance that an ICME sheath will produce resides in predicting the coupling-strength function, $g(\theta)$ (Equation (4)), which multiplies the quantity usually taken to be the dominant measure of potential solar wind geoeffectiveness, the *IEF*. Although $g(\theta)$ varies from 1 when the IMF is south-pointing ($\theta = 0°$) to 0 when the IMF is north-pointing ($\theta = 180°$), its long-term average is about 0.25, based on the formula $\cos^4(\theta/2)$ that yields the greatest correlation coefficient between $VBg(\theta)$ and geomagnetic disturbance indices (Newell *et al.*, 2006). This value, corresponding to $\theta \sim 90°$, is reasonable because on a long-term average the IMF lies in the heliographic equatorial plane, approximately perpendicular to Earth's magnetic dipole.

The problem in predicting $g(\theta)$ in an ICME sheath is that the sheath is usually highly turbulent, and turbulence is inherently unpredictable. For example, McPherron and Siscoe (2004) estimated that the turbulence in the solar wind (not the turbulence in ICME sheaths, which is greater) causes the IMF to alternate randomly between northward tilting and southward tilting (relative to the heliographic equator) about 600 times between the Sun and Earth, or about every 10 minutes. An ICME sheath will compress and speed up the alternations so that the geomagnetic response,which takes typically 15 minutes or longer, acts like a low-pass filter to the IMF fluctuations. The statistical characteristics of IMF turbulence in ICME sheaths has yet to be studied using filters that simulate the magnetospheric response to the *IEF*. Such a project could lead to useful probabilistic forecasts of ICME-sheath disturbances (McPherron and Siscoe, 2004). The problem of forecasting the $g(\theta)$ caused by large-amplitude IMF turbulence in ICME sheaths is unlikely to be solved through deterministic (non-probabilistic) codes of any description.

Of greater importance to the forecaster are systematic southward or northward tilts of the IMF in ICME sheaths since these can bias $g(\theta)$ to high or low values, respectively. Systematic out-of-equatorial tilts can arise from shock deflection and field-line draping around the ICME body (Gosling and McComas, 1987; McComas *et al*, 1989; Wu and Dryer, 1996).

Figure 6 from McComas *et al.* (1989) illustrates their use as a forecast aid. It shows a CME launched from the northern solar hemisphere propagating into an IMF that points towards the Sun. The IMF tilts southward in the region of the ICME sheath that will reach Earth and thus be geoeffective. If the IMF pointed away from the Sun, it would tilt northward in the sheath and not be geoeffective. Extension to CMEs launched from the southern solar hemisphere is obvious. McComas *et al.* (1989) found that 13 of 17 events analyzed (77%) obeyed the draping prediction rule.

The most reliable predictor of a systematic bias in $g(\theta)$ comes from the variation of the tilt of the geomagnetic dipole relative to the ecliptic plane (which combines a 23.5° tilt of the rotation axis with a 11.5° tilt of the dipole relative to the rotation axis and, so, can be as great as 35°). Through a consideration of the geometry of

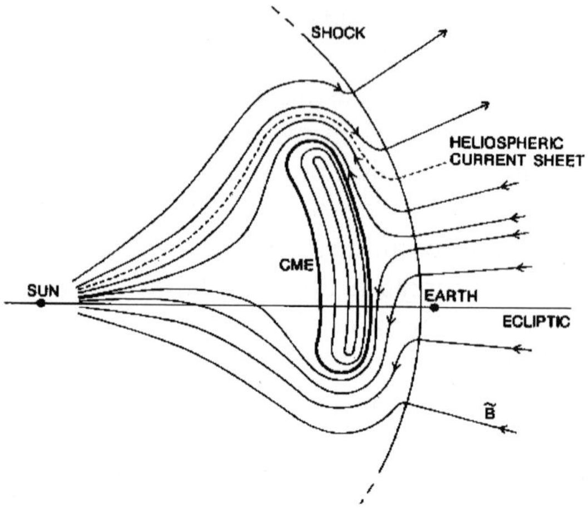

Figure 6. Out-of-equatorial tilting of the IMF cause by transiting of an off-equatorial ICME (from McComas *et al.*, 1989).

the seasonal variation of the dipole tilt relative to the geoeffective component of a Parker-spiral IMF, Russell and McPherron (1973) showed that the tilt-bias in $g(\theta)$ should maximize around the equinoxes (which, by coincidence, is close to where Earth's orbit runs parallel to the heliographic equator, thus minimizing the complicating effect of the 7.25° tilt of the ecliptic relative to it). Then even for the idealized case in which the IMF has no tilt to the equatorial plane, the value of $g(\theta)$ systematically can be considerably greater than its average 0.25 value (more than 0.6 for favorable IMF coupling) or considerably less (under 0.1 for unfavorable IMF coupling). This so-called "Russell-McPherron effect" is predictable from solar observations by the previously-mentioned Wang-Sheeley-Arge model.

3.3. Geomagnetic Disturbances Induced by ICME Bodies

The previous section reduced the problem of forecasting geomagnetic disturbances to the problem of predicting the geoeffective parameters VB, $g(\theta)$ and nV^2 in ICME sheaths. This section looks at how well these parameters can be predicted for ICME bodies. Two preliminary remarks are in order. First, the phenomenology that these parameters display in ICME bodies is completely different than in ICME sheaths; thus, nothing in the previous section regarding phenomenology applies here. Second, "ICME bodies" is a non-unique description. It could mean a magnetic cloud, or a cloud-like structure (having the magnetic but not the thermal signature of a cloud), or a non-cloud-like structure (see Zurbuchen and Richardson, 2006, this volume). Unfortunately, which type actually materializes cannot be predicted from solar observations at present. Only for magnetic clouds and cloud-like structures have disturbance forecast algorithms based on solar observations been developed.

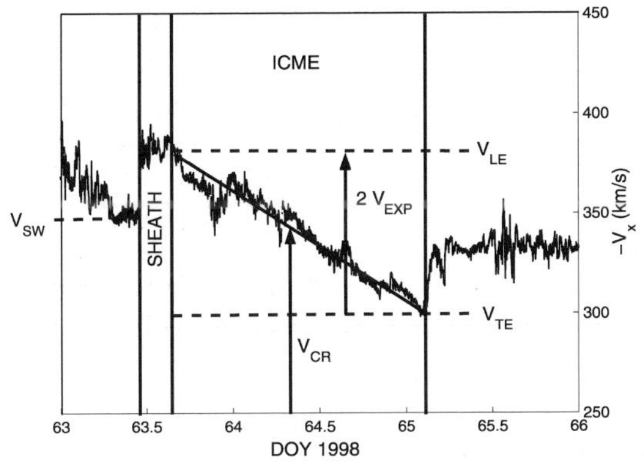

Figure 7. Velocity profile of a cloud or cloud-like ICME and the associated preceding and trailing flows. The figure defines the pre-event solar wind speed, V_{SW}, the leading-edge speed, V_{LE}, the cruise speed, V_{CR}, the trailing-edge speed, V_{TE}, and the expansion speed, V_{EXP} (from Owens *et al.*, 2005).

Estimates of the fraction of ICMEs that fall into the predictable, cloud-or-cloud-like category vary from 14% to 80% (depending on criteria used) with numbers less than 50% dominating, (Richardson and Cane, 2004). Consequently, even before a disturbance forecast algorithm based on solar observations for a cloud-or-cloud-like ICME is brought into play, an initial uncertainty of the order of 50% exists in whether such an algorithm in fact applies, and this existential reality must be incorporated in assigning a reliability tag to the forecast. (As discussed below, however, the situation improves dramatically for shorter range, yet still useful, forecasts.)

Consider then forecast procedures based on solar observations that apply to cloud and cloud-like ICMEs. As in the case of forecast procedures for ICME-sheath disturbances, procedures for ICME-body disturbances pertain to V, B, $g(\theta)$ and n separately (not to their geoeffective combinations), but with an important exception. Regarding n, there is as of yet no prediction procedure except for invoking climatology; for example, the profile of the average value of n through a magnetic cloud (based on a sample of 19 clouds) varies between 10 and 15 protons/cm^3 (Lepping *et al.*, 2003). The variation from this average profile is known to be large, however, especially toward interesting high values; but statistics from a large enough sample to define the extremes is lacking at present.

Forecast algorithms for ICME bodies based on solar observations exist for V, B and $g(\theta)$. Concerning first V, Owens *et al.* (2005) distinguish between the speed of the leading edge of an ICME, V_{LE}, and the 'cruise speed', V_{CR}, by which is meant the speed averaged over the time that the body passes. Figure 7 from the cited paper illustrates the definitions of the two speeds and defines also a 'trailing-edge speed', V_{TE}, and an ICME 'expansion speed', V_{EXP}, (not to be confused with the halo expansion speed discussed in Section 2.1). For the case of a linear velocity profile, as here, V_{CR} is simply the average of V_{LE} and V_{TE}.

The relevance of this work to forecasting is that Owens *et al.* (2005) give empirical relations for the rate at which the speed decreases within the ICME, m_{EXP}, in terms of V_{LE}, which is the same as V_{Max} of the previous section, for which Equation (5) allows predictions from solar observations:

$$m_{\mathrm{EXP}} = 10^{-8}\big(1.19\,V_{\mathrm{LE}}^2 - 954\,V_{\mathrm{LE}} + 284,180\big)\ \mathrm{km/s^2} \tag{7}$$

with a correlation coefficient of 0.9. Thus, the velocity profile through the cloud, $V_{\mathrm{CME}}(t)$, is predictable from solar observations according to

$$V_{\mathrm{CME}}(t) = V_{\mathrm{LE}} - m_{\mathrm{EXP}}\,t \tag{8}$$

The ICME ends when $V_{\mathrm{CME}}(t)$ falls to $V_{\mathrm{TE}} = V_{\mathrm{LE}} - 2V_{\mathrm{EXP}}$, for which Owens *et al.* (2005) provide the empirical relation.

$$V_{\mathrm{EXP}}(\mathrm{km/s}) = 0.266\,V_{\mathrm{LE}} - 70.6\ \mathrm{km/s} \tag{9}$$

Equations (7)–(9) together with (5) constitute a complete forecast algorithm for the velocity within an ICME magnetic cloud. They can also be used to calculate the radial half-thickness of ICME clouds, which yields values between 0.15 AU and 0.2 AU across the observed range of ICME speeds.

The other disturbance-inducing ICME parameter that is (in principle) predictable from solar observations are B and $g(\theta)$. Regarding $g(\theta)$, the state of the art is such that instead of actually predicting $g(\theta)$, one is usually limited to making a binary prediction as to whether the IMF in the ICME cloud points in a direction that favors (southward) or disfavors (northward) strong coupling to the magnetosphere, that is, in effect, whether $g(\theta)$ is close to 1 or to 0.

The forecast procedure in this case is based on the geometry of magnetic flux ropes, which ICME magnetic clouds approximate (e.g., Forsyth *et al.*, Wimmer-Schweingruber *et al.*, Zurbuchen and Richardson, 2006, this volume). The magnetic field in a flux rope has an axial component and a toroidal component (see Figure 2 in Forsyth *et al.*, 2006, this volume). If the axis of the flux rope lies more perpendicular than parallel to the axis of the geomagnetic dipole, the toroidal component dominates in determining $g(\theta)$. In this case the magnetic field will be oriented favorably for coupling in half of the rope and unfavorably in the other half, but there will always be an interval of favorable coupling. The only issue is whether it comes in the leading or trailing half of the cloud. On the other hand, if the axis of the flux rope lies more parallel than perpendicular to the axis of the geomagnetic dipole, the flux rope's axial component dominates in determining $g(\theta)$. An either-or situation results: either the magnetic field in the flux rope is oriented favorably or unfavorably for strong coupling throughout the passage of the cloud over Earth. Faced with on oncoming ICME flux rope, the forecaster is therefore interested in predicting the angle between the flux rope's axis and the axis of the geomagnetic dipole (greater than or less than 45°) and the directions of its axial and toroidal magnetic field components.

An example in which such a forecast procedure has been tested is described by Zhao and Hoeksema (1997). They define the latitude of the flux rope axis to be the angle between it and the ecliptic plane with positive latitudes corresponding to a northward axial magnetic field component. Then latitudes between $+/-45^\circ$ correspond to flux ropes more perpendicular to the dipole axis (they ignored the tilt of the dipole relative to the ecliptic), and latitudes poleward of $+/-45^\circ$ correspond to flux ropes more parallel to the dipole axis. Using data from 23 magnetic flux ropes, they plotted the duration of strong-coupling intervals against cloud-axis latitude. They obtained the expected result, that the duration varies systematically (albeit with appreciable scatter) from close to zero for a latitude of $+90^\circ$ to a maximum value for -90°, corresponding to a $\sim$20-h transit through favorable fields in the entire flux rope. Similarly, they found that the intensity of the geoeffective component of the magnetic field in the clouds also varied systematically with axis latitude in the expected sense from essentially zero to about 20 nT.

The linear fits to the Zhao and Hoeksema data provide the first step needed for a forecast algorithm:

$$D(\text{h}) = (11.49 - 0.12\,L_E) \pm 4.70 \tag{10}$$

$$Bg(\theta)\,(\text{nT}) = (10.76 - 0.10\,L_E) \pm 5.12 \tag{11}$$

where D is duration in hours, $Bg(\theta)$ is the geoeffective intensity in nT (this is the negative of the Zhao and Hoeksema intensity, I, which is modeled after the northward rather than the geoeffective southward component of the IMF), and L_E is the ecliptic latitude (in degrees) of the flux rope axis. Zhao and Hoeksema complete the task of developing a forecast algorithm by relating L_E to a solar observable associated with the release of the CME, a disappearing solar filament, DSF, following the findings of Bothmer and Rust (1997) and Bothmer and Schwenn (1998) that the orientation of magnetic clouds is well-correlated with the orientation of the source filament on the Sun (see also the discussion in Forsyth *et al.*, 2006, this volume). A DSF has a defined axis, the orientation of which relative to the solar equator measured in degrees, *Fo*, determines L_E according to the empirical formula

$$L_E(\text{deg}) = (-1.4 + 0.7Fo) \pm 17.8 \tag{12}$$

Equations (10)–(12) make up a forecast algorithm $Bg(\theta)$ for inside cloud and cloud-like ICMEs. It is, of course, important to carry out a test of the response of the magnetosphere to the $V\,Bg(\theta)$ predicted by Equations (5) and (7)–(12).

3.4. Intermediate-Term Forecasts With L1 Data

Forecasts based on solar observations at the time of the release of the CME offer a one- to four-day advance warning of the oncoming disturbance. Nowcasts based on L1 observations that merely note what is arriving as it arrives give less than one-hour advance warning. There is an intermediate forecast range that uses L1 data together with models to predict what is yet to arrive from what has already arrived. Since a CME disturbance can last more than 24 hours, there is room for useful forecasts in the 10-hour range from L1 observations.

Chen *et al.* (1997) first noted the possibility of making intermediate range forecasts with real-time L1 observations. For example, such observations quickly eliminate the uncertainty (of the order of 50%) regarding whether or not the ICME body is a magnetic cloud (or cloud-like). Chen *et al.* developed a pattern-recognition program that can identify a magnetic flux rope and its orientation after sampling about 20% of it. The remaining 80%, therefore, becomes predictable by fitting to analytical models whose parameters have been determined with data from archived events. The technique already shows successful results and has the capability for improvements by incorporating more aspects of cloud dynamics.

A second forecast procedure of this type has been proposed by Owens *et al.* (2005), in their case based on Equations (7)–(9). Instead of using Equation (5) to determine the leading-edge speed, V_{LE}, from solar observations (and, so, several days in advance), one can measure it when the ICME reaches L1. Then one can instantly calculate from the equations the velocity profile through the ICME and the duration of its passing over Earth. They also note that once the shock arrives, its speed can be calculated instantly from the shock jump relations. Since the shock speed should be the same as the speed of the leading edge of the ICME, which is V_{Max} in Equation (6), from that equation one can then determine B_{Max} in the sheath several hours before the maximum field arrives. Continuing in the same vein, one can use the Chen *et al.* procedure to determine L_E in Equations (10) and (11), to update the forecast obtained from Equation (12). It appears that the possibilities for exploiting intermediate range forecasting with L1 observations are considerable.

4. Summary

Algorithms to predict the arrival time of a CME disturbance (its shock or the ICME itself) from solar observations exist in both empirical, data-based versions and physics-based versions. The empirical algorithms have errors at the 95% confidence level of about 1 day. Physics-based algorithms do a little better but forecast significantly more false alarms. Inhomogeneities in the solar wind through which the shock travels before reaching Earth impose an irreducible uncertainty of the order of 10 hours on any algorithm that does not take them into account. The best hope for improvement in this area is through numerical integrations of the operative equations of motion that self-consistently incorporates the corona, the CME, and the solar wind.

Algorithms that use solar observations to forecast the geoeffective solar wind parameters (VB, $g(\theta)$ and nV^2) in ICME sheaths can be concatenated out of existing formulas that have been developed for other purposes. But the growth of uncertainty that concatenating algorithms entails might reduce their skill relative to climatology essentially to zero or less – the evaluation has not been performed. It would be good to develop and evaluate algorithms that forecast geoeffective quantities directly from solar observations. The problem that strong IMF turbulence in ICME sheaths imposes on forecasting is probably irreducible and will have to be addressed through probabilistic forecast protocols. Nonetheless, under rare conditions with a high potential for a space-weather impact – a super-fast, Earth-directed, Sun-centered halo CME during equinox with a WSA prediction of maximum $g(\theta)$ from the Russell-McPherron effect – a forecast of a strong ICME-sheath disturbance might be made with reasonable confidence. This is a best-case scenario.

Algorithms that use solar observations to forecast the geoeffective solar wind parameters in ICME bodies are beset from the start with an uncertainty (on the order of 50%) whether a forecasted ICME arrival at Earth will bring a predictable-in-principle magnetic cloud ICME or a so-far unpredictable non-cloud ICME. If a cloud ICME arrives, data-based algorithms exist to predict many of its parameters from the time of the CME initiation: the velocity profile through the ICME and the geoeffective component of the magnetic field. These algorithms, however, are based on a prediction algorithm for the speed of the leading edge of the ICME in one case and on the angle that the ICME axis (viewed as a flux rope) makes to the ecliptic plane in the other case. Thus there is also a growth of uncertainties owing to a concatenation of algorithms.

Uncertainty over which type of ICME will materialize and growth of uncertainty owing to concatenations of algorithms can be dramatically reduced by using L1 data to specify crucial input parameters to the forecast codes. The price is a loss in forecast range from more than one day to less than one day.

Acknowledgements

This work was supported in part by the US National Science Foundation under grant ATM-0220396 and by the CISM project which is funded by the STC Program of the National Science Foundation under Agreement Number ATM-0120950.

References

Arge, C. N., and Pizzo, V. J.: 2000, *JGR* **105**, 10,465.

Bothmer, V., and Rust, D. M.: 1997, in: Crooker, N.U., Joselyn, J.A., and Feynman, J. (eds.), *Coronal Mass Ejections*, *Geophys. Monogr. Ser.*, vol. 99, AGU, Washington, D.C., pp. 137–146.

Bothmer, V., and Schwenn, R.: 1998, *Ann. Geophys.* **16**, 1–24.

Cane, H. V., and Richardson, I. G.: 2003, *JGR* **108**, A41156, doi:10.1029/2002JA009817.

Carlowicz, M. J., and Lopez, R. E.: 2002, *Storms from the Sun*, The Joseph Henry Press, Washington, DC.

Chen, J., Cargill, P. J., and Palmadesso, P. J.: 1997, *JGR* **102**, 14,701.

Cliver, E. W., Feynman, J., and Garrett, H. B.: 1990, *JGR* **95**, 17103–17112.

Crooker, N. U.: 2000, *JASTP* **62**, 1071–1085.

Dal Lago, A., Schwenn, R., and Gonzalez, W. D.: 2003, *Adv. Space Res.* **32**, 2637–2640.
Dryer, M.: 1974, *Space Sci. Rev.* **15**, 403–468.
Echer, E., Gonzalez, W. D., and Alves, M. V.: 2006, *Space Weather* **4**, S06001, doi:10.1029/2005SW000200.
Forbes, T. G., Linker, J. A., *et al.*: 2006, *Space Sci Rev.*, this volume, doi: 10.1007/s11214-006-9019-8.
Forsyth, R. J., Bothmer, V., *et al.*: 2006, *Space Sci. Rev.*, this volume, doi: 10.1007/s11214-006-9022-0.
Freeman, J. W.: 2001, *Storms in Space*, Cambridge University Press, Cambridge.
Fry, C. D., Sun, W., Deehr, C. S., Dryer, M., Smith, Z., Akasofu, S.-I., *et al.*: 2001, *JGR* **106**, 20,985–21,001.
Fry, C. D., Dryer, M., Smith, Z., Sun, W., Deehr, C. S., and Akasofu, S.-I.: 2003, *JGR* **108**, A21070, doi:10.1029/2002JA009474.
Gopalswamy, N., Lara, A., Lepping, R. P., Kaiser, M. L., Berdichevsky, D., and O. C. St. Cyr.: 2000, *GRL* **27**(2), 145–148.
Gopalswamy, N., Lara, A., Yashiro, S., Kaiser, M. L., and Howard, R. A.: 2001, *JGR* **106**, A12, 29,207–29,217.
Gosling, J. T., and McComas, D. J.: 1987, *GRL* **14**, 355–358.
Gosling, J. T., McComas, D. J., Phillips, J. L., and Bame, S. J.: 1991, *JGR* **96**, 7831–7839.
Heinemann, M.: 2002, *JASTP* **64**, 315–325.
Hughes, W. J., and Hudson, M. K.: 2004, *JASTP* **64**, 1241–1242.
Joselyn, J. A.: 1995, *Rev. Geophys.* **33**(3), 383–401.
Lepping, R. P., Berdichevsky, D. B., Szabo, A., Arqueros, C., and Lazarus, A. J.: 2003, *Solar Physics* **212**, 425–444.
McComas, D. J., Gosling, J. T., Bame, S. J., Smith, E. J., and Cane, H. V.: 1989, *JGR* **94**, 1465–1471.
McPherron, R. L., and Siscoe, G.: 2004, *Space Weather* **2**, S01001, doi:10.1029/2003SW000003.
Newell, P. T., Soterelis, T., Liou, K., Meng, C., and Rich, F. J.: 2006, *Eos Trans. AGU* **87**(36), Jt. Assem. Suppl., Abstract SM41D-04.
Odenwald, S.: 2001, *The 23rd Cycle*, Columbia University Press New York.
Owens, M. J., Cargill, P. J., Pagel, C., Siscoe, G. L., and Crooker, N. U.: 2005, *JGR* **110**, A01105, doi:10.1029/2004JA010814.
Parker, E. N.: 1963, *Interplanetary Dynamical Processes*, Interscience Publishers.
Richardson, I. G., and Cane, H. V.: 2004, *GRL* **31**, L18804, doi:10.1029/2004GL020958.
Russell, C. T., and McPherron, R. L.: 1973, *JGR* **78**, 92–108.
Russell, C. T., and Mulligan, T.: 2002, *Planet. Space Sci.* **50**, 527.
Schwenn, R., Dal Lago, A. Gonzalez, W. D. Huttunen, E. Cyr, C. 0. St., and S. Plunkett P.: 2001, *Eos Trans. AGU* **82**, 47, Fall Meet. Suppl., Abstract SH12A-0739.
Schwenn, R., Dal Lago, A., Huttunen, E., and Gonzalez, W. D.: 2005, *Ann. Geophys.* **23**, 1033–1059.
Schwenn, R.: 2006, *Living Rev. Solar Phys.* **3**, 2. URL (cited on July 10, 2006): http://www.livingreviews.org/lrsp-2006-2.
Siscoe, G. L., Erickson, G. M., Sonnerup, B. U. Ö., Maynard, N. C., Schoendorf, J. A., Siebert, K. D., *et al.*: 2002, *J. Geophys. Res.* **107**(A6), 1075, doi10.1029/2001JA000109.
Smith, Z., and Dryer, M.: 1990, *Sol. Phys.* **129**, 387–405.
Smith, Z., Dryer, M., Ort, E., and Murtagh, W.: 2000, *JASTP* **62**, 1265–1274.
Srivastava, N., and Venkatakrishnan, P.: 2004, *JGR* **109**, A10103, doi:10.1029/2003JA010175.
Tóth, G., *et al.*: 2005, *JGR* **110**, A12226, doi:10.1029/2005JA011126.
Wimmer-Schweingruber, R. F., Crooker, N. U., et al.: 2006, *Space Science Reviews*, this volume, doi: 10.1007/s11214-006-9017-x.
Wu, C. C., and Dryer, M.: 1996, *JGR* **23**, 1709–1712.
Zhao, X. P., and Hoeksema, J. T.: 1997, *GRL* **24**(23), 2965–2968.
Zurbuchen, T. H., and Richardson, I. G.: 2006, *Space Science Reviews*, this volume, doi: 10.1007/s11214-006-9010-4.

CORONAL MASS EJECTIONS
A Personal Workshop Summary

R. F. WIMMER-SCHWEINGRUBER*
Institut für Experimentelle und Angewandte Physik, Extraterrestrische Physik, Christian-Albrechts-Universität zu Kiel, Kiel, Germany
*(*Author for correspondence: E-mail: wimmer@physik.uni-kiel.de)*

(Received 22 May 2006; accepted in final form 16 June 2006)

Abstract. This workshop summary tries to distill the key difficulties and questions in the art of (I)CME physics and strategies to address them. (I)CMEs are multi-dimensional, multi-parameter, and multi-scale phenomena related to the solar dynamo, corona, and heliosphere. This workshop illustrates the immense progress made in describing and modeling these spectacular energetic solar events, but also shows clear shortcomings in our understanding of them.

Keywords: Coronal mass ejections, flares, solar physics, interplanetary physics, space weather, solar and stellar X-ray luminosity

1. Introduction

Summarizing a workshop that led to the publication of this hefty 500-page volume is a daunting task. So much effort and thought has gone into the individual articles and reports that a summary can hardly do them all justice. From this wealth of material, I have tried to distill the difficulties we face when dealing with coronal mass ejections (CMEs) and their interplanetary manifestations, ICMEs. In one word, (I)CMEs are difficult, because they are "multi-a lot of things": multi-dimensions, multi-parameter, and multi-scale.

It is easy to understand the difficulty with **multi-dimensions**. CMEs are inherently $3 + 1$ dimensional, their spatial properties are linked to their temporal evolution. Obviously, this is difficult, because we are used to drawing two-dimensional sketches and cartoons of CMEs. Figure 2 gives an impression of a dimensional pitfall. Reconnection of the central flux rope with the overlying field lines is topologically impossible in 2-D configuration *a*, but is readily achieved in the 3-D configuration *b*, as shown in *c*. Perhaps less obvious is the problem that many models of CMEs are two (+1) dimensional and that many of our conclusions come from such models. Mikić and Lee (2006, this volume) conclude their introduction to theory and models of CMEs with a list of 10 improvements that are needed to make progress in understanding CME initiation – first of which is the extension of models to 3 (+1) dimensions.

One might be tempted to state that (I)CMEs are even more than 3+1 dimensional. One could argue that e.g. the composition of various parts of an ICME should be

Space Science Reviews (2006) 123: 471–480
DOI: 10.1007/s11214-006-9025-x

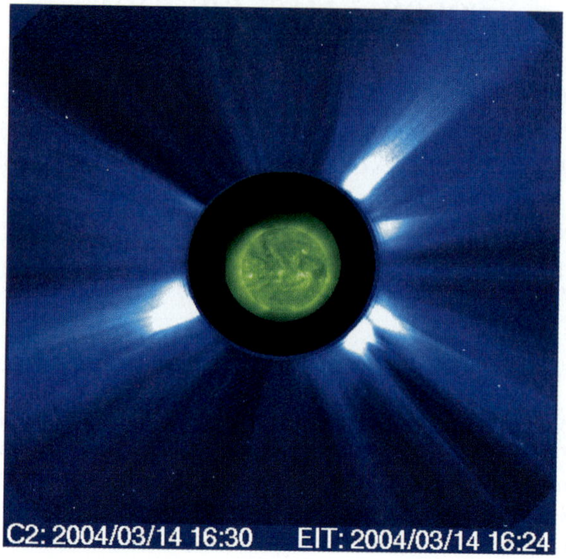

Figure 1. Combined EIT/LASCO image of the "workshop CME" which finally occurred during the final meeting at ISSI. See text for discussion. Image courtesy SOHO (ESA & NASA).

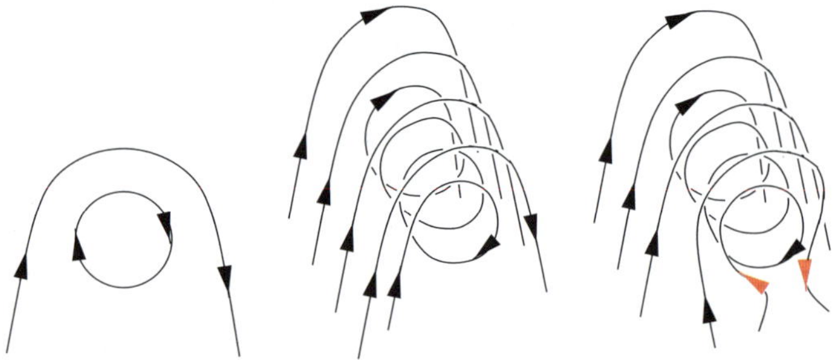

Figure 2. Dimensional pitfall. No reconnection of the central flux rope with the overlying field lines is possible in 2-D configuration a for topological reasons, while it is in 3-D (b and c).

considered as an extra few dimensions, say one per element or even charge state. One could add three extra dimensions for (I)CME speed and we would end up with an (I)CME phase space, but I'm not sure this is really helpful at our present state of nascent understanding, so I prefer to consider these quantities "parameters".

The **multi parameters** that play a role in the behavior of CMEs constitute the next difficulty. To paraphrase Alexander *et al.* (2006, this volume), "In other words, considering parameters one at a time, as is often done for specific events, is inadequate." This is not only true for models, but also for observations. Zurbuchen and Richardson (2006, this volume) suggest that the most practical approach to identify ICMEs is to "examine as many signatures as possible" or available. Since not all signatures agree and all signatures are (almost?) never present simultaneously, this may well be the only solution we have. As we develop more and more ingenious observational methods, the difficulty in absorbing all the possible observations will increase, adding another dimension to the multi-parameter difficulty.

Finally, the physics of (I)CMEs is truly **multi-scale**. For example, the acceleration of particles at ICME-driven shocks requires understanding the overall global structure of the shock and the ambient solar wind, e.g., the magnetic connection to the shock, the intermediate scales such as local curvature, bumps, and dents, on the scale of several ion gyroradii, as well as the microphysics involved in escaping particles generating upstream turbulence on which other particles can scatter. On the Sun, CMEs are driven by or drive small- and large-scale reconnection. Large-scale subsurface flows influence the emergence and appearance of active regions, while small-scale photospheric motions "wiggle" the overlying field lines in the corona which can form large-scale structures that may or may not act as tethers for underlying (buoyant or not) flux ropes. Forbes *et al.* (2006, this volume) illustrate the enormous range of scales in their Figure 1 and Hudson *et al.* (2006, this volume) stress the consequences for modeling.

These few and incomplete examples of the complexity of (I)CMEs demonstrate that "(I)CME-ology" is not an easy field to work in. It is a challenge to understand (I)CMEs and we need to break this challenge down into manageable chunks or questions that we can address.

2. What do We Need to Understand?

There are several ways of dividing up questions that need to be addressed to make progress in the understanding of (I)CMEs. A traditional way is to begin with CME initiation, its propagation through the corona and interplanetary medium, and the consequences of (I)CMEs such as particle acceleration and their influence on cosmic rays. Another way would be to look at some properties or conditions that we believe are crucial to the processes just mentioned, such as the roles of flux emergence, cancellation, or shear, of helicity in (I)CMEs. Other properties or conditions associated with (I)CMEs are flux ropes, seed populations, and their magnetic connection back to the Sun. I have tried to combine both ways inTable I.

The ejection of coronal mass requires energy and we generally agree that it is the magnetic field that supplies this energy. The emergence of magnetic flux alone is insufficient, it appears that flux needs to be sheared, with twisting often injecting helicity and increasing the available energy. How this flux and helicity is injected is unclear. Is it the result of subsurface processes or of photospheric footpoint motions or of rotating sunspots? Both Gopalswamy *et al.* (2006, this volume) and Schwenn *et al.* (2006, this volume) in this volume consider this an important question to answer. Is the increasing helicity important or is it the accompanying increase

TABLE I

Properties or conditions associated with various (I)CME stages.

	Flux emerg./ cancel./shear and helicity	Reconnection/ flares	Flux ropes	Particle populations	Magnetic connection
CME initiation	•	•	•	•	
Coronal propagation	•	•	•	•	•
Interpl. propagation	•	•	•	•	•
Shocks/particles		•	•	•	•

in stored energy that makes active regions with sigmoids more likely to produce CMEs? How does this affect the overall helicity budget of the Sun?

The eruption of CMEs is accompanied by a major reconfiguration of the coronal magnetic field, thereby converting stored magnetic energy into kinetic and thermal energy. However, we still do not fully understand how this conversion occurs (Schwenn *et al.*, 2006, this volume). We all agree it happens by the process of reconnection, but it is very hard to observe because it occurs on such small scales and because of its fundamentally 3-D nature. For instance, we do not understand whether reconnection initiates a CME or whether it is merely a consequence of a CME. Does it trigger a flare or is it triggered by a flare (Pick *et al.*, 2006, this volume; Schwenn *et al.*, 2006, this volume)? Where is the filament with respect to the reconnection site? As the CME moves through the corona into interplanetary space, it interacts with the ambient solar wind and is accelerated or decelerated. Is this the only process or does ongoing reconnection also affect CME propagation speed by converting kinetic energy into thermal energy (Forsyth *et al.*, 2006, this volume)?

A related question is where the flux ropes are formed (Gopalswamy *et al.*, 2006, this volume). Are they already present in the subsurface and then emerge from it, or do they form in the atmosphere? If the latter, do they form before or during the eruption and what is their relation to reconnection? The question whether all CMEs are fluxropes is quite fundamental and of practical purpose (Gopalswamy *et al.*, 2006, this volume). If all CMEs are fluxropes, then our models already incorporate that property. If not, then more work will be required to model non-flux-rope CMEs. The difficulty in settling this question is observational. It is hard to determine whether a CME is a flux rope from remote-sensing observations. Definitely, not all ICMEs are flux ropes (Wimmer-Schweingruber *et al.*, 2006, this volume; Zurbuchen and Richardson, 2006, this volume), but we do not truly understand the relationship between CMEs and ICMEs (Forsyth *et al.*, 2006, this volume). The spacecraft trajectory may simply not be intersecting the flux-rope part of the ICME, or this part has dissolved due to ongoing reconnection, or, indeed, there may well be non-flux-rope ICMEs.

As the CME propagates outwards through the corona, it can drive a shock, depending on its speed and the local magnetosonic speed (Forbes *et al.*, 2006, this volume). These shocks may accelerate particles of various origins in the corona and beyond, but it is unclear from observations where the shocks form (and possibly disappear) relative to the heights where SEPs are accelerated (Cane and Lario, 2006, this volume). Furthermore, the roles of the various seed populations, such as flare suprathermals, shock geometry, and transport processes in determining the properties of individual events are unclear (Klecker *et al.*, 2006, this volume). For instance, the relation of the various systematic dependencies such as m/q, charge-state, and spectral properties, to the local and possibly global plasma and shock properties is unclear. Are flare-accelerated particles, "normal" coronal particles, and possibly inner-source pick-up ions the only seed populations? What do they tell us about shock and hence CME evolution? Are scatter-free particles truly scatter-free, i.e. are the injection delays true delays, implying shock acceleration, or are transport processes affecting flare-produced particles (Cane and Lario, 2006, this volume)? When moving out of the corona into interplanetary space, ICMEs continue to accelerate particles, but intriguingly also can contain particles within. Where do they come from and what do they tell us about ICME topology (Klecker *et al.*, 2006, this volume; Wimmer-Schweingruber *et al.* 2006, this volume)? Even farther out in the heliosphere large ICMEs may compress the solar wind or multiple ICMEs may merge to form Merged or even Global Merged Interaction Regions (MIRs or GMIRs). These regions contribute to the modulation of cosmic rays (Gazis *et al.*, 2006, this volume) by serving as "diffusion barriers". The relation of the constituting ICMEs and these barriers is unclear.

The various particle populations accessible to modern space-based instrumentation have greatly enhanced our possibilities to understand (I)CMEs, be it to detect them in-situ (Wimmer-Schweingruber *et al.*, 2006, this volume; Zurbuchen and Richardson, 2006, this volume; von Steiger and Richardson, 2006, this volume; Gazis *et al.*, 2006, this volume), but also to relate processes involved in CME initiation (Wimmer-Schweingruber *et al.*, 2006, this volume; Zurbuchen and Richardson, 2006, this volume) and particle acceleration (Cane and Lario, 2006, this volume; Klecker *et al.*, 2006, this volume). This workshop certainly drove home the point that there is a richness of information in these measurements that is still waiting to be exploited. Nevertheless, some observations, such as the simultaneous measurement of elevated and low charge states in the bulk plasma of ICMEs, are still puzzling and await explanation (Wimmer-Schweingruber *et al.*, 2006, this volume; Zurbuchen and Richardson, 2006, this volume).

Another group of questions may be summarized under magnetic connection. We have already touched upon the subject of flare-related energetic particles inside flux ropes. How do they get into the interior of the flux-rope ICME (Klecker *et al.*, 2006, this volume; Cane and Lario, 2006, this volume)? Do they leak into the

ICME from the shock it is driving, are they connected along open field lines back to the ejection site, or to some other flaring active region? While the signatures of bidirectional electrons are understood in terms of magnetic connection back to the Sun, those of bidirectional ions are not (Wimmer-Schweingruber *et al.*, 2006, this volume). This magnetic connection may also be a clue to the observation that the kinetic temperature inside ICMEs decreases less quickly with heliocentric distance than that of the solar wind. What keeps on heating ICMEs (Forbes *et al.*, 2006, this volume) once they have been ejected? How long are these open field lines maintained and how are they disconnected? This has immediate implications for the question of the open magnetic flux of the Sun which is strongly influenced by ICMEs (Crooker and Horbury, 2006, this volume), but also on the removal of helicity from the Sun (Hudson *et al.*, 2006, this volume).

3. What is Needed?

In order to further our emerging understanding of (I)CMEs, we need to make progress in the areas of observations/measurements, theory and modeling, and, last but not least, in breaking down our knowledge and understanding to a level manageable for operational use in space weather forecasting (see paper by Siscoe and Schwenn (2006) in this volume). We could also summarize the needs in other ways, along the lines discussed in the introduction. First of all, we need to measure (I)CMEs in 3 (+1) dimensions: This requires 3-D magnetic field measurements in the photosphere, chromosphere, and corona. We are also waiting for the first 3-D images of CMEs from STEREO and are also excited about the opportunities of observing ICMEs while in quadrature with other spacecraft. STEREO, together with Wind, SOHO, and ACE will also allow us multi-point observations of ICMEs and the shocks they drive. Nevertheless, these 3-D measurements are likely to be at the wrong scales or certainly not at all the scales needed. Short of a dedicated multi-scale mission, we should combine the data from Wind, SOHO, ACE, and STEREO with those of ESA's Cluster when it is in the solar wind. In the farther future, we need to investigate the evolution of ICMEs in the inner heliosphere, making use of ESA's Solar Orbiter with its unique orbit, and the multi-point measurements offered by NASA's Sentinels. These missions will no doubt enhance our understanding of (I)CMEs and related phenomena.

On the modeling side, there are probably "only" two wishes, one very difficult, one easier to achieve. The easy goal is to make current and future models more readily available to researchers, thus greatly simplifying data interpretation. The Michigan group is making a promising start in this respect. The other Need (with a capital N) is simply for more realistic models. Almost all papers in this volume call for more realistic modeling. More accurate treatment of the coronal energy balance, inclusion of realistic seed particle populations and composition in models,

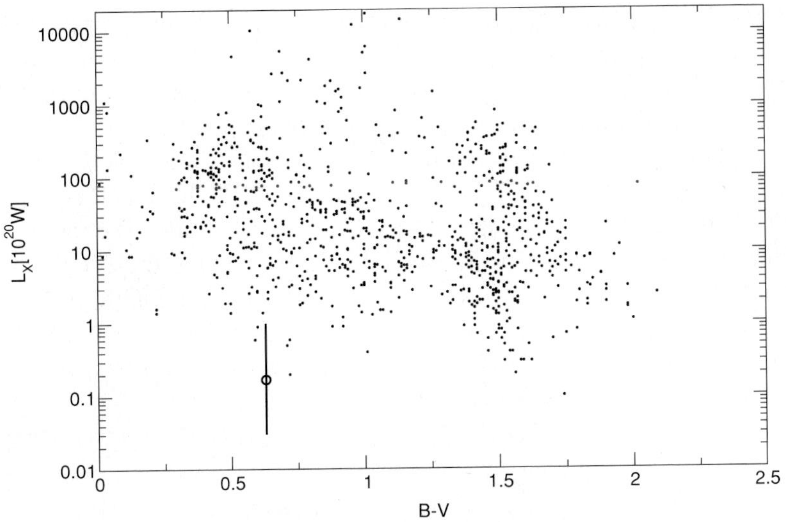

Figure 3. X-ray luminosities of stars in the solar neighborhood vs. B-V color index. The large circle with the vertical solid line indicate the Sun and its X-ray variability. Data: http://vizier.u-strasbg.fr/cgi-bin/VizieR.

relaxation of simplifying assumptions, fully 3-D time-dependent models, inclusion of all relevant scales (kinetic to MHD), self-consistent treatments of CME initiation and evolution, modeling of observable quantities, and inclusion of cross-scale coupling are just a few examples of the many wishes uttered at this workshop. There is still alot of work ahead.

4. The Sun as a Star

The relation of CME initiation and flares is still unclear, yet if we want to apply our understanding of CMEs to other stars and estimate mass loss due to CMEs (e.g. during early evolutionary stages), then we probably need to revert to flares as indicators of CME activity. To me, one of the greatest puzzles in this respect is the extremely low X-ray activity of the Sun when compared to other stars, even solar-type stars.Figure 3 shows X-ray luminosity of a volume-limited Rosat all-sky survey of the solar vicinity vs. B-V color index. The Sun has a B-V index of 0.63, the large circle shows average solar X-ray luminosity, the solid vertical line indicates its variability. The data are from the Vizier Service at Centre de Données astronomiques de Strasbourg (`http://vizier.u-strasbg.fr/cgi-bin/VizieR`). Clearly, the Sun lies at the lower extreme of X-ray activity for all Sun-like stars. Are we just lucky to live with a star that is benigningly X-ray inactive? Or is it a prerequisite for life?

Figure 4. The castle of Elmau in February 2003. Photograph courtesy J. Schmidt.

Figure 5. Group photograph in Elmau. Courtesy J. Schmidt.

5. Final Remarks

The series of workshops leading up to the publication of this volume saw changes in the way we view coronal mass ejections and related coronal and interplanetary phenomena. During the first workshop at the wonderful castle of Elmau (see Figure 4) in the snow-covered Bavarian hills, we were still absorbing the richness of the visual impressions offered by SOHO (and, of course, the more immediate surroundings). For instance, one topic of discussion that came up repeatedly at the first workshop was the relation of slow and fast CMEs to gradual and impulsive solar particle events. Was there a one-to-one correspondence? Was there no relation at all? Today we know that there is a continuum of CME speeds (see, e.g., Figure 1.5 of Schwenn *et al.* (2006, this volume)), then we were still under the impression of a two-class distribution. As we have seen in this workshop, a lot of progress has been made. As is usual in the science business, every question solved generates at least one additional new question – a few selected ones have been mentioned in this summary.

At the last workshop, at the International Space Science Institute, ISSI, we finally got our "workshop CME". It is an utterly unspectacular CME emerging from the upper streamer in the south-western quadrant of Figure 1. May this workshop stand out as more spectacular than its CME!

Acknowledgements

I wish to thank the conveners for organizing a series of three stimulating workshops, especially Horst Kunow for initiating it and pulling it through, the International Space Science Institute, ISSI, for hosting the final one and for its hospitality during the course of writing this summary. Finally, thanks goes to all participants (see Figure 5) for their enthusiasm and for sharing ideas and thoughts during this great series of workshops. This work was supported, in parts, by the Deutsche Forschungsgemeinschaft, DFG.

References

Alexander, D., Richardson, I.G., and Zurbuchen, T.H.: 2006, *Space Sci. Rev.*, (this volume), doi: 10.1007/s11214-006-9013-1.

Cane, H. V., and Lario, D.: 2006, *Space Sci. Rev.*, (this volume), doi: 10.1007/s11214-006-9011-3.

Crooker, N. U., and Horbury, T. S.: 2006, *Space Sci. Rev.*, (this volume), doi: 10.1007/s11214-006-9014-0.

Forbes, T. G., Linker, J. A., Chen, J., Cid, C., Kóta, J., Lee, M. A., *et al.*: 2006, *Space Sci. Rev.*, (this volume), doi: 10.1007/s11214-006-9019-8.

Forsyth, R. J., Bothmer, V., Cid, C., Crooker, N. U., Horbury, T. S., Kecskemety, B., *et al.*: 2006, *Space Sci. Rev.*, (this volume), doi: 10.1007/s11214-006-9022-0.

Gazis, P. R., Balogh, V., Dalla, S., Decker, R., Heber, B., Horbury, T., et al.: 2006, *Space Sci. Rev.*, (this volume), doi: 10.1007/s11214-006-9023-z.

Gopalswamy, N., Mikić, Z., Maia, D., Alexander, D., Cremades, H., Kaufmann, P., *et al.*: 2006, *Space Sci. Rev.*, (this volume), doi: 10.1007/s11214-006-9020-2.

Hudson, H. S., Bougeret, J.-L., and Burkepile, J.: 2006, *Space Sci. Rev.*, (this volume), doi: 10.1007/s11214-006-9009-x.

Klecker, B., Kunow, H., Cane, H. V., Dalla, S., Heber, B., Kecskemety, K., et al.: 2006, *Space Sci. Rev.*, (this volume), doi: 10.1007/s11214-006-9018-9.

Mikić, Z., and Lee, M. A.: 2006, *Space Sci. Rev.*, (this volume), doi: 10.1007/s11214-006-9012-2.

Pick, M., Forbes, T. G., Mann, G., Cane, H., Chen, J., Ciaravella, A., *et al.*: 2006, *Space Sci. Rev.*, (this volume), doi: 10.1007/s11214-006-9021-1.

Schwenn, R., Raymond, J. C., Alexander, D., Ciaravella, A., Gopalswamy, N., Howard, R., et al.: *Space Sci. Rev.*, (this volume), doi: 10.1007/s11214-006-9016-y.

Siscoe, G., and Schwenn, R.: 2006, *Space Sci. Rev.*, (this volume), doi: 10.1007/s11214-006-9024-y.

von Steiger, R., and Richardson, J. D.: 2006, *Space Sci. Rev.*, (this volume), doi: 10.1007/s11214-006-9015-z.

Wimmer-Schweingruber, R. F., Crooker, N. U., Balogh, A., Bothmer, V., Forsyth, R. J., Gazis, P., *et al.*: *Space Sci. Rev.*, (this volume), doi: 10.1007/s11214-006-9017-x.

Zurbuchen, T. H., and Richardson, I. G.: 2006, *Space Sci. Rev.*, (this volume), doi: 10.1007/s11214-006-9010-4.

GLOSSARY

Corona: Outermost layer of the solar atmosphere, characterized by low densities ($<1.0 \times 10^9$ cm^{-3}) and extraordinarily high temperatures ($>1.0 \times 10^6$ K) that extend to several solar radii.

Coronagraph: Telescope for observing the corona by producing an artificial eclipse. It contains an occulting disk which covers the disk of the Sun so that the faint corona may be more easily observed.

Coronal hole: Area of the corona that appear dark in X-rays and ultraviolet light. They are usually located at the poles of the Sun (at activity minima), but can occur at other places as well. The magnetic field lines in a coronal hole extend out into the solar wind rather than coming back down to the Sun's surface as they do in other parts of the Sun.

Coronal Mass Ejection (CME): An observable change in coronal structure that occurs on a time scale between a few minutes and several hours and involves the appearance and outward motion of a new, discrete, and bright white-light feature in the coronagraph field of view.

Coronal streamer belt: Bright region in the corona overlying the solar magnetic equator, consisting of closed field lines at low altitudes and open field lines at high altitudes; source region of the low-speed solar wind.

Corotating Interaction Regions (CIRs): Compression regions formed from the interaction of quasi-stationary high- and low-speed solar wind streams. They are roughly aligned with Archimedean spirals, appear to corotate with the Sun, and typically are bounded by shocks at heliocentric distances greater than about 2 astronomical units.

Cosmic rays: Charged particles of very high energy (up to some 10^{21} eV) that originate outside the solar system, also called Galactic Cosmic Radiation (GCR).

Dimming: Transient dark region observed in X-rays and EUV near CMEs, thought to be due to the lifting of closed field lines during the initial phase of a CME.

Disappearing Filament (DFs): A filament seen in H-alpha light against the solar disk that suddenly fades and disappears completely within tens of minutes. It signals an eruption of the filament and probable launch of a CME. The filament disappears either because it leaves the spectral range of the H-alpha filter due to the Doppler shift or because the material is heated so much that it no longer absorbs H-alpha light.

Doppler shift: A change in the wavelength of radiation received from a source because of its motion along the line of sight. A Doppler shift in the spectrum of an astronomical object is commonly known as a redshift when the shift is towards longer wavelengths (the object is moving away) and as a blueshift when the shift is towards shorter wavelengths (the object is approaching).

Space Science Reviews (2006) 123: 481–484
DOI: 10.1007/s11214-006-9037-6

EIT waves: Bright rings or arcs observed by the Extreme ultraviolet Imaging Telescope (EIT) that sometimes expand around a flaring active region and propagate over a hemisphere of the Sun with speeds of 200–300 km/s.

Energetic particles: Suprathermal ionized particles that are accelerated in the solar system to energies from about 100 eV to several GeV, i.e., to near-relativistic speeds. Those called "Solar Energetic Particles (SEPs)" are accelerated by coronal and interplanetary effects of solar flares and/or CMEs. SEPs are distinct from energetic particles from outside the heliosphere, called "Galactic Cosmic Rays."

Eruptive prominence (EP): Dramatic events observed in H-alpha light above the limb for many years but only recently understood as ejected material that actually leaves the Sun with CMEs. EPs and disappearing filaments are different views of the same physical process.

Expansion speed: Speed of the lateral expansion of a CME. The expansion is measured at the outermost CME edges seen by a coronagraph as projections onto the plane of the sky. The expansion speed can always be uniquely determined, regardless of where the CME is aimed, in contrast to uncertainties in the radial speed caused by projection effects.

Filament: A band-like structure in the corona consisting of cool plasma supported by magnetic fields. Filaments are dark (absorption) structures when seen in the light of the H-alpha line against the bright solar disk but appear bright (as emission structures) when seen over the solar limb, where they are known as prominences.

Flare: A sudden, rapid, and intense variation in brightness in an active region on the Sun. A flare occurs when magnetic energy that has built up in the solar atmosphere is suddenly released. Radiation is emitted across virtually the entire electromagnetic spectrum.

Galactic Cosmic Radiation (GCR): See "Cosmic rays" and "Energetic particles."

Halo CME: A halo of excess brightness completely surrounding the occulting disk of a coronagraph and expanding in all directions from the Sun, caused by the more energetic CMEs that impact the Sun-Earth line, be they moving toward or away from Earth.

H-alpha: Spectral line of neutral hydrogen at 656.3 nm in the red part of the visible spectrum. The H-alpha line is universally used for observations of solar flares, filaments and prominences.

Helicity: The measure of twist in an object, such as the degree of coiling of a magnetic field.

Heliosphere: Region of space containing plasma and magnetic fields of solar origin: a cavity carved in the interstellar medium by the flow of the solar wind.

Heliospheric Current Sheet (HCS): A warped surface in interplanetary space extending from the solar magnetic equator and separating solar wind flows of opposite magnetic polarity. It is often addressed as the Sun's "ballerina skirt".

ICME: Interplanetary counterpart of a CME.

Interplanetary Magnetic Field (IMF): Remnant of the solar magnetic field dragged into interplanetary space by the flow of the solar wind.

Interplanetary medium: Material in between the planets in the solar system. It contains the smaller objects in the solar system, such as asteroids, comets, and meteors, as well as dust , neutral atoms, plasma particles, SEPs and the GCR.

Long Duration Events (LDEs): Slow decrease in the flux of soft X-rays in the aftermath of major X-ray flares, on timescales of days. It is thought that this is a signature of ongoing magnetic reconnection in the aftermath of a CME.

Magnetic reconnection: Sudden interconnection of adjacent magnetic field lines of opposite polarity, causing a substantial change of magnetic topology. Reconnection is fundamental process in plasma physics throughout the universe.

Magnetic sector: A region of unipolar magnetic field in the solar wind. Sectors of opposite magnetic polarity (away from and toward the Sun) are separated by the heliospheric current sheet.

Magnetohydrodynamics (MHD): Fluid dynamics applied to a magnetized medium.

Magnetosphere: The magnetic cavity formed by the geomagnetic field, which shields Earth from the solar wind.

Moreton waves: Bright rings or arcs observed in the H-alpha line that sometimes expand around a flaring active region and propagate over a hemisphere of the Sun with speeds of 440 to 1125 km/s.

Parker spiral: Idealized Archimedean spiral configuration of the IMF in the solar equatorial plane, resulting from solar rotation and the outward flow of the solar wind.

Pick-up ions: Interstellar neutral atoms that drift into the heliosphere and, through ionization, become an integral part of the solar wind flow.

Plane of the sky: Hypothetical plane to which all objects in the sky seem to be projected. For moving objects, we can directly observe and measure only the speed components in the plane of the sky.

Position angle (PA): The apparent latitudinal angular separation of a certain feature (as projected onto the plane of the sky) from the solar north pole, measured counterclockwise.

Post-Eruptive Arcades (PEAs): Transient large-scale loop systems that are often observed in association with CMEs and X-ray flares.

Prominence: See "Filament."

Radio bursts: Electromagnetic waves emitted in the course of explosive events near the Sun, termed type I to type V, depending upon the changes in intensity as a function of frequency and time.

Shock sheath: Compressed ambient gas between a shock and its driver.

Shock wave: A discontinuous, nonlinear change in pressure commonly associated with supersonic motion in a gas or plasma. CMEs with sufficiently high speeds drive shock waves, and CIRs are usually bounded by shock waves at heliocentric distances greater than about 2 astronomical units.

Sigmoid: An s-shaped or inverse s-shaped bright structure on the Sun, as seen in EUV-light or in soft X-rays. Sigmoids often occur near active regions and filaments and may signal incipient CMEs.

Solar activity cycle: Waxing and waning of various forms of solar activity such as sunspots, flares, and CMEs caused by the ~11-year evolution cycle of the Sun's subsurface magnetic fields.

Solar Energetic Particles (SEPs): See "Energetic particles."

Solar wind: An ionized gas, or plasma, that permeates interplanetary space. It exists as a consequence of the supersonic expansion of the Sun's hot outer atmosphere, the solar corona.

Source surface: A hypothetical spherical surface surrounding the Sun at around 2.5–3.25 solar radii, above which all magnetic field lines are treated as strictly radial and open to the outer boundary of the heliosphere.

Stream interface: Boundary separating what was originally fast tenuous plasma from what was originally slow dense plasma within a corotating interaction region.

Sunspot: Dark areas on the photosphere of the Sun with concentrated magnetic flux, typically occurring in bipolar (i.e., two-part with positive and negative poles like a magnet) clusters or groups. They appear dark because they are cooler than the surrounding photosphere.

Sunspot number (SSN): A number traditionally used as an index for solar activity. The SSN is determined from a standardized formula based upon not only the number of sunspots but also their size and the number of groups.

Supergranulation: A large-scale solar convection pattern with a characteristic size of about 30,000 km and a lifetime of about one day.

Termination shock: A discontinuity in the solar wind flow in the outer heliosphere where the speed slows from supersonic to subsonic owing to interaction with the interstellar plasma.

Transition region: A thin layer of the solar atmosphere between the chromosphere and corona where the temperature rises sharply from 20,000 to nearly a million degrees Kelvin.